SCANNING PROBE MICROSCOPY
AND SPECTROSCOPY

Gerd Binnig (left) and Heinrich Rohrer (right) who were awarded the Nobel Prize for their invention of the scanning tunneling microscope.

The investigation and manipulation of matter on the atomic scale have been revolutionized by scanning tunneling microscopy and related scanning probe techniques. This book provides a clear and comprehensive introduction to the field, describing in detail the basic principles and applications of scanning probe microscopy and spectroscopy.

The book consists of two parts. The first deals with the theoretical background of scanning tunneling microscopy, and discusses in detail the design and instrumentation of practical STM systems. Part one concludes with a discussion of the experimental methods employed in scanning force microscopy and other related scanning probe techniques. In part two, the importance and widespread use of local probe techniques are highlighted by a thorough description of their applications in fields such as condensed matter physics, chemistry, biology, and nanotechnology.

Containing 350 illustrations, and over 1200 references, this unique book represents an ideal introduction to the subject for final-year undergraduates in physics or materials science. It will also be invaluable to graduate students and researchers in any branch of science where scanning probe techniques are used.

SCANNING PROBE MICROSCOPY AND SPECTROSCOPY

Methods and applications

ROLAND WIESENDANGER

*Institute of Applied Physics
and Microstructure Research Center
University of Hamburg*

CAMBRIDGE
UNIVERSITY PRESS

Published by the Press Syndicate of the University of Cambridge
The Pitt Building, Trumpington Street, Cambridge CB2 1RP
40 West 20th Street, New York, NY 10011-4211, USA
10 Stamford Road, Oakleigh, Melbourne 3166, Australia

First published 1994

Printed in Great Britain at the University Press, Cambridge

A catalogue record for this book is available from the British Library

Library of Congress cataloguing in publication data

Wiesendanger, R. (Roland), 1961–
Scanning probe microscopy and spectroscopy: methods and
applications / Roland Wiesendanger.
p. cm.
Includes bibliographical references and index.
ISBN 0 521 41810 0. – ISBN 0 521 42847 5 (pbk.)
1. Scanning probe microscopy. 2. Spectroscopy. I. Title.
QH212.S33W54 1994
502'.8'2–dc20   93–31700 CIP

ISBN 0 521 41810 0 hardback
ISBN 0 521 42847 5 paperback

To my parents
Kurt and Elfi Wiesendanger,
to my brother Harald, and
to Martina:

their love always meant light
at the end of the tunnel . . .

and to

Gerd Binnig, Heinrich Rohrer,
Christoph Gerber and Eduard Weibel,
who made the life of many scientists,
including me, even more enjoyable
than before the invention of the
scanning tunneling microscope . . .

To my parents
Kurt and Elfi Wiesendanger,
to my brother Harald, and
to Martina:

their love always meant light
at the end of the tunnel . . .

and to

Gerd Binnig, Heinrich Rohrer,
Christoph Gerber and Eduard Weibel,
who made the life of many scientists,
including me, even more enjoyable
than before the invention of the
scanning tunneling microscope . . .

Contents

ix

Preface

Many high barriers exist in this world – barriers between
nations, races and creeds. Unfortunately, some barriers are
thick and strong. But I hope, with determination, we will
find a way to tunnel through these barriers easily and
freely, to bring the world together . . .

(*L. Esaki*)

In March 1981 G. Binnig, H. Rohrer, Ch. Gerber and E. Weibel at
the IBM Zürich Research Laboratory observed vacuum tunneling of
electrons between a sharp tungsten tip and a platinum sample. Combined with the ability to scan the tip against the sample surface, the
scanning tunneling microscope (STM) was born. Since then, this novel
type of microscopy has continuously broadened our perception about
atomic scale structures and processes. The STM allows one to image
atomic structures directly in real space, giving us the opportunity to
make the beauty of nature at the atomic level directly 'visible'. Moreover, the sharp tip can be regarded as a powerful local probe which
allows one to measure physical properties of materials on a small scale
by using a variety of different spectroscopic methods. However, vacuum
tunneling of electrons is not the only means by which local properties
of matter can be probed. The development of the STM technique has
triggered the invention of a whole family of scanning probe microscopies (SPM) which make use of almost every kind of interaction
between a tip and a sample of which one can think. This class of scanning probe microscopes can now provide information about nanometer-
scale properties of matter which is often inaccessible by any other
experimental technique. Therefore, SPM will play a key role in nanometer-scale science and technology, which we have to understand in
order to develop better and smaller devices. G. Binnig and H. Rohrer
were awarded the Nobel Prize in 1986 for development of the fascinat-
ing STM technique.

This book on scanning probe microscopy and spectroscopy provides an introduction to the experimental methods and theoretical background as well as the various fields of applications of STM/SPM, including condensed matter physics, chemistry, biology, metrology and materials science. The historical background for the development of SPM is described as well as challenges for future research. Since this book is written by a condensed matter physicist, the selected examples of applications are mainly devoted to this field. Besides the invention of STM and SPM, condensed matter physics has seen many significant advances in the 1980s and early 1990s, including the discovery of the quantum Hall effect, of quasicrystals and high-temperature oxide superconductors, progress in the development of organic superconductors and the discovery of the interesting properties of undoped and doped C_{60} and other fullerenes. This concentration of 'top table physics' (N.A. Ashcroft at the APS '87 March meeting) has certainly fascinated as well as surprised the condensed matter physics community. The status of the development of SPM techniques in the various scientific disciplines may further justify the emphasis on condensed matter physics in this book. However, the achievements of SPM in chemistry, biology, metrology and nanotechnology are also described.

Ten years after the invention of STM it is already impossible to present a complete overview of all experimental and theoretical work which has been devoted to this field. It is also impossible to provide a complete list of references. This cannot be the aim of this book because by the time this book appears it would not be 'complete' anyway. The intention is rather to present the historical background of SPM techniques, some of the most important experimental methods together with the theoretical background and to describe representative examples of applications in various fields. A comparison with other competing experimental techniques is often made, to put the role of SPM techniques for understanding nanometer-scale properties of matter into perspective. Most importantly, this book aims at making the reader aware of the impact of SPM techniques on our understanding of atomic scale structures and processes, which in some cases might have been completely changed as a result of the invention and application of SPM.

This book is divided into two parts. The first part provides the historical background of SPM techniques and description of the various experimental methods as well as the theoretical background. This part should be of interest for all readers regardless of their specialized field in science. In the second part, various applications of SPM are described based on the concepts of SPM introduced in part one. The applications are grouped in various topics and subtopics allowing the reader to choose those of his own interest. This book has mainly been written for the following readership.

1. Undergraduate students who are interested in the novel SPM techniques. It would be advantageous, but not absolutely required, to read this book after having attended courses in quantum theory, condensed matter physics and surface physics at the level of textbooks such as *Quantum Mechanics* by Messiah (1969) or *Quantum Theory* by Bohm (1951) and *Solid State Physics* by Ashcroft and Mermin (1976) or Burns (1985).
2. Teachers of courses on STM/SPM in the various scientific disciplines.
3. Postgraduate students who are beginning to do research in the field of STM/SPM.
4. Scientists at research centers or in industry who are starting STM/SPM activities.
5. Scientists outside the SPM community who have become interested in the novel STM/SPM techniques.
6. Specialists who wish to have an overview of the broadened field of STM/SPM.

This book is intended to be attractive to each of these groups of readers.

I would like to thank all my colleagues working in the field of scanning probe microscopy who have collaborated with me at some time or generally shared their knowledge about scanning probe methods with me. In particular, I would like to thank Dr Heinrich Rohrer for his interest in my research work and Professor Harry Thomas for his highly stimulating lectures in theoretical physics at the University of Basel. I also thank Mrs Jacqueline Vetter for her help during the past years. Many thanks are also due to all of the colleagues who provided illustrations of their work for this book. It is also a great pleasure for me to acknowledge the fruitful collaboration with Cambridge University Press. Finally, I wish to thank my parents for their never-ending moral and financial support during my education and Martina for her patience and encouragement during the time this book was written.

Hamburg   *Roland Wiesendanger*

Acronyms

a.c.	alternating current
ADC	analog-to-digital converter
AES	Auger electron spectroscopy
AFM	atomic force microscope/microscopy
APSTM	analytical photon scanning tunneling microscope/microscopy
BCC	body-centered cubic
BCS	Bardeen–Cooper–Schrieffer
BEDT-TTF	bis(ethylenedithio)tetrathiafulvalene
BEEM	ballistic electron emission microscopy
BEES	ballistic electron emission spectroscopy
CCI	constant current image/imaging
CCS	constant current spectroscopy
CCT	constant current topograph/topography
CD	compact disk
CDW	charge density waves
CFI	constant force imaging
CHI	constant height image/imaging
CI	current image
CITS	current imaging tunneling spectroscopy
CSS	constant separation spectroscopy
CVD	chemical vapor deposition
DAC	digital-to-analog converter
DAS	dimer adatom stacking fault
DB	dangling bond
d.c.	direct current
DDB	didodecylbenzene
DFM	differential force microscopy
DNA	deoxyribonucleic acid
DTM	differential tunneling microscopy
EELS	electron energy loss spectroscopy

EFM	electrostatic force microscope/microscopy
EFOM	evanescent field optical microscope/microscopy
ESR	electron-spin resonance
FCC	face-centered cubic
FEIPES	field emission inverse photoelectron spectroscopy
FER	field emission resonances
FESAM	field emission scanning Auger microscope/microscopy
FESEM	field emission scanning electron microscope/ microscopy
FESEMPA	field emission scanning electron microscopy with polarization analysis
FF	far-field
FFM	frictional force microscope/microscopy
FIM	field ion microscope/microscopy
FM	frequency modulation
FS	force spectroscopy
FWHM	full width at half maximum
GIC	graphite intercalation compounds
HCP	hexagonal close-packed
HOMO	highest occupied molecular orbital
HPI	hexagonally packed intermediate (layer)
HREELS	high-resolution electron energy loss spectroscopy
HTSC	high-T_c superconductors/superconductivity
IC	integrated circuit
IET	inelastic electron tunneling
IETS	inelastic electron tunneling spectroscopy
IPES	inverse photoelectron spectroscopy
IR	infrared
IRAS	infrared reflection absorption spectroscopy
LB	Langmuir–Blodgett
LBH	local barrier height
LDA	local density approximation
LDOS	local density of states
LDSTM	laser-driven scanning tunneling microscope/microscopy
LEED	low-energy electron diffraction
LFS	local force spectroscopy
LMIS	liquid metal ion source
LTS	local tunneling spectroscopy
LTSTM	low-temperature scanning tunneling microscope/ microscopy
LUMO	lowest unoccupied molecular orbital
MBE	molecular beam epitaxy
MFM	magnetic force microscope/microscopy
ML	monolayer

NDR	negative differential resistance
NF	near-field
NOS	nitride–oxide–silicon
PAX	photoemission of adsorbed xenon
PCM	point-contact microscope/microscopy
PCS	point-contact spectroscopy
PEEM	photoemission electron microscopy
PES	photoelectron spectroscopy
PLD	periodic lattice distortion
PMMA	polymethylmethacrylate
PODA	polyoctadecylacrylate
PSD	position-sensitive detector
PSTM	photon scanning tunneling microscope/microscopy
PTS	photoassisted tunneling spectroscopy
QMS	quadrupole mass spectrometer
REM	reflection electron microscope/microscopy
RHEED	reflection high-energy electron diffraction
SAFE	STM aligned field emission
SCF	self-consistent field
SCAM	scanning capacitance microscope/microscopy
SCPM	scanning chemical potential microscope/microscopy
SECM	scanning electrochemical microscope/microscopy
SEMPA	scanning electron microscopy with polarization analysis
SERS	surface enhanced Raman spectroscopy
SET	single-electron tunneling
SEM	scanning electron microscope/microscopy
SFA	surface force apparatus
SFEM	scanning field emission microscope/microscopy
SFES	scanning field emission spectroscopy
SFM	scanning force microscope/microscopy
SFS	scanning force spectroscopy
SICM	scanning ion conductance microscope/microscopy
SMM	scanning micropipette microscope/microscopy
SMMM	scanning micropipette molecule microscope/microscopy
SNAM	scanning near-field acoustic microscope/microscopy
SNM	scanning noise microscopy
SNOM	scanning near-field optical microscope/microscopy
SNTM	scanning near-field thermal microscope/microscopy
SOAM	scanning optical absorption microscope/microscopy
SPEM	scanning photoemission microscope/microscopy
SPLEEM	spin-polarized low-energy electron microscopy
SPM	scanning probe microscope/microscopy
SPNM	scanning plasmon near-field microscope/microscopy

SPSTM	spin-polarized scanning tunneling microscope/microscopy
SPSTS	spin-polarized scanning tunneling spectroscopy
SPV	surface photovoltage
SQUID	superconducting quantum interferometer device
STHP	scanning thermal profiler/profilometry
STM	scanning tunneling microscope/microscopy
STOM	scanning tunneling optical microscopy
STP	scanning tunneling potentiometry
STS	scanning tunneling spectroscopy
TAM	tunneling acoustic microscope/microscopy
TED	transmission electron diffraction
TEM	transmission electron microscope/microscopy
TILS	tip-induced localized states
TIR	total internal reflection
TMD	transition metal dichalcogenides
TMT	transition metal trichalcogenides
TS	tunneling spectroscopy
TSMFM	tunneling-stabilized magnetic force microscope/microscopy
TSSFM	tunneling-stabilized scanning force microscope/microscopy
TT	tunneling thermometer
TTF-TCNQ	tetrathiafulvalene-tetracyanoquinodimethane
TTM	tracking tunneling microscopy
UHF	ultra-high frequency
UHV	ultra-high vacuum
UPD	underpotential deposition
UPS	ultraviolet photoelectron spectroscopy
UV	ultraviolet
VDI	variable deflection imaging
VDW	van der Waals
VSS	variable separation spectroscopy
XPS	x-ray photoelectron spectroscopy

Introduction

The general principle of operation of a scanning tunneling microscope (STM) – and related scanning probe microscopies (SPM) as well – is surprisingly simple. In STM a bias voltage is applied between a sharp metal tip and a conducting sample to be investigated (metal or doped semiconductor). After bringing tip and sample surface within a separation of only a few Ångström units[1] (1 Ångström unit (Å) = 0.1 nanometer (nm) = 10^{-10} m), a tunneling current can flow due to the quantum mechanical tunneling effect before 'mechanical point contact'[2] between tip and sample is reached. The tunneling current can be used to probe physical properties locally at the sample surface as well as to control the separation between tip and sample surface. The distance control based on tunneling is very sensitive to small changes in separation between the two electrodes because the tunneling current is strongly (exponentially) dependent on this separation, as we will see later (section 1.2). By scanning the tip over the sample surface while keeping the tunneling current constant by means of a feedback loop, we can follow the surface contours with the tip which – to a first approximation – will remain at constant distance from the sample surface. By monitoring the vertical position z of the tip as a function of the lateral position (x, y), we can get a three-dimensional image $z(x,y)$ of the sample surface. Motion of the tip both laterally and vertically with respect to the sample surface can be realized with sub-atomic accuracy by means of piezoelectric drives. The historical schematic drawing of the STM set-up is shown in Fig. 0.1.

The STM can be viewed as a powerful combination of three important conceptions: scanning, vacuum tunneling and point probing. Scanning

---

[1] The Ångström unit (Å) will often be used rather than the nanometer because it is the most appropriate length unit for dealing with structures on an atomic scale.

[2] The onset of mechanical point contact may not yet be clear conceptually. Possible definitions are provided later (section 1.21).

1

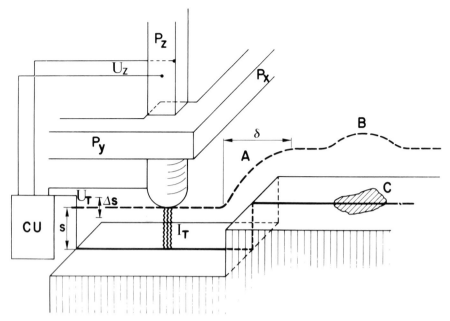

Fig. 0.1. Principle of operation of the STM. (Schematic: distances and sizes are not to scale.) Piezodrives P_x and P_y scan the metal tip over the surface. The control unit (CU) applies the appropriate voltage U_z to the piezodrive P_z for constant tunnel current I_T at constant tunnel voltage U_T. The broken line indicates the z displacement in a scan over a surface step (A) and a chemical inhomogeneity (B) (Binnig *et al.*, 1982b).

as a means for image generation is well known from other microscopies, e.g. scanning electron microscopy, or from television. Electron tunneling had already been used as a powerful experimental technique to probe physical properties of materials a long time before the invention of the STM. However, the tunnel junctions had a planar geometry and an oxide was used to separate the two planar metal electrodes. The tunnel barrier in these metal–oxide–metal tunnel junctions was neither adjustable nor well defined, due to inhomogeneities in the thickness of the oxide layer. Nevertheless, electron tunneling had already proved very important, particularly for probing the superconducting state (section 1.3). The STM makes use of a local geometry which is offered by a point probe tip as one of the two metal electrodes. Point probes have also been used successfully in the past, e.g. in point contact spectroscopy (section 1.7).

   In summary, the scanned point-like tip probing the sample locally via vacuum tunneling of electrons makes STM a unique microscopy. It needs no lenses, unlike other electron microscopes; therefore image distortions due to aberration cannot occur. The electrons involved in STM have an energy of no more than a few electronvolts (eV), in contrast to high-resolution electron microscopes where the electrons have energies of several keV up to MeV, making radiation-induced sample changes and damage very likely. In STM, the electron energies are even smaller than typical energies of chemical bonds, allowing non-destructive atomic resolution imaging. It may be argued that an electron energy E of about 1 eV, as is typical for STM experiments, may not be high enough to resolve individual atoms because the corresponding electron wavelength

$$\lambda = 12.25 \ \text{Å} / \sqrt{E} \simeq 12 \ \text{Å} \tag{0.1}$$

is larger than typical interatomic distances in solids of about 3 Å, which one would like to resolve. However, the STM is operated in the so-called 'near-field' regime where the distance d between tip and sample surface of typically a few Ångström units is comparable to or less than the electron wavelength λ: $d \lesssim \lambda$. In this regime, the spatial resolution which can be achieved is no longer diffraction-limited and is not determined by λ (chapter 3).

In contrast to other electron microscopes and surface analytical techniques using electrons, STM can be operated in air and in liquids as well as in vacuum because there are no free electrons involved in the STM experiment. Therefore, the application of STM is not limited to surface science, but has particularly great potential for *in situ* electro-chemical studies and *in vivo* investigations of biological specimens. STM investigations of the important solid–liquid interface may have even greater impact in the future than STM studies of the solid–vacuum interface because only a very limited number of *in situ* analytical techniques for the solid–liquid interface are available. However, the most significant experimental results during the first ten years of STM have been obtained under ultra-high vacuum (UHV) conditions on well-defined sample surfaces. Here, STM has profited a lot from the knowledge and experience which had been accumulated in the past decades by surface scientists regarding sample surface preparation and UHV technology.

The most important feature of STM, however, might be that this type of microscopy provides local information, ultimately with atomic resolution, directly in real-space, in contrast to diffraction experiments which are traditionally used for determination of the structure of condensed matter. Real-space information is particularly important for the study of non-periodic features such as defects (vacancies, interstitials, impurity sites, steps, dislocations and grain boundaries) and other chemical inhomogeneities. Therefore, STM has the greatest potential for the investigation of complex systems such as multicomponent materials, polycrystalline samples with grains and grain boundaries, composites and nanostructured materials. Other spatial inhomogeneities, e.g. flux vortices in type II superconductors or domain walls in magnetic materials, can also be probed by STM and related SPM techniques. The information gained by local probes, such as STM, and diffraction experiments can be regarded as complementary. Diffraction experiments provide information averaged over macroscopic sample volumes or surface areas and yield mean interatomic distances or lattice constants for crystalline materials with an accuracy which can never be reached by local probes. However, diffraction experiments generally do not provide an accurate picture of the degree of disorder and the nature of defect structures. For instance, it has become clear that a surface showing clear and sharp diffraction spots in low-energy electron diffraction (LEED) experiments can still appear highly defective when imaged by STM. On the other hand, the surface area which one is looking at by STM has been typically quite small (about 1 μm^2 or even less) in early STM investigations. This 'tunnel vision' might sometimes be dangerous when drawing conclusions for the whole surface from information gained at only a few locations of small lateral dimensions. The problem of how representative the obtained STM results are, is at least partly solved by considerably increasing the total scan range of STM/SPM instruments.

The brightest prospects for STM/SPM are offered in the field of nanometer-scale science and technology. As electronic devices become increasingly smaller, there is a strong need for understanding the physical properties of matter on a nanometer scale. Technology at the nanometer level requires nanopositioning and control, nanoprecision machining and reproducible creation of nanometer-scale structures as well as the use and control of super-smooth surfaces. These are tremendous challenges for which STM/SPM can provide appropriate solutions as will be discussed in chapter 8.

The development of STM/SPM has had, however, even greater impact than would appear by considering only those aspects mentioned so far. It should be emphasized that STM/SPM has particularly stimulated interdisciplinary research, mainly for two reasons. Firstly, STM/SPM has found broad applications in many different scientific disciplines such as condensed matter physics, chemistry, biology, metrology and materials science. It is not only the use of the same type of

microscope which provides a joining link between a large number of
scientists from different disciplines but also the existence of many
common projects of interdisciplinary nature for which the application
of the STM/SPM technique plays an important role. Secondly, as solid
state scientists can now probe matter locally down to the atomic level,
an increasing number of common scientific questions exist for scientists
having backgrounds in atomic and molecular physics and chemistry who
are starting from individual atoms and molecules and trying to understand more complex macromolecules (Fig. 0.2).

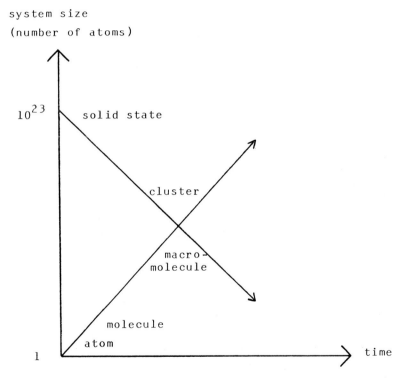

Fig. 0.2. Solid state physics deals with systems containing a large number
(typically of order 10^{23}) of atoms. Traditionally, the collective physical properties of the whole ensemble of atoms have been studied by using macroscopic
measurement techniques. Scanning probe methods now allow probing of physical properties of the solid state down to the scale of individual atoms. On the
other hand, atomic and molecular physics aim at the understanding of the
properties of larger molecules (macromolecules) based on the already known
properties of single free atoms and smaller molecules. The time has now come
at which an intersection of these two developments in research leads to an
increasing common interest within these two originally different branches in
physics and chemistry.

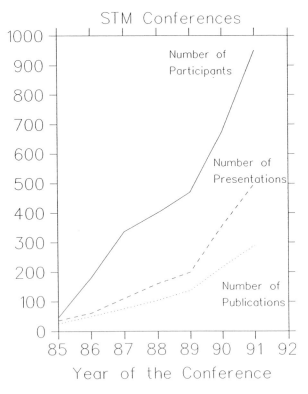

Fig. 0.3. Statistics of STM conferences between 1985 and 1991. Plotted are the number of participants, the number of presentations and the number of publications in the STM conference proceedings.

Furthermore, STM/SPM-type instruments are particularly attractive for universities, where education and research are equally important. Careful design and fine machining of the mechanical parts of the instruments are as important as a profound background in electronics and modern computer automation, including computer graphics. Increasing interest in the STM/SPM technique also exists in industry, where individual fabrication steps can be controlled on the assembly line. Commercially available STM/SPM instruments are generally much less expensive than high-resolution electron microscopes. These and other reasons might be given to explain the fact that ten years after the invention of the STM, several thousand instruments already exist world-wide and are being built. The still increasing interest in the STM/SPM technique is reflected in the statistics of the series of international STM conferences which started in 1985 (Fig. 0.3). Perhaps we should not forget to mention an additional important (non-scientific?) reason why STM/SPM has attracted such a large number of scientists: the fascination of real-space images of atomic structures, sometimes exhibiting all the beauty that nature can offer to us on that scale . . .

Part one

Experimental methods and theoretical background of scanning probe microscopy and spectroscopy

1

Scanning tunneling microscopy (STM)

Before we focus on the different modes of operation of a scanning tunneling microscope (STM) and the information which can be extracted, a historical review of earlier studies of tunneling phenomena is given in the following sections. This will serve as an introduction to tunneling experiments as well as motivation for a variety of experimental methods applied in STM. It will become clear later in this chapter that most of the special effects now studied in STM have already been investigated before by using planar metal–oxide–metal tunnel junctions.

1.1 Historical remarks on electron tunneling

Tunneling is an important mechanism of transport in condensed matter and across artificial junctions. In contrast to other transport mechanisms such as diffusion and drift, which can be described on the basis of classical physics, tunneling can only be understood in terms of quantum theory.

Consider a potential energy barrier and a microscopic particle, e.g. an electron, with an energy smaller than the potential barrier height. From the viewpoint of classical physics, this particle will never be able to traverse this barrier. However, in quantum theory, the wave–particle dualism may in fact allow this electron to traverse the barrier. The wave nature of microscopic particles (de Broglie, 1923) impinging upon a potential energy barrier is expressed in the experimental observation of a finite probability of finding the particles beyond the barrier and they are then said to have tunneled through it. The essential difference between classical theory and quantum theory with respect to tunneling may be illustrated as in Fig. 1.1.

Tunneling phenomena play an important role in solid state physics, nuclear physics and chemical physics, as well as in biology. Some land-

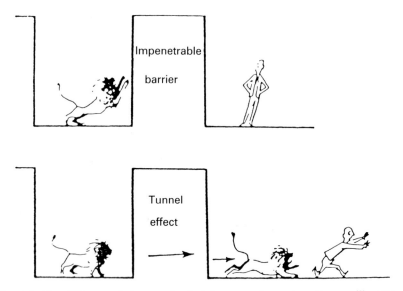

Fig. 1.1. The difference between classical theory and quantum theory, illustrating tunneling through a potential barrier (Bleaney, 1984).

marks in the science of tunneling are summarized in Table 1.1. Among the first applications of tunneling theory have been auto-ionization of excited states of atomic hydrogen in a strong external electric field (Oppenheimer, 1928), field emission from metals in external electric fields (Fowler and Nordheim, 1928), α-decay of heavy nuclei (Gamov, 1928), dissociation of molecules (Rice, 1929), contact resistance between two conductors separated by an insulating layer (Frenkel, 1930; Holm and Meißner, 1932, 1933) and interband tunneling in solids (Zener, 1934). Illustrations of some of these tunneling-based phenomena in potential energy diagrams are presented in Fig. 1.2(*a*)–(*c*).

In the following, we will first review some basic theoretical treatments of electron tunneling before continuing with our historical review of tunneling experiments.

1.2 Theoretical treatment of one-dimensional electron tunneling

To get an overview of the various theoretical treatments of electron tunneling it is important to make the following clear distinctions.

(a) Elastic versus inelastic tunneling. If the energy of the electron is conserved in the tunneling process, i.e. the electron energy is equal in the initial and final states, it is said that the electron has tunneled

Table 1.1. *Landmarks in the science of tunneling (Roy, 1986)*

	Phenomena	Investigators	Year
1	Observation of field emission from metals	Lilienfeld	1922
2	Ionization of hydrogen atoms by electron tunneling	Oppenheimer	1928
3	Explanation of field emission	Fowler and Nordheim	1928
4	Alpha-decay theory	Gamov	1928
		Gurney and Condon	1928
5	Theory of interband tunneling in solids	Zener	1934
6	Field-emission microscope (FEM)	Müller	1937
7	Observation of Zener breakdown	Chynoweth and Mckay	1957
8	Tunneling in degenerate p–n junctions	Esaki	1958
9	Extension of Zener's theory to tunnel diodes	Keldysh	1958
		Price and Radcliffe	1959
		Kane	1961
10	Measurement of energy gap of superconductors	Giaever	1960
11	Perturbation treatment of tunneling	Bardeen	1961
12	Tunneling of Cooper particles	Josephson	1962
13	Experimental verification of the Josephson effect	Anderson and Powell	1963
		Rowell	1963
		Fiske	1964
14	Inelastic tunneling spectroscopy (IETS)	Jaklevic and Lambe	1966
15	Point contact tunneling	Levinstein and Kunzler	1966
		von Molnar *et al.*	1967
16	Experimental observation of Coulomb blockade	Zeller and Giaever	1969
17	Observation of tunneling tails	Lea and Gomer	1970
		Gadzuk and Plummer	1971
18	Spin-polarized tunneling	Tedrow and Meservey	1971
19	Vacuum tunneling and topografiner	Young *et al.*	1971
20	Development of scanning tunneling microscope (STM)	Binnig *et al.*	1982
21	Theory of traversal time for tunneling	Büttiker and Landauer	1982
22	Theory of Coulomb blockade	Ben-Jacob and Gefen	1985
		Averin and Likharev	1986

(a)

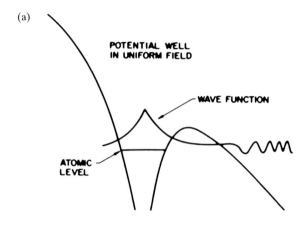

POTENTIAL WELL
IN UNIFORM FIELD

WAVE FUNCTION

ATOMIC
LEVEL

(b)

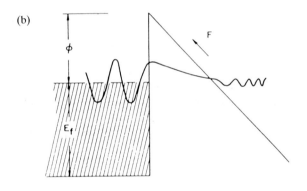

ϕ

F

E_f

(c)

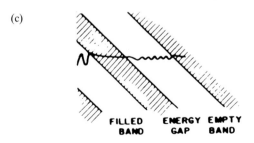

FILLED   ENERGY   EMPTY
BAND   GAP   BAND

Fig. 1.2. (*a*) Ionization of a hydrogen atom in a large external electric field. (*b*) Field emission from metals in an external electric field based on Fowler–Nordheim tunneling. (*c*) Interband tunneling in solids (Esaki, 1969).

elastically. In contrast, if the electron has gained or lost energy due to interaction with elementary excitations such as phonons, magnons or plasmons, the tunneling process is called inelastic. In the following sections, we will first concentrate on elastic tunneling processes, whereas inelastic electron tunneling is treated in section 1.4.

(b) One-dimensional versus three-dimensional potential barriers. Early theoretical treatments of electron tunneling have mainly assumed a one-dimensional potential barrier, which was an appropriate model for the experimentally used planar metal–oxide–metal tunnel junctions. With the replacement of one planar metal electrode by a tip, as in STM, a three-dimensional potential barrier should be used. We will first treat tunneling through one-dimensional potential barriers, whereas the three-dimensional case will be considered later (section 1.11).

(c) Rectangular barriers versus barriers of arbitrary shape. The shape of the potential barrier will first be assumed to be rectangular before treating the general case of a potential barrier of arbitrary shape.

(d) Time-independent versus time-dependent approach. Tunneling phenomena can theoretically be treated in two different ways. One may either use a time-independent approach starting with Schrödinger's time-independent wave equations inside and outside the barrier or a time-dependent perturbation approach starting with Schrödinger's time-dependent wave equation and using Fermi's 'golden rule' of first-order time-dependent perturbation theory.

1.2.1 Elastic tunneling through a one-dimensional rectangular potential barrier using the time-independent approach ('wave-matching method')

We first consider a one-dimensional rectangular potential barrier as shown in Fig. 1.3 and an impinging electron of energy E and mass m. The time-independent Schrödinger equations for the three regions ($j= 1,2,3$) and an *Ansatz* for the corresponding wave functions ψ_j can easily be written.

For region 1:

$$-\frac{\hbar^2}{2m}\frac{d^2\psi_1}{dz^2} = E\psi_1$$

$$\psi_1 = e^{ikz} + Ae^{-ikz} \tag{1.1a}$$

with $k^2 = 2mE/\hbar^2$.

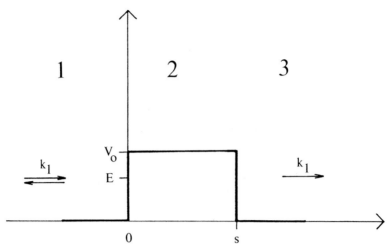

Fig. 1.3. One-dimensional rectangular potential barrier with height V_0 and width s.

For region 2:

$$-\frac{\hbar^2}{2m}\frac{d^2\psi_2}{dz^2} + V_0\psi_2 = E\psi_2$$

$$\psi_2 = B'e^{ik'z} + C'e^{-ik'z} = Be^{-\varkappa z} + Ce^{\varkappa z} \qquad (1.1b)$$

with $\varkappa^2 = -k'^2 = 2m\,(V_0 - E)/\hbar^2$.

For region 3:

$$-\frac{\hbar^2}{2m}\frac{d^2\psi_3}{dz^2} = E\psi_3$$

$$\psi_3 = De^{ikz} \qquad (1.1c)$$

where $\hbar$ is Planck's constant divided by 2π.

We are interested in the barrier transmission coefficient T which is the ratio of the transmitted current density j_t and the incident current density j_i which are given by

$$j_t = \frac{-i\hbar}{2m}\left(\psi_3{}^*(z)\frac{d\psi_3}{dz} - \psi_3(z)\frac{d\psi_3{}^*(z)}{dz}\right)$$

$$= \frac{\hbar k}{m}\,|D|^2$$

$$j_i = \frac{\hbar k}{m} \qquad (1.2)$$

and therefore

$$T = \frac{j_t}{j_i} = |D|^2 \qquad (1.3)$$

By matching the ψ_j and their first derivatives $d\psi_j/dz$ at the discontinuities of the potential $V(z)$ at $z = 0$ and $z = s$ to construct the overall wave function, the coefficient D and therefore T can be derived:

$$T = \frac{1}{1 + (k^2 + \varkappa^2)^2 / (4k^2\varkappa^2)\sinh^2(\varkappa s)} \qquad (1.4)$$

This expression for T is exact. We now consider the limit of a strongly attenuating barrier ($\varkappa s \gg 1$) and obtain

$$T \approx \frac{16\, k^2\, \varkappa^2}{(k^2 + \varkappa^2)^2} \cdot e^{-2\varkappa s} \qquad (1.5)$$

with the decay rate

$$\varkappa = [2m\,(V_0 - E)]^{1/2} / \hbar$$

The dominant contribution to T comes from the factor $\exp(-2\varkappa s)$. As we will see later, the strong exponential dependence of T on the barrier width s and the square root of an effective barrier height $(V_0 - E)^{1/2}$ is typical for tunneling, independent of the exact shape of the barrier. Assuming a barrier width of 5 Å and an effective barrier height of 4 eV, we get a value of about 10^{-5} for the exponential factor. Changing the barrier width by 1 Å then typically leads to a change of the barrier transmission by one order of magnitude. The extreme sensitivity of the tunneling barrier transmission to the barrier width led Binnig, Rohrer and coworkers to the idea that a microscope based on tunneling should provide extremely high spatial resolution (section 1.11).

1.2.2 Elastic tunneling through a one-dimensional rectangular potential barrier using the time-dependent approach

If the barrier transmission T is small, such as in the example provided in the preceding section 1.2.1, a perturbation treatment of tunneling seems very appropriate. However, there is no obvious way of introducing a term in the Hamiltonian which can be treated as small. Instead of introducing states which are the exact solutions of an approximate

Hamiltonian, Bardeen (1961) introduced approximate solutions of the exact Hamiltonian (Fig. 1.4):

$$\psi_l(z) = ae^{-\varkappa z}; \quad z \geqslant 0$$
$$\psi_r(z) = be^{\varkappa z}; \quad z \leqslant s \qquad (1.6)$$

$\psi_1(z)$ has to be matched to the correct solution of the Schrödinger equation for $z \leqslant 0$ but has to decay in the region $z \geqslant s$ instead of satisfying the Schrödinger equation. Similarly, $\psi_r(z)$ has to be matched to the correct solution for $z \geqslant s$ but has to decay in the region $z \leqslant 0$ instead of satisfying the Schrödinger equation. Consequently, ψ_1 is a correct solution of the Hamiltonian H for $z \leqslant s$ and ψ_r is correct for $z \geqslant 0$.

One can then consider an electron being initially in state ψ_1 and compute the transition rate into state ψ_r. We start with the time-dependent Schrödinger equation:

$$H\psi(t) = i\hbar \, \frac{d\psi(t)}{dt} \qquad (1.7)$$

(i is the square root of -1) with

$$H = (H_1 + H_r) + H_T = H_0 + H_T$$
$$H_0 \, \psi_1 = E_1 \, \psi_1$$

where H_1 (H_r) is the Hamiltonian for the left (right) electrode. By substituting the *Ansatz*

$$\psi(t) = c(t)\psi_1 e^{-iE_1 t/\hbar} + d(t)\psi_r e^{-iE_r t/\hbar} \qquad (1.8)$$

into Eq. (1.7), one can identify

$$M_{rl} = \int \psi_r^{*} (H - E_1)\psi_1 \, dz = \int \psi_r^{*} H_T \psi_1 \, dz \qquad (1.9)$$

as the effective tunneling matrix element. H_T is the transfer Hamiltonian which describes electron tunneling from one electrode to the other.

For strongly attenuating potential barriers, as already considered in section 1.2.1, we can use Fermi's 'golden rule' of first-order time-dependent perturbation theory which gives the following expression for the transmitted current:

$$j_t = \frac{2\pi}{\hbar} \, |M_{rl}|^2 \, \frac{dN}{dE_r} \qquad (1.10)$$

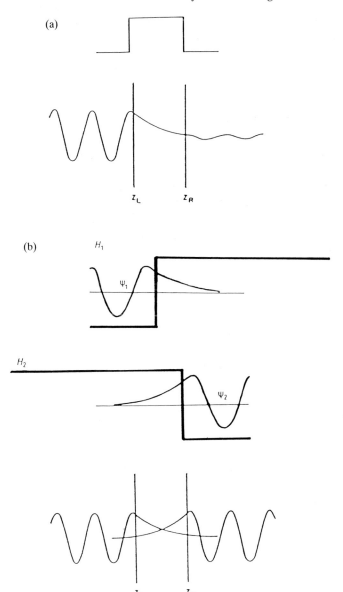

Fig. 1.4. Schematic drawings of the wave functions across a one-dimensional tunnel barrier. In the 'exact' calculation (*a*) an incident state from the left decays exponentially into the barrier and has some probability of appearing on the right. In the transfer Hamiltonian formalism (*b*), initial and final wave functions on either side of the barrier 'transfer' from one side to the other at a finite rate (Kirtley, 1978).

where dN/dE_r is the density of states in the final state. By using the explicit expressions for ψ_l and ψ_r given in Eq. (1.6), one obtains the same result for the barrier transmission in Bardeen's 'transfer Hamiltonian' approach as we have obtained before (Eq. (1.5)). Therefore, the time-independent and time-dependent approach have proven to be equivalent in this case. The transfer Hamiltonian method, however, has the advantage that it is more generally applicable, in particular it is not restricted to one-dimensional tunneling problems.

Based on Eq. (1.10) it is often stated that the tunneling current, as evaluated from Bardeen's transfer Hamiltonian approach, is proportional to the electronic density of states dN/dE_r. However, after calculating the tunneling matrix element M_{rl} explicitly and arriving at the exact expression for the barrier transmission in Eq. (1.4), this expression is no longer simply related to the density of states. Therefore, the interpretation of tunneling experiments as a quantitative probe of the electronic density of states must generally be regarded with great caution. As we will see later (section 1.3.1), there are special cases, such as tunneling in metal–oxide–superconductor junctions, where the tunneling current in fact provides quite accurate information about the density of states. However, in each case, this interpretation of the tunneling current has to be justified theoretically, which was done, for instance, for tunneling into superconductors (Cohen *et al.*, 1962).

1.2.3 Elastic tunneling through a one-dimensional potential barrier of arbitrary shape using the WKB approximation

An extension to potential barriers of arbitrary shape can be made by using either a multistep potential approximation (Ando and Itoh, 1987) or the WKB approximation developed by Wentzel, Kramers and Brillouin in 1926 and described in elementary textbooks of quantum theory. The WKB approximation is based on treating Planck's constant as a small number. This method, though semiclassical, can be applied even for classically forbidden regions where the energy of the particle is smaller than the height of a potential barrier. Therefore we can use it to treat tunneling phenomena. However, easy application of the WKB method is restricted to one-dimensional problems, although the generalization to a multidimensional space is possible (e.g. Das and Mahanty, 1987; Huang *et al.*, 1990).

The following expression for the probability $D(E)$ that an electron can penetrate a one-dimensional potential barrier $V(z)$ of arbitrary shape is found within the WKB approximation:

$$D(E) = \exp\left\{-\frac{2}{\hbar}\int_{s_1}^{s_2}[2m\,(V(z)-E)]^{1/2}\,dz\right\}$$

$$= \exp\left\{-2\int_{s_1}^{s_2}\varkappa(z,E)\,dz\right\} \tag{1.11}$$

where s_1 and s_2 are the classical turning points and $(s_2 - s_1)$ is the width of the barrier. The WKB approximation is adequate only if the energy E is not near or above the top of the barrier ($E \ll V$) and if the sides of the barrier at s_1 and s_2 are gently sloping.

1.2.4 Elastic tunneling in planar metal–insulator–metal junctions

We now focus on planar metal–insulator–metal tunnel junctions which can be described by a one-dimensional potential energy diagram as shown in Fig. 1.5. First, we consider similar metal electrodes with free-electron behaviour. The insulator is treated as if it were a vacuum. We further assume that the tunneling electrodes are essentially in thermal equilibrium. Nonequilibrium effects in tunneling are discussed by Wolf (1985).

A net tunneling current results from the difference $\Delta N = N_1 - N_2$ of the number N_1 of electrons tunneling from the left metal electrode to the right and the number N_2 of electrons tunneling from the right

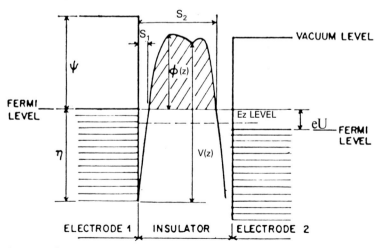

Fig. 1.5. General tunnel barrier between two metal electrodes (Simmons, 1963a).

to the left after applying a bias voltage U on the junction. We first introduce

v_z z component of the electron velocity

$n(v_z)dv_z$ number of electrons per unit volume with a z component of velocity between v_z and $v_z + dv_z$

$D(E_z)$ probability of an electron with energy component $E_z = (mv_z^2)/2$ tunneling through the potential barrier $V(z)$. (A major consequence of the free-electron model is the dependence of D only on E_z)

$f(E)$ Fermi–Dirac distribution function

and

$$E_\parallel = \tfrac{1}{2}mv_\parallel^2 = \tfrac{1}{2}m(v_x^2 + v_y^2)$$

We then obtain

$$N_1 = \int_0^{v_{max}} v_z\, n(v_z)\, D(E_z)\, dv_z = \frac{1}{m}\int_0^{E_{max}} n(v_z)\, D(E_z)\, dE_z$$

$$= \frac{m}{2\pi^2\hbar^3} \int_0^{E_{max}} D(E_z)\, dE_z \int_0^\infty f(E)\, dE_\parallel \tag{1.12}$$

and similarly

$$N_2 = \frac{m}{2\pi^2\hbar^3} \int_0^{E_{max}} D(E_z)\, dE_z \int_0^\infty f(E + eU)\, dE_\parallel \tag{1.13}$$

Therefore

$$\Delta N = \int_0^{E_{max}} D(E_z)\, dE_z \left\{ \frac{m}{2\pi^2\hbar^3} \int_0^\infty \big(f(E) - f(E + eU)\big)\, dE_\parallel \right\} \tag{1.14}$$

where e is the charge of an electron. We can take $D(E_z)$ from Eq. (1.11):

$$D(E_z) = \exp\left\{ -\frac{2}{\hbar} \int_{s_1}^{s_2} [2m(V(z) - E_z)]^{1/2}\, dz \right\}$$

$$= \exp\left\{ -\frac{2\cdot(2m)^{1/2}}{\hbar} \int_{s_1}^{s_2} [E_{F_1} + \phi(z) - E_z]^{1/2}\, dz \right\} \tag{1.15}$$

where we have used

$$V(z) = E_{F_1} + \phi(z) \tag{1.16}$$

It is obvious from the WKB expression for the tunneling probability (Eq. (1.15)) that electrons with a large energy component E_z close to the Fermi energy E_{F_1} of the negatively biased electrode tunnel most effectively. The reason is, of course, that these electrons experience the lowest effective potential barrier height.

Following Simmons (1963a), we introduce a mean potential barrier height ϕ above the Fermi level E_{F_1} of the negatively biased electrode:

$$\bar{\phi} = \frac{1}{\Delta s} \int_{s_1}^{s_2} \phi(z)\, dz \qquad (1.17)$$

with $\Delta s = s_2 - s_1$, and obtain the following expression for the tunneling current density J at zero temperature:

$$J = \frac{e}{4\pi^2 \hbar (\Delta s)^2} \cdot \left\{ \bar{\phi} \exp(-A\bar{\phi}^{\frac{1}{2}} \Delta s) - (\bar{\phi} + eU) \exp(-A(\bar{\phi} - eU)^{\frac{1}{2}} \Delta s) \right\} \quad (1.18)$$

with $A = 2 \cdot (2m)^{1/2}/\hbar$. The first term in Eq. (1.18) can be interpreted as a current density flowing from the left to the right metal electrode whereas the second term represents a current density flowing from the right to the left electrode. The difference between these two results is the net current density J. If no bias voltage is applied ($U = 0$), the current density flowing in either direction is the same and the net current density vanishes.

Simmons (1963a) also gave expressions for the tunnel current density in the following limits.

(a) Low-bias voltage range ($U \approx 0$, i.e. $eU \ll \bar{\phi}$):

$$J = \frac{e^2}{4\pi^2 \hbar^2} \cdot \frac{(2m\bar{\phi})^{\frac{1}{2}}}{\Delta s} \cdot U \exp(-A\bar{\phi}^{\frac{1}{2}} \Delta s)$$

$$= \frac{e^2}{8\pi^2 \hbar} \cdot \frac{A\bar{\phi}^{\frac{1}{2}}}{\Delta s} \cdot U \exp(-A\bar{\phi}^{\frac{1}{2}} \Delta s) \qquad (1.19)$$

As already seen in section 1.2.1, we find again the characteristic exponential dependence of the tunneling current on the width of the potential barrier and the square root of the mean barrier height. In addition, we obtain a linear dependence of the tunneling current on the applied bias voltage in the low-bias limit, i.e. the junction exhibits Ohmic behavior at low bias.

(b) Rectangular barrier of height ϕ_0 and width s and intermediate-bias voltage range $(U < \phi_0/e)$:

$$J = \frac{e}{4p^2\hbar s^2}\left\{\left(\phi_0 - \frac{eU}{2}\right)\exp\left[-\frac{2(2m)^{1/2}}{\hbar}\left(\phi_0 - \frac{eU}{2}\right)^{1/2}s\right]\right.$$

$$\left.-\left(\phi_0 + \frac{eU}{2}\right)\exp\left[-\frac{2(2m)^{1/2}}{\hbar}\left(\phi_0 + \frac{eU}{2}\right)^{1/2}s\right]\right\} \tag{1.20}$$

(c) Rectangular barrier and high-bias voltage range $(U > \phi_0/e)$:

$$J = \frac{2.2\,e^3F^2}{16p^2\hbar\phi_0}\left\{\exp\left[-\frac{4}{2.96\,\hbar eF}(2m)^{1/2}\,\phi_0^{3/2}\right]\right.$$

$$\left.-\left(1 + \frac{2eU}{\phi_0}\right)\exp\left[-\frac{4}{2.96\,\hbar eF}(2m)^{1/2}\,\phi_0^{3/2}\left(1 + \frac{2eU}{\phi_0}\right)^{1/2}\right]\right\} \tag{1.21}$$

where $F = U/s$ is the electric field strength in the insulator.

Simmons (1963b) also considered metal–insulator–metal tunnel junctions with dissimilar metal electrodes having different work functions ϕ_1 and ϕ_2. In this case, an intrinsic electric field is present within the insulator resulting from the contact potential $(\phi_2 - \phi_1)/e$ that exists between two dissimilar electrodes. As a consequence, the potential barrier becomes asymmetric and the $I–U$ characteristic, i.e. the dependence of the tunneling current I on the applied bias voltage U, becomes polarity-dependent. At relatively low bias voltage, greater current flows when the electrode with the lower work function is positively biased. At higher bias voltages, the direction of rectification is reversed, i.e. greater current flows when the electrode with the lower work function is negatively biased.

The temperature-dependence of the tunneling current has been calculated by Stratton (1962), who found the following relation:

$$\frac{j(T)}{j(T=0)} = \frac{\pi ck_BT}{\sin(\pi ck_BT)} \approx 1 + \tfrac{1}{6}(\pi c_1 k_BT)^2 + \ldots \tag{1.22}$$

where k_B is the Boltzmann constant and c is a function of the applied bias voltage U only.

1.2.5 Field emission or Fowler–Nordheim regime and Gundlach oscillations

For very high applied bias voltages $(U > (\phi_0 + E_{F_1})/e)$, Eq. (1.21) reduces to

$$J = \frac{2.2\,e^3 \cdot F^2}{16\,\pi^2 \hbar\,\phi_0} \left\{ \exp\left[-\frac{4}{2.96\,\hbar e F}\,(2m)^{\frac{1}{2}}\,\phi_0^{\frac{3}{2}} \right] \right\} \qquad (1.23)$$

This expression for the current density J is very similar to the Fowler–Nordheim equation which describes electron emission from metals in strong external electric fields, as depicted in Fig. 1.2(b). (In the Fowler–Nordheim equation, the numerator 2.96 is replaced by 3 and the multiplication factor 2.2 does not appear.) According to Eq. (1.23), the bias dependence of the current density in the field emission or Fowler–Nordheim regime can be expressed by

$$J \propto U^2 \exp\left(-\frac{const.}{U} \right) \qquad (1.23b)$$

which is characteristically different from the bias dependence in the low-bias tunneling regime.

Gundlach (1966) additionally predicted that, in the high-bias regime, the tunneling current should exhibit an oscillatory behavior around the classical Fowler–Nordheim bias dependence (Fig. 1.6). He attributed these oscillations to an interference of the incident electron waves and electron waves which are reflected at the sharp barrier–electrode interface. Such oscillations have later been found experimentally, e.g. for metal–oxide–semiconductor tunnel junctions (Maserjian 1974; Maserjian and Zamani, 1982). However, an alternative explanation of these oscillations based on resonant tunneling through localized states in the insulating oxide (see section 1.2.7) has also been proposed (Feuchtwang et al., 1977).

1.2.6 Tunneling including the image potential

To describe tunneling between two metal electrodes correctly, we also have to consider the interaction of the tunneling electrons with the two metal surfaces. This interaction is classically described by the multiple image potential:

$$V_i(z) = -\frac{e^2}{4\pi\varepsilon_0} \left\{ \frac{1}{2z} + \sum_{n=1}^{\infty} \left[\frac{nd}{\left[(nd)^2 - z^2\right]} - \frac{1}{nd} \right] \right\} \qquad (1.24)$$

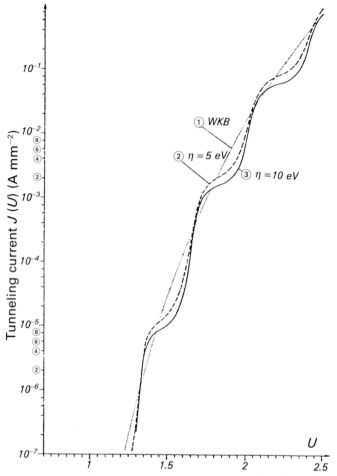

Fig. 1.6. Gundlach oscillations of the tunnel current as a function of the applied bias voltage in the Fowler–Nordheim regime (Gundlach, 1966).

where ε_0 is the dielectric constant and d is the distance between the two image planes. Eq. (1.24) can be approximated by (Simmons, 1963a)

$$V_i(z) = -\frac{2.3\,(\ln 2)\,e^2}{16\,\pi\,\varepsilon_0 d}\cdot\left[\frac{z}{d}\left(1-\frac{z}{d}\right)\right]^{-1} \tag{1.25}$$

The inclusion of the image potential has several consequences (see Fig. 1.7).

1. The corners of the barrier that was originally assumed rectangular are rounded off.

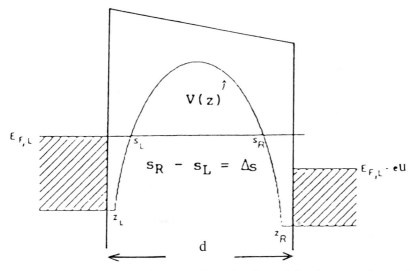

Fig. 1.7. Potential diagram for a one-dimensional metal–insulator–metal tunnel junction including the classical image potential. $V(z)$ is the total potential barrier, $\Delta s = s_R - s_L$ is the effective barrier width at the Fermi energy E_F, and d is the distance between the two image planes (Nguyen *et al.*, 1986).

2. The thickness of the barrier is reduced.
3. As a consequence of the decreased height and width of the barrier, the tunneling current flow between the two electrodes is increased.

It is evident from Eq. (1.25) that the classical image potential cannot describe the interaction between the tunneling electron and the metal electrodes in the vicinity of the two surfaces because the classical image potential diverges to $-\infty$ as the electron approaches either of the two metal surfaces (at $z = 0$ or $z = d$). This is in contrast to the finite ('inner') potential inside a conductor which – for a jellium model – is equal to the sum of the Fermi energy of the metal and the work function of the metal surface. At small distances ($\lesssim 5$ Å) between the electron and a metal surface, quantum theory has to be applied to describe their interaction, including exchange and correlation effects. The classical image potential can then be shown to be the asymptotic limit, as $z \to \infty$, of the quantum mechanical exchange and correlation energy.

Another problem with the application of Eq. (1.25) is the exact location of the two image planes. For the macroscopic scale of classical physics, this is not an important issue. However, as the electron approaches the metal surface to atomic-scale distances, the exact location of the image plane within the classical picture becomes significant.

It is reasonable that the image plane should lie somewhere between the centers of the atoms in the topmost surface layer and their atomic radii, but the precise location will depend on the detailed electronic charge distribution near the metal surface.

Finally, the classical image potential method assumes that the electron charge is stationary, which in fact is not the case. Since the electrons in the metal cannot respond instantaneously to the motion of the electron charge outside, the motion of the image charge is retarded, leading to dynamical corrections to the classical image potential.

The limitations of the classical image charge method as well as their extension have recently been reviewed (Jennings and Jones, 1988). Vacuum tunneling between two metal surfaces, where the separation of the electrodes can be varied continuously, can provide important information about the classical image potential as well as the transition to the quantum regime.

1.2.7 Band structure effects in elastic tunneling

At the beginning of section 1.2.4, we made the assumption that the two metal electrodes can be described in terms of a free-electron model and the insulator was treated as if it were a vacuum. We now discuss the influence of the band structure on the tunneling current, where we can distinguish between three major effects (Duke, 1969).

1.2.7.1 Influence of the electrodes' band structure on tunneling probability

One consequence of the free-electron model is the dependence of the tunneling probability D only on the energy component E_z. More generally, D has to be considered as a function of the total energy E of the tunneling electron and the wave vector component $k_{||}$ parallel to the junction, which are both conserved in the tunneling process. The quantities E, E_z, $E_{||}$ and $k_{||}$ are related by

$$E_z = E - E_{||}(k_{||}) \tag{1.26}$$

From a phenomenological point of view, within the effective mass approximation, $E_{||}(k_{||})$ is given by

$$E_{||}(k_{||}) = E_{||}(k_{||}) = \frac{\hbar^2 k_{||}^2}{2m^*} \tag{1.27}$$

where m^* is the effective electron mass which is different from the free-electron mass.

The dependence of D upon E and $k_\parallel$ (i.e. upon E and $E_\parallel$ within the effective-mass approximation) has to be taken into account explicitly in calculations of the tunneling current. Band structure effects are particularly important for the case of tunnel junctions with at least one semiconductor electrode (sections 1.5 and 1.6) where the tunneling probability D may change substantially at the onset of tunneling into a 'new' band. However, band structure effects can also be observed in metal–oxide–metal tunnel junctions (Jaklevic and Lambe, 1973).

1.2.7.2 Influence of the electrodes' band structure on effective tunneling barrier height

Another band structure effect is the dependence of the effective tunneling barrier height on $k_\parallel$ of the tunneling electron which results from the $k_\parallel$ conservation law. In the free-electron model, the barrier height is simply given by

$$V(z) - E_z = \frac{\hbar^2}{2m} \varkappa^2 \tag{1.28}$$

To take band structure effects into account, we have to use the relation Eq. (1.26). Within the effective-mass approximation we obtain for the effective barrier height

$$V(z) - \left(E - E_\parallel(k_\parallel)\right) = \frac{\hbar^2}{2m} \varkappa^2 \tag{1.29}$$

The decay rate $\varkappa$ is given by

$$\varkappa^2 = k_\parallel^2 + \frac{2m^*}{\hbar^2}\left[V(z) - E\right] \tag{1.30}$$

For a given energy E the decay rate therefore increases with larger $k_\parallel$. As a consequence, electrons in states with a large parallel wave vector component $k_\parallel$ tunnel much less effectively than electrons in states with small $k_\parallel$.

1.2.7.3 Influence of the insulator's band structure on tunneling current

The band structure of a single-crystal insulator used as tunneling barrier can also influence the tunneling current. This is seen from the WKB expression for the tunneling probability (Eq. (1.11)):

$$D(E) = \exp\left(-2\int_{s_1}^{s_2} \varkappa(E)\, dz\right) \tag{1.31}$$

Neglecting the dependence of the decay constant $\varkappa$ on $k_{\parallel}$ (Eq. (1.30)) to a first approximation, we have

$$\varkappa^2 (z,E) = \frac{2m^*}{\hbar^2}\left[V(z)-E\right] \qquad (1.32)$$

The z-dependence of $\varkappa$ results from the z-dependence of the potential $V(z)$ which itself is caused by the presence of an electric field within the insulator built up by the inner contact potential and the externally applied bias voltage. For a uniform electric field, $V(z)$ is proportional to z (trapezoidal barrier approximation) and the tunneling probability can be expressed by

$$\ln D(E) = \frac{-2s}{\chi_1 - \chi_2 + eU}\int_{E_1}^{E_2}\varkappa(E)\,\mathrm{d}E \qquad (1.33)$$

where the initial and final state energies of the tunneling electron, E_1 and E_2, as well as the positions of the Fermi levels of the two metal electrodes, χ_1 and χ_2, are measured with respect to the bottom of the insulator's conduction band (Fig. 1.8). Since tunneling occurs predominantly with electron energies close to E_{F_1} (section 1.2.4), the relation (1.33) can be approximated by

$$\ln D(E) \approx \frac{-2s}{\chi_1 - \chi_2 + eU}\int_{-\chi_1}^{-\chi_2+eU}\varkappa(E)\,\mathrm{d}E \qquad (1.34)$$

Based on measurements of the bias voltage dependence of the tunneling current in metal–insulator–metal tunnel junctions with a single-crystalline insulating barrier, one can extract $\varkappa(E)$ by 'inverting' the relation Eq. (1.34). Since $\varkappa$ can be interpreted as the imaginary part of a propagation wave vector $k' = i\varkappa$ (Eq. (1.1)) within the insulator, the one-electron dispersion relation $E(k')$ within the forbidden energy gap of the insulator can be derived (Kurtin *et al.*, 1970). Tunneling can therefore be used as a direct probe of the exponentially damped electronic states within the energy gap of insulators.

1.2.8 Resonant tunneling

We now consider a one-dimensional rectangular double barrier as shown in Fig. 1.9. The transmission coefficient T for elastic tunneling through such a double barrier can be calculated either by using the wave matching procedure within the time-independent approach

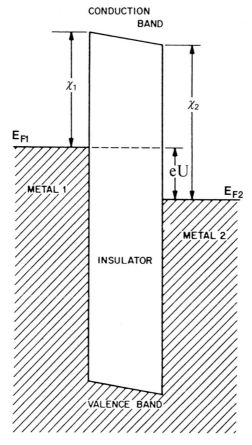

Fig. 1.8. Energy diagram of a metal–insulator–metal tunnel junction (Mead, 1969).

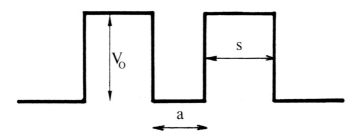

Fig. 1.9. The one-dimensional double tunnel barrier.

(section 1.2.1, see Burstein and Lundquist (1969)) or by using the transfer Hamiltonian approach (section 1.2.2, see Payne (1986)). One obtains

$$T = \frac{2k^2\varkappa^2}{(k^2+\varkappa^2)^2} \cdot \left(\frac{a}{2}+\frac{1}{\varkappa}\right)^{-1} \cdot e^{-2\varkappa s} \qquad (1.35)$$

with $k^2 = 2mE/\hbar^2$ and $\varkappa^2 = 2m(V_0 - E)/\hbar^2$.

The exponentially dependent part of the transmission coefficient which describes the barrier attenuation is equal to the case of a single rectangular tunneling barrier (Eq. (1.5)). By looking carefully at the amplitude of the transmission coefficient, one can additionally find a condition

$$\left[(k^2 - \varkappa^2) \cdot \sin(ka) - 2k\varkappa\cos(ka)\right] = 0 \qquad (1.36)$$

for which the transmission coefficient becomes of the order of unity. There is no barrier attenuation in this case, there is only reflection due to mismatch at the various edges of the double barrier potential. This is the situation where 'resonant tunneling' can be observed.

Resonant transmission through potential barriers is very familiar to solid state physicists. Within the Kronig–Penney model of a one-dimensional crystal, which consists of a series of identical rectangular potential barriers and square wells, one gets allowed bands of perfect transmission separated by forbidden bands of attenuation.

An experimental realization of the one-dimensional double potential barrier shown in Fig. 1.9 is given by semiconductor heterostructures, e.g. GaAs –Al_xGa_{1-x} As heterostructures fabricated by molecular beam epitaxy (MBE) where resonant tunneling can be observed directly (Chang *et al.*, 1974; Sollner *et al.*, 1983).

Resonant tunneling can also occur via localized states in the oxide barrier. These localized states, capable of trapping electrons, originate either from defects which are usually present in the oxide layer or from intentionally doped impurities (Koch and Hartstein, 1985). An early theoretical treatment based on a one-dimensional model already predicted that the tunneling current flow through oxide films containing traps is increased relative to the case of trap-free oxide layers and that the current increase is greatest when the trapping level is near the Fermi energy of the metal (Penley, 1962). The resonance effect at the energy levels of the trap is illustrated by a calculation based on a rectangular potential barrier with a square well in its center (Fig. 1.10).

A realistic theoretical treatment of resonant tunneling via localized

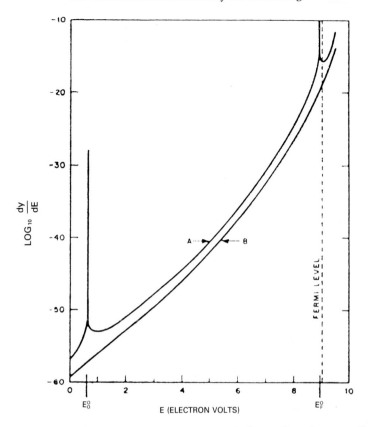

Fig. 1.10. Tunnel current per unit energy interval of tunneling electrons. Curve A: rectangular barrier with square well trap (total barrier thickness 43 Å). Curve B: rectangular barrier without trap (total barrier thickness 43 Å) (Penley, 1962).

states in the insulating barrier region has to be based on a three-dimensional model. For resonant tunneling via a series of localized states, the following distance dependence of the transmission coefficient is found (Knauer *et al.*, 1977; Halbritter, 1982, 1985):

$$T \propto \exp(-\varkappa s) \tag{1.37}$$

indicating a much slower decrease of resonant tunneling current with barrier thickness s than in direct tunneling, for which

$$T \propto \exp(-2\varkappa s) \tag{1.38}$$

Therefore, resonant tunneling through a series of localized states should dominate for sufficiently thick tunnel barriers.

1.2.9 Traversal time for tunneling

When a particle with a given energy has tunneled through a potential barrier, how much time did it, on average, spend in the barrier region? To find an answer to this question is important both from a theoretical viewpoint, particularly in many-body problems of tunneling which have been reduced to approximate single-particle problems, e.g. the classical image potential method, as well as from an experimental viewpoint, e.g. for determination of the practical speed limit in quantum devices based on tunneling phenomena.

Unfortunately, an expression for the traversal time for tunneling ('tunneling time') cannot be found in an easy and unique way. The reason is that time is not an observable in quantum theory and therefore no time operator exists. However, an expression for the tunneling time can be found based on a *Gedankenexperiment*.

Following Büttiker and Landauer (1982), we can consider tunneling through a time-modulated potential barrier. At low barrier modulation frequencies ω, corresponding to modulation periods long relative to the time during which the electron interacts with the barrier, the electron experiences an effectively static barrier during its traversal. At modulation frequencies high relative to the reciprocal traversal time, the electron experiences many series of oscillation and tunnels through a time-averaged potential barrier, but can do it inelastically by losing or gaining modulation quanta (see section 1.4). As the modulation frequency is varied, the crossover between the two types of behavior occurs when $\omega\tau \simeq 1$, where τ is the interaction time of a transmitted electron.

In the WKB limit one finds

$$\tau^{BL} = \int_{s_1}^{s_2} \frac{m}{\hbar \varkappa(z)}\, dz = \int_{s_1}^{s_2} \left(\frac{m}{2[V(z)-E]} \right)^{1/2} dz \qquad (1.39)$$

where s_1 and s_2 are the classical turning points and τ^{BL} is known as the Büttiker–Landauer tunneling time. For a rectangular barrier of height V_0 and width s, Eq. (1.39) reduces to

$$\tau^{BL} = \frac{ms}{\hbar \varkappa} \qquad (1.40)$$

Assuming a barrier width of 5 Å and an effective barrier height of 4 eV, we get a value of about 4×10^{-16} s for τ^{BL}. The tunneling time is therefore generally several orders of magnitude shorter than the average time between two successive tunneling events given by $\Delta t = e/I$, where I denotes the tunneling current. On the other hand, the tunneling time can be of the same order of magnitude as the typical response time of electrons in metals, which is given by the inverse of the plasma frequency ω_p $(1/\omega_p \simeq 10^{-16} - 10^{-15}$ s). Therefore, dynamical corrections to the classical image potential as discussed in section 1.2.6 can become important.

Another *Gedankenexperiment* based on the Larmor precession of a spin in a homogeneous magnetic field also yields an expression for the tunneling time (Baz', 1967a, 1967b; Rybachenko, 1967). Consider an incident spin-$\frac{1}{2}$ particle, such as an electron, polarized in the z direction and an infinitesimal magnetic field $B = B_0\hat{y}$ covering the barrier region and aligned parallel to this barrier (taken as the y direction). The time spent in the magnetic field region should be proportional to the averaged spin component $<s_x>$ of the transmitted electron. The 'Larmor time' for transmission is then given by

$$\tau^L = \lim_{w_L \to 0} \langle s_x \rangle / (-\tfrac{1}{2}\hbar w_L) \qquad (1.41)$$

where $\omega_L = g\mu_B B_0/\hbar$ with the Bohr magneton μ_B and the gyromagnetic ratio g. Büttiker (1983) has re-examined the use of the Larmor precession as a clock to measure the traversal time and showed that, for an opaque rectangular barrier, the traversal time in the Baz'–Rybachenko *Gedankenexperiment* becomes $ms/\hbar\varkappa$ which is equal to the Büttiker–Landauer time given in Eq. (1.40).

More recent theoretical considerations are in agreement with the Büttiker–Landauer expression for the tunneling time (e.g. Fertig, 1990; Guéret, 1991). However, some controversy still exists (Hauge and Støvneng, 1989).

1.3 Elastic electron tunneling experiments with planar metal–insulator–metal junctions

We now focus on some important early experimental studies of tunneling phenomena in planar metal–insulator–metal junctions starting around 1960. Such junctions were fabricated according to a procedure illustrated in Fig. 1.11. As bottom metal electrode, aluminum was often preferred because it forms a relatively homogeneous Al_2O_3 oxide layer which served as the tunneling barrier. The oxide thickness was typically chosen to be about 30 Å. The growth of a reasonably homogeneous oxide layer was most critical, particularly for excluding the presence of

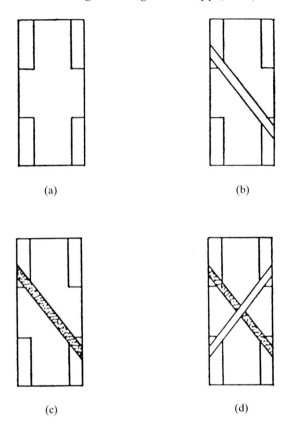

Fig. 1.11. Preparation of planar metal–oxide–metal tunnel junctions. (*a*) Glass slide with indium contacts. (*b*) A metal strip has been deposited across the contacts. (*c*) The metal strip has been oxidized. (*d*) A second metal film has been deposited across the first metal film, forming a metal–oxide–metal sandwich (Giaever and Megerle, 1961).

pinholes which would form short circuits. The bottom and top metal electrodes were typically less than 1 μm thick and about 1 mm wide, leading to a junction area of about 1 mm². Depending on the oxide thickness, the resistance of the 1 mm² junction varied between 10^{-2} Ω and 10^7 Ω.

1.3.1 Giaever tunneling

1.3.1.1 Tunneling between two normal metals

We first consider tunneling between two normal metals which is illustrated in the energy diagram of Fig. 1.12. The insulating layer is treated

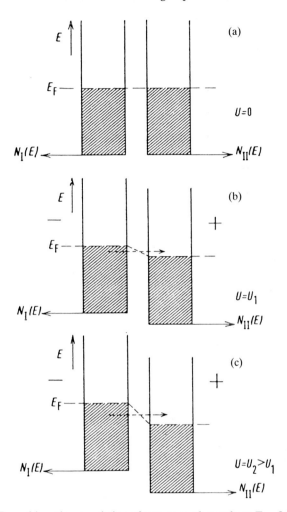

Fig. 1.12. Tunnel junction consisting of two normal metals at $T = 0$ K (Buckel, 1984).

as if it were a vacuum. According to Eq. (1.10) the transmitted current from an occupied state of the left electrode to an unoccupied state of the right electrode is given by

$$j_t = \left(\frac{2\pi}{\hbar}\right)|M_{rl}|^2 \, n_r \, (1 - f_r)$$ (1.42)

where M_{rl} is the tunneling matrix element, n_r is the density of states on the right side and $(1 - f_r)$ the probability that the state on the right side is empty. (For Eq. (1.10) we assumed that $(1 - f_r) = 1$).

To get the total current j_{rl} from the left to the right electrode for an applied positive bias voltage U on the right electrode, we have to sum over the occupied states on the left side and obtain

$$j_{rl} \propto \int_{-\infty}^{+\infty} |M_{rl}|^2 n_l (\mathscr{E}) n_r (\mathscr{E} + eU) f_l (\mathscr{E}) \left[1 - f_r (\mathscr{E} + eU) \right] d\mathscr{E} \quad (1.43)$$

where n_l is the density of states on the left side, $\mathscr{E}$ is the energy measured from the Fermi level ($\mathscr{E} = E - E_F$) and f_l the probability that the state on the left side is occupied. $f_{l,r} (\mathscr{E})$ is given by

$$f_{l,r} (\mathscr{E}) = \frac{1}{e^{\mathscr{E}/k_B T} + 1} \quad (1.44)$$

where k_B is the Boltzmann constant and T is the temperature. After subtracting a similar expression for the current j_{lr} from the right to the left side, we get the total net current j:

$$j = j_{rl} - j_{lr}$$
$$\propto \int_{-\infty}^{+\infty} |M|^2 n_l (\mathscr{E}) n_r (\mathscr{E} + eU) \left[f_l (\mathscr{E}) - f_r (\mathscr{E} + eU) \right] d\mathscr{E} \quad (1.45)$$

Note that $M_{rl} = M_{lr} = M$.

In a small energy window around the Fermi level E_F we can assume that the tunneling matrix element M is almost independent of the energy E:

$$|M|^2 \approx \text{constant} \quad (1.46)$$

For the tunneling current j_{NN} between two normal metals at zero temperature, we therefore obtain in the limit of small applied bias voltage

$$j_{NN} \propto n_l (E_F) n_r (E_F) \cdot eU \quad (1.47)$$

i.e. the current is linearly proportional to the applied bias voltage (Ohmic behavior) as found earlier (Eq. (1.19)).

1.3.1.2 Tunneling between normal metal and superconductor

We now consider tunnel junctions with one electrode being a superconductor (Fig. 1.13). The density of states n_s of a superconductor with weak electron–phonon coupling can be taken from the Bardeen–

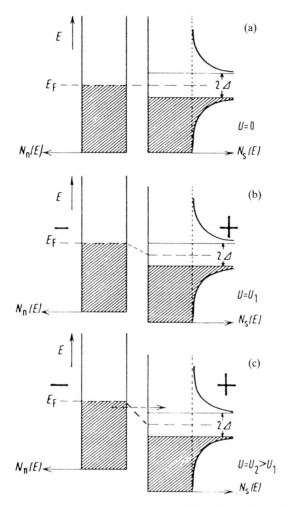

Fig. 1.13. Metal–insulator–superconductor tunnel junction at $T = 0$ K (Buckel, 1984).

Cooper–Schrieffer (BCS) theory of superconductivity (Bardeen *et al.*, 1957):

$$n_{\mathrm{s}}(\mathscr{E}) = \begin{cases} n_{\mathrm{N}}(0) \cdot \dfrac{|\mathscr{E}|}{\left(\mathscr{E}^2 - \Delta^2\right)^{1/2}} & \text{for } |\mathscr{E}| \geq \Delta \\[2ex] 0 & \text{for } |\mathscr{E}| < \Delta \end{cases} \qquad (1.48)$$

where n_N is the density of states in the normal state, $\mathscr{E}$ is measured from the Fermi energy and Δ is half the energy gap, which is independent of energy in the BCS theory. The tunneling current j_{SN} between a metal in the normal state and a metal in the superconducting state is then given by

$$j_{SN} \propto n_1\,(E_F)\,n_r\,(E_F) \cdot \int_{-\infty}^{+\infty} \frac{|\mathscr{E}|}{\left(\mathscr{E}^2 - \Delta^2\right)^{\frac{1}{2}}} \cdot \left[f_1(\mathscr{E}) - f_r(\mathscr{E} + eU)\right] d\mathscr{E} \quad (1.49)$$

$$|\mathscr{E}| \geq \Delta$$

If we differentiate this expression with respect to the applied bias voltage U, we obtain

$$\frac{dj_{SN}}{dU} \propto \int_{-\infty}^{+\infty} \frac{|\mathscr{E}|}{\left(\mathscr{E}^2 - \Delta^2\right)^{\frac{1}{2}}} \left[\frac{(1/k_B T)\exp\left[(\mathscr{E} + eU)/k_B T\right]}{\left(\exp\left[(\mathscr{E} + eU)/k_B T\right] + 1\right)^2} \right] d\mathscr{E} \quad (1.50)$$

$$|\mathscr{E}| \geq \Delta$$

At $T = 0$, the second factor in the integral degenerates into a delta function and we get

$$\left(\frac{dj_{SN}}{dU}\right)_{T=0} \propto \begin{cases} \dfrac{|eU|}{\left[(eU)^2 - \Delta^2\right]^{\frac{1}{2}}} & \text{for } |eU| \geq \Delta \\[2mm] 0 & \text{for } |eU| < \Delta \end{cases} \quad (1.51)$$

i.e. the voltage dependence of the tunneling conductance at $T = 0$ resembles the density of states of the superconductor (Cohen *et al.*, 1962). Integration of Eq. (1.51) yields the bias dependence of the tunnel current:

$$j_{SN} \propto \begin{cases} \left[(eU)^2 - \Delta^2\right]^{\frac{1}{2}} & \text{for } |eU| \geq \Delta \\[2mm] 0 & \text{for } |eU| < \Delta \end{cases} \quad (1.52)$$

For a finite temperature well above the transition temperature of the superconductor, the exponential tail of the Fermi distribution leads to a finite current also for $|eU| < \Delta$. However, there remains a sudden rise of current for $|eU| \approx \Delta$.

The current–voltage characteristics for normal metal–insulator–

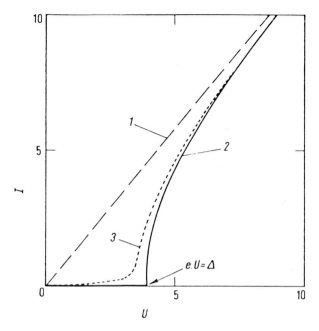

Fig. 1.14. Current–voltage characteristics for different tunnel junctions: (1) normal metal/normal metal (Fig. 1.12), (2) normal metal/superconductor at $T = 0$ K (Fig. 1.13), (3) normal metal/superconductor for $0 < T \ll T_{\mathrm{c}}$ (Buckel, 1984).

normal metal and normal metal–insulator–superconductor tunnel junctions are shown in Fig. 1.14.

If a superconductor with strong electron–phonon coupling is chosen, e.g. Pb, the measured conductance–voltage characteristics exhibit additional structure which is not described by Eq. (1.51) with an energy-independent gap. In this case, the Eliashberg theory for strong coupled superconductors has to be applied. This predicts that the phonon density of states $g(\omega)$ is reflected in the superconductor's electronic density of states via the energy-dependent gap function $\Delta(E)$ (Eliashberg, 1960). Therefore, since tunneling into superconductors directly probes the electronic density of states, information about the spectrum of phonons effective in producing Cooper pairs can be derived as well. More specifically, the second derivative of the current–voltage characteristic is closely related to the Eliashberg function $\alpha^2 g(\omega)$, where α^2 describes the strength of the electron–phonon coupling. From a measurement of $\mathrm{d}^2 I/\mathrm{d}U^2(U)$, the Eliashberg function can be derived directly (McMillan and Rowell, 1965).

1.3.1.3 Single-particle tunneling between two superconductors

Tunneling in superconductor–oxide–superconductor junctions can be described in terms of a single-particle picture which allows us to consider the potential energy diagram as shown in Fig. 1.15.

For the tunneling current j_{ss} between two weakly coupled superconductors with energy gaps Δ_1 and Δ_2, we obtain the following expression (Nicol *et al.*, 1960):

$$j_{ss} \propto \int_{-\infty}^{+\infty} \frac{|\mathscr{E}|}{\left(\mathscr{E}^2 - \Delta_1^2\right)^{1/2}} \cdot \frac{|\mathscr{E} + eU|}{\left[\left(\mathscr{E} + eU\right)^2 - \Delta_2^2\right]^{1/2}} \left[f(\mathscr{E}) - f(\mathscr{E} + eU)\right] d\mathscr{E} \qquad (1.53)$$

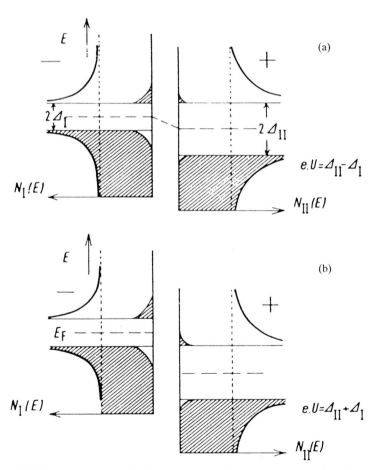

Fig. 1.15. Superconductor–insulator–superconductor tunnel junction (Buckel, 1984).

There is a logarithmic singularity in the current at a bias voltage of approximately $U = \pm|\Delta_2 - \Delta_1|$ and a finite discontinuity of the current at $U = (\Delta_2 + \Delta_1)/e$ even at $T \neq 0$.

The corresponding current–voltage characteristics for $T = 0$ and $T > 0$ are presented in Fig. 1.16. The maximum of the tunneling current at bias voltage $U = (\Delta_2 - \Delta_1)/e$ is explained by the fact that all single electrons of the left superconductor can tunnel into the high density of unoccupied states of the right superconductor. On increasing the bias voltage further, the tunneling current drops because the density of states of the right superconductor decreases. As a consequence, a negative resistance region appears in the current–voltage characteristic. At bias voltage $U = (\Delta_2 + \Delta_1)/e$, a strong increase of the tunneling current can be observed because the high densities of states of both superconducting electrodes are now contributing to the current.

Single-particle tunneling between two superconductors can also be described within the picture of Cooper pairs (Buckel, 1984). The strong increase of the tunneling current at $U = (\Delta_2 + \Delta_1)/e$ is then explained by the fact that the energy $(\Delta_2 + \Delta_1)$ is required to break a Cooper pair into a single particle in the left superconductor and a single particle

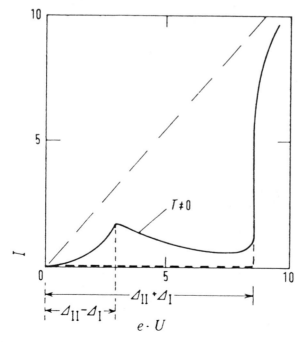

Fig. 1.16. Current–voltage characteristic for a superconductor–insulator–superconductor tunnel junction for $0 < T < T_c$ (Buckel, 1984).

in the right superconductor, which corresponds to the transition of a single particle through the insulating layer.

Tunneling experiments using planar metal–oxide–superconductor or superconductor–oxide–superconductor tunnel junctions as pioneered by Giaever and coworkers (Giaever, 1960a, 1960b, 1974; Fischer and Giaever, 1961; Giaever and Megerle, 1961) represent a powerful method for the determination of the energy gap in superconductors. In principle, any physical property which depends on the electronic density of states near the Fermi level, such as the specific heat, infrared absorption, ultrasonic attenuation or electron tunneling, can be used to determine this energy gap. However, only electron tunneling has the potential to probe the superconductor energy gap on a local scale (section 4.3.1). In addition, for the case of strongly coupled superconductors, tunneling experiments directly yield information about the spectrum of phonons effective for producing the Cooper pairs, as already mentioned above. Tunneling can therefore be regarded as the most detailed available probe of the superconducting state.

1.3.2 Josephson tunneling

In superconductor–oxide–superconductor tunnel junctions with thin oxide layers (10–20 Å), tunneling of Cooper pairs can be observed in addition to single-particle tunneling, as first theoretically predicted by Josephson (1962, 1965, 1974) and experimentally confirmed by Anderson and Rowell (1963).

In Fig. 1.17 a current–voltage characteristic of a 'Josephson junction' consisting of two identical superconductors is shown. In contrast to single-particle tunneling (Fig. 1.14), a supercurrent I_s exists for zero bias U_s across the junction. This Josephson (direct) current can be increased to a maximum value $I_{s_{max}}$ by increasing the bias U_0 in the external circuit (see Fig. 1.17). A further increase of U_0 leads to a finite bias U_s across the junction and the current jumps from $I_{s_{max}}$ to I^*, where I^* is determined by the resistance R in the external circuit. For this state of finite bias U_s across the junction, Josephson predicted the existence of coherent current oscillations in the junction:

$$I_s = I_{s_{max}} \cdot \sin(2\pi v_J t) \tag{1.54}$$

with a frequency v_J given by

$$v_J = \frac{2e}{h} U_s \tag{1.55}$$

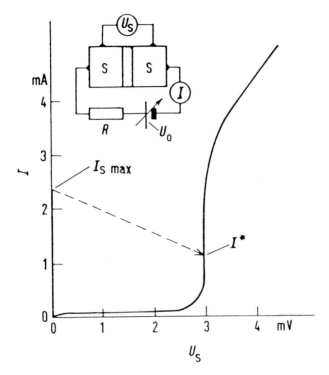

Fig. 1.17. Current–voltage characteristic of a Josephson tunnel junction consisting of two identical superconductors. The resistance R determines the behavior of the tunnel junction at the point of instability (Buckel, 1984).

Experimental observation of the d.c. and a.c. Josephson effects provided direct proof for phase correlation in the superconducting state (Anderson and Rowell, 1963).

A change of the phase relation between the two superconductors by means of an externally applied magnetic field parallel to the junction leads to a characteristic dependence of the maximum value $I_{s_{max}}$ of the Josephson direct current on the magnetic field B:

$$I_{s_{max}}(B) = I_{s_{max}}(0) \cdot \frac{\sin(\pi\phi / \phi_0)}{(\pi\phi / \phi_0)} \tag{1.56}$$

where $\phi_0 = h/2e$ is the elementary magnetic flux quantum and ϕ is the total flux generated by the magnetic field in the tunnel junction. It can easily be seen from Eq. (1.56) that $I_{s_{max}}$ becomes zero for $\phi = n\phi_0$ (n is an integer). The dependence of $I_{s_{max}}$ on B is shown in Fig. 1.18.

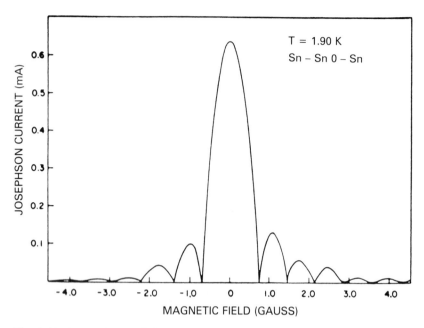

Fig. 1.18. Dependence of the maximum Josephson current on the strength of the magnetic field applied parallel to the insulating layer of a metal–insulator–metal tunnel junction (Mercereau, 1969).

By constructing double junctions, one gets interference structures in the $I_{s_{max}}(B)$ dependence which are highly sensitive to slight changes of B. Such 'superconducting quantum interferometer devices' (SQUIDs) are commonly used for high-precision measurements of magnetic fields.

The a.c. Josephson effect (Eq. (1.54) and (1.55)) has also found an important application. Since frequency measurements can be performed with high accuracy, Eq. (1.55) provides a key for ultra-precise voltage measurements leading to a voltage standard. Many other applications exist which have been summarized by Solymar (1972).

Finally, it should be emphasized that the important role of the thin oxide layer in planar superconductor–oxide–superconductor Josephson junctions is to provide a weak coupling ('weak link') between the two superconductors. However, such weak links can also be realized by pressing a sharp tip of a superconducting material against another superconductor while the area of contact is kept small. This realization of a weak link is more closely related to the STM-based geometry.

For their important contributions to the science of tunneling phenomena, Giaever and Josephson were awarded the Nobel Prize in 1973, together with L. Esaki who introduced the tunnel diode (section 1.5).

1.3.3 Coulomb blockade and single-electron tunneling

We now consider tunnel junctions with an extremely small capacitance, where the macroscopic charging energy $e^2/2C$ of the capacitance C is much larger than the thermal energy k_BT and also larger than the quantum fluctuation energy $\hbar/\tau$ with $\tau = RC$, R being the resistance of the tunnel junction:

$$k_B T < \frac{\hbar}{RC} < \frac{e^2}{2C} \tag{1.57}$$

In this case, tunneling of a single electron results in noticeable re-charging of the junction capacitance, so that the probability of other tunneling events is drastically affected.

If the absolute value of the charge Q on an isolated junction is less than $e/2$, tunneling of an electron in either direction will increase the charging energy, resulting in a blocking of tunneling events ('Coulomb blockade' of tunneling). As a consequence, a 'Coulomb gap' $\Delta_c = 2(eU_c)$ opens up in the current–voltage characteristic of such a tunnel junction with a threshold voltage $U_c = e/2C$ below which no tunneling current can flow ($I = 0$). In addition, there exists a voltage offset $U_{off} = \pm e/2C$ of the large-current asymptotes from the origin. As the absolute value of the charge $|Q|$ on the tunnel junction exceeds $e/2$, tunneling becomes favorable and a single electron can tunnel. After the tunnel event, $|Q|$ is again less than $e/2$, so that tunneling is again blocked until a current source has recharged the junction above $e/2$. Therefore, steps in the current–voltage characteristic occur at bias voltages given by

$$U_n = \pm (2n+1) \frac{e}{2C} \qquad (n \text{ integer}) \tag{1.58}$$

leading to the 'Coulomb staircase' of tunneling. A simulated *I–U* characteristic together with its first derivative are shown in Fig. 1.19.

The most interesting consequence of the Coulomb blockade is time correlation between tunneling events (Ben-Jacob and Gefen, 1985; Averin and Likharev, 1986). In a current-biased tunnel junction fulfilling condition (1.57), the electrons are expected to tunnel one at a time, at more or less fixed time intervals. This time-correlated single-electron tunneling gives rise to coherent voltage oscillations with a frequency ν_c fundamentally related to the current I by

$$\nu_c = \frac{I}{e} \tag{1.59}$$

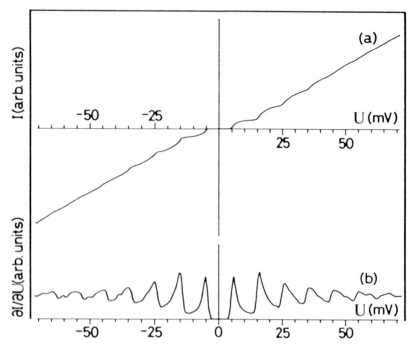

Fig. 1.19. (*a*) *I–U* characteristic and (*b*) d*I*/d*U–U* characteristic calculated from Monte Carlo simulations for $C_1 = 3.2 \times 10^{-18}$ F, $C_2 = 1.6 \times 10^{-17}$ F and $R_2/R_1 = 5$ (van Bentum *et al.*, 1988).

Since frequency measurements can be performed with high accuracy, experimental observation of these voltage oscillations in single-electron tunneling (SET) offers a way to ultra-precise current measurements, possibly leading to an improved current standard.

The realization of SET experiments is guided by the requirement (1.57); i.e. the capacitance of the tunnel junction has to be small and the temperature generally low. Of course, the temperature range depends on how small the capacitance can actually be made. For a capacitance of the order of $C \simeq 10^{-15}$ F, a cryogenic temperature of $T < 1$ K is required. However, if C is as small as 10^{-18} F or even less, SET can be observed even at room temperature. Such a small capacitance can only be realized if the tunnel junctions are made small. These are often called 'mesoscopic tunnel junctions'.

For a single mesoscopic tunnel junction, one has to be aware of the fact that the parasitic capacitance of the leads, which is usually much larger than the junction capacitance itself, has to be added, thereby decreasing the charging energy considerably and making the observation of SET difficult. Therefore, SET can best be studied either in two serially coupled small tunnel junctions or in multijunction arrays.

The first experimental observation of the Coulomb blockade was made with two planar tunnel junctions in series where the middle electrode consisted of isolated metal particles with a radius less than 100 Å embedded in an oxide matrix (Zeller and Giaever, 1969). In this case, the small capacitance was related to the small size of the metal particles. Owing to advances in the field of microfabrication, it has become possible to fabricate mesoscopic tunnel junctions and even multijunction arrays by electron beam lithography, where single-electron charging effects as well as SET oscillations can clearly be studied (Fulton and Dolan, 1987; Delsing *et al.*, 1990). Furthermore, mesoscopic double tunnel junctions with adjustable capacitances in the 10^{-18} F range can be realized by using a sharp metal tip, like in STM, and tunneling into a small isolated metal particle on top of an oxidized metal base electrode (van Bentum *et al.*, 1988, 1989; Wilkins *et al.*, 1989). Finally, it should be noted that Coulomb blockade of Cooper pair tunneling is also possible and has been observed experimentally (Fulton *et al.*, 1989; Geerligs *et al.*, 1990).

1.3.4 Spin-polarized tunneling

1.3.4.1 Tunneling between a ferromagnet and a superconductor
So far we have not considered the spin of the tunneling electrons. In fact, the tunneling current can become spin-dependent, as first shown by Tedrow and Meservey (1971) using planar ferromagnet–oxide–superconductor tunnel junctions. In these experiments, determination of the spin polarization of the tunneling current was based on the discovery of the Zeeman splitting of the quasiparticle density of states of a superconducting thin film in a strong parallel magnetic field (Meservey *et al.*, 1970). This Zeeman splitting leads to two BCS-type density-of-states curves shifted in energy by $\pm \mu H$ (μ is the electron magnetic moment, H is the applied magnetic field) with respect to the density-of-states curve in the absence of a magnetic field (Fig. 1.20, top part). Within the Stoner model of band ferromagnetism, the two parts (spin up and spin down) of the spin-dependent density of states of the ferromagnetic counterelectrode are shifted relative to each other by the exchange energy (Fig. 1.21) leading to a different density of states at the Fermi level for the two different spin directions. To a first approximation, we can assume the spin-dependent density of states to be constant within a small energy window around the Fermi level (Fig. 1.20,

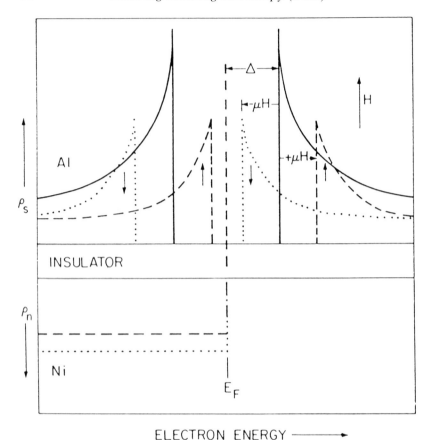

ELECTRON ENERGY ──────▶

Fig. 1.20. Tunneling density-of-states diagram for an Al–oxide–Ni tunnel junction showing for the Al electrode a BCS density-of-states split into spin-up (increased in energy by μH) and spin-down (decreased in energy by μH) parts and an assumed predominance of spin-down (magnetic moment parallel to the field) carriers at the Fermi surface of Ni. Arrows on density of states refer to spin direction (Tedrow and Meservey, 1971).

bottom part). If we now examine the full density-of-states diagram for the ferromagnet–oxide–superconductor tunnel junction in an applied parallel magnetic field (Fig. 1.20), it is obvious that the difference in the density of states at the Fermi level for the two different spin species of the ferromagnetic electrode will lead to an asymmetry in the current–voltage and conductance–voltage characteristics, which is indeed observed experimentally (Fig. 1.22).

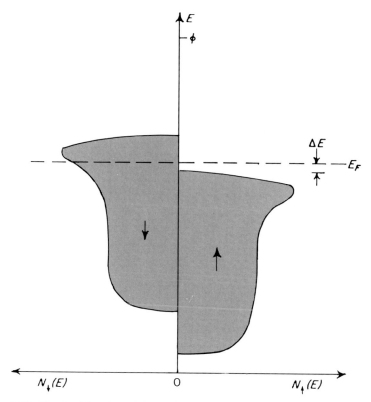

Fig. 1.21. Simple d band model of a ferromagnet. The majority spins reside in a filled band slightly below the Fermi energy (Wolf, 1985).

We can define a spin polarization P of the tunneling current by

$$P \equiv \frac{N_{\uparrow} - N_{\downarrow}}{N_{\uparrow} + N_{\downarrow}} \qquad (1.60)$$

where $N_{\uparrow}$ $(N_{\downarrow})$ is the number of tunneling electrons with magnetic moment aligned parallel (antiparallel) to the direction of the magnetic field. By neglecting spin–orbit scattering in the superconducting elec-trode, which is a good first approximation for sufficiently thin films (50 Å), and by assuming the absence of spin scattering in the tunneling barrier, Tedrow and Meservey (1973) showed that the polarization P can be determined experimentally from the relationship

$$P = \frac{(\sigma_{d} - \sigma_{a}) - (\sigma_{c} - \sigma_{b})}{(\sigma_{d} - \sigma_{a}) + (\sigma_{c} - \sigma_{b})} \qquad (1.61)$$

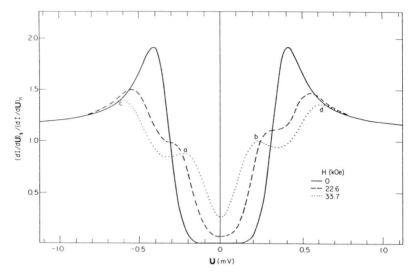

Fig. 1.22. Normalized conductance of an Al–Al$_2$O$_3$–Ni tunnel junction measured as a function of the voltage applied to the Al film for three values of applied magnetic field. The asymmetry of the conductance peaks a, b, c and d (H = 33.7 kOe) results from polarization of the Ni carriers (Tedrow and Meservey, 1971).

where σ_i is the measured conductance at point i (i=a,b,c,d) of the conductance–voltage curve (Fig. 1.22).

Since the tunneling current in these experiments is mainly dominated by the mobile itinerant d$_i$ electrons of the ferromagnetic electrode which obtain their polarization through exchange interaction with the localized d$_l$ electrons, Tedrow and Meservey measured the polarization of these d$_i$ electrons in a thin shell within 1 meV of the Fermi surface. In contrast, photoelectrons are dominated by localized electrons within a much larger energy window (typically of the order of electronvolts) below the Fermi level. Therefore, spin polarizations as measured by spin-polarized tunneling and spin-resolved photoemission can generally not be compared with each other (Stearns, 1977).

1.3.4.2 Tunneling between two ferromagnets

Spin-dependent tunneling is also found in planar ferromagnet–insulator–ferromagnet tunnel junctions (Julliere, 1975; Maekawa and Gäfvert, 1982). To explain this experimental observation, Slonczewski (1988, 1989) considered a tunnel junction with two ferromagnetic electrodes where the directions of the internal magnetic fields differ by an angle θ (Fig. 1.23). Within a free-electron model and in the limit of small applied bias voltage, Slonczewski found the following expression

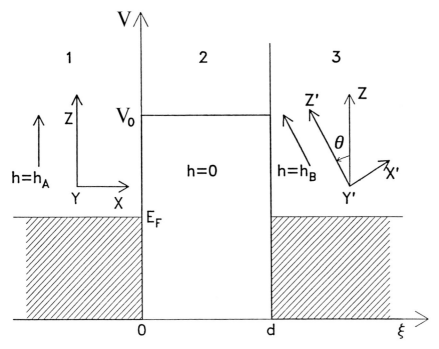

Fig. 1.23. Schematic potential diagram for two metallic ferromagnets separated by an insulating barrier. The molecular fields h_A and h_B within the magnets form an angle θ (Slonczewski, 1989).

for the conductance σ of the ferromagnet–insulator–ferromagnet tunnel junction for the case of two identical ferromagnetic electrodes:

$$\sigma = \sigma_{fbf} \, (1 + P_{fb}^2 \cos \theta) \qquad |P_{fb}| \leq 1 \qquad (1.62)$$

Here P_{fb} is the effective spin polarization of the ferromagnet–barrier interface and σ_{fbf} is a mean conductance which is proportional to $\exp(-2\varkappa s)$. If the two ferromagnetic electrodes are different, the conductance becomes

$$\sigma = \sigma_{fbf'} \, (1 + P_{fb} P_{f'b} \cos \theta) \qquad (1.63)$$

For the two special cases of parallel and antiparallel alignment of the internal magnetic field directions, one finds

$$\sigma_{\uparrow\uparrow} = \sigma_{fbf'} \, (1 + P_{fb} P_{f'b})$$
$$\sigma_{\uparrow\downarrow} = \sigma_{fbf'} \, (1 - P_{fb} P_{f'b}) \qquad (1.64)$$

Finally, we obtain

$$\frac{\sigma_{\uparrow\uparrow} - \sigma_{\uparrow\downarrow}}{\sigma_{\uparrow\uparrow} + \sigma_{\uparrow\downarrow}} = P_{fb}P_{f'b} =: P_{fbf'} \tag{1.65}$$

where $P_{fbf'}$ is the effective polarization for the whole tunnel junction.

1.3.4.3 Tunneling between two non-ferromagnetic electrodes

Spin polarization of the tunneling current can be observed even in the case of two non-ferromagnetic electrodes in zero applied field if a ferromagnetic barrier, e.g. EuS, is used (Moodera *et al.*, 1988; Hao *et al.*, 1990). EuS is a ferromagnetic semiconductor with a Curie temperature of $T_c = 16.6$ K and a bulk band gap of 1.65 eV. Below T_c, the conduction band is split by the ferromagnetic exchange interaction ($\Delta E_{ex} = 0.36$ eV at 4 K). Therefore, if EuS is used as a barrier in tunnel junc-

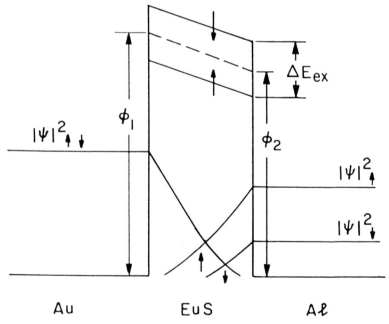

Fig. 1.24. The spin-filter model for a ferromagnetic EuS tunnel barrier. The dashed line represents the tunnel barrier height at temperatures above T_c, and the solid lines the tunnel barrier heights for spin-up and spin-down electrons (as indicated by the arrows) for $T \ll T_c$. The tunneling probabilities for each spin-polarization are shown schematically as the overlap of the wave functions (Hao *et al.*, 1990).

tions, the barrier height will be split for the two spin directions below T_c (Fig. 1.24):

$$\phi_{\uparrow,\downarrow} = \phi_0 \mp \Delta E_{ex}(T) / 2 \tag{1.66}$$

where ϕ_0 is the average barrier height above T_c. As a consequence, the probability of tunneling for spin-up electrons is much higher than for spin-down electrons because the spin-up electrons see a barrier height $\phi_\uparrow$ which is lower by ΔE_{ex} compared with the barrier height $\phi_\downarrow$ seen by the spin-down electrons. A tunnel barrier consisting of a ferromagnetic semiconductor can therefore act as a very effective 'spin filter' below T_c. In fact, spin polarization as high as 85% has been observed in metal–EuS–metal tunnel junctions (Hao et al., 1990). If one of the metal electrodes is a superconductor, e.g. Al, the ferromagnetic ordering of EuS below T_c additionally leads to an exchange-induced Zeeman splitting of the quasiparticle density of states of the superconducting electrode even in zero applied magnetic field. The spin-filter effect of EuS barriers can be used to provide a low-energy spin-polarized electron source whereas the zero-field Zeeman-split density of states in the superconductor provides a spin-selective detector for tunneling electrons without any external magnetic field.

In view of the geometry given in STM, it is interesting that earlier field-emission studies of EuS-coated tungsten tips also showed a high degree of spin polarization of the field-emitted electrons below T_c (Müller et al., 1972; Kisker et al., 1976, 1978; Baum et al., 1977).

1.4 Inelastic electron tunneling in planar metal–insulator–metal junctions

So far, we have only considered elastic tunneling processes, in which the energy of the tunneling electrons is conserved. We now focus on inelastic tunneling where the electron energy is changed due to interaction of the tunneling electrons with elementary excitations. Here, we can distinguish between inelastic tunneling processes due to excitations within the insulating barrier and inelastic tunneling requiring interactions in the electrodes.

1.4.1 Inelastic tunneling involving excitation within the insulating barrier

In Fig. 1.25(a) an energy diagram for $T = 0$ is shown, illustrating elastic and inelastic tunneling processes. In the case of inelastic tunneling, the electron loses a quantum of energy $\hbar\omega_0$ to some mode within the insulat-

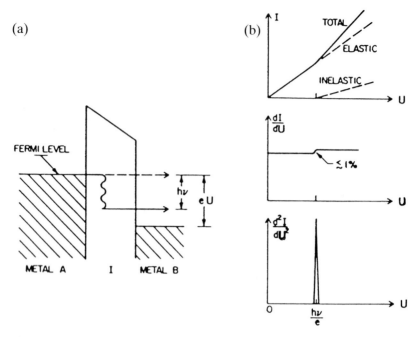

Fig. 1.25. Tunneling electrons can excite a molecular vibration of energy $h\nu$ only if $eU > h\nu$. For smaller voltages, there are no vacant final states for the electron to tunnel into. Thus the inelastic current has a threshold at $U = h\nu/e$. The increase in conductance at this threshold is typically <1%. A standard tunneling spectrum, d^2I/dU^2 versus U, accentuates this small increase; the step in dI/dU becomes a peak in d^2I/dU^2 (Hansma, 1977).

ing barrier. According to the Pauli exclusion principle, the final state after the inelastic tunneling event must be initially unoccupied as depicted in Fig. 1.25(a).

We can now discuss the bias voltage dependence of the tunneling current (Fig. 1.25(b)). Starting from zero applied bias voltage U, the elastic tunneling current will increase linearly, proportional to U (Eq. (1.34)). As long as the applied bias voltage is sufficiently small ($U < \hbar\omega_0/e$, where ω_0 is the lowest energy mode in the barrier), inelastic tunneling processes cannot occur due to the Pauli exclusion principle. At the threshold bias $U_0 = \hbar\omega_0/e$, the inelastic channel opens up, and the number of electrons which can use the inelastic channel will increase linearly with U (Fig. 1.25(b)). Therefore, the total current, including both elastic and inelastic contributions, has a kink at $U_0 = \hbar\omega_0/e$. In the conductance (dI/dU) versus voltage curve, the kink becomes a step at U_0. Since the fraction of electrons which tunnel inelastically is tiny (typically 0.1–1%), the conductance increase at U_0 due to onset of the inelastic tunnel channel is too small to be conveniently observed. There-fore, the second derivative (d^2I/dU^2) is measured, which exhibits a peak at U_0. Of course, the insulating barrier may have more than one mode ω_0 which can be excited in the tunneling process. Each mode ω_i contrib-utes a peak in the second derivative $d^2I/dU^2(U)$ at the corresponding bias voltage $U_i = \hbar\omega_i/e$ so that $d^2I/dU^2(U)$ represents the spectrum of possible excitations. Inelastic electron tunneling can therefore be regarded as a special kind of electron energy loss spectroscopy.

The peaks in the measured spectrum can be characterized by their positions, widths and intensities. As we have already seen, the peak positions are determined by the energy conservation law and are inde-pendent of the mechanism of interaction between the tunneling electron and mode ω within the barrier. However, small shifts in the peak posi-tions are sometimes observed due to interactions with the metal elec-trodes. The peak widths result from three different contributions: the natural width Γ of the mode ω, thermal broadening which is of the order of $k_B T$, and additional broadening due to the experimental meas-urement technique which is based on bias voltage modulation. The peak intensities depend on the details of the mechanism of interaction between the tunneling electrons and the possible modes within the bar-rier and are generally difficult to calculate accurately. For further discus-sion, we have to specify the modes within the barrier. We can distin-guish between excitations of vibrational modes of molecules that were doped into the junction and excitations of collective modes in the insu-lating barrier, such as phonons or magnons.

1.4.1.1 Excitation of vibrational modes of molecules

Inelastic electron tunneling spectroscopy (IETS) has been shown to be a powerful technique for measuring the vibrational spectra of molecules, which can be used for identification of these molecules or molecular fragments based on the characteristic frequencies of functional groups (Jaklevic and Lambe, 1966; Lambe and Jaklevic, 1968; Hansma, 1977, 1982; Wolfram, 1978; Adkins and Phillips, 1985).

The vibrational modes have been proved to be relatively insensitive to incorporation of the molecules into the tunnel junctions by comparing tunneling with infrared and Raman spectra. We can distinguish between two possible types of interaction between the tunneling electron and the molecule. For molecules with a permanent dipole moment, we can consider the Coulomb interaction between the tunneling electron and the dipole moment of the molecule. The size of the conductance increase due to inelastic tunneling can then be calculated to be propor-tional to the square of the dipole matrix element of the corresponding

Table 1.2. *Comparison of experimental techniques for surface vibrational spectroscopy (Hansma, 1982).*

	Inelastic electron tunneling	Transmission infrared	Reflection-absorption infrared	Raman spectroscopy	High-resolution electron energy loss spectroscopy	Inelastic neutron scattering
Spectral range (cm^{-1})	240–8000	1500–3300	1500–3300	1–4000	300–5000	16–2000
Resolution (cm^{-1})	1–10	1–10	1–10	1–10	40–80	4–200
Sample area (mm^2)	1	10^6	100	10–10^6	10	10^8
Sensitivity (% monolayer)	0.1	0.1	0.5	1	0.1	1
Substrate	Metal oxide	Metal oxide	Metal, metal oxide	High surface area or roughened Ag, Au, Cu	Metal, semiconductor, insulator	Graphite, metal oxide, powdered metals
Adsorbate	Many	Many	Few, mainly CO	Few, pyridine and related compounds	Many	Hydrogen, hydrocarbons
Maximum pressure	Sample prep. $\simeq 1$ atm	>1 atm	10 Torr	>1 atm	10^{-5} Torr	1 atm
Theory	Electron–dipole, evolving	Photon–dipole, well established	Photon–dipole, well established	Surface-enhanced Raman, evolving	Electron–dipole, evolving	Neutron–nucleus, well established
Orientational selection rule	Yes, weak	Weak, if any	Yes	No	Yes, depending on mechanism	No

vibrational mode, which also determines the peak intensity of 'infrared active modes' in infrared absorption spectra (Scalapino and Marcus, 1967). For molecules without a permanent dipole, interaction can occur via the polarizability of the molecule. The electron induces a dipole moment in the molecule and interacts with the induced dipole. The associated vibrational modes are called 'Raman active' because they can be observed in Raman spectroscopy. IETS is sensitive to both infrared and Raman active modes.

By comparing IETS with other experimental techniques for surface vibrational spectroscopy (Table 1.2), the following features of IETS appear remarkable.

1. The spectral range of IETS (typically 30–1000 meV corresponding to 240–8000 cm^{-1}) is relatively wide. The upper limit has been pushed even further, which requires prevention of dielectric breakdown in the barrier. Thus, even electronic transitions of molecules have become accessible.
2. The spectral resolution of IETS is comparable to that of other vibrational spectroscopies. However, cryogenic temperatures and long measurement times are typically required to achieve good resolution by decreasing the effects of thermal and instrumental line width broadening. The resolution in inelastic tunneling spectra can be optimized by using superconducting electrodes because the presence of the gap and the sharply peaked density of states of the superconductors both act to increase the resolution (Lambe and Jaklevic, 1968).
3. The sample area can be tiny for IETS while the sensitivity of IETS is high (about 0.1% of a monolayer). Therefore, vibrational spectra of a minute quantity of organic molecules can be obtained by IETS.
4. IETS has no strong selection rules in contrast, for instance, to infrared reflection absorption spectroscopy (IRAS). However, there is a slight dependence of the coupling strength of the tunneling electrons to vibrational modes on the orientation as well as on the location of the dipoles in the insulating barrier. For dipoles located near the metal–insulator interface, as is usually the case, this coupling is stronger for vibrational modes oscillating perpendicular to the interface rather than parallel to it. The coupling strengths reverse for dipoles located in the middle of the barrier.

There are also some weaknesses of IETS.

1. The junction geometry, particularly the molecular layer, is relatively ill-defined.

2. The top metal electrode can cause shifts in the peak positions in inelastic tunneling spectra or may even damage the molecular layer. These two most serious limitations of IETS can be removed by using metal–vacuum–metal tunnel junctions as in STM.

3. IETS requires cryogenic temperatures to keep thermal line width broadening sufficiently small.

4. The theory for IETS, based on the WKB approximation (Scalapino and Marcus, 1967), the transfer Hamiltonian formalism (Hansma, 1977; Kirtley and Soven, 1979) or other methods, is not yet sufficiently advanced to predict even the relative peak intensities in inelastic tunneling spectra accurately.

Besides intentionally doping the tunnel junction with molecules to perform vibrational spectroscopy, it often happens that a small concentration of magnetic impurities is introduced unintentionally into the junction during preparation. As a consequence, a zero-bias anomaly appears in the IETS spectra. In the absence of a magnetic field, this zero-bias anomaly is explained by elastic scattering, similar to the Kondo effect in metals containing a small concentration of magnetic impurities (Appelbaum, 1966). However, with a magnetic field applied, the magnetic levels split in energy and the interaction with the tunneling electrons can become inelastic due to possible spin-flip scattering. As a result, splitting of the zero-bias peak is observed.

1.4.1.2 Excitation of collective modes in the insulating barrier

Tunneling electrons can also excite collective modes in the insulating barrier, for instance, collective vibrational modes. (The expression 'collective vibrational modes' is preferred instead of phonons because the insulating barrier typically has amorphous structure.) The IETS spectra can therefore provide information about the vibrational density of states $g(\omega)$ of the barrier material in addition to other experimental techniques such as inelastic neutron scattering, infrared and Raman spectroscopy (Phillips, 1984; Adkins and Phillips, 1985). However, there exists no simple relationship between $g(\omega)$ and the measured spectrum $d^2I/dU^2(U)$. By writing

$$\frac{d^2I}{dU^2}(U) \propto |M(\omega)|^2 g(\omega) \qquad (1.67)$$

the matrix element $M(\omega)$ will depend on the coupling between the tunneling electrons and the collective vibrational modes as well as on the coherence between scattering from different atoms which is deter-

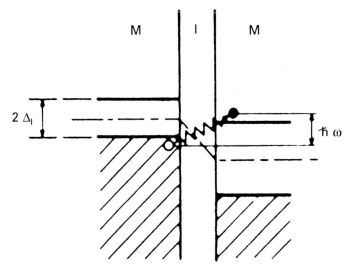

Fig. 1.26. Schematic illustration of photon (phonon) absorption by the tunneling electron. This process is most analogous to phonon-assisted tunneling in semi-conductors. It occurs for $\hbar\omega > \Delta_L$ $(\Delta_L < \Delta_R)$ (Duke, 1969).

mined by the relative phase of the displacements of neighboring atoms. As a consequence, attention has been focused mainly on the peak positions in the IETS spectra rather than on deducing $g(\omega)$ from the measured spectra.

Consideration of the interaction of tunneling electrons with phonons (or photons) is particularly important for explaining phonon-(photon-)assisted tunneling between two superconductors in a 'single-step' process as illustrated in Fig. 1.26. This is the only type of mechanism which can yield a single-particle tunneling current at zero temperature for values of the applied bias voltage below $U = [\min(\Delta_1,\Delta_2)]/e$.

Tunneling electrons can also interact with collective magnetic modes (magnons), as seen in inelastic tunneling experiments using a thin crystalline film of insulating NiO as the tunneling barrier. The coupling of tunneling electrons to magnons is found to be much stronger than the coupling to phonons (Tsui *et al.*, 1971).

1.4.2 Inelastic tunneling involving excitation in the metal electrodes or metal–insulator interfaces

Inelastic tunneling can also be induced by coupling of the tunneling electrons to collective modes in the metal electrodes, e.g. to metal phonons. A distinction between inelastic tunneling based on coupling to modes in the barrier and to modes in the metal electrodes can be made by varying either the metal used for the electrodes or the insulator used for the tunneling barrier. In addition, metal phonons are restricted to energies below 50 meV whereas vibrational modes in insulating oxides can have energies up to 150 meV.

The interaction of tunneling electrons with phonons (or photons) within the metal electrodes can lead to phonon-(photon-)assisted tunneling between two superconductors in a 'two-step' process as illustrated in Fig. 1.27.

Tunneling electrons can also interact with collective plasma oscillation (plasmon) modes (Tsui, 1969). As a consequence of inelastic tunneling excitation of optically coupled surface plasmon modes, broadband light emission in the visible regime from the tunnel junction can be observed with a high-frequency cut-off ω_{co} which is dependent only on the applied bias voltage U through the quantum relation $\hbar\omega_{co} = |eU|$ (Lambe and McCarthy, 1976; Rendell *et al.*, 1978; Adams *et al.*, 1979). The process of light emission by inelastic tunneling is illustrated in Fig. 1.28. For effective light emission it is important that the junction surface is rough, allowing decoupling of the energy stored in the surface plasmons from the tunnel junction by their radiative decay. However, even then the radiation efficiency, i.e. the probability that a tunneling

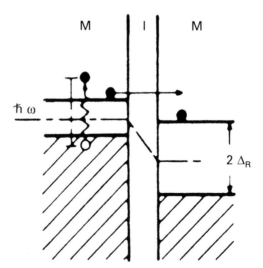

Fig. 1.27. Schematic illustration of photon (phonon) assisted tunneling between superconductors via the two-step process of (1) 'photo-excitation' and (2) elastic tunneling. This process can occur only for $\hbar\omega \geqslant 2\Delta_L$ ($\Delta_L < \Delta_R$) (Duke, 1969).

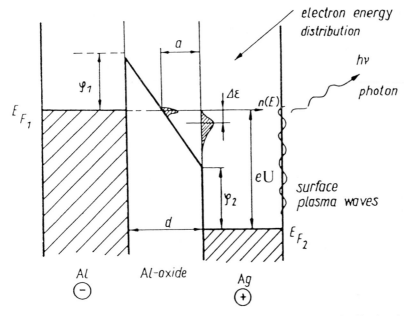

Fig. 1.28. Schematic energy band diagram and electron energy distribution in a thin film metal–oxide–metal tunnel junction in the case when $eU > \phi_1, \phi_2$ (Kroó *et al.*, 1980).

electron emits a photon, is typically no more than 10^{-5}–10^{-4}, which is low relative to the probability for inelastic tunneling (typically 10^{-3}–10^{-2}). On the other hand, this radiation efficiency is high enough to allow observation of the emitted light directly with the naked eye if the tunneling current is of the order of 1 mA. Therefore, though the radiation efficiency may be low compared with the probability for inelastic tunneling, the emitted light as a secondary effect is much more easily detected than the small changes in tunneling conductance (typically 0.1–1%) due to the opening of inelastic channels.

1.5 Semiconductor (p–n) junctions: Esaki tunnel diodes

So far we have concentrated on planar metal–insulator–metal tunnel junctions which are prepared by evaporation of the electrode materials to form polycrystalline films with an oxide growth step in between (section 1.3). However, tunnel junctions can also be realized by a variety of different ways. In this section, we focus on semiconductor p–n tunnel junctions. The p–n tunnel junctions can be fabricated by different methods including the grown junction method, the alloy junction method, and the diffused junction method (Burstein and Lundquist, 1969). The tunnel barrier in a p–n junction is formed by the depletion zone. A space charge region is left, with a width w which can be controlled by the acceptor and donor concentrations, N_A and N_D, of the p- and n-type sides of the junction according to the relationship

$$w \propto \left(\frac{1}{N_A} + \frac{1}{N_D} \right)^{1/2} \tag{1.68}$$

If the current flow across the p–n junction is to be dominated by tunneling, the junction must be narrow ($w \leqslant 100$ Å). Therefore, according to relation (1.68) the concentrations N_A and N_D have to be increased, in fact up to values where both sides of the junction become degenerately doped.

The current–voltage characteristics of such narrow p–n tunnel junctions of degenerately doped semiconductors were studied in detail by Esaki (1958, 1974), who made two important observations.

1. The p–n tunnel diode is a 'backward rectification diode', being more conductive in the reverse than in the forward direction.
2. A negative resistance is observed for a particular bias voltage window in the forward direction.

A typical forward current–voltage characteristic of an Esaki tunnel diode is shown in Fig. 1.29, which can be analyzed in terms of interband tunneling based on the energy diagrams shown in Fig. 1.30. Since both sides of the junction are degenerately doped, the Fermi energies are located well inside the valence or conduction band. For zero applied bias voltage, the two Fermi energies are at the same level (Fig. 1.30(a)). By applying a reverse bias U_R, valence electrons from the p-doped side can tunnel into empty states in the conduction band of the n-doped side (Fig. 1.30(b)). At low forward bias, electrons in the conduction band of the n-doped side tunnel into empty states in the valence band of the p-doped side (Fig. 1.30(c)). However, as the forward bias is increased, the bottom of the conduction band of the n-doped side will pass the top of the valence band of the p-doped side and direct elastic tunneling will be prohibited (Fig. 1.30(d)). Therefore, the tunneling current decreases, leading to the negative resistance regime in the current–voltage characteristics shown in Fig. 1.29. The excess current is explained by inelastic tunneling processes (section 1.4) where tunneling electrons emit phonons as indicated in Fig. 1.30(d). The first observation of inelastic tunneling processes was in fact made by using Esaki

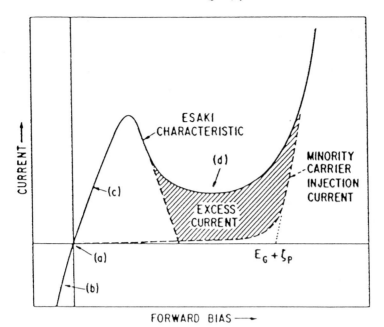

Fig. 1.29. A typical forward current–voltage characteristic of an Esaki tunnel diode (Logan, 1969).

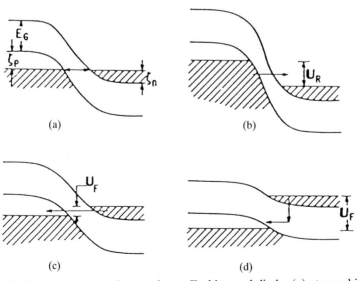

Fig. 1.30. Potential energy diagram for an Esaki tunnel diode, (*a*) at zero bias, (*b*) at reverse bias, (*c*) at small forward bias and (*d*) at large forward bias. In each case the arrow indicates the tunneling path (Logan, 1969).

tunnel diodes (Holonyak *et al.*, 1959). At even higher forward bias, the current flow will be dominated by the injection current, whereby electrons and holes are excited over the potential barrier.

The Esaki tunnel diode was the first type of junction in which a negative resistance effect was observed. Similar current–voltage characteristics can be obtained with metal–oxide–semiconductor tunnel junctions (Esaki and Stiles, 1966) where the origin of the negative resistance is quite different from that in the Esaki tunnel diode. Negative resistance is also observed in resonant tunneling (section 1.2.8) using semiconductor heterostructures (Chang *et al.*, 1974).

1.6 Metal–semiconductor junctions: Schottky barrier tunneling

If a metal is brought into contact with a heavily doped semiconductor, a Schottky barrier is formed as illustrated in Fig. 1.31. In this case, the barrier is formed by the semiconductor depletion region from which charge is transferred to the metal or to interface states. A uniform space charge region is left in which the electrostatic potential varies quadratically with distance z, leading to a barrier potential $\phi(z)$ of the form

$$\phi(z) = \frac{e^2 N_\mathrm{D}(s-z)^2}{2\varepsilon} + eU - E_\mathrm{F} \qquad \text{for } 0 < z < s \qquad (1.69)$$

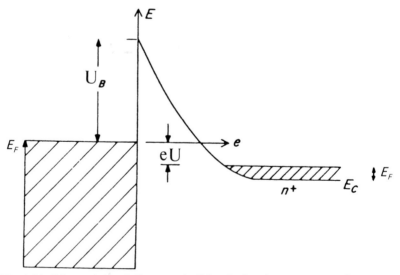

Fig. 1.31. Schematic band diagram of a Schottky barrier contact to a degenerate (metallic) n-type semiconductor (Wolf, 1985).

where N_D is the donor concentration (in the n-type case), ε the dielectric constant and E_F the Fermi energy of the semiconductor. The barrier thickness s can be controlled by the donor concentration N_D and the applied bias voltage U according to the relationship

$$s = \left[\frac{2\varepsilon \, (U_B + E_F / e - U)}{eN_D} \right] \tag{1.70}$$

The barrier height $\phi_0 = eU_B$ at the metal–semiconductor interface can be varied by the choice of contact metal.

The tunneling problem for a Schottky barrier described by Eq. (1.69) with the Hamiltonian

$$H_b = -\frac{\hbar^2}{2m} \frac{d^2}{dz^2} + \phi(z) \tag{1.71}$$

in the barrier region ($0 < z < s$) can be exactly solved (Wolf, 1985) by using the wave-matching method (section 1.2.1). If the barriers are sufficiently thin and the temperature is sufficiently low, then the current flow will be primarily due to electron tunneling. For forward biasing a Schottky barrier, as illustrated in Fig. 1.31, one obtains in the low-temperature limit (Burstein and Lundquist, 1969)

$$j \propto \exp \, (U / U_{00}) \tag{1.72}$$

whereas for the reverse bias one obtains

$$j \propto \exp \left[-\frac{2 \phi_0^{3/2}}{3 \, (eU_{00}) \, (\phi_0 - eU)^{1/2}} \right] \tag{1.73}$$

which is essentially equivalent to the Fowler–Nordheim relation (Eq. (1.23)) because the electric field F at the interface is proportional to the square root of the effective barrier height.

For thicker barriers and higher temperatures, tunneling of thermally excited carriers becomes important. In this case, one obtains for forward bias

$$j \propto \exp \, (U / U_0) \tag{1.74}$$

with $U_0 = U_{00} \coth \, (eU_{00}/k_B T)$ and for reverse bias

$$j \propto \exp \left(-U / U_0' \right) \qquad\qquad (1.75)$$

with

$$U_0' = \frac{U_{00}}{(e U_{00} / k_B T) - \tanh \left(e U_{00} / k_B T \right)}$$

In the low-temperature limit, the parameter U_0 in Eq. (1.74) tends to the constant value U_{00} which corresponds to the pure field-emission case (Eq. (1.73)). In the high-temperature limit, U_0 tends toward $k_B T/e$ which would be expected for pure thermionic emission.

Schottky barrier junctions can be realized in different ways. Planar metal–semiconductor junctions with an abrupt and contamination-free interface are obtained by cleavage of a semiconductor single crystal in the presence of an evaporating metal under ultra-high vacuum conditions. As an alternative, a semiconductor probe tip, cut and etched, for instance, from GaAs, can be brought into contact with a metal surface (von Molnar, *et al.*, 1967; Thompson, 1968; Thompson and von Molnar, 1970; Güntherodt *et al.*, 1982). Clean probe tips are obtained by an *in situ* sputter cleaning method which removes the surface contamination layer by low-energy bombardment in a chemically inert environment. Point contact superconductor–oxide–superconductor tunnel junctions involving anodized or oxidized metal tips were introduced earlier to study the superconductor energy gap for single crystals locally (Levinstein and Kunzler, 1966; Levinstein *et al.*, 1967).

In contrast to the Giaever-type planar metal–oxide–metal tunnel junctions prepared by the evaporation technique, point contact junctions can easily be formed with single crystals, compounds and intermetallics, thereby extending the applicability of the tunneling probe technique to important classes of materials. For Schottky-barrier type junctions, the tunneling barrier is also better defined than the oxide barriers which tend to be spatially and chemically non-uniform. On the other hand, the thickness of a Schottky barrier depends on the applied bias voltage (Eq. (1.70)), making a quantitative analysis of measured current–voltage characteristics more difficult.

1.7 Point contact spectroscopy (PCS)

In this section we want to focus on electron transport through small metallic constrictions such as point contacts between two conductors. Although the transport mechanism is no longer electron tunneling, point contact spectroscopy is closely related to tunneling spectroscopy with respect to the experimental techniques involved as well as the information which can be extracted from the experimental results (Yanson, 1974, 1983; Jansen *et al.*, 1980, 1987; van Kempen, 1982).

If we consider metallic point contacts, we can distinguish between two fundamentally different types.

(a) Point contacts for which the electron mean free path l is considerably smaller than the radius a of the contact area: $l \ll a$. In this case, we are in the diffusive regime of electron transport with a linear current–voltage characteristic according to Ohm's law. The resistance R of such point contacts in the so-called Maxwell limit is given by

$$R_M = \frac{\rho}{2a} \qquad (1.76)$$

where ρ is the specific resistivity of the metal.

(b) Point contacts for which $l \gg a$. In this case, we are in the ballistic regime of electron transport. The resistance in this so-called Knudsen limit has been calculated by Sharvin:

$$R_S = \frac{4\rho l}{3\pi a^2} \qquad (1.77)$$

Since ρ is inversely proportional to l, the Sharvin resistance R_S does not depend on the mean free path l, in contrast to R_M.

Between the two limits $l \ll a$ and $l \gg a$, the resistance can be expressed by

$$R = \frac{4\rho l}{3\pi a^2} + \Gamma(K)\frac{\rho}{2a} \qquad (1.78)$$

with $\Gamma(K) \approx 1$. $K = l/a$ is the Knudsen number. The voltage dependence of R is given by

$$\frac{dR}{dU} = \frac{\Gamma(K)\rho l}{2a} \cdot \frac{d(1/l)}{dU} \qquad (1.79)$$

The total scattering length l of the electrons is determined by elastic scattering processes with impurities (l_{imp}) and inelastic scattering with elementary excitations, in particular with the phonons (l_{ep}). If the scattering processes are independent, the total scattering length l is given by

$$\frac{1}{l} = \frac{1}{l_{imp}} + \frac{1}{l_{ep}} \tag{1.80}$$

Since only l_{ep} will be energy-dependent, we find

$$\frac{d(1/l)}{dU} = \frac{d\left(1/l_{ep}(\mathcal{E})\right)}{dU} = \frac{1}{v_F} \cdot \frac{d\tau_{ep}^{-1}(\mathcal{E})}{dU} \tag{1.81}$$

where v_F is the Fermi velocity and $\tau_{ep}(\mathcal{E}) = l_{ep}(\mathcal{E})/v_F$ is the electron–phonon scattering time. Therefore, the voltage derivative of the resistance becomes

$$\frac{dR}{dU} = \frac{\Gamma(K)\rho l}{2av_F} \cdot \frac{d\tau_{ep}^{-1}(\mathcal{E})}{dU} \tag{1.82}$$

In bulk normal metals the electrons cannot gain enough energy between scattering events to make the energy dependence of the scattering time significant, i.e. $dR/dU = 0$. As a consequence, the linear relationship between current and voltage, expressed by Ohm's law, is obeyed for a very wide range of currents and voltages. In contrast, by applying a voltage to a point contact, sufficiently high current densities of the order of 10^{10} A cm^{-2} can be obtained for which $dR/dU \neq 0$ and nonlinearities in the current–voltage characteristics become visible.

At low temperatures, in the limit $T = 0$, the relaxation rate $1/\tau_{ep}(\mathcal{E})$ for the electron–phonon interaction is given by

$$\tau_{ep}^{-1}(\mathcal{E}) = \frac{2\pi}{\hbar} \int_0^{\mathcal{E}=eU} \alpha^2 g(\omega)\, d\omega \tag{1.83}$$

where $\alpha^2 g(\omega)$ is the Eliashberg function, being the product of the phonon density of states $g(\omega)$ and the squared matrix element α for the electron–phonon interaction, averaged over the Fermi sphere. We finally find an important relation between dR/dU and $\alpha^2 g(\omega)$:

$$\frac{dR}{dU}(U) = \frac{\pi\Gamma(k)\rho l e}{a\hbar v_F}\, \alpha^2 g(eU) \tag{1.84}$$

According to this equation, we can directly determine the Eliashberg function by measuring the voltage derivative of the resistance at low temperatures. We have mentioned earlier (section 1.3.1) that tunneling into superconductors allows one to derive $\alpha^2 g(\omega)$ for the superconducting metals. Point contact spectroscopy extends the possibility of an experimental determination of $\alpha^2 g(\omega)$ also to normal metals with a sufficiently large mean free path of the electrons such that $l \gtrsim a$ can be fulfilled.

As discussed in section 1.4, inelastic scattering of tunneling electrons by phonons or other excitations also leads to an influence of the electron–phonon interaction on the current–voltage characteristics. However, there is a significant difference between inelastic tunneling spectroscopy and point contact spectroscopy. In tunneling spectroscopy, inelastic scattering leads to the opening of new channels for electrons to tunnel through the barrier. As a consequence, the conductance increases, i.e. the resistivity decreases. In contrast, electrons flowing through a point contact can return through the contact after a scattering event. This back flow leads to an increase of the resistance.

It is clear that the point contact spectroscopy technique can be applied not only to study the electron–phonon interaction, but also the interaction of electrons with other excitations, e.g. with magnons (Verkin *et al.*, 1979). Furthermore, point contact spectroscopy is used to probe mixed valence and heavy-fermion compounds (Bussian *et al.*, 1982; Moser *et al.*, 1985).

Point contacts can be realized in two different ways. Yanson (1974), who discovered and exploited point contact spectroscopy, has used planar metal–oxide–metal tunnel junctions prepared by the evaporation technique. A metallic short circuit due to a pinhole in the oxide film, which is introduced either accidentally or intentionally, serves as a point contact (Fig. 1.32(*a*)). Alternatively, a sharp needle obtained by electrochemical etching of a wire can be pressed against the surface of a flat piece of metal. By carefully adjusting the pressure by means of a differential screw mechanism (Fig. 1.32(*b*)), the contact area and resistance can be properly controlled (Jansen *et al.*, 1977). While the point contacts in evaporated thin film structures have the advantage of higher mechanical stability, the pressure-type point contacts allow investigation of a large variety of samples, including studies of anisotropies of single crystals, as well as the transition between the tunneling and point contact regimes (Blonder *et al.*, 1982; Blonder and Tinkham, 1983).

By using a double point contact arrangement, the electron–surface interaction can be probed from the inside of a material by transverse electron focusing (Tsoi, 1974; van Son *et al.*, 1987). Electrons are injected into a single crystal through one point contact (emitter) and are then deflected by means of a magnetic field to another contact

(a)

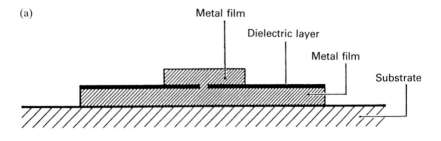

(b)

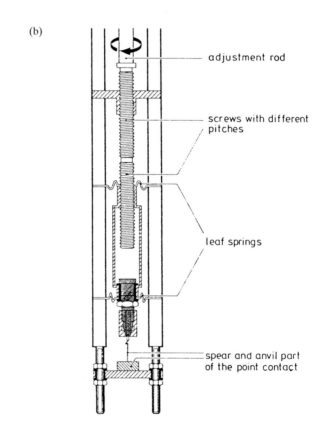

Fig. 1.32. (*a*) Schematic view of a metal–insulator–metal tunnel junction with a short circuit between the two metal films. (*b*) Schematic view of a pressure-type point contact. By means of a differential screw mechanism the spear can be moved towards the anvil in order to adjust a contact (Jansen *et al.*, 1980).

(collector). By studying the collector signal as a function of the magnetic field, the coefficient for specular reflection of the electrons at the metal surface can be determined.

Finally, ballistic point contacts in semiconductors have also become an active field of research. By measuring the conductance as a function of the width of the contact defined in the two-dimensional electron gas of a GaAs–AlGaAs heterostructure, quantized steps of $2e^2/h$ were found in zero magnetic field, which can be explained by the quantized transverse momentum in the point-contact region (van Wees *et al.*, 1988).

1.8 Vacuum tunneling before the invention of the STM

We now come back to the investigation of tunneling phenomena. As we have already mentioned in a number of previous sections, several limitations of the planar oxide tunnel junctions can be recognized. An important drawback is the existence of inhomogeneities in the junction on a 'nanoscopic scale', which are significant because the total thickness of the oxide layer is typically 3 nm or even less. An oxide junction might look as depicted in Fig. 1.33(*a*). Owing to roughness of the electrode–oxide interfaces there is a significant variation of the oxide thickness on the nanometer scale, leading to strong inhomogeneities in tunneling current flow as a result of the strong exponential distance dependence of the tunneling current. Additional inhomogeneities in the chemical composition of the oxide layer can make the situation even worse. Though there have been significant advances in the fabrication of planar oxide tunnel junctions with relatively smooth oxide layers (e.g. Imamura and Hasuo, 1991), recognizable inhomogeneities still remain, as seen by cross-sectional transmission electron microscopy (Fig. 1.33(*b*)).

An alternative is complete removal of the oxide layer and choosing vacuum as the tunnel barrier. Clean electrode surfaces can then be prepared and carefully characterized without disturbance by an attached barrier layer. In addition, dielectric screening and other effects caused by the presence of an oxide layer are absent in vacuum tunnel junctions. Moreover, the electrode spacing can be changed very conveniently in vacuum tunnel junctions, which additionally allows study of the distance dependence of the tunneling current. Finally, a comparison of experimental results with tunneling theory assuming a vacuum barrier is better motivated.

Though vacuum tunneling has long been considered superior to tun-

Oxide Junction

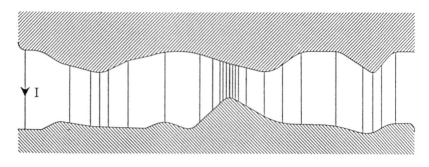

(b)

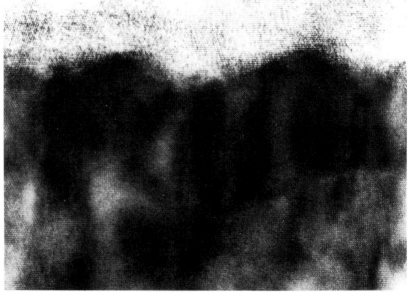

100 Å

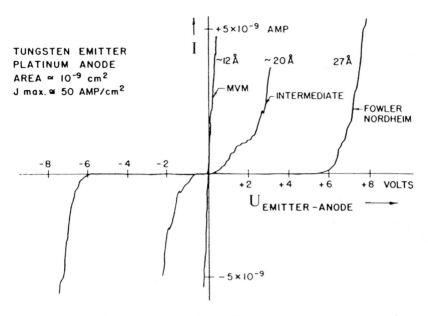

Fig. 1.34. *I–U* characteristics for three different emitter-to-surface spacings. Note the linear metal–vacuum–metal (MVM) tunneling characteristic (Young *et al.*, 1971).

neling through oxide layers, there was a serious obstacle to overcome before its experimental realization. This main obstacle was described by Giaever in his Nobel Prize Lecture (Giaever, 1974): 'To be able to measure a tunneling current the two metals must be spaced no more than about 100 Å apart, and we decided early in the game not to attempt to use air or vacuum between the two metals because of problems with vibration'. The vibration coupled into the tunnel junction leads to temporal variations of the tunnel barrier width, which result in temporal instabilities of the tunneling current.

The first observation of vacuum tunneling was therefore not made until 1971, when Young, Ward and Scire succeeded in measuring the current–voltage characteristics (Fig. 1.34) in the tunneling as well as in the field emission and intermediate regime (Young *et al.*, 1971). A

Fig. 1.33. (*a*) Schematic microscopic structure of a planar metal–oxide–metal tunnel junction. (*b*) Cross-sectional TEM lattice image of a planar Nb/AlO$_x$–Al/Nb tunnel junction fabricated on a single-crystalline silicon substrate (Imamura and Hasuo, 1991).

linear *I–U* characteristic was found for metal–vacuum–metal tunneling as expected (section 1.2.4 and 1.3.1). In this experiment, a vacuum tunnel junction was established by bringing an electrochemically etched tungsten field emitter tip very close (about 12 Å) to a platinum surface

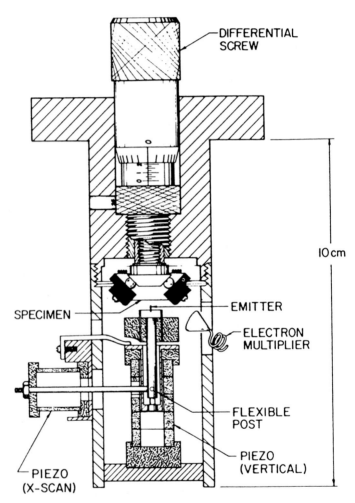

Fig. 1.35. Schematic drawing of the topografiner. A differential screw is used as a coarse adjustment to bring the specimen close enough to the emitter so that it is within the range of the vertical (*z*) piezoelectric element. The *x*-scan piezo-drive deflects the emitter support post so as to scan the emitter in one direction. The orthogonal (*y*) piezo-drive is not shown. The specimen is clamped between copper blocks to permit heating. An electron multiplier permits detection of secondary electrons (Young *et al.*, 1972).

by means of a differential screw for coarse adjustment and a (z) piezo-electric drive for fine adjustment (Fig. 1.35). A low-vibration stand with a vertical period of 1 Hz and an acoustic shield were used for stabilization of the tunnel barrier.

Moreover, the instrument, named 'topografiner', had two additional orthogonal (x and y) piezodrives, allowing raster scanning of the field emitter tip relative to the sample surface. By using a servo system (Fig. 1.36) containing the z piezodrive, the field emitter tip could be kept at a constant distance of a few hundred Ångström units above the surface during scanning. By recording the vertical height z of the tip, i.e. by recording the servo signal as a function of the lateral position (x,y), a topographic map $z(x,y)$ of the surface was obtained (Young, 1971; Young et al., 1972). The topografiner, operated in the field emission regime with relatively large tip-to-sample spacing, achieved a vertical resolution perpendicular to the surface of about 30 Å and a lateral resolution in the plane of the sample surface of about 4000 Å. A successful combination of scanning and vacuum tunneling to a scanning tunneling microscope (STM) failed at that time due to problems with mechanical vibrations on the Ångström unit level as well as deficiencies concerning the servo system. As a consequence, development of the STM was delayed by another ten years. However, the potential of such a microscope for studying single and multiple steps on crystal surfaces, the adsorption of gases and processes involving excitations at surfaces, as well as for metrological applications was already recognized in 1971 or even earlier (Young, 1966).

Improvement of the stability of vacuum tunnel junctions was an important issue for several other investigations of vacuum tunneling after the pioneering work of Young, Ward and Scire. Excellent stability was achieved with a thermal drive apparatus allowing two lead electrodes to be brought within 10–20 Å of each other by using the differential contraction of two metals as the temperature is varied (Thompson and Hanrahan, 1976).

An extensive study of vacuum tunneling between two gold balls was performed by Teague (1978). The electrode holder and micropositioning assembly including a differential screw mechanism and piezoelectric drives is shown in Fig. 1.37. For the first time, the tunneling potential barrier height was deduced from the slope of measured log current–distance (log I–s) characteristics (Fig. 1.38). In addition, the possible influence of attractive forces, van der Waals type and electrostatic forces, on the tunneling characteristics was already being discussed at that time.

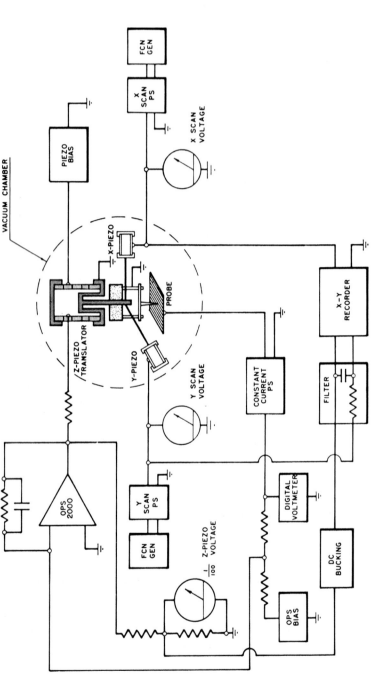

Fig. 1.36. Block diagram for the topografiner electrical circuitry. The *x–y* recorder was frequently replaced with a memory oscilloscope allowing more rapid scan rates (Young *et al.*, 1972).

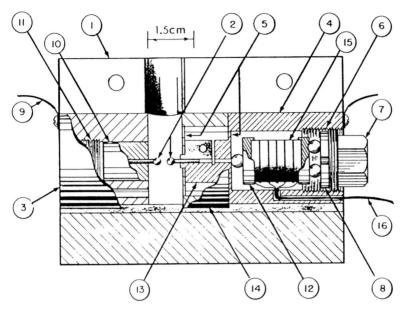

Fig. 1.37. Electrode holder and micropositioning assembly for gold–vacuum–gold tunneling experiments. (1) Housing for electrode holders. (2) Gold electrodes with spherical ends. (3) Copper mount for differential-screw mechanism. (4) Polyimide film. (5) Diaphragm flexures. (6) Threads. (7) Prestressing nut. (8) Hardened steel washer. (9) Electrical connections to gold electrodes. (10) Clamp-holder for left electrode. (11) Threads. (12) Hardened steel end caps. (13) Clamp-holder for right electrode. (14) Support for outer diaphragm flexure. (15) Piezoelectric displacement element. (16) Electrical connections to piezoelectric element (Teague, 1978).

Very high stability in vacuum tunneling experiments (after the invention of the STM) was achieved with squeezable electron tunnel junctions (Fig. 1.39) which can be adjusted purely mechanically without using piezoelectric drives (Moreland *et al.*, 1983).

All the vacuum tunneling studies mentioned above after the early work of Young, Ward and Scire have in common that they were performed with tunnel junctions being adjustable in one dimension only, due to the lack of an additional scanning unit. Therefore, topographic imaging or spatially resolved measurements were not possible.

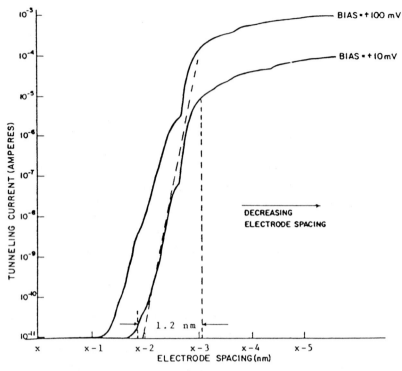

Fig. 1.38. Experimental log current–spacing characteristics. The sloping dashed line corresponds to a work function of approximately 3.0 eV (Teague, 1978).

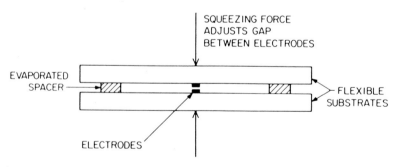

Fig. 1.39. Schematic illustration of a squeezable electron tunnel junction showing two electrodes supported by flexible substrates separated by thin-film spacers (Moreland *et al.*, 1983).

1.9 The birth of STM

The successful combination of vacuum tunneling and scanning capability was achieved early in the 1980s when Binnig, Rohrer, Gerber, and Weibel at the IBM Rüschlikon laboratory succeeded in the development of the scanning tunneling microscope (STM). The early design of the STM included a piezoelectric tripod scanner and a piezoelectric walker, the 'louse' to allow for the approach of sample and tip (Fig. 1.40). During the night of 16 March, 1981, the first log I–s characteristics showing an exponential dependence of the tunnel current I on the tip–surface separation s (Fig. 1.41) were obtained by using a tungsten tip and a platinum sample (Binnig *et al.*, 1982a, c). This date has subsequently been identified with the birth of the STM (Binnig and Rohrer, 1987). However, the real breakthrough for the STM instrument as developed by Binnig, Rohrer, Gerber, and Weibel came with the first atomic resolution image of the Si(111) 7×7 surface obtained in autumn 1982 and shown in Fig. 1.42 (Binnig and Rohrer, 1982; Binnig *et al.*, 1983a). This first atomic resolution STM image of a well-defined surface structure made the STM different and superior compared with all other vacuum tunneling instruments developed before and perhaps it would be more appropriate to identify this event with the birth of STM.

The atomic resolution capability of STM directly in real space for surfaces of conducting materials has become the most important feature of STM. The extremely high lateral as well as vertical resolution makes

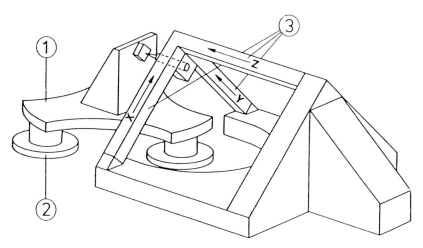

Fig. 1.40. 'Louse' with piezoelectric base plate (1) and three feet (2) allowing the sample to be approached towards the tip which is mounted on a piezoelectric tripod scanner (3) (Binnig and Rohrer, 1982).

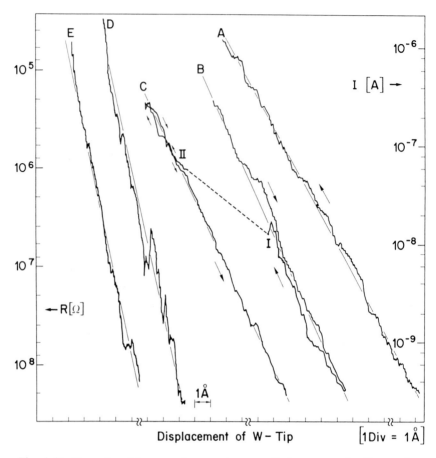

Fig. 1.41. Tunnel resistance and current versus displacement of a Pt sample with respect to the tip for different sample and tip conditions (Binnig *et al.*, 1982a).

STM an outstanding microscopy (Fig. 1.43). In addition, the atomic resolution capability in real space has several important consequences.

1. Surface structures, either given by nature or artificially created, e.g. by MBE, can be monitored and controlled on the atomic scale.

2. Based on the atomic resolution capability, the tip can be positioned with atomic accuracy above a pre-selected atomic site and a local experiment can be performed.

3. The capability of monitoring surface structures with atomic resolution combined with the ability to position the tip with atomic accuracy above pre-selected surface sites finally leads to the ability to

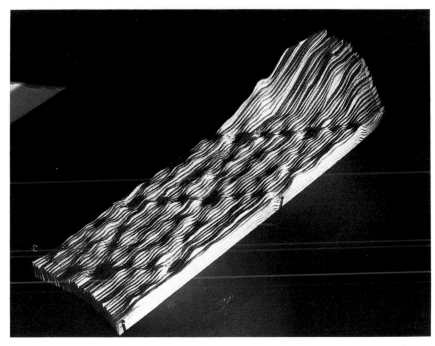

Fig. 1.42. STM topograph of the 7×7 reconstruction of the Si(111) surface, showing two complete rhombohedral unit cells (Binnig and Rohrer, 1982).

perform direct and controlled manipulation at the atomic level, offering the opportunity to create atomic-scale devices.

There can be no doubt that the birth of STM has marked the beginning of a novel field of Ångström unit scale research and technology which will perhaps dominate developments in the 21st century in a way similar to that in which the invention of the transistor and integrated circuits did in the 20th century.

1.10 STM design and instrumentation

In the following, we first focus on STM design and instrumentation before discussing various modes of operation as well as the investigation of 'special effects' in STM. This section is not intended to provide the reader with full information on how to build a reliably working STM. Rather a summary of some general design and instrumentation considerations (Pohl, 1986; Kuk and Silverman, 1989) is given. We can distinguish between the mechanical construction of the STM, the electronics

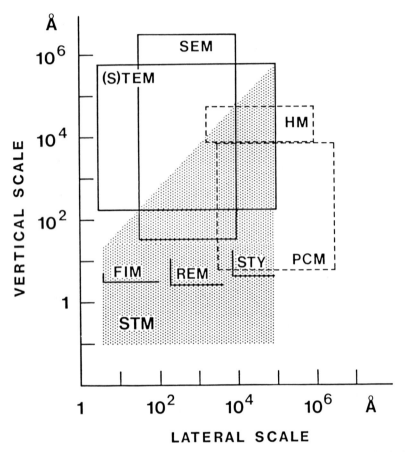

Fig. 1.43. Comparison of resolutions of different microscopes. STM: shaded area. HM: high-resolution optical microscope, PCM: phase-contrast microscope, (S)TEM: (scanning) transmission electron microscope, SEM: scanning electron microscope, REM: reflection electron microscope, and FIM: field ion microscope (Binnig and Rohrer, 1982).

and the computer automation for data acquisition and processing. We will also pay particular attention to the most critical component of the STM, i.e. the sensor tip, its preparation and characterization.

1.10.1 Vibration isolation

As already mentioned in section 1.8, vibration has been one of the most serious problems on the way to successful operation of an STM. The vibration of a tunnel unit can result, for instance, from vibration in the building or acoustic noise. To obtain a vertical resolution of 0.01 Å (1 pm) in STM, which is desired for high-resolution applications, a stability of the tip-to-sample spacing at the level 0.1 pm is required. This is at least six orders of magnitude smaller than typical floor vibration amplitudes (0.1–1 μm). The required stability can only be achieved by combination of an effective vibration isolation system and a rigid design of the STM instrument.

We first discuss damping systems for vibration isolation. They can be either passive or active. However, most STM instruments are used in combination with a passive damping system and we therefore restrict our discussion to those. Passive damping can be achieved either by using viscoelastic materials, e.g. rubber or Viton, or metal springs combined with some damping system. Viscoelastic materials are most effective against large-amplitude shock and high-frequency vibration. However, if they are used as spacers separating metal plates, these spacers often have a relatively large compression stiffness resulting in a rather high (10–100 Hz) resonance frequency. Such high resonance frequencies are not desirable because below or at the resonance frequency the damping system enhances perturbations instead of reducing them (Fig. 1.44). Therefore, the resonance frequency of the damping system should be as low as possible. Metal springs usually have a lower resonance frequency (typically 1–6 Hz) which depends only on the stretched length of the spring. On the other hand, metal springs provide only little damping and therefore have to be combined with viscoelastic materials or with an eddy-current damping system consisting of copper elements and permanent magnets (e.g. CoSm).

For their first design in 1981, Binnig and coworkers used magnetic levitation on a superconducting bowl of lead combined with eddy-current damping for vibration isolation which, however, worked only at liquid helium temperature. (High-T_c superconductors were not yet known at that time.) The second-generation STM included a two-stage spring system combined with Viton elements (Binnig and Rohrer, 1982). The spring system was later also combined with eddy-current damping. It is easily verified that a two-stage spring system provides more efficient vibration isolation than a one-stage system (Park and Quate, 1987). For the 'pocket-size' STM (Gerber *et al.*, 1986), a stack of metal plates separated by Viton spacers and springs was used for vibration isolation. This allowed considerable reduction in size of the whole STM unit compared with the two-stage spring system. These 'home-built' vibration isolation systems are often combined with commercially available vibration isolation systems using pneumatic feet.

Apart from the external damping systems discussed so far, one can

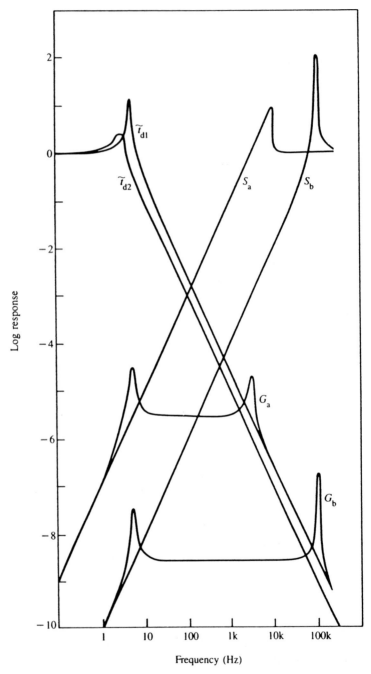

Fig. 1.44. Frequency characteristics of damping ($\tilde{t}_{d1}$, $\tilde{t}_{d2}$), stiffness (S_a, S_b), and combination of $\tilde{t}_{d1}$ with S_a, S_b (G_a, G_b) (Pohl, 1986).

make use of the inherent structural damping of a rigid STM instrument. To prevent a disturbing influence of the low-frequency vibration left by the external damping system on the stability of the tip-to-sample spacing, the first resonance frequency of the STM unit should be as high as possible (Fig. 1.44) which means that the construction of the tunnel unit has to be very stiff and it should be small. The total instrument response including the external damping system and the inherent structural damping of the STM unit is obtained by multiplying the two corresponding transfer functions (Fig. 1.44). It can be seen from Fig. 1.44 that, by combining an external vibration isolation system with a rigid STM design, a reduction of external vibration amplitudes by a factor of 10^{-6}–10^{-7} under optimized conditions can indeed be achieved, which is required for high-resolution STM operation, as mentioned at the beginning.

To achieve the highest possible stability it is best to operate the STM in a room with acoustic shielding located in the basement, where floor vibrations are considerably reduced. If the STM is to be operated in an ultra-high vacuum (UHV) chamber, the whole UHV system can be mounted on commercially available pneumatic feet with a resonance frequency of 1 Hz or even less. Inside the UHV system, a two-stage spring system combined with eddy-current damping and an additional stack of metal plates separated by Viton spacers on the inner platform can provide a very efficient external damping (Fig. 1.45). Together with the inherent structural damping of a rigid STM design, it is possible to reach a stability at the 0.1 pm level.

1.10.2 Positioning devices

The design of the STM has to allow for three-dimensional movement of the sensor tip and, ideally, three-dimensional positioning of the sample. Three-dimensional movement of the tip is always achieved by using piezoelectric drives, which are described in the next section 1.10.3. For sample positioning, several different solutions exist.

The minimum requirement for a sample positioning device is to move the sample in one dimension from millimeter to submicrometer distances towards the sensor tip. The step sizes for this movement have to be smaller than the total range of the z piezodrive for the tip in order to avoid accidental contact between tip and sample during the approach. In addition, two-dimensional positioning of the sample in the surface (x–y) plane with a smallest reproducible step size below the total (x–y) scan range of the scanning unit for the tip is desired, particularly for

Fig. 1.45. The STM unit used in the author's laboratory is mounted on top of a stack of five stainless steel plates separated by Viton rubber elements. This stack itself is supported from the inner platform of a two-stage spring system combined with eddy-current damping. The entire microscope (STM unit plus vibration isolation system) can be mounted into a UHV chamber through the bottom flange by using an elevator (Wiesendanger *et al.*, 1990a).

investigations of inhomogeneous sample surfaces where inequivalent surface regions above the micrometer length scale exist.

(a) Sample movement in STM was first realized by using a *piezoelectric walker*, the 'louse', shown in Fig. 1.40 (Binnig and Rohrer, 1982). The louse allowed for two-dimensional positioning of the sample relative to the tip based on a sequence of clamping and piezo expansion or contraction. Several variations of the original louse design have been developed (e.g. Mamin *et al.*, 1985; Uozumi *et al.*, 1988). Another kind of one-dimensional piezoelectric translation device based on clamping and piezo expansion or contraction is the commercially available 'inchworm'. Three inchworms can be combined to allow full three-dimensional sample positioning (Takata *et al.*, 1989b).

(b) As an alternative to the piezoelectric walkers, *magnetic walkers* for sample positioning have also been developed (e.g. Smith and Elrod, 1985; Corb *et al.*, 1985; Ringger *et al.*, 1986). Steps can be realized, for instance, by applying voltage pulses to a coil with a permanent magnet inside. The electromagnetic forces push the magnet forward or backward depending on the polarity of the voltage pulse. Movement of the magnet can be translated into movement of the sample. An STM design based on this kind of sample positioning is shown in Fig. 1.46 (Wiesendanger *et al.*, 1990 a,b).

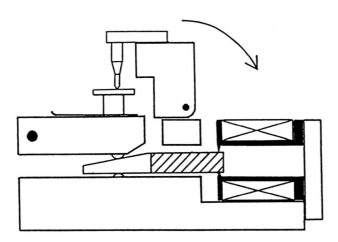

Fig. 1.46. Schematic drawing of a STM unit used in the author's laboratory for surface science studies under UHV conditions. An electromagnetically driven wedge transforms its horizontal movement into vertical movement of a sample stage which carries the clamped sample holder with the horizontally mounted sample. The scanhead rests on three ball bearings and can be flipped backwards, giving easy access to the sample and tip. The scale bar is 1 cm (Wiesendanger *et al.*, 1990a, 1990e).

(c) Sample movement can also be realized by *inertial sliding* based on controlled change between gliding and sticking of an object on a surface. The motion is initiated by a piezoelectric element excited by a sawtooth electrical waveform (Pohl, 1987; Anders *et al.*, 1987; Niedermann *et al.*, 1988; Lyding *et al.*, 1988b; Frohn *et al.*, 1989).

(d) Purely mechanical positioning of the sample can be achieved by using a *screw pushing against a reduction lever* as shown in Fig. 1.47

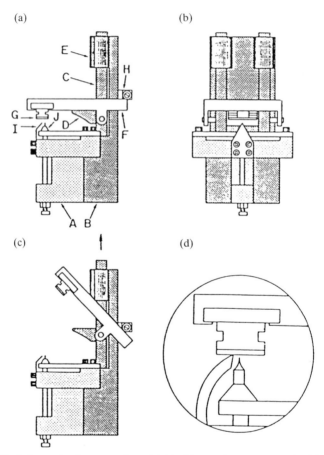

Fig. 1.47. Schematic diagram of an UHV STM unit. Indicated in (*a*) are (A) the macor block onto which the *x*, *y*, *z* piezo scanners are mounted, (B) the microscope base plate, (C) carriage rods, (D) stop, (E) ball bushing assembly, (F) connecting arms to sample, (G) sample and sample holder, (H) catch, (I) foot, and (J) probe tip mounted in collet on the *x*–*y* piezoelectric drives. In (*c*) is shown the microscope with the sample retracted while in (*d*) is shown a close-up during tunneling (Demuth *et al.*, 1986a).

(Demuth *et al.*, 1986a, 1986b; Coombs and Pethica, 1986; Kaiser and Jaklevic, 1988) or against a single spring or double spring system as shown in Fig. 1.48 and Fig. 1.49 respectively (Smith and Binnig, 1986; Fein *et al.*, 1987; Wiesendanger *et al.*, 1990a). For a differential spring system, reduction of motion is produced by the difference in spring constants between two springs. The use of purely mechanical positioning devices can lead to very rigid STM designs with a high first resonance frequency as desired.

1.10.3 Scanning units

Scanning units allowing for three-dimensional tip motion can be realized using piezoelectric drives which can be built up by bars, tubes or bimorphs.

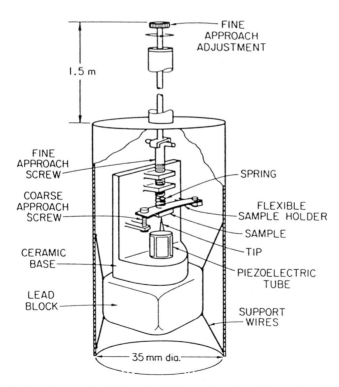

Fig. 1.48. An ultra-small STM unit which can be dipped into a liquid-helium storage dewar for fast turn-around low-temperature experiments (Smith and Binnig, 1986).

(a)

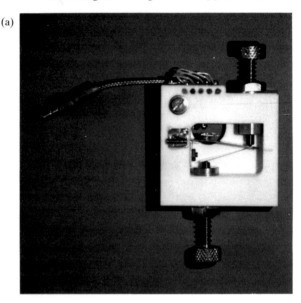

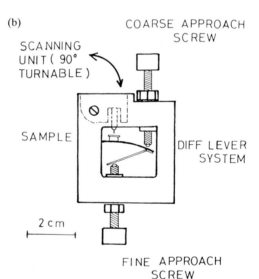

Fig. 1.49. (*a*) Photograph and (*b*) schematic drawing of a rigid STM unit. The approach mechanism is based on a differential lever system consisting of two levers with different spring constants. The sample, which is mounted on the lever with the higher spring constant, can be moved towards the tip by reducing the tension of this lever with the coarse approach screw. By bending the lever with the lower spring constant with the fine approach screw, a high reduction of sample movement can be achieved. The piezoelectric scanner with the tip holder is fixed to a rotatable scan-head. By rotating the scan unit by 90°, the tip and sample can easily be exchanged (Wiesendanger, *et al.*, 1990a).

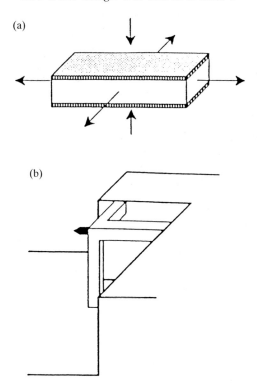

Fig. 1.50. (*a*) Piezoelectric bar. (*b*) Piezoelectric tripod scanner consisting of three piezoelectric bars perpendicular to each other.

(a) If a voltage U_p is applied to opposite electrodes of a *piezoelectric bar* of thickness h and length l (Fig. 1.50(*a*)), the resulting length change Δl is given by

$$\Delta l = d_{31} \frac{l}{h} U_p \qquad (1.85)$$

where d_{31} is the relevant piezoelectric coefficient. By combining three piezoelectric bars via a piezoelectric tripod (Fig. 1.50(*b*)), full three-dimensional movement of the tip can be achieved (Binnig and Rohrer, 1982; Okayama *et al.*, 1985).

(b) If a voltage U_p is applied between the inner and outer electrodes of a *piezoelectric tube* of length l and wall thickness h (Fig. 1.51(*a*)), the resulting length change Δl is given by

$$\Delta l = d_{31} \frac{l}{h} U_p \qquad (1.86)$$

(a)

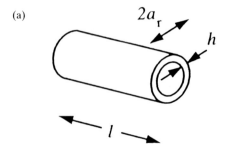

(b)

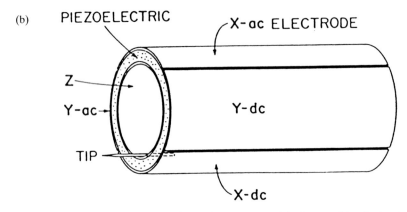

Fig. 1.51. (*a*) Piezoelectric tube. (*b*) Tube scanner with the outside electrode sectioned into four equal areas parallel to the axis of the tube. As a voltage is applied to a single outside electrode the tube bends away from that electrode. Voltage applied to the inside electrode causes a uniform elongation. A small a.c. signal and a large d.c. offset can be separated on electrodes 180° apart (Binnig and Smith, 1986).

The higher sensitivity of tubes compared with bars of about the same size results from the fact that the wall thickness of the tubes is much less than the thickness of the bars. Three piezoelectric tubes can again be combined to form a tripod scanner allowing for three-dimensional movement of the tip.

As an alternative, a compact single-tube scanner can be used for full three-dimensional movement if the outer electrode is divided into four segments (Fig. 1.51(*b*)). Pairs of opposite outer electrode segments can then be excited by a voltage leading to bending modes of the tube fixed at one end and thereby allowing for *x–y* movement in addition to *z* movement by expansion or contraction along the tube axis (Germano, 1959; Binnig and Smith, 1986). If the voltage is applied symmetrically to the opposite outer electrode segments, i.e. two equal and opposite voltages on two opposite quadrants, the deflection of the tube in the x (y) direction is given by

$$\Delta x\,(\Delta y) = 2\sqrt{2}\,d_{31}\,\frac{l^2}{\pi D h}\,U_{x(y)} \tag{1.87}$$

where D is the inner diameter of the tube (Chen, 1992a). If the voltage is applied antisymmetrically to the opposite outer electrode segments, i.e. one voltage on a single quadrant, the deflection is given by

$$\Delta x\,(\Delta y) = \sqrt{2}\,d_{31}\,\frac{l^2}{\pi D h}\,U_{x(y)} \tag{1.88}$$

which is exactly half of the deflection obtained for the symmetric case (Eq. (1.87)).

(c) For a *piezoelectric bimorph* of length l and thickness h (Fig. 1.52), clamped at both ends, the deflection is given by (Pohl, 1986)

$$\Delta x = \tfrac{3}{8}d_{31}\left(\frac{l}{h}\right)^2 U_p \tag{1.89}$$

The sensitivity is higher compared with piezoelectric bars by a factor of the order of l/h.

The design of a piezoelectric scanning unit has to be optimized with respect to the following issues.

1. High resonance frequencies and high scan speed. For stable STM operation it is of advantage to have high resonance frequencies as discussed in section 1.10.1. Single-tube scanners offer the highest

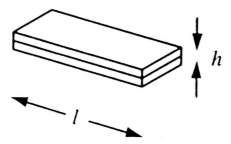

Fig. 1.52. Piezoelectric bimorph element.

resonance frequencies and allow for very compact and rigid STM designs. Based on high resonance frequencies, the scan speed can be increased accordingly without the risk of exciting vibrations. Video scan rates can be realized offering scanning electron microscope (SEM)-type imaging, which is useful for studying dynamic processes at surfaces, such as diffusion.

2. High sensitivity and large scan areas. A high sensitivity of the piezoelectric drives is of advantage for increasing the maximum scan area for a given applied voltage. Large scan areas are useful for imaging larger scale surface structures as well as for checking how representative the surface structures observed on a smaller scale are. On the other hand, if a large scan area is not required, the output voltage of the electronics for the piezoelectric drives can be reduced accordingly, with a higher sensitivity of the piezodrives.

3. Low cross-talk between x, y, and z piezodrives. Cross-talk between the three orthogonal directions of movement is present for both tripod and single-tube scanner designs and can become significant, particularly for large displacements. Cross-talk is undesirable for distortion-free imaging. However, it can be used for calibrating the scanner sensitivity *in situ*, e.g. inside an UHV chamber or a cryostat (Chen, 1992b).

4. Low nonlinearities, hysteresis and creep. Piezoelectric hysteresis can cause image distortions which become particularly significant for large-area scans. Nonlinearities result from an increased scanner sensitivity with increased scan size. In addition, piezoelectric elements show a slow logarithmic creep after a fast voltage change described by

$$\frac{\Delta l}{l} = \alpha + \beta \ln t \qquad (1.90)$$

(α, β are constants, t is time). Piezoelectric creep therefore leads to image distortions on a longer time scale due to the logarithmic dependence on time. These image distortions caused by hysteresis and creep can be reduced by appropriate choice of the piezoelectric ceramics and electronically by a capacitor insertion method (Kaizuka and Siu, 1988; Kaizuka, 1989). Alternatively, image distortions may be eliminated by real-time feedback scan correction or post-imaging software image correction. This method can be applied if the position of the piezoelectric scanner is accurately measured and controlled, for instance, by using an integrated heterodyne interferometer (Stemmer *et al.*, 1988), a capacitance-based position monitor

(Griffith *et al.*, 1990), or an integrated optical beam displacement sensor (Barrett and Quate, 1991a).

5. Low thermal drifts. Thermal drifts in STM result from slow temperature variations ΔT and the difference between the thermal expansion coefficients α of the different materials used:

$$\frac{\Delta l}{l} = \Delta \alpha \cdot \Delta T \qquad (1.91)$$

Thermal drifts can be reduced considerably, if materials with similar thermal expansion coefficients are chosen, e.g. piezoelectric ceramic ($\alpha \approx 7 \times 10^{-6}$ K^{-1}) in contact with Macor ($\alpha \approx 9 \times 10^{-6}$ K^{-1}) rather than piezoelectric ceramic in contact with stainless steel ($\alpha \approx 16 \times 10^{-6}$ K^{-1}). Additionally, thermal drifts can be reduced by a symmetric design of the scanning unit, which can be achieved either with piezoelectric bars (van der Walle *et al.*, 1985) or piezoelectric tubes (Lyding *et al.*, 1988b).

1.10.4 Electronics

The STM electronics has to be designed at least as carefully as the mechanical parts of the STM in order to avoid electrical noise and pickup or feedback loop instabilities, to mention only some of the problems which can arise. A block diagram of an STM electronic system is shown in Fig. 1.53.

The tip (or sample) bias voltage (typically between 1 mV and 4 V) can be supplied by a battery or computer-controlled digital-to-analog converter (DAC). After approaching tip and sample, the tunneling current is measured by a preamplifier with a gain of 10^6–10^9 V A^{-1}. The tunneling current signal, which is exponentially dependent on the separation between tip and sample, can then be linearized by a logarithmic amplifier to improve the dynamic range. This is particularly important for relatively rough surfaces, whereas for atomically flat surfaces, a linear approximation of the exponential dependence of the tunneling current can be used and the logarithmic amplifier is no longer necessary. The measured tunneling current is then compared with the demanded current (typically between 10 pA and 10 nA) and the resultant error signal is fed into the feedback amplifiers which usually consist of a proportional (P) amplifier and an analog integrator (I). The differential part (D) of the familiar PID feedback loop system is not used for STM because it can cause instabilities. The feedback signal is finally fed into a high-voltage amplifier which generates the amplified voltage signal

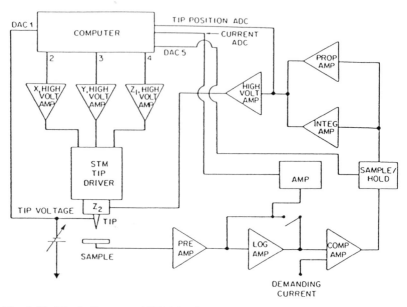

Fig. 1.53. Block diagram of STM feedback circuit (Kuk and Silverman, 1989).

applied to the z piezoelectric drive. Scanning the tip in the x–y plane is achieved by using either computer-controlled DACs or synchronized ramp generators with a subsequent signal amplification by high-voltage amplifiers.

A sample-and-hold amplifier which can be inserted in the feedback loop circuit in front of the PI amplifiers allows the error signal to be set to zero during a hold period so that the z piezodrive voltage does not change and the separation between tip and sample is kept fixed. During this hold period, the bias voltage U can be ramped and the tunneling current can be measured, resulting in I–U characteristics at fixed tip–sample separation. Consequently, spectroscopic information can be obtained (section 1.13).

1.10.5 Computer automation

A full computer-automated STM system includes data acquisition, data analysis, data processing and data visualization. From the previous section it is clear that the computer used for data acquisition should have at least four to five DACs and at least two analog-to-digital converters (ADCs). The data acquisition software should allow for real-time plane subtraction correcting for possible sample tilt, as well as for easy on-line change of the experimental parameters, such as applied bias voltage, tunneling current, feedback gain and scan windows. The data analysis software should allow, for instance, for drawing arbitrary line sections, selecting sub-areas within the total scan window and for calculating two-dimensional Fourier transforms and autocorrelation functions. The data processing software may include interpolation and smoothing routines, a selection of filters including median, Wiener and Fourier filters, as well as other useful routines, e.g. statistical differencing (Wilson and Chiang, 1988). Finally, a variety of means of data visualization should exist, including a line scan representation, a grey-scale top-view representation, an illuminated surface representation with shading effects, as well as three-dimensional renderings.

1.10.6 Sensor tip preparation and characterization

After having discussed the general principles of STM design and instrumentation, we now focus on perhaps the most critical component of an STM, i.e. the sensor tip, its preparation and characterization.

1.10.6.1 General considerations

There are several requirements for an STM tip. The importance of each requirement depends, however, on the application.

1. For atomic resolution STM studies on relatively flat sample surfaces, the STM tip should terminate in a single atom but the macroscopic shape of the tip is of little importance. Since the tunneling current is exponentially dependent on the tip–surface separation, the tip atom closest to the sample surface always gives the major contribution to the tunneling current. Another tip atom 3 Å more distant from the surface than the front tip atom would make a tiny contribution of only about 0.1% to the total current (see section 1.2.1). However, it can happen that two or more atoms at the front of the tip have nearly the same distance to the sample surface, leading to double or multiple tip imaging artifacts (Park *et al.*, 1987; Mizes *et al.*, 1987; Albrecht *et al.*, 1988b).

2. For topographic STM studies of larger scale structures on relatively rough surfaces, the macroscopic shape of the tip also becomes important (Fig. 1.54). A tip with a relatively large cone angle fails to penetrate into deep and narrow grooves on the sample surface which leads to a smoothing of topographic surface features in STM images (Gimzewski *et al.*, 1985; Reiss *et al.*, 1990a; Musselman and Russell, 1990). An attempt was made to calculate the true surface

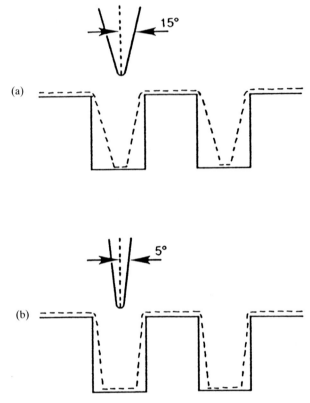

Fig. 1.54. Effect of tip geometry on measured STM profile of grooves 0.75 μm wide and 1 μm deep using: (*a*) tip with 500 Å radius of curvature and 15° cone half angle, (*b*) tip with 500 Å radius of curvature and 5° cone half angle (Musselman and Russell, 1990).

topography based on STM images by taking the finite size of the tip into account (e.g. Reiss *et al.*, 1990b; Keller, 1991). The existence of several minitips at the front end of the tip can cause additional imaging artifacts due to switching of the tunneling location from one minitip to another when scanning a rough surface (e.g. van Loenen *et al.*, 1990a). For this case, a reconstruction of the true surface topography becomes considerably more sophisticated. The preparation of the tip should therefore be optimized with regard to its geometry for STM studies of rough surfaces.

3. The chemical composition at the front end of the tip is of great importance as well. Oxide or insulating contamination layers of thickness several nanometers can prevent vacuum tunneling, leading

to mechanical contact between tip and sample. The tunneling current is then flowing through the oxide or contamination layer rather than through a vacuum barrier. For atomic resolution STM studies, it is even important what kind of atom and orbital at the front end of the tip dominates the tunneling current because this has a significant influence on measured atomic corrugation amplitudes and spectroscopic results (Park *et al.* (1988); sections 1.11.1 and 1.13.6).

1.10.6.2 Choice of the tip material

The choice of an appropriate tip material depends on the specific application.

1. Since contact between tip and sample surface can occur during scanning, it is necessary to choose a relatively hard material for the tip. Tungsten tips have been most widely used for STM studies, particularly under UHV conditions. Alternatively, molybdenum and iridium tips have been tried as well. Very hard tips which are highly resistant against accidental or intentional contact with the sample surface have been fabricated from titanium carbide single crystals (Yata *et al.*, 1989) and from ion-implanted diamond (Kaneko and Oguchi, 1990; Miyamoto *et al.*, 1991).
2. For STM studies in air or under poor vacuum conditions, it is favorable to choose an inert material for the tip, e.g. a noble metal such as platinum or gold. However, these materials are relatively soft and the tips can easily be damaged as a result of accidental contact with the sample surface. Therefore, Pt–Ir alloys of increased hardness have mainly been used as tip material for STM studies in air.
3. For magnetic-sensitive STM studies, a variety of magnetic tips have been prepared, e.g. from CoCr (Allenspach *et al.*, 1987), CrO_2 (Wiesendanger *et al.*, 1990c), Ni (Johnson and Clarke, 1990), Fe (Wiesendanger *et al.*, 1991a, 1992a) and Cr (Wiesendanger *et al.*, 1991a).

1.10.6.3 Tip preparation methods

Several different methods for sensor tip preparation exist (Melmed, 1991). These methods have sometimes been combined, leading to novel tip preparation procedures.

1. Sharp tips with radius of curvature 10 nm or even less can be prepared by electrolytical etching of a wire, as is well known from field ion microscopy (Müller and Tsong, 1969). In Fig. 1.55 an experimental set-up for electrolytical etching of an STM tip from a tungsten wire is shown (Bryant *et al.*, 1987). The choice of etching solution

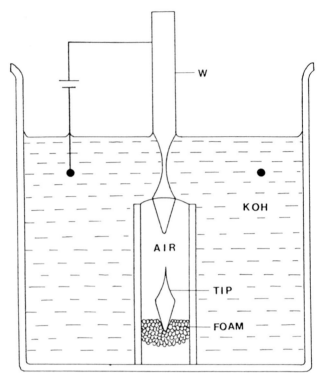

Fig. 1.55. A pretapered tungsten wire (W) is held by a micromanipulator with end protected in air column while the shank is etched in 3–5% aqueous KOH. After dropping off, the lower end is caught by the foam to avoid damage of the tip (Bryant *et al.*, 1987).

depends on the tip material (Müller and Tsong, 1969). For etching tungsten tips, either KOH or NaOH solutions are appropriate. Etching of the shank proceeds until the tensile strength of the notched region can no longer sustain the weight of the small lower end. Tensile shearing of the shank while the lower end drops off produces fine asperities at the front end of the tip with radii much smaller than the radius of the macroscopic tip shape. Alternative set-ups for electrolytical etching have been proposed which, however, all make use of the drop-off technique (e.g. Ibe *et al.*, 1990; Lemke *et al.*, 1990).

2. Though electrolytical etching can lead to very sharp tips, the electro-chemical reactions during etching involve the formation of an unwanted oxide layer which can be several nanometers thick (Garnaes *et al.*, 1990). The oxide layer can be removed after the

electrolytical etching procedure by ion milling, which additionally helps to further decrease the radius of curvature of the tips (Biegelsen *et al.*, 1987; Biegelsen *et al.*, 1989; Morishita and Okuyama, 1991). Focused ion beam milling (Fig. 1.56) leads to tip radii

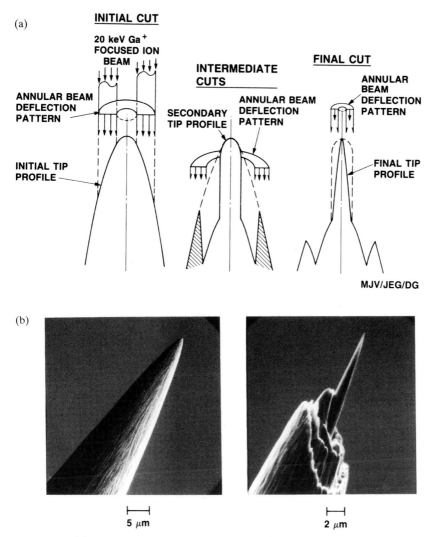

Fig. 1.56. (*a*) Focused ion beam milling steps: (1) Initial cut made using an annular beam deflection pattern (0.9 μm internal diameter, 4.0 μm outer diameter). (2) Intermediate shoulder cleaning step (approximately 2.0 μm i.e., 6.0 μm outer diameter). (3) Final cut (0.46 μm internal diameter, 1.5 μm outer diameter). (*b*) SEM images of a tungsten tip after electrochemical etching (left) and after focused ion beam machining (right) (Vasile *et al.*, 1991).

as small as 4 nm and cone angles of as little as 10° (Vasile *et al.*, 1991).

3. For topographic STM studies in air or moderate vacuum on atomic-ally flat sample surfaces, Pt–Ir tips are often used, which are simply prepared by mechanically cutting a Pt–Ir wire. This tip preparation procedure avoids the formation of a thick oxide layer as typically present after electrolytical etching. Though tiny asperities usually exist at the front end of mechanically cut tips, which guarantee atomic resolution on atomically flat surfaces, the macroscopic radius of curvature of these tips is huge and the tips are irregularly shaped, preventing useful STM studies of rough surfaces.

4. Alternatively, a combination of electrolytical etching and mechanical cutting can be used for tip preparation (Yata *et al.*, 1989). First, a wire is thinned by electrolytical etching until a constriction with a diameter of about 50 μm is formed. By pulling apart the two ends

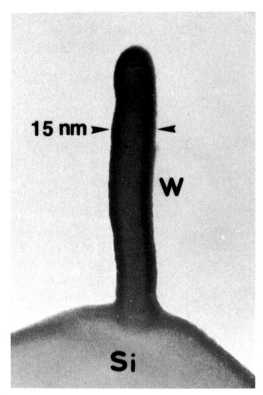

Fig. 1.57. SEM image of a W rod of 15 nm diameter. The rod was fabricated on a Si particle, using a focused electron beam, 3 nm in diameter (Ichihashi and Matsui, 1988).

of the wire, sharp tips can be obtained due to tensile shearing. Thick oxide layers are avoided because the wire is not etched through. On the other hand, the macroscopic shape of the tip is considerably improved compared with the conventional wire cutting procedure. The second step in the combined etching–cutting technique for tip preparation can even be performed under UHV conditions, thus avoiding any contamination of the obtained tip (Wiesendanger *et al.*, 1991a).

5. For STM studies of extremely rough surfaces with deep and narrow grooves, it is desirable to have extremely small and narrow tips. Such tips can be fabricated by electron beam-induced chemical vapor deposition of tungsten on top of a substrate tip in a transmission electron microscope (TEM) as shown in Fig. 1.57 (Ichihashi and Matsui, 1988).

6. Microfabrication techniques are particularly useful for fabricating twin probes (Tsukamoto *et al.*, 1991) or multiple tip arrays, which are important for metrological applications of the STM, as well as for introducing parallelism into STM reading and writing.

7. The ultimate goal, controlled preparation of well-defined single-atom tips, can be reached by using field-ion techniques. Under controlled conditions in a field-ion microscope (FIM), a well-defined single-atom tip can be prepared by a combination of ion sputtering, annealing, and field evaporation of a (111)-oriented tungsten tip (Fink, 1986, 1988). The single atom at the front of the tip is bonded to a trimer which itself is bonded to the third layer consisting of seven atoms (Fig. 1.58). Single-atom tips can also be prepared by heating a (111)-oriented tungsten tip in the presence of a strong electric field (Fink, 1986; Binh and Marien, 1988). Surface diffusion under the electric field gradient leads to a building-up process tending towards the equilibrium tip shape, which has a single-atom termination. Unfortunately, the atomic structure at the front end of an STM tip can easily change during scanning due to tip–surface interaction. Therefore, the effective tip for tunneling is generally not known even if one starts with a well-defined single-atom tip.

1.10.6.4 Tip characterization methods

The detailed atomic structure at the front end of the tip can be studied and controlled only by FIM. TEM is useful to characterize the tip shape at the nanometer level and to reveal the existence of possible oxide layers. SEM is usually used to examine tip shapes at the micrometer scale whereas optical microscopes allow for a quick check of freshly prepared tips. The chemical composition of tips can be studied either

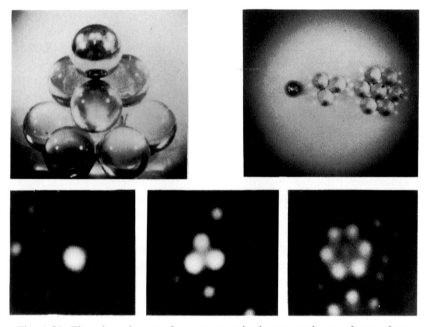

Fig. 1.58. First three layers of a monoatomic tip are made up of one, three, and seven atoms. Top: ball model of monoatomic tip. Bottom: successive removal by the field shows field-ion pattern of first three layers of the tip (Fink, 1986).

by energy-dispersive x-ray analysis in combination with SEM or TEM investigations, or by point Auger electron spectroscopy (AES) analysis.

1.10.7 Combined analysis instruments

The combination of STM instruments with other analysis techniques is useful from two different view-points.

1. The tip and sample surface preparation as well as the location of the tip above the surface can be controlled by using additional micro-scopical and analytical techniques.
2. The local information gained from STM experiments can be combined with the additional information from other surface analytical techniques, which are often complementary. A complete picture is usually obtained only by combining several different techniques. For instance, many previously unknown surface structures of semi-conductors and metals have been solved since the invention of the STM. However, it has usually been combination of STM with other

surface analytical techniques which has allowed full determination of the surface structure.

1.10.7.1 Combined STM–FIM instruments

It has already become clear in the previous section 1.10.6 that detailed information about the tip's atomic structure can only be obtained by FIM. Therefore, the combination of STM and FIM is important for studying the role of the tip's atomic structure in STM (Kuk and Silverman, 1986; Kuk *et al.*, 1988a; Michely *et al.*, 1988; Sakurai *et al.*, 1989). However, detailed investigations of the influence of the tip's atomic structure on the STM results is often prevented by tip changes, which frequently occur during scanning.

1.10.7.2 Combined STM–SEM/REM and combined STM–optical microscope instruments

A combination of STM with a scanning electron microscope (SEM) or reflection electron microscope (REM) is useful because the magnification of SEM can easily be changed to very low values, thereby extending the imaged surface regions from a microscopic to a macroscopic scale. Consequently, the STM tip can be guided to preselected surface locations of inhomogeneous sample surfaces under SEM control (Gerber *et al.*, 1986; Ichinokawa, *et al.*, 1987; Anders *et al.*, 1988; Vázquez *et al.*, 1988; Fuchs and Laschinski, 1990; Kuwabara *et al.*, 1989; Iwatsuki *et al.*, 1991). Inhomogeneous surface structures exist, for instance, for polycrystalline samples exhibiting grains and grain boundaries, for semiconductor heterostructures, or for microcircuits. Biological specimens are often studied with combined STM–optical microscope instruments to facilitate the search for the biological material deposited on a flat substrate (e.g. Guckenberger *et al.*, 1988; Stemmer *et al.*, 1988). However, as the maximum scan range of STM instruments becomes considerably increased, up to the millimeter scale, the need for combined STM–SEM or STM–optical microscope systems is somewhat reduced.

1.10.7.3 Combined STM–surface analysis instruments

Control of sample surface preparation as well as additional surface characterization can be performed under UHV conditions with well-known surface analytical techniques, such as low-energy electron diffraction (LEED), Auger electron spectroscopy (AES), x-ray photoelectron spectroscopy (XPS), ultraviolet photoelectron spectroscopy (UPS) and ion scattering spectroscopy (ISS), to mention only a few. Several combined STM–surface analysis instruments have already been built (Fig. 1.59) and successfully applied to surface science problems (e.g. Chiang

(a)

(b)

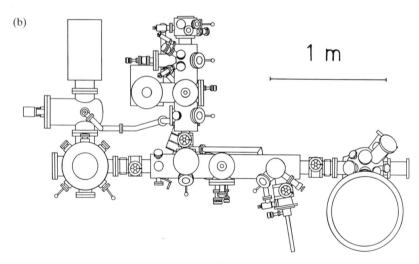

1 m

Fig. 1.59. (*a*) Photograph and (*b*) schematic drawing of an UHV surface analysis system (NANOLAB-I) consisting of a surface analysis chamber, a transfer, a preparation and a separate chamber containing the STM (from right to left). Preparation facilities include resistive and electron beam heating, ion etching, evaporation, gas inlet and high-pressure gas reaction. Conventional surface analysis techniques such as LEED, SEM/SAM, XPS and UPS are used for additional characterization of the surfaces and for getting complementary information (Wiesendanger *et al.*, 1990a, 1990e).

et al., 1988a; Wiesendanger *et al.*, 1990b). Other combinations of STM with surface analysis techniques exist, even outside UHV, which help to complete the information from the sample surface under investigation.

1.11 Topographic imaging by STM

We now focus on various modes of operation of an STM and the information which can be extracted. It should be emphasized right at the beginning that generally no single, distinct imaging mode for a specific sample (and tip) property exists and that the separation of the various properties reflected in any type of image obtained in a specific operation mode is by no means trivial (Rohrer, 1990).

1.11.1 Constant current imaging (CCI)

The first and most widely used mode of STM operation is the constant current imaging (CCI) mode as illustrated in Fig. 1.60 (Binnig and Rohrer, 1982). As already described in section 1.10.4, a feedback loop adjusts the height of the tip during scanning so that the tunneling current flowing between tip and sample is kept constant. The height (z) adjustment is performed by applying an appropriate voltage U_z to the z piezoelectric drive while the lateral tip position (x,y) is determined by the corresponding voltages U_x and U_y applied to the x and y piezoelectric drives. Therefore, the recorded signal $U_z(U_x,U_y)$ can be translated into the 'topography' $z(x,y)$, provided that the sensitivities of the three orthogonal piezoelectric drives are known.

The principle of the CCI mode sounds rather simple. However, the interpretation of the obtained contour map $z(x,y)$ is not at all trivial. Though the contour map $z(x,y)$ is often referred to as the 'topographic image' of the sample surface, it is generally not solely the arrangement of atoms on a surface which determines $z(x,y)$. The contour map $z(x,y)$ rather reflects a constant current surface according to the experimental measurement procedure. (We first assume the ideal case where the feedback response is indefinitely fast to keep the current constant at all times.) To interpret the contour map $z(x,y)$ appropriately, we have to evaluate the contributions of specific sample (and tip) properties to the tunneling current.

1.11.1.1 Theoretical expression for the tunneling current and its interpretation

In section 1.2 we have evaluated the tunneling probability for a one-dimensional potential barrier based either on a time-independent treat-

(a) Constant Current Mode

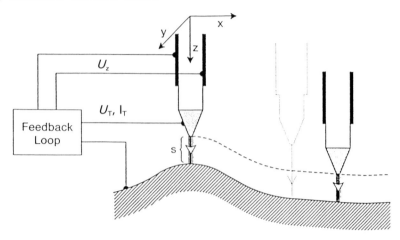

(b) CONSTANT CURRENT MODE

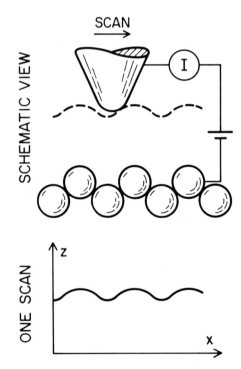

Fig. 1.60. Schematic illustration of the constant current mode of STM operation (Binnig and Rohrer, 1987; Hansma and Tersoff, 1987).

ment (wave-matching method) or on a time-dependent treatment (Bardeen's transfer Hamiltonian approach). For the three-dimensional case to be considered for STM, the method of matching wave functions across the interface becomes an extremely difficult problem, whereas Bardeen's transfer Hamiltonian approach, based on a perturbative treatment of tunneling, is still applicable and provides an appropriate insight into the physics of the tunneling process as well as a connection to specific tip and sample surface properties.

Within Bardeen's formalism, the tunneling current I can be evaluated in first-order time-dependent perturbation theory according to

$$I = \frac{2\pi e}{\hbar} \sum_{\mu,\nu} \left\{ f(E_\mu)[1 - f(E_\nu + eU)] - f(E_\nu + eU)[1 - f(E_\mu)] \right\}$$
$$\cdot |M_{\mu\nu}|^2 \, \delta(E_\nu - E_\mu) \tag{1.92}$$

where $f(E)$ is the Fermi function, U is the applied sample bias voltage, $M_{\mu\nu}$ is the tunneling matrix element between the unperturbed electronic states ψ_μ of the tip and ψ_ν of the sample surface, and E_μ (E_ν) is the energy of the state ψ_μ (ψ_ν) in the absence of tunneling. The delta function describes the conservation of energy for the case of elastic tunneling. The essential problem is the calculation of the tunneling matrix element which, according to Bardeen (1961), is given by

$$M_{\mu\nu} = \frac{-\hbar^2}{2m} \int d\mathbf{S} \cdot (\psi_\mu^* \nabla \psi_\nu - \psi_\nu \nabla \psi_\mu^*) \tag{1.93}$$

where the integral has to be evaluated over any surface lying entirely within the vacuum barrier region separating the two electrodes. The quantity in parentheses can be identified as a current density $j_{\mu\nu}$. To derive the matrix element $M_{\mu\nu}$ from Eq. (1.93), explicit expressions for the wave functions ψ_μ and ψ_ν of the tip and sample surface are required. Unfortunately, the atomic structure of the tip is generally not known, as has been discussed in section 1.10.6. Therefore, a model tip wave function has to be assumed for calculation of the tunneling current.

Tersoff and Hamann (1983, 1985), who first applied the transfer Hamiltonian approach to STM, used the simplest possible model for the tip with a locally spherical symmetry (Fig. 1.61). In this model, the tunneling matrix element is evaluated for an s-type tip wave function, whereas contributions from tip wave functions with angular dependence (orbital quantum number $l \neq 0$) have been neglected. Tersoff and

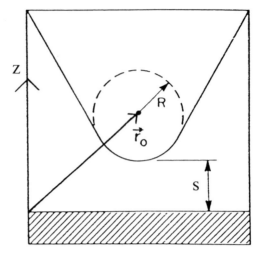

Fig. 1.61. Schematic picture of tunneling geometry in the Tersoff–Hamann model. The probe tip is assumed to be locally spherical with radius of curvature R, where it approaches nearest the surface (shaded). Distance of nearest approach is s. The center of curvature of the tip is labeled r_0 (Tersoff and Hamann, 1983).

Hamann first considered the limits of low temperature and small applied bias voltage for which the tunneling current becomes

$$I = \frac{2\pi e^2}{\hbar} U \sum_{\mu,\nu} |M_{\mu\nu}|^2 \, \delta(E_\nu - E_F) \cdot \delta(E_\mu - E_F) \qquad (1.94)$$

where E_F is the Fermi energy. Within the s-wave approximation for the tip, the following expression for the tunneling current can finally be obtained:

$$I \propto U \cdot n_t(E_F) \cdot \exp(2\varkappa R) \cdot \sum_\nu |\psi_\nu(r_0)|^2 \, \delta(E_\nu - E_F) \qquad (1.95)$$

with the decay rate $\varkappa = (2m\phi)/\hbar$ where ϕ is the effective local potential barrier height, $n_t(E_F)$ is the density of states at the Fermi level for the tip, R is the effective tip radius, and r_0 is the center of curvature of the tip. The quantity

$$n_s(E_F, r_0) = \sum_\nu |\psi_\nu(r_0)|^2 \, \delta(E_\nu - E_F) \qquad (1.96)$$

can be identified with the surface local density of states (LDOS) at the Fermi level E_F, i.e. the charge density from electronic states at E_F, evaluated at the center of curvature r_0 of the effective tip. The STM images obtained at low bias in the constant current mode therefore represent contour maps of constant surface LDOS at E_F evaluated at the center of curvature of the effective tip, provided that the s-wave approximation for the tip can be justified. Since the wave functions decay exponentially in the z direction normal to the surface towards the vacuum region:

$$\psi_v(r) \propto \exp(-\varkappa z) \qquad (1.97)$$

we find that

$$|\psi_v(r_0)|^2 \propto \exp[-2\varkappa(s+R)] \qquad (1.98)$$

where s denotes the distance between the sample surface and the front end of the tip (Fig. 1.61). Therefore, the tunneling current, given by Eq. (1.96), becomes exponentially dependent on the distance s as expected:

$$I \propto \exp(-2\varkappa s) \qquad (1.99)$$

Unfortunately, the simple interpretation of constant current STM images as given by Tersoff and Hamann is no longer valid for high bias or for tip wave functions with angular dependence. In the following, we first focus on the effect of a finite bias on the expression for the tunneling current given by Eq. (1.92) and Eq. (1.93). The applied bias voltage enters through the summation of states which can contribute to the tunneling current. Additionally, a finite bias can lead to a distortion of the tip and sample surface wave functions ψ_μ and ψ_v as well as a modification of the energy eigenvalues E_μ and E_v (Chen, 1988). The derivation of these distorted tip and sample surface wave functions and energy eigenvalues under the presence of an applied bias is, however, a difficult problem. Therefore, as a first approximation, the undistorted zero-voltage wave functions and energy eigenvalues are usually taken. Consequently, the effect of a finite bias U only enters through a shift in energy of the undistorted surface wave functions or density of states relative to the tip by an amount of eU. Under this approximation, we may use the following expression for the tunneling current as a generalization of the result of Tersoff and Hamann for the low-bias limit:

$$I \propto \int_0^{eU} n_t(\pm eU \pm \mathscr{E}) \cdot n_s(\mathscr{E}, r_0) \, d\mathscr{E} \qquad (1.100)$$

where $n_t(\mathcal{E})$ is the density of states for the tip and $n_s\,(\mathcal{E},r_0)$ is the density of states for the sample surface evaluated at the center of curvature r_0 of the effective tip. All energies are measured with respect to the Fermi level. We can now make the following approximation motivated by a generalization of Eq. (1.96) together with Eq. (1.98):

$$n_s(\mathcal{E},r_0) \propto n_s(\mathcal{E}) \cdot \exp\left\{-2(s+R)\left[\frac{2m}{\hbar^2}\left(\frac{\phi_t + \phi_s}{2} + \frac{eU}{2} - \mathcal{E}\right)\right]^{\frac{1}{2}}\right\} \quad (1.101)$$

where a WKB-type expression for the decay rate $\varkappa$ in the exponential term has been used. ϕ_t (ϕ_s) denotes the tip (sample surface) work function. We finally obtain

$$I \propto \int^{eU} n_t(\pm eU \mp \mathcal{E}) \cdot n_s(\mathcal{E}) \cdot T(\mathcal{E},eU)\,\mathrm{d}\mathcal{E} \quad (1.102)$$

with an energy- and bias-dependent transmission coefficient $T(\mathcal{E},eU)$ given by

$$T(\mathcal{E},eU) = \exp\left\{-2(s+R)\left[\frac{2m}{\hbar^2}\left(\frac{\phi_t + \phi_s}{2} + \frac{eU}{2} - \mathcal{E}\right)\right]^{\frac{1}{2}}\right\} \quad (1.103)$$

By writing Eq. (1.102) and Eq. (1.103), matrix element effects in tunneling are expressed in terms of a modified decay rate $\varkappa$ including a dependence on energy E and applied bias voltage U. The expression (1.103) for the transmission coefficient neglects image potential effects (section 1.2.6) as well as the dependence of the transmission probability on parallel momentum (section 1.2.7). This can be included by an increasingly more accurate approximation for the decay rate $\varkappa$.

We now focus in more detail on the validity of the s-wave approximation used in the Tersoff–Hamann theory. As discussed in section 1.10.6, tungsten and platinum–iridium tips are most widely used in STM experiments. For these materials, the density of states at the Fermi level is dominated by d states rather than by s states (Table 1.3). Consequently, it is natural to consider the role of d states in STM. It has been argued that d states decay much faster than s states and therefore should have negligible influence on the tunneling current because the transmission probability would be highly reduced for d states compared with s states. However, as pointed out by Chen (1990a), the faster decay of d states is inferred from free atoms whereas the situation for d-type surface states on top of bulk tungsten may be completely differ-

Table 1.3. *Electronic density of states at the Fermi level for common tip materials (Chen, 1990a).*

Material	W	Pt	Ir
s state	3.1%	0.77%	0.94%
d state	85%	98%	96%

ent. For instance, a highly localized metallic d_{z^2} surface state is found both experimentally and theoretically on a W(100) surface which extends much further into the vacuum than do the atomic 6s and 5d states (Chen, 1990a). First-principles calculations of the electronic states of several kinds of tungsten clusters used to model the STM tip also reveal the existence of dangling-bond states near the Fermi level at the apex atom which can be ascribed to d_{z^2} states (Ohnishi and Tsukada, 1989, 1990).

Evaluation of the tunneling current according to Eq. (1.92) and Eq. (1.93) now requires calculation of the tunneling matrix element for tip wave functions with angular dependence ($l \neq 0$). Chen (1990a) has shown that generally the tunneling matrix element can simply be obtained from a 'derivative rule'. The angle dependence of the tip wave function in terms of x, y and z has to be replaced according to

$$
\begin{aligned}
x &\to \partial / \partial x \\
y &\to \partial / \partial y \\
z &\to \partial / \partial z
\end{aligned}
\tag{1.104}
$$

where the derivatives have to act on the sample surface wave function at the center of the apex atom. For instance, the tunneling matrix element for a p_z tip state is proportional to the z derivative of the sample surface wave function at the center of the apex atom at r_0. Table 1.4 summarizes the obtained expressions for the tunneling matrix element according to the derivative rule.

In terms of a microscopic view of the STM imaging mechanism illustrated in Fig. 1.62 (Chen, 1991a), a dangling-bond state at the tip apex atom is scanned over a two-dimensional array of atomic-like states at the sample surface. Overlap of the tip state with the atomic-like states on the sample surface generates a tunneling conductance which depends on the relative position of the tip state and the sample state. The atomic corrugation Δz depends on the spatial distribution as well as on the type of tip and sample surface states.

Table 1.4. *Tunneling matrix elements (Chen, 1990a).*

State	$M \propto$ value at r_0
s	ψ
p $[z]$	$\dfrac{\partial \psi}{\partial z}$
p $[x]$	$\dfrac{\partial \psi}{\partial x}$
p $[y]$	$\dfrac{\partial \psi}{\partial y}$
d $[zx]$	$\dfrac{\partial^2 \psi}{\partial z \, \partial x}$
d $[zy]$	$\dfrac{\partial^2 \psi}{dz \, \partial y}$
d $[xy]$	$\dfrac{\partial^2 \psi}{\partial x \, \partial y}$
d $[z^2 - \frac{1}{3}r^2]$	$\dfrac{\partial^2 \psi}{\partial z^2} - \frac{1}{3}\varkappa^2 \, \psi$
d $[x^2 - y^2]$	$\dfrac{\partial^2 \psi}{\partial x^2} - \dfrac{\partial^2 \psi}{\partial y^2}$

The wave functions for different tip states are listed in Table 1.5. Assuming atomic-like states at the sample surface as well, the conductance distribution $\sigma(r)$ can be evaluated based on the derivative rule for obtaining the matrix elements (Chen, 1991a). Table 1.6 lists the results for three different types of tip and sample surface states respectively with azimuthal quantum number $m = 0$. For s-like tip and sample surface states, a conductance distribution proportional to exp $(-2\varkappa r)$ is obtained in agreement with Eq. (1.99). However, for all other combinations of tip and sample surface states, a dependence of the conductance distribution on cos $\theta = z/r$ appears. It is also evident from Table 1.6 that, on interchanging tip and sample surface states, the conductance distribution remains the same. This has been called the 'reciprocity principle' in STM (Chen, 1990b). If, for example, a d_{z^2} tip state is scanning over s-like atomic sample surface states as illustrated in Fig. 1.63, the obtained constant current image no longer represents a contour map of constant LDOS at E_F, but represents a charge-density contour of a fictitious surface with a d_{z^2} state on each surface atom. Generally, for non-s-wave tip states, the tip apex atom follows a contour, determined by the derivatives of the sample surface wave func-

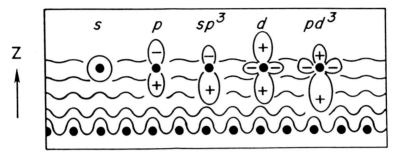

Fig. 1.62. Microscopic view of the STM imaging mechanism (Chen, 1991a).

Table 1.5. *Tip wave functions (Chen, 1990a).*

State	Wave functions
s	$C(\varkappa\rho)^{-1}\exp(-\varkappa\rho)$
p_z	$C[(\varkappa\rho)^{-1}+(\varkappa\rho)^{-2}]\exp(-\varkappa\rho)\cos\theta$
p_x	$C[(\varkappa\rho)^{-1}+(\varkappa\rho)^{-2}]\exp(-\varkappa\rho)\sin\theta\cos\phi$
p_y	$C[(\varkappa\rho)^{-1}+(\varkappa\rho)^{-2}]\exp(-\varkappa\rho)\sin\theta\sin\phi$
d_{z^2}	$C[(\varkappa\rho)^{-1}+3(\varkappa\rho)^{-2}+3(\varkappa\rho)^{-3}]\exp(-\varkappa\rho)(\cos^2\theta-\tfrac{1}{3})$
d_{xz}	$C[(\varkappa\rho)^{-1}+3(\varkappa\rho)^{-2}+3(\varkappa\rho)^{-3}]\exp(-\varkappa\rho)\sin(2\theta)\cos\phi$
d_{yz}	$C[(\varkappa\rho)^{-1}+3(\varkappa\rho)^{-2}+3(\varkappa\rho)^{-3}]\exp(-\varkappa\rho)\sin(2\theta)\sin\phi$
d_{xy}	$C[(\varkappa\rho)^{-1}+3(\varkappa\rho)^{-2}+3(\varkappa\rho)^{-3}]\exp(-\varkappa\rho)\sin^2\theta\sin(2\phi)$
$d_{x^2-y^2}$	$C[(\varkappa\rho)^{-1}+3(\varkappa\rho)^{-2}+3(\varkappa\rho)^{-3}]\exp(-\varkappa\rho)\sin^2\theta\cos(2\phi)$

tions, which exhibits much stronger atomic corrugation than the contour of constant surface LDOS at E_F.

While Bardeen's transfer Hamiltonian approach is based on a perturbative treatment of tunneling, which is appropriate for weakly overlapping electronic states of the two electrodes, it is also possible to evaluate the tunneling current from non-perturbative treatments of the tunneling process without the restriction to low transmissivities. For instance, a theoretical approach based on scattering of electrons by rough surfaces has been used to evaluate an expression for the tunneling current non-perturbatively (García *et al.*, 1983; Stoll *et al.*, 1984). More

Table 1.6. *Conductance distribution function for different tip states and sample states (as single localized surface states). The relation* $\cos\theta = z/r$ *is implied (Chen, 1990a).*

Tip state	Sample state	Conductance distribution	Apparent radius
s	s	$\exp(-2\varkappa r)$	z
s	p	$\cos^2\theta \exp(-2\varkappa r)$	$z/[1+1/\varkappa z]$
s	d	$(\cos^2\theta - \frac{1}{3})^2 \exp(-2\varkappa r)$	$z/[1+3/\varkappa z]$
p	s	$\cos^2\theta \exp(-2\varkappa r)$	$z/[1+1/\varkappa z]$
p	p	$\cos^4\theta \exp(-2\varkappa r)$	$z/[1+2/\varkappa z]$
p	d	$(\cos^2\theta - \frac{1}{3})^2 \cos^2\theta \exp(-2\varkappa r)$	$z/[1+4/\varkappa z]$
d	s	$(\cos^2\theta - \frac{1}{3})^2 \exp(-2\varkappa r)$	$z/[1+3/\varkappa z]$
d	p	$(\cos^2\theta - \frac{1}{3})^2 \cos^2\theta \exp(-2\varkappa r)$	$z/[1+4/\varkappa z]$
d	d	$(\cos^2\theta - \frac{1}{3})^4 \exp(-2\varkappa r)$	$z/[1+6/\varkappa z]$

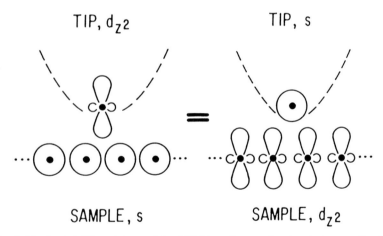

Fig. 1.63. An intuitive picture of the STM imaging mechanism in terms of a d_{z^2} tip state, in the light of the reciprocity principle (Chen, 1990b).

generally, a non-perturbative expression for the tunneling current can be derived from a Green function formalism based on Keldysh's theory of non-equilibrium processes (Keldysh, 1965). This alternative non-perturbative theoretical treatment of tunneling was found to be important for quantitative assessment of the accuracy of approximate tunneling theories such as the perturbative transfer Hamiltonian formalism (Lucas *et al.*, 1988; Noguera, 1988, 1989, 1990).

Apart from being a perturbative tunneling theory, the transfer Hamiltonian approach as used by Tersoff and Hamann suffers from the fact that assumptions for the tip and sample surface wave functions have to be made in order to derive the tunneling current. Alternatively, Lang (1985, 1986a) has calculated the tunneling current between two planar metal electrodes with adsorbed atoms where the wave functions for the electrodes have been obtained self-consistently within density-functional theory. However, relatively easy application of self-consistent density-functional calculations is limited to simple models for the electrodes, e.g. a jellium model for the metal surfaces with adsorbed atoms.

In Fig. 1.64 the calculated current density distribution from a single sodium atom adsorbed at its equilibrium distance on one of the two metal electrodes is shown (Lang, 1985). The plot illustrates how spatially localized the tunneling current is. By scanning one adsorbed atom (taken as the tip) past another adsorbed atom (taken as the sample), the vertical tip displacement versus lateral position can be evaluated under the constant current condition (Lang, 1986a). In Fig. 1.65, constant current scans at low bias of a sodium tip atom past three different sample adatoms (Na, S and He) are shown. Most striking is the negative tip displacement for adsorbed helium. The closed valence shell of helium is very much lower in energy with respect to the Fermi level and its only effect is to polarize metal states away from E_F, thereby producing a decrease in the Fermi-level state density. This results in a reduced tunneling current flow, i.e. a negative tip displacement in a constant current scan. The present example illustrates nicely that 'bumps' or 'holes' in 'topographic' STM images may not correspond to the presence or absence of surface atoms respectively – sometimes even the reverse is true.

In Fig. 1.66 a comparison is made between the tip-displacement curve obtained for a sodium tip atom and a sodium adatom as sample with contours of constant LDOS at E_F and constant total density of states. For the chosen example, all three curves are quite close to each other. This, however, depends on the specific adatoms used for the tip and sample. A direct comparison between an experimentally obtained constant current scan and a calculated contour of constant local state density at the Fermi level using the adatom-on-jellium model was made for the case of xenon atoms adsorbed on a Ni(110) surface (Eigler *et al.*, 1991a). A reasonable agreement between experiment and theory was

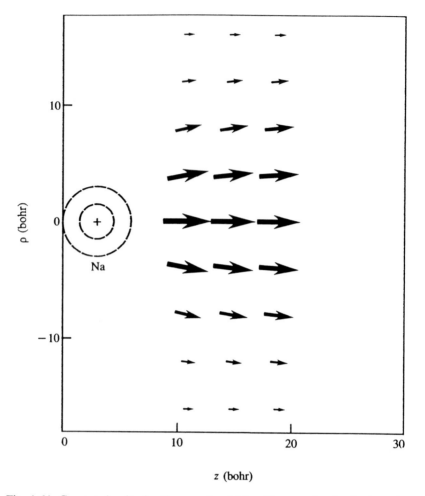

Fig. 1.64. Current density for the case in which a Na atom is adsorbed on the left electrode. Length and thickness of the arrow are proportional to ln (ej/j_0) evaluated at the spatial position corresponding to the center of the arrow (1 bohr = 0.529 Å) (Lang, 1986c).

found as shown in Fig. 1.67. The xenon adatom appears as a protrusion in a constant-current scan because xenon binds at relatively large distances from metal surfaces and the size of the relevant Xe 6s orbital is comparatively large so that it extends considerably further out into the vacuum than do the bare-surface wave functions.

It is clear that the applicability of the adatom-on-jellium model is limited. Nevertheless, important insight into the interpretation of constant current STM images was obtained by calculations based on this

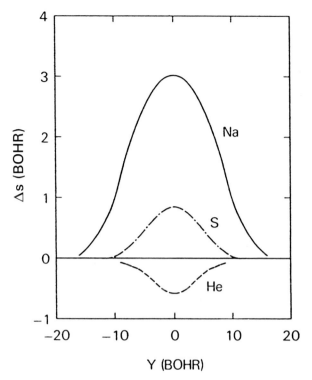

Fig. 1.65. Change in tip distance versus lateral separation for constant tunnel current. The tip atom is Na; sample adatoms are Na, S and He (Lang, 1986a).

model which, in some respects, can be regarded as complementary to the Tersoff–Hamann theory for the STM.

1.11.1.2 Spatial resolution in constant current STM images

The definition of spatial resolution in STM raises problems for two main reasons (Tersoff, 1990). First, STM is inherently a nonlinear imaging technique. Therefore, the usual definition of resolution in terms of convolution with an instrument function cannot be applied directly. Second, to define the resolution it is necessary to know what the instrument should ideally measure. However, this is not obvious because STM is not only a microscopical but also a spectroscopical analysis technique, i.e. the obtained images can depend critically on the applied bias voltage for specific choices of tip and sample surface (section 1.13). The spatial resolution, regardless of its definition, will therefore depend on the experimental conditions including the status of the tip, the electronic properties of the sample surface and the applied bias.

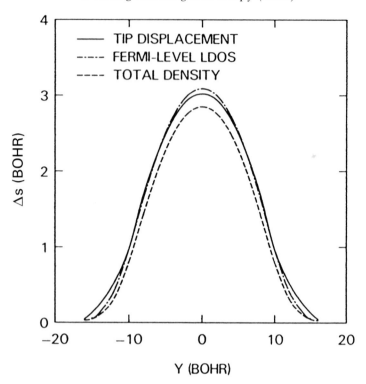

Fig. 1.66. Comparison of tip-displacement curve from Fig. 1.65 for Na adatom sample and Na tip with contours of constant Fermi-level local density-of-states and constant total density (Lang, 1986a).

We first consider tunneling experiments between free-electron-like metal electrodes at low bias. Following Tersoff and Hamann (1985), we define a corrugation amplitude, or briefly, corrugation, Δ by

$$\Delta := z_+ - z_- \tag{1.105}$$

where z_+ and z_- denote the extremal values of the z displacement of the tip in a constant current scan. This corrugation Δ decreases exponentially with distance z from the surface:

$$\Delta \propto \exp(-\gamma z) \tag{1.106}$$

where the decay rate γ is very sensitive to the surface lattice constant a because it depends quadratically on the corresponding Fourier component G:

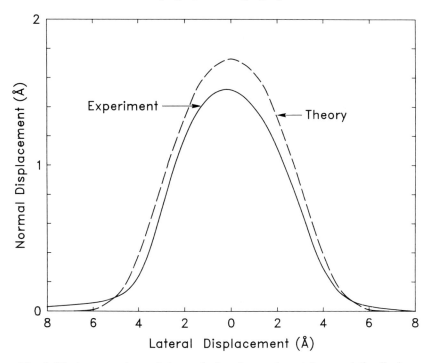

Fig. 1.67. A comparison of theoretical and experimental normal tip displacement (Å) versus lateral tip displacement (Å) curve for Xe adsorbed on a metal surface (Eigler *et al.*, 1991a).

$$\gamma \approx \tfrac{1}{4} G^2 / \varkappa \qquad (1.107)$$

with $\varkappa^2 = (2m\phi)/\hbar^2$. Consequently, only the lowest non-zero Fourier component determines the corrugation at sufficiently large distances. Tersoff and Hamann (1983, 1985) argued that suppression of higher Fourier components in their expression for the tunneling current between a spherical tip of radius R and a sample surface at a distance s from the front end of the tip is equivalent to a spatial resolution determined by $[(R + s)/\varkappa]^{1/2}$. This result was confirmed by Stoll (1984) who calculated the decrease of the observed corrugation Δ of equi-current surfaces with increasing separation s within the framework of the scattering theoretical approach (Stoll *et al.*, 1984):

$$\frac{\Delta}{\Delta_0} = \exp\left\{ -\tfrac{1}{4} \left[\frac{2\pi}{a} \cdot \left(\frac{R+s}{\varkappa} \right)^{1/2} \right]^2 \right\}$$

$$= \exp\left\{ -\gamma (R+s) \right\} \qquad (1.108)$$

where a surface structure with a weak sinusoidal one-dimensional corrugation of amplitude Δ_0 and period $a = 2\pi/G$ was assumed.

Alternatively, the lateral resolution can be defined by the diameter L_{eff} of a circle which has a constant current density j equal to that obtained in the direction of the tip axis, i.e. j at $R = 0$, and results in the same current $I(s)$ as that flowing through the entire tip–vacuum–sample junction (García *et al.*, 1983; García, 1986):

$$\pi \left(\frac{L_{eff}}{2} \right)^2 j \, (R = 0) = I(s) \qquad (1.109)$$

Again, L_{eff} can be well approximated by

$$L_{eff} \approx \left[(R + s)/\varkappa \right]^{1/2} \qquad (1.110)$$

According to this expression, the lateral resolution in STM is determined by geometrical quantities R and s, rather than by the wavelength of the tunneling electrons. This is characteristic for so-called near-field microscopes which are operated at distances between probe and sample surface that are small compared with the wavelength. For STM, typical tip–surface separations are 3–10 Å whereas the wavelength of tunneling electrons typically varies in the range 12–120 Å for an applied bias voltage of 0.01–1 V.

The expression (1.110) for the lateral resolution in constant current STM images implies that high spatial resolution is obtained at small tip–surface separation, i.e. at low tunneling resistance, and with a small radius of curvature of the effective tip, both of which have been verified experimentally. The dependence of the measured corrugation on the radius of curvature of the effective tip was studied by combined STM–FIM experiments where the obtained STM results could directly be correlated with the size of the effective tip as revealed by FIM (Kuk and Silverman, 1986; Kuk *et al.*, 1988a). The relationship (1.108) was verified for two periodic surface structures, Au(100) 5×1 and Au(110) 2×1, with reasonable agreement between the experimental and theoretical values for γ (Fig. 1.68). As a direct consequence, measured absolute values for the corrugation Δ are meaningless if the microscopic structure of the tip is not known.

Unfortunately, the expression (1.110) for the spatial resolution in constant current STM images can only be applied for relatively large-wavelength structures on reconstructed metal surfaces such as Au(100) 5×1 and Au(110) 2×1 where the surface periodicity is about 14 Å and 8.2 Å respectively. For atomic-scale surface structures the validity of the expression (1.110) obtained within a macroscopic spherical tip

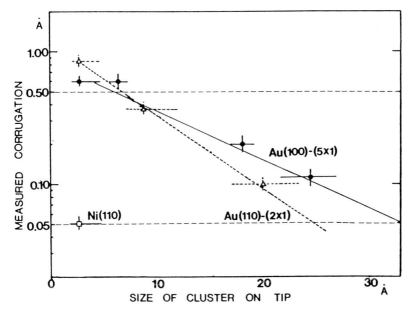

Fig. 1.68. Dependence of measured corrugation on size of cluster on tip in the Au(110) (2×1) and Au(100) (5×1) systems (Kuk *et al.*, 1988a).

model breaks down. In particular, within the spherical tip model it is not possible to explain the experimentally observed atomic resolution on close-packed metal surfaces, such as Au(111) (Hallmark *et al.*, 1987) and Al(111) (Wintterlin *et al.*, 1989). Baratoff (1984) was the first to point out that the spatial resolution might be considerably increased compared with the expression (1.110) if tunneling occurs via localized surface states or dangling bonds. More recently, Chen (1990b, 1991a) has systematically investigated the influence of different tip orbitals on the spatial resolution within a microscopic view of STM. The calculated enhancement of the tunneling matrix element by tip states with $l \neq 0$ as shown in Fig. 1.69 leads to increased sensitivity to atomic-size features with large wavevector. For instance, a p_z tip state acts as a quadratic high-pass filter, whereas a d_{z^2} tip state acts as a quartic high-pass filter. Consequently, the resolution of STM can be considerably higher than predicted within the s-wave tip model. The spontaneous switching of the resolution often observed in or between atomic-resolution STM images (Fig. 1.70) can be explained by the fact that a very subtle change in the tip involving a change of the effective orbital can induce a tremendous difference in STM resolution. In particular, Chen has shown that the experimentally observed atomic corrugation on the close-

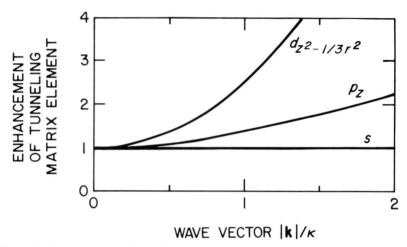

Fig. 1.69. Enhancement of tunneling matrix elements by p and d tip states. The enhancement of the tunnel current is the square of the enhancement of the matrix element (Chen, 1990a).

packed Al(111) surface as well as its dependence on the tip–sample distance can accurately be explained by assuming a d_{z^2} tip state whereas an s-wave tip state, which would trace contours of constant surface LDOS, fails to explain the experimental observations (Fig. 1.71).

In conclusion, it is the orbital at the front end of the tip which mainly determines the spatial resolution in STM. A p_z orbital typical for elemental semiconductors or a d_{z^2} orbital from d-band metals are most favorable. Therefore, 'tip sharpening procedures' have to aim at bringing such favorable orbitals to the front of the tip (Chen, 1991a).

It is interesting that our theoretical understanding of the STM imaging mechanism, as discussed in this section, is closely related to the problem of matching macroscopic and microscopic views of STM, i.e. matching the 'band' and the 'bond' pictures. Within the macroscopic view of STM leading to the LDOS interpretation of constant current STM images, it is not possible to explain the observed atomic resolution on close-packed metal surfaces. On the other hand, the microscopic view of STM is based on orbital symmetry considerations without referring to the particular surface electronic structure. STM can be regarded as an outstanding experimental technique also because it can help to unify the two complementary macroscopic and microscopic pictures.

(a)

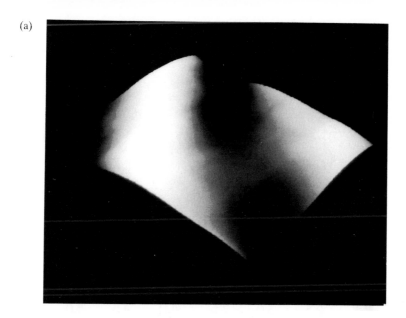

(b)

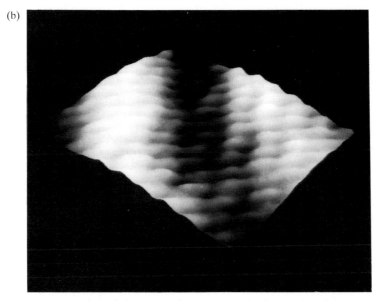

Fig. 1.70. (*a*) Perspective view of a constant current STM image of a 2 nm × 2 nm surface area on Au(111) showing stripes due to a reconstruction of the Au(111) surface. An additional corrugation due to the atomic surface structure is totally absent ($U = +0.03$ V, $I = 1$ nA). (*b*) Surface area of the same size as that shown in (*a*). The additional corrugation due to the atomic surface structure can now clearly be observed, although the image was obtained with similar tunneling parameters ($U = +0.05$ V, $I = 1$ nA) (Wiesendanger *et al.*, 1990b).

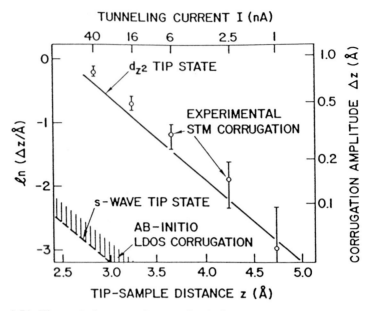

Fig. 1.71. Theoretical corrugation amplitude for an s and a d_{z^2} tip state, on a close-packed metal surface with $a = 2.88$ Å and $\phi = 3.5$ eV. The orbitals on each metal atom on the sample are assumed to be of 1s type (Chen, 1991a). Measured STM corrugation amplitudes are from the data of Wintterlin *et al.* (1989).

1.11.2 Constant height imaging (CHI) or variable current mode

A significant drawback of the constant current imaging (CCI) mode lies in the finite response time of the feedback loop which limits the scan speed and data acquisition time considerably. In principle, the tip can be scanned faster than the feedback can respond (Bryant *et al.*, 1986a,b). Consequently, the feedback can only maintain an average tunneling current and cannot respond to the higher frequency components modulating the tunneling current which result from atomic-scale surface features. Alternatively, the feedback can even be switched off completely. The modulation of the tunneling current with the full sensitivity of the exponential dependence of the current on tip–surface spacing then reflects the atomic-scale 'topography' provided that the response of the preamplifier (section 1.10.4) is still faster than the scan speed (Fig. 1.72). This 'constant-height imaging' (CHI) or 'variable current' mode can be used to collect STM images at real-time video rates, offering the opportunity to observe dynamic atomic-scale processes at surfaces, such as surface diffusion.

The performance of the STM is usually also improved in this fast scan mode because the microscope becomes insensitive to most low-

CONSTANT HEIGHT MODE

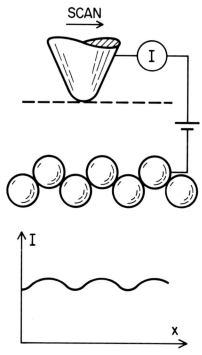

Fig. 1.72. Schematic illustration of the constant height mode of STM operation (Hansma and Tersoff, 1987).

frequency disturbances including mechanical vibration, electronic noise and low-frequency ($1/f$) noise inherent in the tunneling process which is often conjectured to arise from diffusion of molecular species in the spatial gap between tip and sample surface. In addition, piezoelectric hysteresis of the scanning unit appears to be reduced and thermal drifts cause less image distortions.

The useful upper limit of the scan speed in CHI mode is given by the lowest mechanical resonance frequency of the STM unit. Furthermore, application of the CHI mode is limited to surfaces that are nearly atomically flat over the imaged surface regions where the amplitudes of topographic features with high spatial frequencies have to be smaller than the tip–surface spacing itself. A significant drawback of the CHI mode compared with the CCI mode lies in the fact that vertical height (z scale) information is not directly available. This information could only be extracted from CHI data provided that the absolute value of the effective local tunneling barrier height would be known as a function of location (section 1.12). Usually, topographic STM data on a larger scale are collected using the CCI mode whereas the CHI mode should then be used after 'zooming' to selected smaller surface areas of sufficient flatness.

1.11.3 Differential tunneling microscopy (DTM)

Another way of minimizing the effect of noise in STM images, resulting from mechanical vibration, electronic and $1/f$ noise, is offered by vibrating the tip parallel to the sample surface (x direction) at a frequency v_0 above the feedback response frequency but below the first mechanical resonance frequency (Abraham *et al.*, 1988a, 1988b; Stoll and Gimzewski, 1991). The conventional topography can then be measured simultaneously with a differential image $\partial I(x,y)/\partial x|_{I,U}$, corresponding to the amplitude of current modulation at v_0 obtained by lock-in detection. A significant improvement in the signal-to-noise ratio is achieved in the differential image from which the conventional topographic STM image can be regained by Fourier-domain integration.

1.11.4 Tracking tunneling microscopy (TTM)

For the modes of STM operation discussed so far, the tip is raster scanned over the sample surface by means of triangular voltage ramps applied to the lateral (x,y) motion piezoelectric drives. Alternatively, the STM can be operated in various tracking modes (Pohl and Möller, 1988). The voltage ramps are replaced by small a.c. voltages causing the tip to vibrate in both x and y directions with frequencies v_x and v_y larger than the cut-off frequency of the feedback loop system but considerably smaller than the first mechanical resonance frequency. The response of the tunneling current then determines the driving voltages U_x and U_y for the x and y piezoelectric elements, producing a small lateral displacement of the tip which has to be small compared with the surface structures under investigation. The paths of tip motion can be selected, e.g. the tip can be forced to follow paths of steepest inclination or equal surface topographic height by adapting the electronic circuitry accordingly. Alternatively, the tip can be locked to a surface extremity for an indefinite time.

1.11.5 Scanning noise microscopy (SNM)

Tunneling current noise is usually considered to be a basic limitation for the performance of a STM and it is therefore tried to eliminate it as far as possible (section 1.10). However, it has been shown that current noise can be exploited for topographic imaging at zero bias (Möller *et al.*, 1989a). At bias voltages above a few millivolts, the current noise typically shows a $1/f$ spectral distribution over a range of 1 Hz to 100 kHz, whereas white noise is found at zero bias. Owing to its exponential distance dependence, this white noise at zero bias can be exploited to control the tip–sample separation instead of the tunneling current itself, using the electronic feedback for maintaining constant current noise. This topographic imaging technique at vanishing bias voltage avoids high electric fields and current densities present in a biased tunnel junction and is therefore particularly suitable for investigations of delicate samples.

1.11.6 Nonlinear alternating-current tunneling microscopy

The topographic STM modes of operation discussed so far can only be applied for conducting samples. To extend the applicability of STM to insulators, a nonlinear a.c. technique was introduced (Kochanski, 1989). This technique involves tunneling of a small number of electrons onto and off the surface to be studied during alternate half-cycles of a high-frequency (GHz) a.c. driving voltage. The position of the STM tip is controlled by the a.c. tunneling current. The third harmonic of the current is used rather than the first harmonic to eliminate the effects of stray capacitances. The nonlinearities necessary to produce the third harmonic can be caused, for instance, by Coulomb blockade (section 1.3.3) or by the energy-dependence of the electronic density of states near the Fermi level.

1.12 Local tunneling barrier height

In section 1.11.1 we saw that the tunneling current I in STM depends exponentially on the tip–surface separation s (Eq. (1.99)):

$$I \propto \exp\left(-2\varkappa s\right)$$

with a decay rate $\varkappa$ given by (Eq. (1.95))

$$\varkappa = \left(2m\phi\right)^{1/2} / \hbar$$

where ϕ is an effective local potential barrier height. Tersoff and Hamann (1983, 1985) assumed that ϕ is laterally uniform, i.e. ϕ does not depend on the lateral position (x,y), and that ϕ is equal to the surface work function. Lateral variations of the tunneling current were explained in terms of lateral variations of the surface local density of states (LDOS). In the following, we shall focus in more detail on the meaning of ϕ as well as its local and spatially resolved measurement in STM.

Motivated by Eq. (1.99), an apparent local barrier height (LBH) in STM is usually defined by

$$\phi_A[eV] = \frac{\hbar^2}{8m}\left(\frac{d\ln I}{ds}\right)^2 \approx 0.95\left(\frac{d\ln I}{ds\,[\text{Å}]}\right)^2 \qquad (1.111)$$

For large tip–surface separation outside the effective range of image forces (section 1.2.6) it is clear that ϕ has to approach the surface local work function ϕ_s which is defined as the work needed to remove an electron from the solid's Fermi level to a position somewhat outside the surface where image force effects can be neglected (Forbes, 1990). However, for small tip–surface separations (5–10 Å), image potential effects certainly have to be considered. Binnig *et al.* (1984a) assumed a one-dimensional potential barrier and included the image potential by writing

$$\phi(d) = \phi_0 - \frac{\alpha}{d} \qquad (1.112)$$

where ϕ_0 is the average work function of the sample surface and the tip ($\phi_0 = (\phi_s + \phi_t)/2$), d is the distance between the two image planes, which is different from the distance s between the two metal electrodes modeled by jelliums ($d \approx s - 1.5$ Å), and $\alpha \approx 10$ eV Å. We can now calculate the distance dependence of the tunneling current by using the model potential described by Eq. (1.112) and obtain

$$\frac{d\ln I}{ds} = -\frac{2\,(2m)^{\frac{1}{2}}}{\hbar}\cdot\phi_0^{\frac{1}{2}}\left[1+\frac{\alpha^2}{8\phi_0^2 d^2}+\mathcal{O}\left(\frac{1}{d^3}\right)\right] \qquad (1.113)$$

Interestingly, the first-order term in $1/d$, although present in the potential $\phi(d)$, cancels exactly in the expression for d ln I/ds. The second-order term in $1/d$ ($\alpha^2/8\phi_0^2 d^2$) usually contributes only a few percent of the zero-order term and can therefore be neglected to a first approximation. As a consequence, we find

$$\frac{d\ln I}{ds} \approx \text{const.} \qquad (1.114)$$

and $\phi_A \approx \phi_0 = \text{const}$. This means that the presence of the image potential does not show up in the distance-dependence of the tunneling current although the absolute values of the current are drastically affected by the presence of the image potential. The distance-independence of the apparent barrier height deduced from the current–distance relation (Fig. 1.73a) has been verified experimentally as well as by additional theoretical work (Payne and Inkson, 1985; Coombs *et al.*, 1988a).

As already discussed in section 1.2.6, the classical image potential can no longer be valid as the two metal electrodes are separated by only a few Ångström units or even by a distance approaching the point contact value. Instead, the quantum-mechanical exchange and correlation potential has to be used. Lang (1988) has calculated the apparent barrier height as a function of tip–surface separation for distances less than 7 Å (Fig. 1.73b) based on the adatom-on-jellium model for both tip and sample surface (section 1.11.1). As can be seen in Fig. 1.73b, the apparent barrier height, though almost constant and equal to ϕ_0 for tip–surface separations above 5 Å, decreases gradually towards zero as point contact is approached, where complete collapse of the barrier is

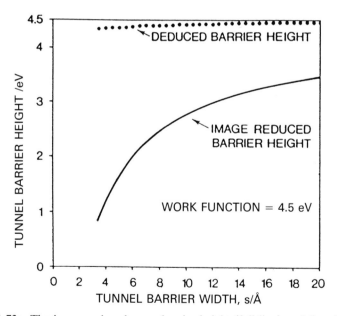

Fig. 1.73a. The image-reduced mean barrier height (full line) and the effective barrier height deduced from the current–distance relation for this barrier (dotted line). The work function used in the calculation is 4.5 eV. It can be seen that the deduced barrier height is always within 0.2 eV of the work function despite the collapse of the image-reduced barrier (Coombs *et al.*, 1988a).

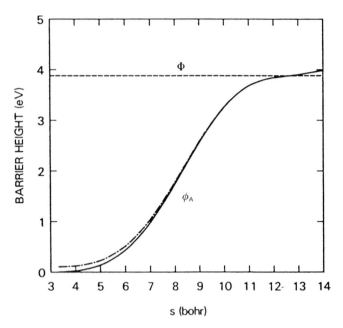

Fig. 1.73b. Apparent barrier height ϕ_A for two jellium electrodes, one representing the sample, the other, with an adsorbed Na atom, representing the tip. The work function ϕ for the sample electrode by itself is shown for comparison. The tip–sample separation is denoted by s (1 bohr = 0.529 Å) (Lang, 1988).

expected. A similar behavior was found in subsequent theoretical work (Pitarke *et al.*, 1989).

Finally, as already discussed in section 1.2.7, the apparent barrier height measured in STM can be drastically affected by band structure effects, particularly in tunneling experiments on semiconductors or semimetals. According to Eq. (1.30), band structure effects lead to an increase of the apparent barrier height given by

$$\phi_A = \phi_0 + \frac{\hbar^2}{2m} k_{\parallel}^2 \tag{1.115}$$

where $k_{\parallel}$ is the wave vector component parallel to the surface plane.

1.12.1 Local tunneling barrier height measurements at fixed surface locations

According to Eq. (1.111), the apparent barrier height ϕ_A can be determined locally by measuring the slope of ln *I*–*s* characteristics at a fixed sample bias voltage U and at a fixed sample surface location. To demonstrate vacuum tunneling it is necessary to obtain reasonably high values for ϕ_A of several electronvolts in addition to establishing the exponential dependence of the current on the tip–surface separation (Fig. 1.41).

Alternatively, the apparent barrier height can be deduced from the slope of local ln U–s characteristics in a low applied bias voltage range and at a fixed tunneling current. In the low-bias regime, the tunnel junction exhibits Ohmic behavior, as was found earlier (Eq. (1.95)):

$$I \propto U \cdot \exp\left(-2 \varkappa s\right) \tag{1.116}$$

Therefore, we obtain

$$\phi_A = \frac{\hbar^2}{8m}\left(\frac{d \ln U}{ds}\right)^2 \tag{1.117}$$

at constant current.

1.12.2 Spatially resolved local tunneling barrier height measurements

After having discussed the distance dependence of the apparent barrier height $\phi_A(s)$ at a fixed surface location (x,y) in section 1.12.1, we now focus on spatial variations of the apparent barrier height $\phi_A (x,y)$ at constant tip–surface separation s. We shall distinguish between two limiting cases, the 'macroscopic' and 'microscopic' limits.

In the macroscopic limit, the tip–surface distance is larger than 5 Å so that the measured apparent barrier height ϕ_A reflects the average value of the tip and sample surface work functions. For a given work function ϕ_t of the tip, spatial variations of $\phi_A (x,y)$ can be attributed to spatial variations of the surface local work function $\phi_s (x,y)$ and $\phi_A (x,y)$ can be regarded as a 'characterization parameter' of the local surface work function. However, the absolute value of $\phi_A (x,y)$ is generally not equal to $\phi_s (x,y)$. Furthermore, we exclude atomic resolution of surface features in the macroscopic limit.

In contrast, in the microscopic limit we shall concentrate on interpretation of the spatially resolved apparent barrier height in atomic-resolution STM data.

1.12.2.1 Macroscopic limit

Since in the macroscopic limit spatial variations of $\phi_A (x,y)$ reflect spatial variations of the surface local work function $\phi_s (x,y)$, we first have to discuss possible origins of such variations.

The local work function as defined at the beginning of section 1.12 can be split into two contributions, a 'chemical' and an 'electrical' contribution (Forbes, 1990). The chemical component of ϕ_s is determined by the chemical nature and structure of the solid only, whereas the electrical component of ϕ_s depends on the chemical nature of the solid as well as on the surface crystallographic orientation. The electron charge distribution at the surface gives rise to an electric dipole layer which leads to an electric potential difference between the exterior and interior of the solid. Since the nature of the dipole layer depends on the crystallographic face, a variation of the local work function with crystallographic orientation results. It should be noted that the Fermi level within the solid remains uniform by definition as long as thermodynamic equilibrium exists for the electrons within the solid. However, the electrostatic potentials outside each of the different crystallographic faces are different. Consequently, even a neutral solid is surrounded by a system of electric fields, known as 'patch fields' which extend out into the space surrounding the solid with an effective range comparable to the dimensions of the different crystallographic facets. These patch fields are responsible for the spatial variations of the local work function.

The absolute value of the surface work function is mainly determined by chemical effects contributing several electronvolts whereas electric effects contribute typically between 0.1 and 1 eV depending, for instance, on the crystallographic orientation (Smoluchowski, 1941). Particularly for inhomogeneous samples, various small surface areas or 'patches' with a different local work function can exist. For instance, in the case of polycrystalline samples, variations of the local work function can result from a different chemical composition of different grains or from a different chemical composition of the grains and the grain boundaries. Additionally, a polycrystalline sample usually exhibits different crystallographic orientations of the grains at the surface, leading to spatial variations of the local work function as well. Furthermore, spatial variations of the local work function can result from non-uniform adsorption of atoms and molecules on surfaces. Charge transfer between adsorbates and solid surface can produce or influence an electric dipole layer at the surface, thereby changing the value of the local work function through patch-field effects.

Several experimental methods have been developed in the past, some before the invention of the STM, to detect or map spatial variations of the local surface work function; for instance, by using an electron beam scanning technique (Haas and Thomas, 1963, 1966), by ultraviolet photoemission electron microscopy (PEEM) (e.g. Bauer *et al.*, 1989), or by scanning photoemission microscopy (SPEM) (Rotermund *et al.*, 1989). These techniques are able to map variations of the local work function of macroscopic surface areas with a spatial resolution down to the micrometer scale. On the other hand, the technique of photoemission of adsorbed xenon (PAX) is sensitive to differences in the local work function between individual atomic surface sites, but integrates over macroscopic surface areas without providing any lateral resolution (Küppers *et al.*, 1979). Mapping of the apparent barrier height in STM can provide information about the spatial variation of the local work function with a lateral resolution down to the atomic scale. However, the absolute value of the local work function ϕ_s (x,y) can generally not be extracted from the measured apparent barrier height ϕ_A (x,y).

The experimental determination of ϕ_A (x,y) is based on modulating the tip–surface distance s by Δs while scanning at a constant average current $\bar{I}$, with a modulation frequency ν_0 higher than the cut-off frequency of the feedback loop but smaller than the first mechanical resonance frequency of the STM unit (Binnig and Rohrer, 1983). The modulation of ln I at ν_0, which is in-phase with the modulation of the tip–surface separation, can be measured by a lock-in amplifier simultaneously with the corresponding constant current topograph and directly yields a signal proportional to the square root of the apparent barrier height via the relation

$$\frac{\Delta \ln I}{\Delta s} = - \frac{2\,(2m)^{1/2}}{\hbar} \phi_A^{1/2} \qquad (1.118)$$

The apparent barrier height obtained in this way is not measured at a constant tip–surface separation s. Scanning at constant average current (and at constant applied bias voltage) implies that the product $(\phi_A)^{1/2}{\cdot}s$ is kept constant, rather than s. However, since the spatial variation of ϕ_A is usually small (about 10% or less of the absolute value of ϕ_A), and ϕ_A enters only under the square root, the spatial variation of ϕ_A (x,y) is usually measured almost at constant tip–surface separation s.

Spatially resolved measurements of the apparent barrier height in the macroscopic limit have mainly been used to map chemical inhomogeneities at surfaces (e.g. Binnig and Rohrer, 1983; Wiesendanger *et al.*, 1987a, 1987b). Spatial variations of the apparent barrier height originating from different crystallographic facets of a crystallite embedded in an amorphous matrix have been detected as well (Wiesendanger *et al.*, 1987b). The lateral resolution in the apparent barrier height images has always been found to be comparable to the lateral resolution in the corresponding topographic images.

1.12.2.2 Microscopic limit

The spatial variation of the apparent barrier height can, of course, also be measured simultaneously with atomic-resolution constant-current images by using the modulation technique described above. It was shown that atomic resolution can also be obtained in the barrier height images of metal surfaces (e.g. Marchon *et al.*, 1988a), surfaces of layered materials (e.g. Anselmetti *et al.*, 1988), and semiconductor surfaces (e.g. Villarrubia and Boland, 1989). Usually, the barrier height images exhibit an improved signal-to-noise ratio compared with the corresponding topographic constant current images due to the lock-in detection technique.

In the microscopic limit, where variations of the apparent barrier height appear on an atomic scale, these variations can no longer be related to variations of the local surface work function because the work function is clearly defined only in the macroscopic limit. The work function was originally introduced as a property of the surface as a whole. A distinction can then be made between the total and the local work function. The total work function is the work needed to remove an electron from the Fermi level of a solid to infinity. This concept is usually used in theoretical work together with the assumption of having a flat surface of infinite extent. On the other hand, the concept of the local work function, defined as the work needed to remove an electron from the solid's Fermi level to a position somewhat outside the surface where image force effects can be neglected, is practically more important. However, if we approach the atomic scale, even the concept of the local work function is no longer useful because the electron remains within the effective range of the classical image force or even of quantum mechanical exchange and correlation.

In the microscopic quantum mechanical regime, it is certainly more appropriate to relate the measured apparent barrier height with the decay rates of the wave functions describing the sample surface and the tip. Lateral variations of the measured apparent barrier height then have to be interpreted as lateral variations in the decay rate of the surface wave function, an aspect which was not addressed in the early theoretical treatments of the STM (e.g. Tersoff and Hamann, 1983, 1985). We already know that, with increasing tip–surface separation s, the measured surface atomic corrugation Δ in constant current STM images is smoothed out exponentially (Eq. (1.108)). This can only occur if the decay rate $\varkappa_p$ above a local protrusion in the topography is larger than the decay rate $\varkappa_d$ above a local depression. Consequently, the apparent barrier height above a local topographic protrusion has to be larger than the barrier height above a local depression. Therefore, atomically resolved apparent barrier height images closely reflect corresponding topographic constant current images. If the measured signal $\Delta \ln I/\Delta s$ is directly displayed, there is a contrast reversal because $(\Delta \ln I/\Delta s) \propto - \phi^{1/2}$ (Eq. (1.118)).

1.12.3 Anomalous barrier heights

Measured values for the apparent barrier height can be subject to anomalies caused either by the experimental conditions or the measurement technique itself.

1.12.3.1 Geometry-induced anomalies

For surfaces which are not atomically flat, topographic features have an influence on the barrier height images obtained by the modulation technique described in the previous section. Experimentally, the tip–surface distance s is modulated by applying an additional a.c. voltage on the z piezodrive. The resulting modulation Δz of the vertical tip position is equal to the modulation Δs of the tip–surface distance only if the z-direction coincides with the direction of the normal vector for the local surface element. In general, however, there is an angle φ between these two directions and we find (Fig. 1.74):

$$\Delta s = \Delta z \cdot \cos (\varphi) \tag{1.119}$$

Therefore

$$\frac{\Delta \ln I}{\Delta s} = \frac{1}{\cos (\varphi)} \cdot \frac{\Delta \ln I}{\Delta z} = - A\phi_A^{1/2} \tag{1.120}$$

where $A = 2(2m)^{1/2}/\hbar$. By using the z modulation technique we measure

$$\frac{\Delta \ln I}{\Delta z} = -A\phi_A^{1/2} \cos (\varphi) \tag{1.121}$$

which includes the geometry-dependent factor ($\cos \varphi$) (Binnig and Rohrer, 1983). Significant variations of the measured apparent barrier height

$$\phi_{A,m}^{1/2} = - \frac{1}{A} \frac{\Delta \ln I}{\Delta z} = \phi_A^{1/2} \cdot \cos (\varphi) < \phi_A^{1/2} \tag{1.122}$$

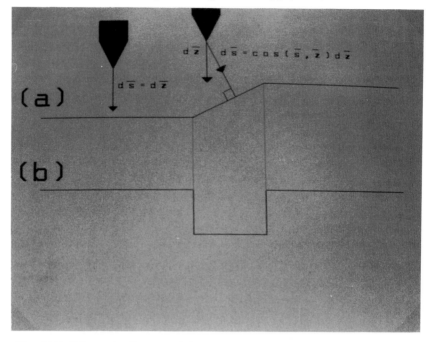

Fig. 1.74. Schematic diagram of the topographic contribution on STM barrier height images. (*a*) Topographic profile. (*b*) Corresponding cos (*s*, *z*) profile, where (*s*, *z*) is the angle formed by the gradient vector of the surface at each point (*s*) and the modulation direction (*z*). Notice that when a step is scanned in topography, the corresponding cos (*s*, *z*) profile is a depression (Gómez-Rodríguez *et al.*, 1989).

due to geometric non-uniformities can be expected particularly at step edges as illustrated in Fig. 1.75, or at grain boundaries (Gómez-Rodriguez *et al.*, 1989; Golubok and Tarasov, 1990).

1.12.3.2 Surface elasticity-induced anomalies

Particularly in STM experiments performed in air with contaminated tips and sample surfaces, anomalously low (< 1 eV) barrier heights have often been measured. Coombs and Pethica (1986) explained this observation by assuming that tip and sample are in contact somewhere. This contact behaves in a linear elastic fashion, thereby changing the calibration for the vertical tip displacement being some fixed fraction α of the z piezodrive movement. The exponential dependence of the current on separation is still preserved, but a lower apparent barrier height is measured because the actual tip displacement Δs is less than the z piezodrive movement Δz:

$$\Delta s = \alpha \cdot \Delta z \qquad \alpha < 1 \qquad (1.123)$$

(a)

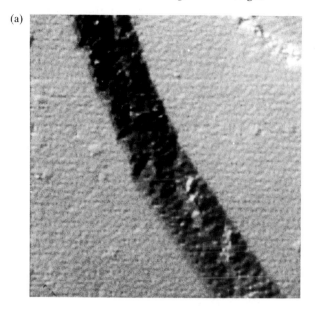

(b)

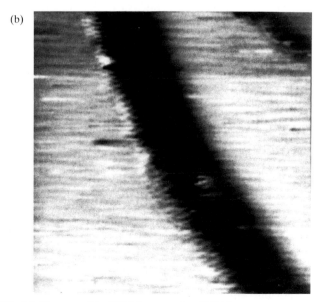

Fig. 1.75. (*a*) Topographic STM image of a Si(111) surface with a (113) step edge. (*b*) Corresponding local barrier height image showing the topographic induced lowering of the barrier height at the step edge (Wiesendanger, unpublished results).

Therefore

$$\phi_{A,m}^{1/2} = -\frac{1}{A}\frac{\Delta \ln I}{\Delta z} = -\frac{1}{A}\,\alpha\,\frac{\Delta \ln I}{\Delta s}$$
$$= \alpha \cdot \phi_A^{1/2} < \phi_A^{1/2} \qquad\qquad (1.124)$$

In general, repulsive forces acting between tip and sample lead to a decrease of the measured apparent barrier height $\phi_{A,m}$ whereas attractive forces lead to increased $\phi_{A,m}$ values. This dependence of the measured apparent barrier height on forces acting between tip and sample becomes significant particularly for soft elastic samples, such as biological specimens, where measurement of the spatial variation of $\phi_{A,m}$ can be used to probe the spatial variation of surface elasticity with a contrast often improved compared with topographic constant current images.

1.13 Tunneling spectroscopy (TS)

In section 1.11.1 we saw that in addition to the distance- and apparent barrier height-dependence of the tunneling current there also exists a bias-dependence which we will now focus on further. For tunneling between metal electrodes in the low-bias (millivolt) limit, the tunneling current is found to be linearly proportional to the applied bias voltage (Eq. (1.95)). For higher bias (volts) and particularly for semiconductor samples, the bias-dependence of the tunneling current generally does not exhibit Ohmic behavior and the constant current STM images can depend critically on the applied bias (Lang, 1987a). Studying this bias-dependence in detail allows extraction of various spectroscopic information of high spatial resolution, ultimately down to the atomic level. The spectroscopic capability of STM combined with its high spatial resolution is perhaps the most important feature of STM and has been applied widely, particularly for investigation of semiconductor surfaces (Hamers, 1989a; Tromp, 1989; Feenstra, 1990). In principle, STM has to be considered primarily as a spectroscopic method due to its inherent bias-dependence.

We consider a simplified one-dimensional potential energy diagram at zero temperature for the system consisting of the tip (left electrode) and the sample (right electrode) which are separated by a small vacuum gap (Fig. 1.76). For zero applied bias (Fig. 1.76(b)) the Fermi levels of tip and sample are equal at equilibrium. When a bias voltage U is applied to the sample, the main consequence is a rigid shift of the energy levels downward or upward in energy by an amount $|eU|$,

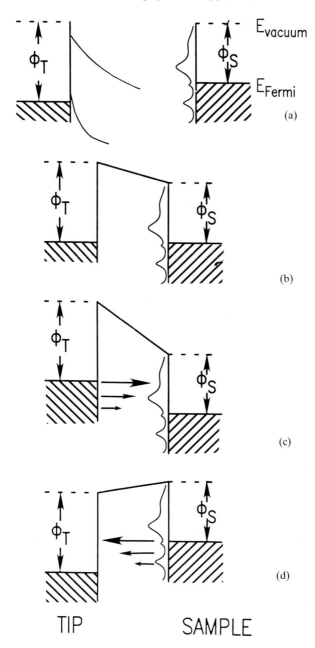

TIP   SAMPLE

Fig. 1.76. Energy level diagrams for sample and tip. (*a*) Independent sample and tip. (*b*) Sample and tip at equilibrium, separated by a small vacuum gap. (*c*) Positive sample bias: electrons tunnel from the tip to the sample. (*d*) Negative sample bias: electrons tunnel from the sample into the tip (Hamers, 1989a).

depending on whether the polarity is positive (Fig. 1.76(*c*)) or negative (Fig. 1.76(*d*)). (As discussed in section 1.11.1, we neglect the distortions of the wave functions and the energy eigenvalues due to the finite bias to a first approximation.) For positive sample bias, the net tunneling current arises from electrons that tunnel from the occupied states of the tip into unoccupied states of the sample (Fig. 1.76(*c*)), whereas at negative sample bias, electrons tunnel from occupied states of the sample into unoccupied states of the tip. Consequently, the bias polarity determines whether unoccupied or occupied sample electronic states are probed. We further see that the electronic structure of the tip enters as well, as is also obvious from Eq. (1.102) for the tunneling current:

$$I \propto \int_0^{eU} n_t(\pm eU \mp \mathscr{E}) \, n_s(\mathscr{E}) T(\mathscr{E}, eU) \, d\mathscr{E}$$

By varying the amount of the applied bias voltage, one can select the electronic states that contribute to the tunneling current and, in principle, measure the local electronic density of states. For instance, the current increases strongly if the applied bias voltage allows the onset of tunneling into a maximum of the unoccupied sample electronic density of states. Therefore, the first derivative dI/d$U(U)$ reflects the electronic density of states to a first approximation. However, we also have to consider the energy- and bias-dependence of the transmission coefficient $T(\mathscr{E}, eU)$. Since electrons in states with the highest energy 'see' the smallest effective barrier height, most of the tunneling current arises from electrons near the Fermi level of the negatively biased electrode. This has been indicated in Fig. 1.76 by arrows of differing size. The maximum in the transmission coefficient $T(\mathscr{E}, eU)$ given in Eq. (1.103) can be written as (Feenstra, 1990)

$$T_{max}(U) = \exp\left\{-2(s + R)\left[\frac{2m}{\hbar^2}\left(\frac{\phi_t + \phi_s}{2} - \frac{|eU|}{2}\right)\right]^{1/2}\right\} \quad (1.125)$$

The bias-dependence of the transmission coefficient typically leads to an order-of-magnitude increase in the tunneling current for each volt increase in magnitude of the applied bias voltage. Since the transmission coefficient increases monotonically with the applied bias voltage, it contributes only a smoothly varying 'background' on which the density-of-states information is superimposed.

As an important consequence of the dominant contribution of tunneling from states near the Fermi level of the negatively biased electrode, tunneling from the tip to the sample (Fig. 1.76(c)) mainly probes the sample's empty states with negligible influence of the tip's occupied states. On the other hand, tunneling from the sample to the tip is much more sensitive to the electronic structure of the tip's empty states which often prevents detailed spectroscopic STM studies of the sample's occupied states (Klitsner *et al.*, 1990). Griffith and Kochanski (1990a) have calculated a spectroscopic dI/dU–U curve for the hypothetical case when the tip and sample are assumed to have an uniform density of states, with a 10% ripple added for identification purposes (Fig. 1.77). It can clearly be seen from Fig. 1.77 (lower part) that the calculated spectroscopic curve dI/dU(U) reflects the slow wiggles characteristic for the sample's electronic density of states only for positive sample bias voltage whereas, at negative sample bias, the spectroscopic curve shows the fast wiggles characteristic for the tip's electronic density of states.

Spectroscopic information from bias-dependent STM measurements can be obtained in a number of ways based on different experimental techniques which will now be described and discussed.

1.13.1 Voltage-dependent STM imaging

The simplest way of obtaining spectroscopic information consists of sequentially recording conventional constant current topographs (CCT) at different applied bias voltages followed by comparison of the experimental results. However, thermal drift, creep of the piezodrives and tip changes during scanning often make direct comparison between successive STM images difficult. Alternatively, the applied bias voltage can be changed on a line-by-line basis during data acquisition of a single STM image, providing two (or more) interleaved images at different bias voltages which are often chosen with a different polarity.

Voltage-dependent STM imaging as a spectroscopic method relies on the fact that, at a particular chosen bias voltage, only the electronic states between the Fermi levels of the tip and the sample can contribute to the tunneling current. This method often provides a quick and simple way to assess whether there are any interesting differences between different bias voltages, particularly at different polarities (e.g. Stroscio *et al.*, 1988). The most serious limitation of this experimental procedure lies in the fact that constant current topographs generally reveal both geometric and electronic structure information which cannot easily be separated. Therefore, additional experimental techniques have been introduced to extract the electronic structure information more clearly, preferably independent of the surface geometric structure.

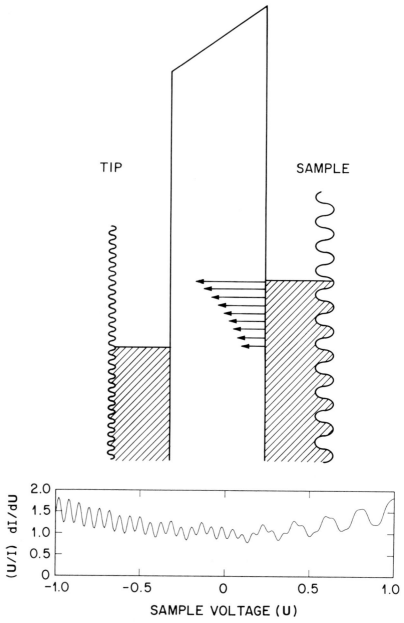

Fig. 1.77. Schematic illustration of a spectroscopic tunneling experiment. The upper portion of the figure shows energy levels as a function of position; the tip is on the left (biased 1 V positive), and the sample is on the right. Hatched areas correspond to filled states below the Fermi levels; the length of the arrows is proportional to the barrier transmission factor at a given energy, and the density of states of the tip and sample are shown by the wiggly vertical lines. This diagram corresponds to an experiment attempting to measure the sample density of states 1 V below its Fermi level. The lower portion displays the computed spectroscopic characteristic of the system (displayed as $(dI/dU)/(I/U)$). This curve contains information about both the tip (fast wiggles) and the sample (slow wiggles) (Griffith and Kochanski, 1990b).

1.13.2 Scanning tunneling spectroscopy (STS) at constant current

In scanning tunneling spectroscopy (STS), a high-frequency sinusoidal modulation voltage is superimposed on the constant d.c. bias voltage applied between tip and sample. The modulation frequency is chosen higher than the cut-off frequency of the feedback loop which keeps the average tunneling current constant. By recording the tunneling current modulation, which is in-phase with the applied bias voltage modulation, by means of a lock-in amplifier, a spatially resolved spectroscopic signal dI/dU can be obtained simultaneously with the constant current image (Binnig *et al.*, 1985a,b; Becker *et al.*, 1985a; Baratoff *et al.*, 1986). Based on the expression (1.102) for the tunneling current and by assuming $dn_t/dU \approx 0$, we obtain (Hamers, 1989a):

$$\frac{dI}{dU}(U) \propto en_t(0)\, n_s(eU)\, T(eU,eU)$$

$$+ \int_0^{eU} n_t(\pm eU \mp \mathscr{E})n_s(\mathscr{E}) \cdot \frac{dT(\mathscr{E},eU)}{dU}\, d\mathscr{E} \qquad (1.126)$$

At a fixed location, the increase of the transmission coefficient with applied bias voltage is smooth and monotonic, as mentioned earlier (section 1.12). Therefore, structure in dI/dU as a function of U can usually be attributed to structure in the state density via the first term in Eq. (1.126). However, interpretation of the spectroscopic data dI/dU as a function of position (x,y) is more complicated. As discussed in section 1.12.2, the apparent barrier height above a local topographic protrusion is larger, i.e. the transmission coefficient is smaller, than above a local topographic depression. This spatial variation in the transmission coefficient shows up in spatially resolved measurements of dI/dU as a 'background' that is essentially an 'inverted' constant current topography. Therefore, spectroscopic images corresponding to the spatial variation of dI/dU obtained in the constant current mode in fact contain a superposition of topographic and electronic structure information.

There are also other drawbacks of the STS technique. As in the case of voltage-dependent STM imaging, complete investigation of the surface electronic structure as a function of energy and position requires many repeated measurements over the same surface area at different bias voltages, which is often prevented by tip instabilities during scanning and by thermal drift. Furthermore, measurements of dI/dU (Eq. (1.126)) at constant average tunneling current contain a strongly voltage-dependent background originating from the voltage-dependence of the transmission coefficient. In the low-bias limit, where the tunnel junction exhibits Ohmic behavior, we simply find

$$\frac{dI}{dU} = \frac{I}{U} \qquad (1.127)$$

For a constant tunneling current I, the quantity dI/dU therefore diverges like $1/U$ as U approaches zero. This $1/U$ background, on which the desired electronic structure information is superimposed, makes it difficult to observe structure in the dI/dU data at low bias voltages, thereby preventing investigation of the electronic structure near the Fermi level.

Another problem arising in STS measurements lies in the fact that the tip–sample separation varies with applied bias voltage if the tunneling current is kept constant. With increasing voltage and distance, lateral resolution is degraded. On the other hand, with decreasing bias voltage, the tip plunges toward the sample surface in order to maintain a constant tunneling current if a semiconductor energy gap is present. Consequently, investigation of the electronic structure near the Fermi level, which is often of primary importance, has to be excluded. In practice, useful STS data on semiconductors is usually not obtained below 1 V.

Many problems in STS discussed above, such as the $1/U$ divergence, can be solved by operating the STM at a constant tunneling resistance U/I instead of a constant tunneling current I (Kaiser and Jaklevic, 1986). Electronic structure information can then be obtained close to the Fermi level as well.

1.13.3 Local I–U measurements at constant separation

Another way to eliminate the $1/U$ divergence and additionally the influence of the z-dependence of the transmission coefficient is based on measuring local I–U curves at a fixed tip–sample separation. This can be achieved by breaking the feedback circuit for a certain time interval at selected surface locations by means of a sample-and-hold amplifier (section 1.10.4) while local I–U characteristics are recorded (Feenstra *et al.*, 1986, 1987a; Stroscio *et al.*, 1986). The I–U characteristics are usually repeated several times at each surface location and finally signal-averaged. Since the feedback loop is inactive while sweeping the applied bias voltage, the tunneling current is allowed to become extremely small. Therefore, band-gap states in semiconductors, for instance, can be probed without difficulties. The first derivative dI/dU can be obtained from the measured I–U curves by numerical differentiation. The dependence of the measured spectroscopic data on the value of the tunneling conductance I/U can be compensated by normalizing the differential conductance dI/dU to the total conductance I/U. Feenstra *et al.*, (1987a) showed that the normalized quantity $(dI/dU)/(I/U)$ = $(d \ln I)/(d \ln U)$ reflects the electronic density of states reasonably well by minimizing the influence of the tip–sample separation. This conclusion was supported by calculations of Lang (1986b) based on the adatom-on-jellium model. However, the close resemblance of the $(d \ln I/d \ln U)$–U curve to the electronic density of states is generally limited to the position of peaks while peak intensities can differ significantly.

1.13.4 Current imaging tunneling spectroscopy (CITS)

The measurements of local I–U curves at constant tip–sample separation can be extended to every pixel in an image, which allows performance of atomically resolved spectroscopic studies (Hamers *et al.*, 1986a; Hamers 1989a). The ability to probe the local electronic structure down to atomic scale has great potential, for instance, for investigation of surface chemical reactivity on an atom-by-atom basis (section 5.1).

The method, denoted current imaging tunneling spectroscopy (CITS), also uses a sample-and-hold amplifier to alternately gate the feedback control system on and off (Fig. 1.78(a)). During the time of active feedback, a constant stabilization voltage U_0 is applied to the sample and the tip height is adjusted to maintain a constant tunneling current. When the feedback system is deactivated, the applied sample bias voltage is linearly ramped between two preselected values and the I–U curve is measured at a fixed tip height. Afterwards, the applied bias voltage is set back to the chosen stabilization voltage U_0 and the feedback system is reactivated. By acquiring the I–U curves rapidly while scanning the tip position at low speed, a constant current topograph and spatially resolved I–U characteristics can simultaneously be obtained. To increase the possible scan speed and to decrease the amount of data to be stored, one can predefine a coarse grid of pixels in the image at which local I–U curves will be measured (Fig. 1.78(b)).

Based on the spatially resolved I–U information, a 'current image' (CI) at some chosen bias voltage U_1 different from the stabilization voltage U_0 can be formed. Unfortunately, these current images depend on the contour that the tip follows during scanning, which is determined by the stabilization voltage U_0. Essentially, these current images represent the spatial distribution of the difference in the tunneling current

(a)

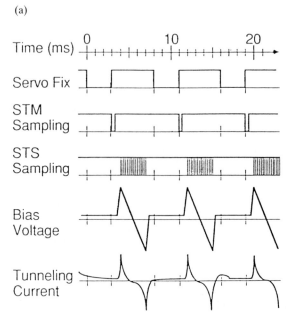

Fig. 1.78. (*a*) Timing chart of control signals in CITS measurements (Bando *et al.*, 1988). (*b*) STM topograph of the unoccupied states of a Si(111) (7×7) surface (sample bias +2 V). The atoms imaged are the top-layer Si adatoms. The grid encompasses a 14 Å × 14 Å area of this surface for which tunneling spectra have been obtained. The 100 tunneling spectra are plotted in the dI/dU form. Such spectral maps allow one not only to obtain the energies of the occupied (negative bias) and unoccupied (positive bias) states of particular atomic sites, but also to obtain information on the spatial extent of their wave functions (Avouris and Lyo, 1990).

measured at U_1 and U_0. This difference $I(U_1) - I(U_0)$ can arise, for instance, from a surface state at energy eU_1 (Fig. 1.79, curve (*a*)). In this case, the CI can be interpreted as a spatial map of the surface state itself. However, a difference $I(U_1') - I(U_0)$ can also arise from the voltage-dependence of the transmission coefficient (Berghaus *et al.*, 1988b; Stroscio *et al.*, 1988; Feenstra, 1990), as illustrated in Fig. 1.79, curve (*b*). In this case, the CI would erroneously be interpreted as a spatial map of a surface state at an energy eU_1'. Therefore, one has to be cautious in the interpretation of current images. This conclusion may even hold for differences between current images. In general, a complete spectrum of the surface states must first be examined before assigning specific 'topographic' features to individual surface states.

(b)

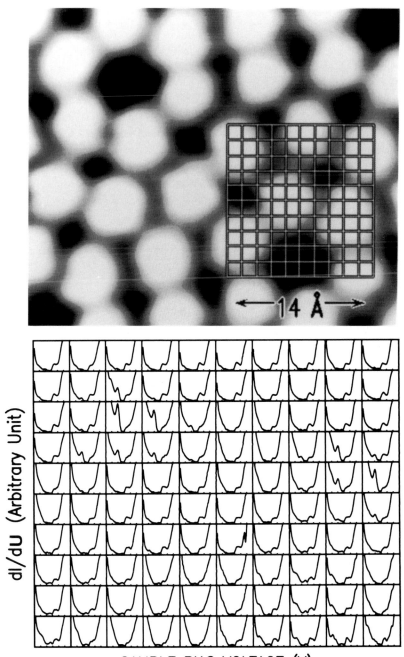

dI/dU (Arbitrary Unit)

−30 3   SAMPLE BIAS VOLTAGE (V)

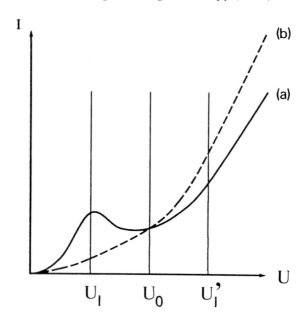

Fig. 1.79. Schematic illustration of current–voltage characteristics at two different spatial locations (a) and (b) on a surface. The currents are equal at the
voltage U_0. At U_1, a surface state feature appears at location (a), and it can be
imaged at that voltage. An image formed at the voltage U'_1 will show a maximum at spatial location (b), associated with the varying background level of
the current (Feenstra, 1990).

1.13.5 Variable-separation spectroscopy

The constant-separation spectroscopy (CSS) discussed in section 1.13.3
and 1.13.4 has the advantage of eliminating the $1/U$ divergence problem
in the constant current spectroscopy (CCS) discussed in section 1.13.2,
thereby allowing scanning of the applied voltage through zero bias. On
the other hand, constant current spectroscopy has the intrinsic advantage of offering a large dynamic range because the conductivity always
maintains large, measurable values as a function of bias if the current
is kept constant. By using the constant-separation spectroscopy method,
the I–U curves usually have to be measured at various values of the
tip–sample separation in order to achieve a large dynamic range, which
is needed particularly for investigations of wide band-gap materials. To
eliminate this problem, a variable-separation spectroscopy (VSS) was
introduced in which the tip–sample separation is continuously varied by
a linear ramp while the applied sample bias voltage is scanned (Feenstra
and Mårtensson, 1988; Mårtensson and Feenstra, 1989; Feenstra, 1990).

The separation is reduced for decreasing applied sample bias voltage and enlarged for increasing bias. A modulation technique is used to measure the differential conductivity dI/dU which is finally normalized to the total conductivity I/U as motivated in section 1.13.3.

1.13.6 Assessment of tunneling spectroscopy compared with other spectroscopical techniques

By using the experimental methods described above, the STM can provide spectroscopic information with an unprecedented spatial resolution, ultimately at the atomic level, as was first demonstrated for the Si(111) 7×7 surface (Hamers *et al.*, 1986a). Apart from spatially resolved spectroscopy, it is always possible to perform local spectroscopy where the tip is positioned above a preselected surface site, e.g. an atomic defect, and an $I–U$ curve is measured which characterizes the electronic structure of the selected surface site only. Such site-specific information is generally not provided by area-integrating surface analytical techniques. Since tunneling spectroscopy is sensitive to both occupied and unoccupied electronic states, one might expect to perform a combined photoemission and inverse photoemission experiment with a STM instrument which additionally allows for atomic resolution. However, there are also some limitations of tunneling spectroscopy which will now be discussed.

1.13.6.1 Fundamental weaknesses of tunneling spectroscopy (TS)

1. As already discussed in section 1.11.1, STM (and therefore also TS) senses only those electronic states that protrude into the vacuum region and significantly overlap with the tip wave functions. Consequently, STM and TS probe only the tails of the surface wave functions, protruding several Ångström units into the vacuum region. If one defines an experimental information depth measured from the top surface layer towards the bulk of the solid, this information depth would have negative values for STM and TS. This is in strong contrast to the positive information depth, for instance, in photoemission experiments, Auger electron spectroscopy, or energy-dispersive x-ray microanalysis. Electronic states which are completely localized between the first and second surface layer or electronic states that are localized too close to the atomic cores of the top surface layer are invisible in STM and TS. However, in many cases, e.g. for fully understanding surface reconstructions, it is important also to get subsurface electronic structure information.

2. An important feature of angle-resolved photoemission and inverse photoemission experiments is the ability to provide information about the band structure of the sample surface. In contrast, STM is more closely related to angle-integrated photoemission and has no ability to resolve the electron wave vector in its initial state (Tromp, 1989). However, there exists some kind of k-selection rule in STM because states with $k_{||} = 0$ decay slowest into the vacuum and contribute most strongly to the tunneling current signal (section 1.2.7).

3. Perhaps the most serious limitation of STM and TS is the lack of chemical sensitivity. In contrast to x-ray photoelectron spectroscopy and Auger electron spectroscopy, which probe the element-specific core levels, STM and TS are sensitive to the valence states only, because the bias range in tunneling experiments is limited to $\pm \phi/e$. If the bias voltage were to exceed the apparent barrier height, we would enter the field emission regime (section 1.20). Since the valence states are drastically affected by chemical bond formation in compound solids, it is generally impossible to extract element-specific information from conventional STM and TS experiments. However, in special cases where information about the charge transfer between surface atoms already exists, STM can be used for atom-selective imaging, as first demonstrated for the GaAs(110) surface (Feenstra *et al.*, 1987a,b). Charge transfer from Ga to As results in an occupied electronic state centered at the As sites and an empty state centered at the Ga sites. Consequently, STM selectively images the As atoms at negative bias and the Ga atoms at positive bias. For the general case of an arbitrary multicomponent solid, however, the details of charge transfer between different atomic species are usually not known and therefore no assignment of 'topographic' features in STM images to particular atomic species can be made.

1.13.6.2 Tip artifacts in tunneling spectroscopy

Another weakness of tunneling spectroscopy lies in the fact that the information obtained always reflects a convolution of electronic states at the sample surface and at the tip apex, as is evident from expression (1.102) for the tunneling current. To exclude a strong influence of the tip's electronic structure on tunneling spectroscopy results, it is desirable to use tips with a featureless electronic structure within the energy window accessible by TS. While this is the case for many bulk metals, the electronic structure at the apex of atomically sharp tips can have very prominent features. It was indeed found that the most reliable spectroscopic data of a sample surface are obtained with relatively blunt tips while the sharpest, highest-resolution tips are the most likely to yield unreliable spectroscopic information (e.g. Feenstra *et al.* 1987a; Klitsner *et al.*, 1990). Tip electronic structure effects are not only visible in the *I–U* characteristics but can, of course, also drastically influence the appearance of topographic constant-current images, STS images, or current images obtained by the CITS method (Tromp *et al.*, 1988; Park *et al.*, 1988; Klitsner *et al.* 1990; Pelz, 1991). This can most nicely be seen in STM images where a tip change has occurred during data acquisition as illustrated in Fig. 1.80. Such tip changes can frequently and spontaneously occur during scanning, induced, for example, by transfer of an atom from the sample surface to the tip or from the tip to the surface, or by rearrangement of the apex atoms at the tip. As a consequence, the effective tip orbital usually changes, which can have a drastic influence on spatial resolution (section 1.11.1) as well as on the electronic structure of the tip apex, leading to tip-dependent STM and TS images. Particularly for atomic resolution STM images, it is not clear *a priori* which STM image obtained at a particular sample bias reflects the properties of the sample surface, its geometric and electronic structure, most closely. Careful spectroscopic STM studies as well as comparison with spectroscopic information provided by other surface analytical techniques are usually required for interpretation of atomic-resolution STM data. The fact that reliable tunneling spectroscopy data can usually be obtained only with relatively blunt tips, represents, of course, a severe limitation on the applicability of atomically resolved tunneling spectroscopy techniques. However, the spatial resolution of tunneling spectroscopy achieved even with relatively blunt tips is still unprecedented compared with other surface spectroscopical techniques.

1.13.6.3 *Energy resolution in tunneling spectroscopy*

Since the spatial resolution of STM is exceptionally high, one might expect that the uncertainty principle plays a role in determining the energy resolution of atomically resolved tunneling spectroscopy (Kuk and Silverman, 1989). If we assume a spatial uncertainty of $\Delta x \approx 2$ Å, the momentum uncertainty would be $\Delta k \approx 0.5$ Å^{-1} because $\Delta x \cdot \Delta k \approx 1$. Although the crystal momentum in a solid is not a physical momentum, a momentum resolution limit may roughly be estimated in this way. By considering the band structure of the solid surface, it is clear that poor momentum resolution results in poor energy resolution. Therefore, better resolution will certainly be obtained with tips that are not atomically sharp, which again favors relatively blunt tips for performing tunneling spectroscopy.

In conclusion, the requirements for the tip depend critically on the specific application. Atomically sharp tips, preferably with a p_z or d_{z^2}

Fig. 1.80. (*a*) Gray-scale CCT of the clean Si(111) (7×7) surface. (*b*) Subsequent CCT of the same area. (*c*)–(*h*) CITS images acquired simultaneously with (*b*), imaged at $U = +2.25, +1.75, +1.25, -1.25, -1.75$ and -2.25 V, respectively (Pelz, 1991).

orbital at the apex, are desirable for atomic-resolution 'topographic' STM imaging (section 1.11.1). On the other hand, relatively blunt tips provide the most reliable tunneling spectroscopy data with a better energy resolution. By using *in situ* 'tip sharpening' (Chen, 1991a) and 'tip blunting' (Chen, 1992c) procedures, it is possible to fulfill both requirements with a single tip.

1.14 Spin-polarized scanning tunneling microscopy (SPSTM)

So far, we have considered the dependence of the tunneling current on the tip–sample separation s, the local barrier height ϕ and the applied sample bias voltage U:

$$I = I(s, \phi, U) \tag{1.128}$$

Accordingly, we have discussed the corresponding modes of STM operation: 'topographic' imaging, local barrier height imaging and tunneling spectroscopy. However, we have not yet considered the spin of the tunneling electrons and the additional spin-dependence of the tunneling current if magnetic electrodes are involved (section 1.3.4):

$$I = I(s, \phi, U, \uparrow) \tag{1.129}$$

By using this spin-dependence of the tunneling current to obtain magnetic contrast, we arrive at another mode of STM operation, denoted as spin-polarized STM (SPSTM).

For SPSTM experiments, an additional requirement for the sensor tip exists: it should provide a highly efficient source or detector for spin-polarized electrons. This requirement can be fulfilled by using well-known spin-filter phenomena.

1. The Zeeman-split density of states of a superconductor can act as a very efficient spin-filter (Meservey, *et al.*, 1970), as already discussed in section 1.3.4. However, the use of this spin-filter requires low-temperature SPSTM experiments.
2. Ferromagnetic semiconductors or insulators used, for instance, as spin-dependent tunneling barriers (Müller *et al.*, 1972; Moodera *et al.*, 1988) can act as efficient spin-filters as well. For the case of EuS (Curie temperature $T = 16.7$ K), this would again require low temperatures.
3. Optically pumped semiconductors, such as GaAs, provide well-known sources for spin-polarized electrons. Unfortunately, the

polarization is typically well below 50%. It can be increased above the theoretical limit of 50% for GaAs by using semiconductor heterostructures, such as GaAs–AlGaAs superlattices (e.g. Omori *et al.*, 1991; Ciccacci *et al.*, 1987).

4. Half-metallic ferromagnets (de Groot *et al.*, 1983) and half-metallic ferrimagnets or antiferromagnets (de Groot, 1991) offer highly efficient spin-filters at room temperature. These materials exhibit a metallic-like density of states for one spin direction and a semiconductor-like density of states for the opposite spin direction. Consequently, a spin polarization of 100% is theoretically predicted for the conduction electrons at the Fermi level. The Mn-based Heusler alloys XMnSb (with X = Ni, Co, Pt), CrO_2, Fe_3O_4 as well as other compounds, were theoretically predicted to be half-metallic ferromagnets. For some of these materials, such as CrO_2, a very high spin polarization of nearly 100% was indeed found experimentally (Kämper *et al.*, 1987).

5. Ferromagnetic materials, such as Fe, Ni and Co, or antiferromagnetic materials, such as Cr, are also used as tip materials for SPSTM experiments.

Generally, the degree of spin polarization of electrons from such tips used in SPSTM studies is not yet known very precisely and is expected to depend critically on the tip preparation procedure.

1.14.1 Concepts of SPSTM

SPSTM experiments can be performed either without or with an applied external magnetic field. The two different types of experiments will now be discussed separately. We shall focus on room-temperature SPSTM studies using two magnetic electrodes.

1.14.1.1 SPSTM experiments without external magnetic field

Tunneling between two ferromagnetic electrodes separated by a non-magnetic tunneling barrier was treated by Slonczewski (1988, 1989) as already discussed in section 1.3.4. For this calculation, free-electron behavior of the spin-polarized conduction electrons inside each ferromagnet was assumed and the limit of vanishing sample bias voltage was considered. An effective spin polarization $P_{fbf'}$ of the whole ferromagnet (f)–barrier (b)–ferromagnet (f′) tunnel junction can be defined as the product of the effective spin polarizations P_{fb} and $P_{f'b}$ of the two ferromagnet–barrier couples as introduced by Slonczewski (1989):

$$P_{fbf'} = P_{fb} \cdot P_{f'b} \qquad (1.130)$$

In section 1.3.4 we found that the effective polarization $P_{fbf'}$ can be expressed by (Eq. (1.65))

$$P_{fbf'} = \frac{\sigma_{\uparrow\uparrow} - \sigma_{\uparrow\downarrow}}{\sigma_{\uparrow\uparrow} + \sigma_{\uparrow\downarrow}}$$

where $\sigma_{\uparrow\uparrow}(\sigma_{\uparrow\downarrow})$ denotes the tunneling conductance for the case of parallel (antiparallel) alignment of the magnetization directions within the two ferromagnetic electrodes. An experimental determination of the quantity on the right-hand side of Eq. (1.65) by means of a SPSTM offers a way to derive the effective polarization $P_{fbf'}$ locally with a spatial resolution comparable to that of topographic STM images and therefore ultimately on the atomic scale.

The spin-dependence of the tunneling current in SPSTM experiments with two magnetic electrodes in zero external magnetic field was demonstrated by using a ferromagnetic CrO_2 sensor tip and a Cr(001) test surface (Wiesendanger *et al.*, 1990c). The topological antiferromagnetic order of the Cr(001) surface (Blügel *et al.*, 1989) with alternately magnetized terraces separated by monatomic steps was confirmed. In addition, a local effective polarization of the CrO_2–vacuum–Cr (001) tunnel junction was derived.

In Fig. 1.81 a schematic drawing of a ferromagnetic tip scanning over alternately magnetized terraces separated by monatomic steps of height h is shown. The magnetization direction within the tip is assumed to be aligned along the tip axis due to shape anisotropy. However, the exist-

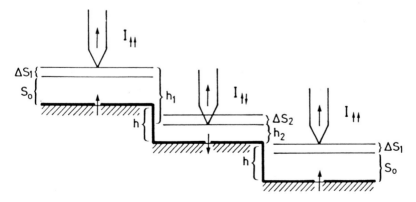

Fig. 1.81. Schematic drawing of a ferromagnetic tip scanning over alternately magnetized terraces separated by monatomic steps of height h. An additional contribution from SP tunneling leads to alternating step heights $h_1 = h + \Delta s_1 + \Delta s_2$ and $h_2 = h - \Delta s_1 - \Delta s_2$ (Wiesendanger *et al.*, 1990c).

ence of closure domains at the front end of the tip could lead to a tip magnetization with a significant component parallel to the sample surface plane. For the alternately magnetized terraces of the Cr (001) surface, only the out-of-plane component of the surface magnetization along the assumed direction of the tip magnetization has been depicted for clarity. A significant in-plane component of the surface magnetization may exist as well. In any case, spin-polarized tunneling experiments mainly sense the surface magnetization component aligned parallel to the tip magnetization direction because of the factor $(\cos \theta)$ in Eq. (1.63) where θ denotes the angle between the two magnetization directions of tip and sample surface.

If the ferromagnetic tip were to be scanned at a constant distance s_0 from the sample surface, the tunneling current would alternate between $I_{\uparrow \uparrow} = I_0(1+P)$ and $I_{\uparrow \downarrow} = I_0(1-P)$, where P denotes the effective polarization of the tunnel junction given by

$$P = \frac{I_{\uparrow\uparrow} - I_{\uparrow\downarrow}}{I_{\uparrow\uparrow} + I_{\uparrow\downarrow}} \qquad (1.131)$$

If the SPSTM experiments are performed at a constant current, the feedback system has to adjust the distance of the tip from the sample surface accordingly, being either $s_1 = s_0 + \Delta s_1$ or $s_2 = s_0 - \Delta s_2$. As a consequence, the measured monatomic step heights alternate between $h_1 = h + \Delta s_1 + \Delta s_2$ and $h_2 = h - \Delta s_1 - \Delta s_2$ (Fig. 1.81). In the low-bias limit, considered by Slonczewski (1989), the following relationship between the effective polarization P and the measurable quantities, $\Delta s = \Delta s_1 + \Delta s_2$ as well as the local tunneling barrier height ϕ, can be found to a first approximation (Wiesendanger *et al.*, 1990c):

$$P = \frac{\exp(A \phi \Delta s) - 1}{\exp(A \phi \Delta s) + 1} \qquad (1.132)$$

with $A = 1.025 \text{ eV}^{-1/2} \text{ Å}^{-1}$.

The SPSTM experiments with the Cr(001) surface directly probed the topological antiferromagnetic order of this surface and allowed derivation of a local effective polarization, whereas conventional spin-resolved surface analytical techniques usually provide information averaged over macroscopic surface areas and therefore average over many terraces which are typically only several hundred Ångström units wide. However, more detailed information about the tip, including the magnetization direction and the degree of spin polarization, is required for quantification of experimental SPSTM results.

The first SPSTM studies of the Cr(001) surface were performed on a nanometer scale. The CrO_2 tips used in these experiments were too blunt, preventing atomic resolution studies. Atomic resolution in SPSTM experiments was demonstrated on a magnetite ($Fe_3O_4(001)$) test surface, where the two different magnetic ions Fe^{2+} and Fe^{3+} on the Fe B-sites in the Fe–O(001) planes could be distinguished by using a sharp Fe sensor tip prepared *in situ* (Wiesendanger *et al.*, 1992b, 1992c). The observation of magnetic contrast at the atomic level was attributed to the different spin configurations being $3d^5 \uparrow 3d \downarrow$ for Fe^{2+} and $3d^5 \uparrow$ for Fe^{3+}. At room temperature, fluctuations of the $3d \downarrow$ electrons among the Fe B-sites are rapid in the bulk, and the electrical conductivity is about $100 \ \Omega^{-1} \ cm^{-1}$, which is relatively high for an oxide. Below the Verwey transition at about 120 K, these charge fluctuations cease and a static periodic arrangement of the Fe^{2+} and Fe^{3+} ions sets in, driven by the Coulomb interaction. Consequently, the electrical conductivity decreases significantly. The Verwey transition can be considered as an example of Wigner crystallization in three dimensions. The SPSTM experiments on the $Fe_3O_4(001)$ surface performed with a Fe sensor tip provided strong evidence for the existence of a Wigner glass state at the (001) surface of magnetite even at room temperature. A static arrangement of the two different Fe ions was observed with short-range order exhibiting a periodicity equal to that present in the bulk below the Verwey transition (Fig. 1.82). The reduced coordination and band narrowing at the surface are likely to increase the order–disorder transition temperature for the $3d \downarrow$ electrons considerably, which would explain the experimental observations.

Similar to the SPSTM studies of the Cr(001) surface, more detailed information about the tip is required for quantification of the experimental results, e.g. to relate the measured corrugation between Fe^{2+} and Fe^{3+} sites with the local magnetization of the tip and the sample surface.

1.14.1.2 SPSTM experiments with external magnetic field

With an external magnetic field applied additionally, the magnetization of the sample can be modulated periodically, for instance, from parallel to antiparallel alignment relative to the tip magnetization direction. Consequently, a portion of the tunneling current is predicted to oscillate at the same frequency, with an amplitude linearly proportional to the average tunneling current. The advantage of this experimental procedure lies in the fact that lock-in detection techniques can be used,

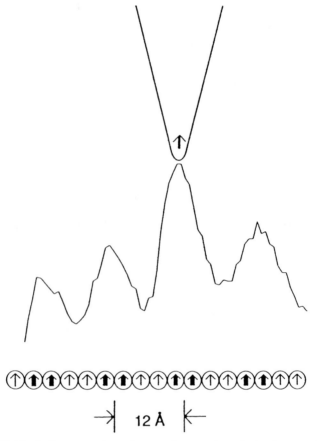

Fig. 1.82. Single line scan along the atomic row of octahedrally coordinated Fe B-sites within the Fe–O(001) plane of magnetite. The observed magnetic contrast on the atomic scale has been attributed to the different spin configurations (i.e. different magnetic moments) of Fe^{2+} and Fe^{3+} (Wiesendanger *et al.*, 1992b, 1992c).

resulting in an improvement of the signal-to-noise ratio. Initial experiments were performed in air by using a single-crystal Ni tip and a permalloy (Ni–Fe alloy) sample which were likely to be in contact (Johnson and Clarke, 1990). Values for the local effective polarization P were derived in stationary experiments without scanning. However, in principle the magnetic field can be modulated at a frequency v_0 well above the cut-off frequency of the feedback loop and the corresponding amplitude of the current oscillation at frequency v_0 can be recorded with a lock-in amplifier simultaneously with the constant current topograph. The spatially resolved lock-in signal then provides a map of the effective spin polarization.

1.14.2 Spin-polarized scanning tunneling spectroscopy (SPSTS)

Thus far, we have considered only the low-bias limit, in which the effective polarization of the tunnel junction can be expressed as a product of the effective polarizations of the two magnetic electrodes (Eq. (1.130)). This is no longer valid for a finite sample bias voltage where spin-resolved density-of-states effects have to be taken into account. A model calculation of the effective polarization assuming a simplified spin-dependent tunneling probability $T(s_1, s_2) = \delta_{s_1, s_2}$, where s_1 and s_2 denote the spin directions in the initial and final states, shows that the effective polarization becomes bias-dependent and can even change its sign as a function of bias (Fig. 1.83), as was found for a CrO_2–vacuum–Cr (001) tunnel junction (Wiesendanger *et al.*, 1991b). If the spin-resolved density-of-states (DOS) of one electrode (e.g. the tip) were known, a measurement of the bias-dependence of the effective polarization could provide information about the spin-resolved DOS of the other electrode (e.g. the sample), leading to a spin-polarized scanning tunneling spectroscopy (SPSTS) method.

Furthermore, distance-dependence of the effective polarization is expected because the decay rates for s, p and d states usually differ, resulting in different contributions to the effective polarization at differ-

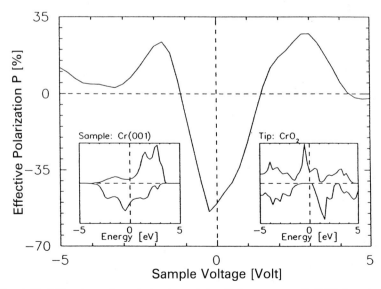

Fig. 1.83. Effective spin polarization of the CrO_2–vacuum–Cr(001) junction as a function of sample bias voltage. The insets show the spin-up (upper part) and spin-down (lower part) density-of-states of CrO_2 and the Cr(001) surface (Wiesendanger *et al.*, 1991b).

ent tip – surface separations. As for conventional tunneling spectroscopy, it is important to realize that STM probes only the tails of the wave functions of those electronic states that have significant overlap with the tip. Therefore, a direct comparison of SPSTM/SPSTS and spin-resolved photoelectron spectroscopy results, for instance, is generally not possible.

1.14.3 Other spin-sensitive STM experiments

Spin-sensitive STM experiments may be performed with non-magnetic tips as well. For instance, the precession of individual paramagnetic spins on a partially oxidized Si(111) surface in a constant magnetic field induces a modulation in the tunneling current at the Larmor frequency (Manassen *et al.*, 1989). The spin precession can influence the tunneling probability, for instance, via dipole–dipole or spin–spin exchange interaction between the localized spin on the paramagnetic center and the delocalized electrons. It was proposed that spin precession does not provide a direct contribution to the tunneling current at the Larmor frequency, but induces a time correlation between tunneling electrons with opposite spins due to interaction of the localized spin with the spins of the delocalized electrons, which would explain the experimental observations (Molotkov, 1992).

The measured radio-frequency signal from the individual paramagnetic spin center was shown to be localized over distances less than 1 nm (Manassen *et al.*, 1989). In contrast, electron spin resonance (ESR) measures the precession of an induced macroscopic magnetization in an external magnetic field which is a result of the alignment of many (at least 10^{10}) electron spins. Valuable spectroscopic information about the properties of individual spins is lost as a result of inhomogeneous line broadening.

1.15 Inelastic electron tunneling (IET) in STM

In section 1.4 we discussed the fact that, besides the dominant elastic electron tunneling process, for which the electron energy is equal in the initial and final states, inelastic tunneling can occur if the tunneling electrons couple to some modes ω in the tunnel junction. The tunneling current then exhibits an additional dependence on these modes ω:

$$I = I(s, \phi, U, \omega) \qquad (1.133)$$

A major drawback of inelastic electron tunneling spectroscopy (IETS) with planar metal–oxide–metal tunnel junctions has been the total lack of microscopic characterization of the junction region. Replacing the oxide layer by vacuum and the top planar metal electrode by a sharp metal tip, capable of locally probing preselected, well-characterized surface sites and additionally providing a map of the spatial variations in the coupling of the tunneling electrons to the modes ω by recording IET spectra during raster scanning of the tip over the sample surface, would be the ultimate goal for IETS.

Unfortunately, the changes in tunneling conductance $\Delta\sigma/\sigma$ as a result of the opening of additional inelastic tunneling channels are typically rather small, being 0.1–1% for planar oxide tunnel junctions (section 1.4). For the STM, conductance changes of 1–10% were theoretically predicted (Binnig *et al.*, 1985b; Persson and Demuth, 1986; Persson and Baratoff, 1987; Baratoff and Persson, 1988; Persson, 1988). The values for $\Delta\sigma/\sigma$ are generally larger for IET through a vacuum barrier than with an oxide barrier because of the absence of screening by the dielectric constant of the oxide. The smaller values of about 1% were obtained by assuming a long-range dipole coupling between the tunneling electrons and the modes ω, which might not be justified because of the close proximity of the STM tip to the sample surface of only a few Ångström units. By assuming a short-range *resonant* coupling, a conductance *decrease* of up to 10% was predicted at the onset of inelastic tunneling.

In any case, the relative stability of the tunneling current has to be better than 1% to obtain reasonable IET spectra with the STM. In addition, low temperatures are required to keep thermal line-width broadening in the IET spectra of the order of $k_B T$ (section 1.4) small compared with the energy $\hbar\omega$ of the modes ω being typically a few millielectronvolts.

1.15.1 Phonon spectroscopy by IET

Electrons tunneling between tip and sample can create phonons at the interface between the conductor and the tunneling barrier. The emission of phonons is believed to take place within a few atomic layers of the interface. Therefore, an atomically well-defined interface is expected to be important, motivating IET experiments with the STM.

The opening of additional inelastic tunneling channels usually leads to a step-like increase in conductivity dI/dU or, equivalently, to peaks in the second derivative d^2I/dU^2 as a function of bias (section 1.4). Low-temperature experiments performed with a rigid STM using a tungsten tip and a graphite sample indeed revealed a spectrum of peaks in d^2I/dU^2–U characteristics where the positions of the peaks corresponded closely to the energies of the phonons of the graphite sample and the tungsten tip (Smith *et al.*, 1986a). The measured increase in conductance at the phonon energies was of the order of 5%. By analogy with elastic scanning tunneling spectroscopy (section 1.13), spectroscopic imaging in IET can be performed by recording d^2I/dU^2 at a particular phonon energy while scanning the tip over the sample surface. This method allows one to map spatial variations of the phonon spectra, caused by spatial variations in the coupling between the tunneling electrons and the phonons, on the atomic scale (Smith *et al.*, 1986a).

1.15.2 Molecular vibrational spectroscopy by IET

As discussed in section 1.13.6, it is generally impossible to identify chemical species by elastic tunneling spectroscopy. In contrast, IETS can yield vibrational spectra of molecules adsorbed on a surface exhibiting features characteristic of particular molecular functional groups (section 1.4). By using low-temperature STM, a vibrational spectrum of an individual adsorbed molecule can be obtained after positioning the tip over the preselected adsorbate. It is even possible to form a map showing the sites within a molecule where particular resonances occur.

For sorbic acid adsorbed on graphite, a spectrum of strong peaks was observed in the first derivative dI/dU instead of the expected second derivative d^2I/dU^2 (Smith *et al.*, 1987b). The energies of the peaks corresponded approximately to the vibrational modes of the molecule. The measured increase in conductivity at the molecular resonances was as much as a factor of ten, which is at least two orders of magnitude larger than expected.

IET spectra were also obtained by using crossed electrodes of very fine gold wire as an alternative adjustable microscopic tunnel junction (Gregory, 1990). For hydrocarbons, the measured increase in conductivity at the C–H stretch peak was about 10%. However, spatially resolved information cannot be gained by using this type of junction.

1.15.3 Photon emission by IET

Inelastic tunneling events which are normally buried in the dominant elastic tunnel current background can also be studied indirectly via radiative decay processes (Gimzewski *et al.*, 1988, 1989; Coombs *et al.*, 1988b; Berndt *et al.*, 1991a, 1991b). The tunneling electrons cannot directly generate photons because energy and momentum conservation laws forbid such processes. However, they can excite surface plasmons localized in the region close to the tip. Owing to the conservation of parallel momentum, surface plasmons on perfectly smooth surfaces could not radiate. If, however, the translational invariance along the surface is broken by roughness or irregularities, radiative decoupling of the energy stored in the surface plasmons becomes possible. For planar tunnel junctions, the surfaces and interfaces therefore must be rough as mentioned in section 1.4.2.

In the case of STM, the presence of the tip already breaks the translational invariance and leads to considerable enhancement of the rate of light emission for frequencies near the eigenfrequency of the localized interface plasmon mode which is sensitive to the radius of curvature of the tip, being larger for a smaller radius (Johansson *et al.*, 1990; Johansson and Monreal, 1991; Smolyaninov *et al.*, 1991). This enhancement results from an amplification of the electromagnetic field in the region below the tip due to the localized interface plasmon mode. The photon creation efficiency is typically of the order of 10^{-4}–10^{-3} photons per tunneling electron which is high enough to allow detection of the emitted photons by a photon detector placed near the tip–sample region of a STM operated at a constant tunneling current of 1 nA–1 μA, as illustrated in Fig. 1.84. Optical frequency light emitted when the STM is operated in the low-bias tunneling mode, i.e. $eU < \phi$, can be detected by means of a photomultiplier. The photon detector is usually placed over a viewport of the UHV chamber resulting in a photon acceptance angle of approximately 0.1 sterad. Typical measured count rates are 50 cps μA^{-1}.

To increase the photon detection efficiency it was proposed to use one metal-covered end of a quartz fiber as STM tip while the other end

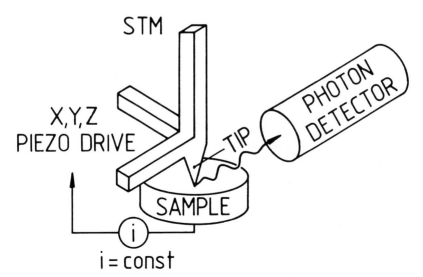

Fig. 1.84. Schematic drawing of the experimental arrangement used for STM stimulated photon emission (Coombs *et al.*, 1988b).

of the fiber cable is glued to the photomultiplier (Smolyaninov *et al.*, 1990). Photon spectra were obtained by means of a fiber optics cable attached to the entrance slit of a spectrometer equipped with a holographic grating and an optical multichannel analyzer (Gimzewski *et al.*, 1989). The high-energy threshold of the spectral distribution was found to equal the sample bias voltage as expected. A Geiger–Müller-type photon detector is used for the detection of ultraviolet photons emitted from the tip–sample region if the STM is operated in the field emission mode, i.e. $eU > \phi$ (section 1.20).

The emitted light can experimentally be investigated using various spectroscopic methods (Gimzewski *et al.*, 1988).

(a) In isochromat spectroscopy the photon energy is kept constant while the sample bias voltage is scanned (Fig. 1.85(a)).
(b) In fluorescence spectroscopy the sample bias voltage is held fixed while the spectral distribution of the emitted photons is measured using a spectrometer (Fig. 1.85(b)). The local geometry of STM

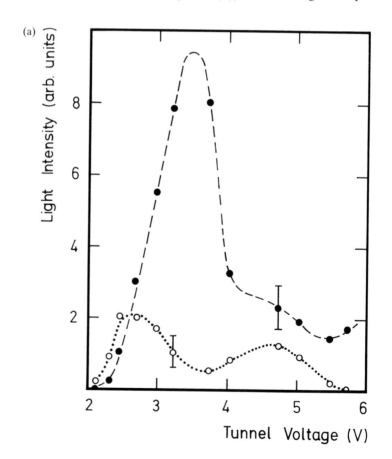

(b)

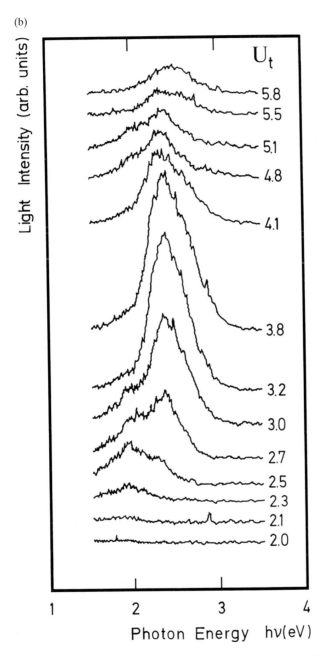

Fig. 1.85. (*a*) Photon intensity of two resonance structures at $h\nu = 1.9$ eV (open circles) and $h\nu = 2.4$ eV (filled circles) as a function of tunnel voltage for a polycrystalline silver film. Dotted lines are guides for the eye only. (*b*) Optical spectra recorded at constant tunnel current at a series of tunnel voltages as indicated. Note maxima at $h\nu = 1.9$ and 2.4 eV (Gimzewski *et al.*, 1989).

might allow one to obtain a photon spectrum even of an individual adsorbed molecule providing chemical-specific information (Smolyaninov *et al.*, 1990).

(c) In spatially resolved photon yield mapping a selected feature in the photon spectrum can be used to generate a map of the intensity of emitted photons at a particular photon energy as a function of the spatial (*x,y*) coordinates, leading to a scanning tunneling optical microscopy (STOM) with high lateral resolution at the nanometer or even subnanometer level (Coombs *et al.*, 1988b). In principle, STOM is not limited to a UHV environment, but can be envisaged equally well in air and in liquids. Finally, STOM can be extended by exploiting the emitted electromagnetic radiation from the infrared to soft x-ray regime if the STM is operated with a correspondingly wide sample bias voltage range.

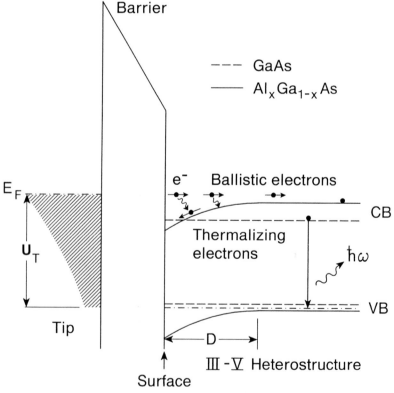

Fig. 1.86. Model of the tunnel injection process of electrons on GaAs and AlGaAs. The strong band bending on AlGaAs induces quantum mechanical electron reflection and trapping near the surface, thus reducing the fraction of electrons injected into the bulk. *D* is the width of the depletion zone (Alvarado *et al.*, 1991).

1.16 Tunneling-induced luminescence microscopy and spectroscopy

Besides light emission from a STM tunnel junction as a result of inelastic electron tunneling processes (section 1.15.3), photons can also be emitted due to electron–hole recombination in semiconductors which can be studied by tunneling-induced luminescence microscopy and spectroscopy (Abraham *et al.*, 1990; Alvarado *et al.*, 1991). For GaAs–AlGaAs heterostructures, the observed luminescence was explained in terms of a three-step model (Fig. 1.86):

1. injection of electrons by tunneling from the tip into the sample surface;
2. transport of electrons from the surface into the bulk of the material including ballistic transport, diffusion, thermalization of hot electrons to the bottom of the conduction band or into an acceptor level, as well as trapping; and finally
3. radiative recombination.

The significant difference between the luminescence intensities for GaAs and AlGaAs, as found experimentally (Fig. 1.87), was attributed

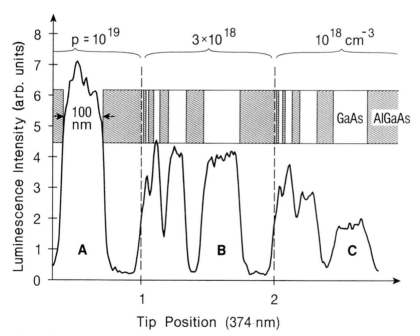

Fig. 1.87. Luminescence intensity profile of a multiple quantum well structure. The inset shows the related structure. A, B, C denote 100 nm wells with doping levels of $p = 10^{19}$, 3×10^{18} and 10^{18} cm^{-3}, respectively. The tunneling voltage is $U = 2.5$ V (Alvarado *et al.*, 1991).

to strong band bending at the surface of AlGaAs in contrast to the GaAs surface (Fig. 1.86). This strong band bending can induce quantum mechanical electron reflection and subsequent confinement of the electrons in the depletion zone where the probability for direct radiative recombination is very low because of the reduced number of holes available. Consequently, the luminescence intensity signal decreases if the tip is positioned above AlGaAs regions. Trapping of electrons by contaminants or defects at the surface can reduce the fraction of electrons injected into the bulk by tunneling as well. Therefore, reduction of luminescence intensity is also observed above defective surface regions.

Luminescence maps of a multiple quantum well GaAs–AlGaAs heterostructure, obtained by measuring the recombination radiation at a particular photon energy by means of a photomultiplier tube while the tip is raster scanned over the sample surface, showed that spatial resolution down to the nanometer level can be achieved. Tunneling-induced luminescence microscopy and spectroscopy can therefore be used to study spatial variations in the recombination rate at high lateral resolution. In addition, the energy of bulk bands as well as transport parameters, e.g. the thermalization and diffusion length of minority electrons, can be determined.

Finally, by measuring the circular polarization of the recombination luminescence light, it is possible to extract information about the spontaneous polarization of the conduction electrons in GaAs associated with the energy-dependent spin splitting of the conduction band (Alvarado and Renaud, 1992). If the electrons have tunneled from a ferromagnetic tip into the conduction band of GaAs, information about the initial polarization of these electrons can be deduced as well.

1.17 STM with laser excitation

Instead of studying the electromagnetic radiation emitted by a STM (section 1.15.3 and 1.16), one can also investigate the interaction of incident laser radiation with the tunnel junction. Several mechanisms may influence the tunneling current and bias as a result of this interaction.

1.17.1 Thermal effects

The illumination of the tip–sample region of a STM by a focused laser beam causes heating of both tip and sample inducing thermal expansion (Amer *et al.*, 1986; Grafström *et al.*, 1991). The photothermal displacement effect, which is more noticeable for the tip being more efficiently heated than the sample, can directly influence the tunneling current via a change in the tip–sample separation. In addition, thermoelectric EMFs, arising from differential heating of the tunnel junction by the absorbed radiation, can contribute to the tunneling bias, leading to additional thermocurrents. Since the tip is more effectively heated than the sample, a thermoelectric current flows even if the tip and sample materials are identical. To minimize thermal effects, the power density of the incident laser radiation should be kept as low as possible.

1.17.2 Rectification, frequency mixing and laser-driven STM

Point-contact metal–insulator–metal (MIM) diodes have been used for a long time to rectify RF signals and to generate harmonics and difference frequencies of injected microwave and laser radiation for high-precision frequency measurements. However, since an oxide layer has usually been used as insulator for MIM diodes, the point contacts are not well-defined and repeatable. Using a STM instead of a MIM diode offers several advantages related to replacement of the oxide layer by a vacuum gap. The vacuum gap is better defined and, equally important, can be precisely controlled by adjusting the tip–sample separation.

A STM tunnel junction can act as a diode as a result of nonlinearities in the I–U characteristic (Cutler *et al.*, 1987; Krieger *et al.*, 1990). Laser radiation is efficiently coupled into this diode by means of the STM tip which represents a long-wire antenna for the electromagnetic radiation. The oscillating antenna currents induce an a.c. voltage U_i across the tunnel junction in addition to the externally applied bias voltage U_e. The nonlinearity in the I–U characteristic of the STM tunnel junction is essential for generating new frequency components of the current, e.g. at zero frequency (rectification) and at the difference frequency of two simultaneously injected laser signals. The tip finally serves again as an antenna for transmitting a signal at the difference frequency.

Laser frequency mixing using a STM was demonstrated in a wide frequency range, from megahertz (Arnold *et al.*, 1987, 1988) to gigahertz (Krieger *et al.*, 1990; Völcker *et al.*, 1991a, 1991b), and even up to terahertz (Krieger *et al.*, 1989). For the megahertz range, the mixing signals were detected by analyzing the frequency spectrum of the current (Fig. 1.88(a)), whereas for the gigahertz range a pyramidal microwave horn mounted directly above the STM as part of a microwave circuit was used for detecting the mixing signals emitted by the tip (Fig. 1.88(b)). It was found that the intensity of the mixing signal increases with laser power and with decreasing tip–sample separation. By measuring the mixing-signal intensity while the STM tip is raster scanned over

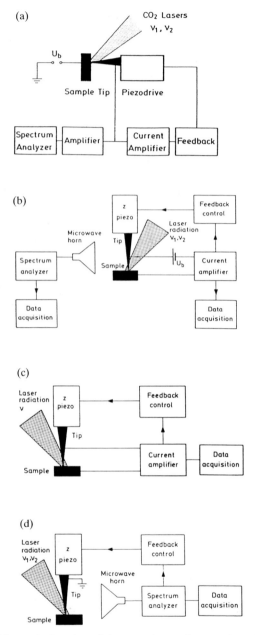

Fig. 1.88. (*a*) Schematic drawing of the experimental set-up for mixing two laser frequencies in the tunnel junction of an STM (Arnold *et al.*, 1987). (*b*) Schematic drawing of the experimental set-up for generation and detection of laser difference frequencies in the tunnel junction with the STM as part of a microwave circuit. (*c*) Schematic drawing of a laser-driven STM whose tip–sample distance is controlled by the rectified current. Note the absence of a bias voltage source in the tunneling circuit. (*d*) Schematic drawing of a laser-driven STM whose tip–sample distance is controlled by the difference-frequency signal of two injected laser signals. Note the absence of any electrical connection to the sample (Völcker *et al.*, 1991b).

the sample surface, spatial variations of the difference-frequency signal intensity on the atomic scale were recorded (Völcker *et al.*, 1991b).

A rectified signal, i.e. a d.c. current at zero external bias voltage, was also detected when the STM was illuminated with laser radiation. This rectified signal can be used to control the tip–sample separation even in the absence of an external bias voltage source (Fig. 1.88(*c*)). Spatially resolved measurements with atomic resolution were demonstrated with a laser-driven STM (LDSTM) whose tip–sample separation was controlled by the rectified current (Völcker *et al.*, 1991b).

By using the emitted radiation in laser frequency mixing experiments for tip–sample distance control and imaging, it is necessary neither to have an external bias voltage source nor a d.c. current, allowing removal of all electrical connections to the sample (Fig. 1.88(*d*)). Atomic resolution was indeed achieved in spatially resolved measurements of the difference-frequency signal intensity without any electrical connection to the sample (Völcker *et al.*, 1991b). This mode of operation of the laser-driven STM also allows study of insulators similar to nonlinear alternating-current tunneling microscopy (section 1.11.6).

We now focus on interpretation of the difference-frequency signal. The total tunneling current I can be expressed as a power series in terms of the laser-induced a.c. voltage U_i:

$$I = I\,(U_e) + \left(\frac{\partial I}{\partial U}\right) U_i + \frac{1}{2}\left(\frac{\partial^2 I}{\partial U^2}\right) U_i^2 + \dots \qquad (1.134)$$

where U_e is the externally applied bias voltage, and the derivatives are evaluated at U_e. By writing

$$U_i = U_i^0 \cdot \cos(\omega_i t) \qquad i = 1, 2 \qquad (1.135)$$

with $\omega_i(i=1,2)$ being the two frequencies of the incident laser radiation, it is seen that both the rectified current and the current $I_{\Delta\omega}$ at the difference frequency $\Delta\omega = \omega_1 - \omega_2$ are proportional to the second derivative (d^2I/dU^2) at the externally applied bias voltage (Krieger *et al.*, 1990). Since the radiated power $P_{\Delta\omega}$ at the difference frequency grows as $I^2_{\Delta\omega}$, we finally obtain

$$P_{\Delta\omega}(U_e) \sim \left(\frac{d^2 I}{dU^2}\right)^2 P_{\omega_1} P_{\omega_2} \qquad (1.136)$$

where P_{ω_1} and P_{ω_2} denote the laser power at frequencies ω_1 and ω_2. Spatially resolved measurements of the difference-frequency signal intensity therefore reflect the spatial variations of the second derivative of the I–U characteristic. An important application of the laser-driven STM lies in spectroscopic studies of adsorbate-covered surfaces. Adsorbate-specific resonances can first be excited and subsequently detected by atomically resolved measurements of $P_{\Delta\omega}$ or, equivalently, of d^2I/dU^2. A fingerprint for molecular functional groups can therefore be obtained similarly to inelastic electron tunneling spectroscopy (section 1.15.2).

The frequency mixing experiments up to the terahertz range further show that the STM responds to considerably higher frequencies than usually exploited in STM imaging either in the constant-current mode (section 1.11.1) or in the variable-current mode of operation (section 1.11.2) where atomic data rates are typically below 1 MHz. The fundamental limit of the frequency response of the STM is determined by the tunneling time of the electrons (section 1.2.9) and the RC time constant of the tunnel junction. The junction resistance typically varies between 1 MΩ and 10 GΩ whereas the capacitance of the microscopic tunnel junction can be estimated to be about 10^{-18} F. However, the capacitance of the leads, for instance, is appreciably higher and therefore limits the frequency response of the tunneling current in a measuring electronic circuit in practice (Arnold *et al.*, 1988). The extremely small response time of STM as a nonlinear device of microscopic dimensions can be used for ultrafast time resolution experiments (Hamers and Cahill, 1990, 1991).

1.17.3 Generation and detection of surface plasmons

Besides excitation of surface plasmons by inelastic electron tunneling (section 1.15.3), plasmons at metal surfaces can also be generated by means of an incident laser beam (Fig. 1.89) and subsequently detected by their influence on the tunneling current (Möller *et al.*, 1991a; Kroó *et al.*, 1991). The excitation of surface plasmons was shown to be an efficient method of coupling laser radiation into the tunnel junction of a STM leading to an increase of current flow between tip and sample. This current rise can either result from a rectification of the a.c. surface plasmon field (Möller *et al.*, 1991a) or from a thermal effect (Kroó *et al.*, 1991). The decay of surface plasmons by electron excitation or photon emission causes local heating which can induce a change of the tunneling current by thermal expansion or by generation of a thermoelectric current (section 1.17.1).

The spatial distribution of the intensity of the surface plasmon-induced current signal revealed variations on a nanometer scale (Möller

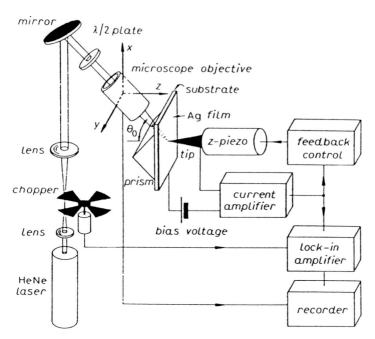

Fig. 1.89. Schematic drawing of the experimental set-up for the decay-length measurement of surface plasmons using an STM (Kroó *et al.*, 1991).

et al., 1991a). Large-area scans were used to determine the decay length of surface plasmons by measuring the dependence of the laser-induced signal on the lateral distance between the tip and the laser focal spot (Kroó *et al.*, 1991).

1.17.4 Photovoltaic effects and photoassisted tunneling spectroscopy

If an appropriate semiconductor is chosen as sample in a laser-excited STM, a surface photovoltage may result from the photoexcited electron–hole pairs. The absorption of incident laser radiation with an energy at or above the semiconductor band gap creates electron–hole pairs within the absorption depth. These photoexcited electron–hole pairs subsequently become separated by the electric field of the semiconductor space-charge layer in a near-surface region (Fig. 1.90). The spatial separation of charge carriers finally generates a voltage differ-ence between the surface and the bulk which is denoted as surface photovoltage (SPV). As a consequence of this SPV effect, a current is observed at zero applied sample bias (Fig. 1.91). This zero-bias current

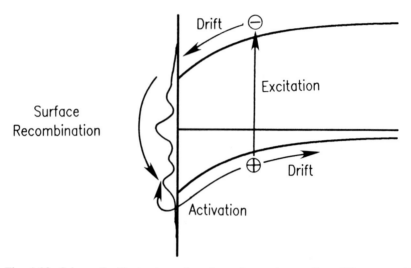

Fig. 1.90. Schematic illustration of surface photovoltage effect (Hamers and Markert, 1990b).

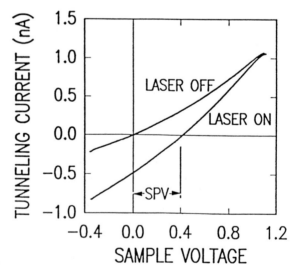

Fig. 1.91. Tunneling *I–U* curves measured on clean p-type Si(111) (7×7) in the absence and presence of optical illumination. The stabilization voltage is +1.2 V (Hamers and Markert, 1990b).

is negative for p-type semiconductors as a result of downward surface band bending causing photoelectrons to accumulate at the surface (Fig. 1.90). If a positive bias equal to the SPV is externally applied, the tunneling current can be made zero. By recording the bias necessary to counter the tunneling current as a function of tip location, a spatially resolved map of the SPV is obtained. In contrast, conventional photovoltage measurements by means of Kelvin probes or photoemission spectroscopy average over macroscopic surface areas. The sign of the SPV changes for n-type semiconductors and can therefore be used to determine the kind of doping.

Different experimental methods were applied to measure the spatial variations of the SPV, including feedback gating techniques (Hamers and Markert, 1990a, 1990b; Kuk *et al.*, 1990a, 1991), a double modulation technique (Cahill and Hamers, 1991a, 1991b), as well as a numerical fit technique (Kochanski and Bell, 1992). Upon elimination of error sources in the SPV measurements, it was found that the photovoltages are uniform within 10 mV on a 100-Å length scale, despite the presence of adsorbates and defects (Cahill and Hamers, 1991b; Kochanski and Bell, 1992).

Apart from SPV measurements, the photoassisted tunneling current at a particular sample bias voltage can be determined as a function of the wavelength of the incident laser radiation leading to a photoassisted tunneling spectroscopy (PTS) method for semiconductors and semiconductor quantum well structures (Akari *et al.*, 1991; Qian and Wessels, 1991a, 1991b).

1.18 Scanning tunneling potentiometry (STP)

Another mode of STM operation, denoted as scanning tunneling potentiometry (STP), has been introduced to study the potential distribution on a sample surface simultaneously with its topography (Muralt and Pohl, 1986; Muralt *et al.* 1986). In STP, two electrodes are attached to the sample (Fig. 1.92) and a potential difference $\Delta U = U_2 - U_1$ is applied across the sample (in the x direction). Consequently, a potential distribution $U(x,U_1,U_2)$ is built up. If the STM were operated with a d.c. bias for the tunneling current as usual, the tunnel voltage would become position-dependent. In the constant-current mode it would then be impossible to distinguish between topographic surface features and features resulting from variations of the tip–surface separation in response to spatial variations of tunneling voltage.

A separation of surface topography and spatial potential variations can, however, be achieved if an a.c. tunneling voltage $U_T = U_T^\circ \sin(\omega t)$

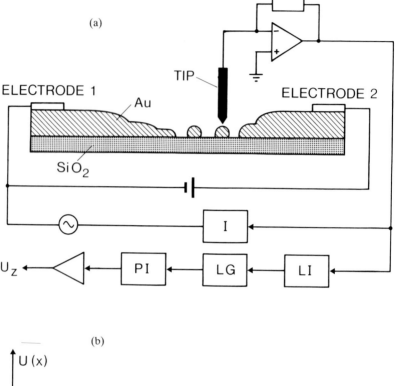

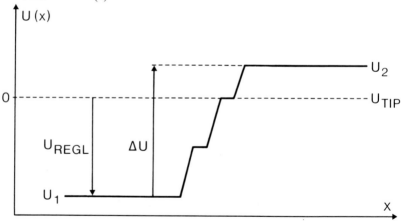

Fig. 1.92. (*a*) Schematic diagram of sample and feedback circuitry for scanning tunneling potentiometry. LI: lock-in amplifier, LG: logarithmic amplifier, PI: control circuitry. U_z is the voltage applied to the z piezoelectric element. (*b*) Schematic diagram of potential distribution. The local potential $U(x)$ at the site of the tip is shifted to ground potential by potential regulation (Muralt *et al.*, 1986).

is used for tip–sample distance control and applied to both sample electrodes in addition to the d.c. voltage ΔU between the two electrodes. The d.c. part of the tunneling current is regulated to zero by shifting the sample potential by the amount U_{Regl}. Therefore, U_{Regl} is a direct measure of the local potential at the tip site. The two sample electrode potentials are

$$U_1 = U_{\text{Regl}} + U_{\text{T}}^{\circ} \sin(\omega t)$$

$$U_2 = U_{\text{Regl}} + U_{\text{T}}^{\circ} \sin(\omega t) + \Delta U \qquad (1.137)$$

Since no steady current between tip and sample is needed for measurement of the potential U_{Regl}, the STP method works similarly to an impedance bridge, with the advantage of zero-current detection. Consequently, potentiometric measurements can be performed on highly resistive samples as well.

In Fig. 1.92 the STP control system is shown schematically. The distance is regulated for constant a.c. component of the tunneling current by means of a lock-in amplifier (LI) in series with a logarithmic amplifier (LG) and control circuitry (PI), providing the topographic signal. The local potential is adjusted for vanishing d.c. component of the tunneling current with the help of an integrator I defining the second feedback circuit. Its output is equal to U_{Regl} and provides the potentiometric signal which is obtained simultaneously with the topographic signal. Typical voltage noise levels have been of the order of 10–100 mV using this STP measurement technique.

Alternatively, the topographic and potentiometric signals can be measured sequentially by means of a 'gated' feedback circuit (sections 1.10.4 and 1.13.3) as illustrated in Fig. 1.93 (Kirtley *et al.*, 1988a, 1988b). Voltage noise levels as low as 10 μV can be achieved by using this method. A further improved low-noise STP control system can attain even sub-microvolt sensitivity (Pelz and Koch, 1989). Finally, a scanning noise potentiometry (SNP), derived from scanning noise microscopy (section 1.11.5), avoiding any external voltages across the sample, was introduced with a sensitivity in the microvolt range (Möller *et al.*, 1991b).

STP has found several applications.

The resistance of a metal is believed to result from a series of small voltage drops occurring when the carriers collide with imperfections in the lattice potential. STP offers a unique method for direct experimental confirmation and spatial mapping of the assumed microscopic potential fluctuations and can therefore contribute substantially to our micro-

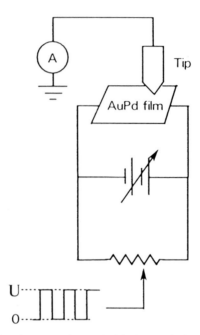

Fig. 1.93. Schematic diagram of sample-biasing technique used for scanning tunneling potentiometry (Kirtley *et al.*, 1988a).

scopic understanding of local electron transport properties. For instance, polycrystalline metal films show abrupt steps in surface potential due to scattering from grain boundaries in these films (Kirtley *et al.*, 1988a, 1988b). Similar potential steps were found in STP studies of granular films of high-temperature superconductors (Kent *et al.*, 1989).

STP can be applied to semiconductors as well. In this case, a rectifier *I–U* characteristic exists and a pure a.c. tunnel voltage leads to a d.c. component of the tunneling current. As a result, the type of conductivity (n- or p-type) can easily be determined (Muralt and Pohl, 1986). STP can also be used to map the spatial extent of the space-charge as well as the electron–hole recombination region in semiconductor p–n junctions and heterojunctions with nanometer resolution. For instance, it was shown how the thickness of the recombination layer changes as a function of the applied bias (Muralt, 1986; Muralt *et al.*, 1987). Spatial variations in the surface photovoltage resulting from spatial variations in the local recombination rate (section 1.17.4) were also studied using STP (Hamers and Markert, 1990a,b; Kuk *et al.*, 1990a).

As for all scanning probe microscopy methods, one has to be aware of possible tip-related artifacts in STP (Pelz and Koch, 1990a; Rohrer, 1990). On rough sample surfaces, such as granular thin films, a discon-

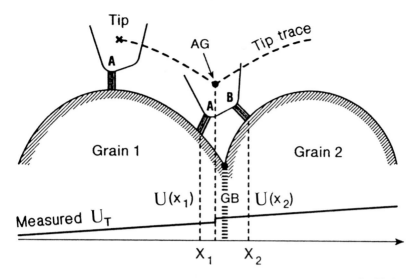

Fig. 1.94. Apparent voltage drop at an apparent grain boundary (AG) in a current-carrying granular structure. The true grain boundary is at GB with no voltage drop across it (Rohrer, 1990).

tinuity in the potential distribution can be mimicked by switching the tunneling location from one grain to another (Fig. 1.94). The artificially abrupt jumps in the measured potential distribution generally occur at topographic valleys which are usually identified as grain boundaries. The tip artifacts in STP images are often even more pronounced than in the corresponding topographic STM images and have therefore been suggested as a valuable means to detect the presence of tip artifacts in STM experiments (Pelz and Koch, 1990a).

1.19 Ballistic electron emission microscopy (BEEM) and spectroscopy (BEES)

As discussed in section 1.13, STM and tunneling spectroscopy probe the tails of the surface wave functions protruding into the vacuum region. Subsurface information can only be gained in those cases where an influence of subsurface properties on the tails of the surface wave functions exists. Ballistic electron emission microscopy (BEEM) and spectroscopy (BEES) as STM-related methods have been introduced to allow spatially resolved investigations of buried interfaces which are located far below the surface (Kaiser and Bell, 1988).

Metal–semiconductor interfaces have most widely been studied by the BEEM technique since its invention. Similarly to STP, BEEM is based on a three-terminal configuration (Fig. 1.95). An STM tip is

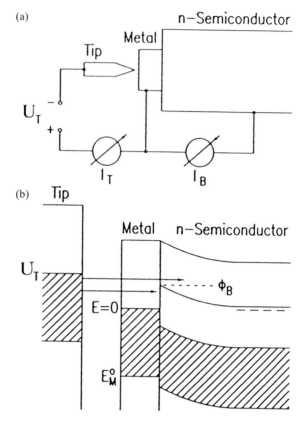

Fig. 1.95. Schematic view (*a*) and energy diagram (*b*) of the BEEM experiment. Electrons tunnel from the tip into the thin metal film, resulting in the tip current I_T. Those which reach the metal–semiconductor interface ballistically may cross it if their energy exceeds the height of the Schottky barrier ϕ_B to form the BEEM current I_B (Prietsch and Ludeke, 1991b).

positioned close to the surface of a metal–semiconductor heterojunction. Electron tunneling from the tip to the metal base electrode leads to injection of ballistic electrons into the base electrode which is typically about 100 Å thick. Since the injected ballistic electrons typically have attenuation lengths greater than 100 Å, they may propagate through the metal layer and thus allow probing of the subsurface metal–semiconductor interface. Experimentally, the collector current I_c flowing between the metal base electrode and the semiconductor collector electrode is measured as a function of the applied tunnel bias voltage between the STM tip and the metal base electrode (Fig. 1.96). For a

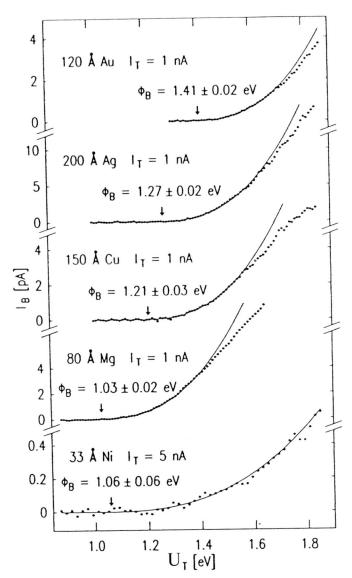

Fig. 1.96. Representative BEEM spectra of various metal/GaP(110) interfaces, taken with constant tunnel current I_T at fixed tip positions (dots). The lines represent fit curves, resulting in the given Schottky barrier heights (Prietsch and Ludeke, 1991b).

tip–base tunnel bias voltage U less than the base–collector barrier height U_B, where eU_B is the Schottky barrier height of the metal–semiconductor interface as introduced in section 1.6, the ballistic electrons cannot enter the collector electrode and I_c remains zero (Fig. 1.96). However, as the tip–base tunnel bias U is increased above U_B, ballistic electrons are allowed to enter the collector electrode and the base–collector current I_c increases considerably. The measured spectrum $I_c(U)$ provides information about the ballistic-electron transport properties of the base film, the interface electronic structure, including the Schottky barrier height, and the quantum-mechanical reflection of electrons at the interface.

As a first approximation, the measured BEEM spectrum $I_c(U)$ can be fitted by a simple one-dimensional theory (Kaiser and Bell, 1988). Assuming that the ballistic-electron mean free path is independent of the energy E, the density of states in the tip and the metal base electrode is constant over the energy range to be considered, and the energy-dependent transmission probability of ballistic electrons arriving at the interface with a barrier height eU_B can be approximated by a step function θ, the following expression for the collector current I_c as a function of tunnel bias voltage U at a constant tunneling current I_T can be found (Kaiser and Bell, 1988):

$$I_c(U) = \alpha I_T \int dE \left[f(E) - f(E+eU) \right] \cdot \theta \left[E - (E_F - eU + eU_B) \right] \qquad (1.138)$$

where $f(E)$ is the Fermi function. The quantity α is a measure of ballistic-electron attenuation due to scattering in the metal base layer and is bias-independent. A fit of the measured spectra based on Eq. (1.138) with the parameters α and U_B is quite satisfactory (Fig. 1.96) and yields a value for the Schottky barrier height eU_B. Owing to the localized injection of ballistic electrons by the STM tip, BEEM allows determination of the Schottky barrier height locally whereas other experimental techniques, such as photoemission, photoresponse and current–voltage measurements, only provide values for the Schottky barrier height averaged over a large interface region due to their lack of spatial resolution. BEEM even permits direct spatially resolved mapping of heterostructure interface properties by measuring the spatial variation in the collector current I_c at fixed tunnel bias voltage U and tunneling current I_T during scanning of the STM tip over the heterostructure surface. Simultaneous acquisition of BEEM and constant current STM images can be performed.

While the high spatial resolution of STM becomes plausible by considering the close proximity of the STM tip and the sample surface in addition to the strong distance-dependence of the tunneling current, a high spatial resolution of BEEM might not be expected because the ballistic electrons first have to propagate through a metal base layer 100 Å thick before arriving at the metal–semiconductor interface. However, the conservation of electron transverse momentum across the interface in the absence of scattering facilitates high spatial resolution in BEEM images as well. An electron with large transverse momentum $k_{||}$ cannot enter the semiconductor because there are no states available satisfying the $E(k_{||})$ relationship for the electron. Therefore, a critical angle φ_c for electron propagation in the metal base electrode exists outside of which electrons are reflected back at the metal–semiconductor interface and do not contribute to the collector current I_c. This critical angle φ_c is given by (Bell and Kaiser, 1988):

$$\sin^2 \varphi_c = \frac{m_{||}^*}{m} \cdot \frac{U - U_B}{U + E_F} \qquad (1.139)$$

where $m^*_{||}$ denotes the electron effective mass parallel to the interface within the semiconductor whereas m denotes the free-electron mass. An important consequence of transverse-momentum conservation at the interface and the existence of a critical angle φ_c is a focusing effect which provides high spatial resolution of interface properties, typically of the order of 10 Å for a metal base layer 100 Å thick (Fig. 1.97).

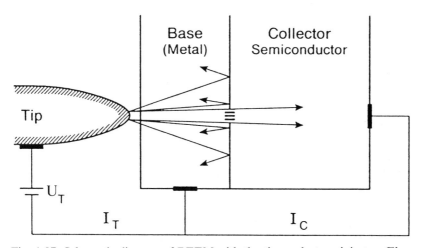

Fig. 1.97. Schematic diagram of BEEM with the tip as electron injector. Electrons reaching the base–collector (metal–semiconductor) interface ballistically can be reflected because their energy is insufficient or because of perpendicular momentum mismatch (Rohrer, 1990).

Since only electrons with small transverse momenta in the metal base electrode can be collected, scattering in the metal layer mainly leads to fewer electrons being collected, rather than to a decrease of the lateral resolution.

An important extension of the BEEM method is offered by varying the kind of doping of the semiconductor collector electrode. An n-type semiconductor serves as a collector for ballistic electrons, as illustrated in Fig. 1.95, while a p-type semiconductor will collect holes leading to a ballistic-hole spectroscopy of interfaces (Bell *et al.*, 1990; Hecht *et al.*, 1990). For the case of a p-type semiconductor collector and a negative metal base bias (positive tip bias), electrons tunneling from the metal base electrode into the tip leave holes behind, which may propagate ballistically through the base electrode to the subsurface metal–semiconductor interface. Transmission through the interface can occur if the hole energy, measured with respect to the base conduction-band minimum, is less than the threshold defined by the valence-band maximum. By measuring $I_c(U)$ spectra, it is therefore possible to probe the valence-band structure of subsurface metal–semiconductor interfaces. A positive metal base bias (negative tip bias) will result in injection of 'hot' electrons into the metal base layer. Some of these hot electrons undergo inelastic energy-loss processes via scattering with equilibrium conduction-band electrons in the base electrode. As a result of energy and momentum transfer from the incident hot electrons to equilibrium electrons, the latter are excited to energies above the Fermi level, leaving a distribution of holes in the base layer. These holes may propagate ballistically to the metal–semiconductor interface and may enter the valence band of the semiconductor under conservation of transverse momentum. Since the holes have been generated in electron–electron scattering events in this case, ballistic hole spectroscopy can provide valuable information about these scattering processes as well (Bell *et al.*, 1990).

BEEM and BEES data can be subject to anomalies, as are all other tip-based microscopies and spectroscopies. For instance, it has been found that the contrast observed in BEEM images can be dominated by surface topography, particularly for relatively rough metal-base surfaces where large topographic gradients are present (Prietsch and Ludeke, 1991a, 1991b). The electrons injected from the tip into the metal base layer have a predominant wave vector in the direction of the local surface normal (Fig. 1.98). For the injection geometry illustrated in Fig. 1.98(*a*), the values of $k_\parallel$ are smaller than for the injection geometry illustrated in Fig. 1.98(*b*). Consequently, the interface transmission coefficient for the two cases differs, being reduced for electrons with higher $k_{\parallel}$ as a result of transverse momentum conservation at the interface as discussed above. Therefore, the collector current will decrease for tip locations with a large surface gradient, resulting in a close correspondence of the simultaneously acquired BEEM and STM images (Prietsch and Ludeke, 1991a). Another artifact in BEES data can be caused by an insulating tip contamination layer leading to a shift of the measured $I_c(U)$ spectra as a result of the band gap of the contaminant (Prietsch and Ludeke, 1991b). Well-defined tip and surface preparation and a clean environment are therefore important for BEEM and BEES measurements.

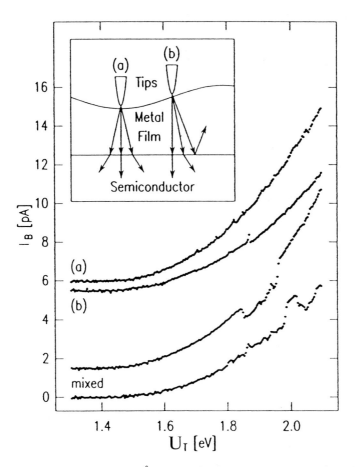

Fig. 1.98. BEEM spectra of 120 Å Au/GaP(110), taken at different tip positions with $I_T = 1$ nA. The inset shows the effect of variations in injection geometry on spectral intensities (Prietsch and Ludeke, 1991b).

1.20 Scanning field emission microscopy (SFEM) and spectroscopy (SFES)

Similarly to the topografiner (section 1.8), the STM can also be operated in the field emission regime (section 1.2.5) with an applied sample bias voltage U exceeding the local tunneling barrier height ϕ. For nondestructive operation, the tip–sample separation s has to be increased accordingly with increasing bias voltage to keep the electric field strength below the threshold for field evaporation (Müller and Tsong, 1969). As a result of the tip–sample separation being larger than for the tunneling regime, lateral resolution in scanning field emission microscopy (SFEM) is reduced, being typically at the nanometer level.

Besides topographic imaging at constant field emission current, a modulation technique, as similarly used for measurements of the local tunneling barrier height (section 1.12.2), can provide maps of the local field enhancement factor (Niedermann *et al.*, 1990). In addition, spectroscopic investigations in the field emission regime revealed the existence of field emission resonances, as predicted by Gundlach (section 1.2.5), offering the opportunity to perform electron interferometry at surfaces.

1.20.1 Field emission resonances (FER) and electron interferometry at surfaces and interfaces

For a sample bias voltage $U > \phi/e$, the tunneling barrier in a one-dimensional potential energy diagram is triangular, rather than trapezoidal in shape (Fig. 1.99). The electrons no longer have to tunnel through the total spatial gap s between tip and sample surface, but only through a barrier of reduced width s':

$$s'(U) = s \cdot (\phi / eU) \qquad (1.140)$$

A region between the triangular barrier and the positively biased electrode (either tip or sample) exists where the energy of the electrons is higher than the potential $V(z)$. The boundary conditions at the electron's classical turning point ($z = s'$) and the potential step at the surface of the positively biased electrode ($z = s$) lead to the formation of standing waves (Fig. 1.99), which depend on the detailed shape of the tunneling barrier and the electron energy. By varying the electron energy, the transmission probability exhibits resonances due to the nodal character of these standing wave states. Consequently, spectroscopic measurements of dI/dU as a function of the applied bias voltage U, using, for instance, the bias-modulation technique (section 1.13.2),

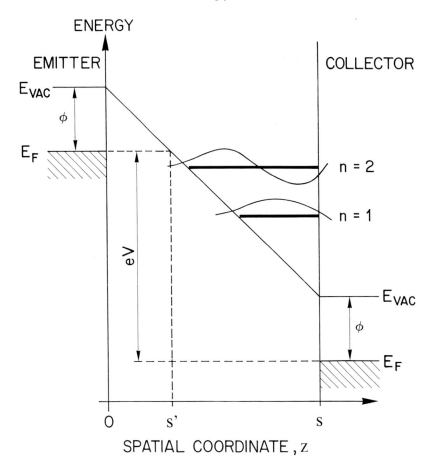

Fig. 1.99. Energy diagram for field emission. The field emission current depends upon the overlap of emitter and collector wave functions in the tunnel barrier. The collector wave functions can have resonantly enhanced amplitudes outside the collector surface when standing-wave conditions in the vacuum are fulfilled and this leads to field emission resonances in tunnel current (Coombs and Gimzewski, 1988).

show oscillations (Fig. 1.100) reflecting the field emission resonances associated with the standing wave states (Becker *et al.*, 1985d; Binnig *et al.*, 1985a). Each successive oscillation in the dI/d$U(U)$ curve incorporates an additional standing wave. The standing-wave interference effects are not associated with strongly resonant energy levels corresponding to long-lived states because the electrons spend only a short time in the vacuum gap as a result of the relatively small reflection coefficient at the surface of the positively biased electrode.

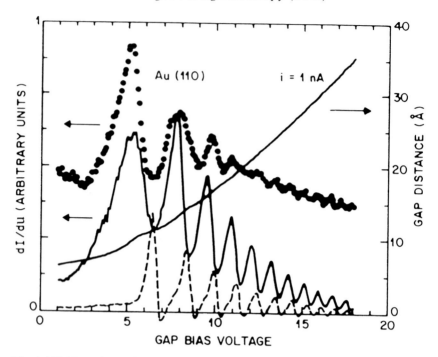

Fig. 1.100. Experimental curves of dI/dU (closed circles) and gap distance versus bias voltage. The oscillatory solid curve is the theoretical barrier penetration factor. The dashed curve omits the image-potential contribution (Becker *et al.*, 1985d).

The measured oscillations in the dI/dU–U characteristics can be fitted by numerical integration of the Schrödinger equation. As a result of this fitting procedure, various parameters of interest can be obtained, e.g. the effective tunneling area as well as the absolute tip–sample separation s, which is very difficult to determine otherwise (Becker *et al.*, 1985d; Feenstra *et al.*, 1987a; Leavens and Aers, 1986). The electron standing waves can therefore be used for electron interferometry, allowing calibration of the tip position on the atomic scale. Since the detailed shape of the tunneling barrier also depends on the shape of the image potential and the tip shape, quantitative analysis of the measured oscillations can additionally yield information about the image potential contribution (Becker *et al.*, 1985d; García *et al.*, 1986), as well as the tip, e.g. its effective radius of curvature, its roughness and its composition (Bono and Good, 1987; Pitarke *et al.*, 1990).

Early spectroscopic observations of the field emission resonances

(FER) using the conventional bias modulation method showed that the oscillations are damped out typically after 4–12 periods. It was proposed that the rapid damping of most FER oscillations results from the roughness of the tip (Bono and Good, 1987). By increasing the modulation voltage in proportion to the applied bias voltage, Coombs and Gimzewski (1988) succeeded in observing also higher-index (n) states up to $n \approx 40$, covering gap spacings s up to about 80 Å. It was also found that the detailed structure in the FER spectra is remarkably sensitive to the lateral tip position above the sample surface, which was attributed to inhomogeneities of the field emitter tip.

Spatially resolved electron interferometry at a heterojunction interface, based on a quantum-size effect in electron transmission through a thin adlayer, was demonstrated for a metal-on-semiconductor system (Kubby *et al.*, 1990). A resonance in the dI/dU–U characteristic was found, showing a dependence on the structural integrity of the overlayer which has to be continuous on a length scale of the electron's de Broglie wavelength. The resonance was attributed to a quantum-size effect associated with the discrete levels of an electron confined within the thin overlayer. Since a dependence on the specularity of the reflections from both the surface and the buried semiconductor–overlayer interface exists, localized studies of buried interfaces, ultimately with atomic spatial resolution, have become possible. Electron interferometry can therefore complement BEEM experiments (section 1.19) for investigation of buried interfaces (Kubby *et al.*, 1990, 1991).

1.20.2 Field emission scanning microscopy

As a result of the interaction of the primary field-emitted electrons from the tip with the sample, secondary electrons as well as other particles are created, which can be detected and analyzed, yielding additional information about the sample surface. Consequently, we arrive at novel microscopes based on STM technology which have well-known counterparts in traditional surface science equipment.

1.20.2.1 Field emission scanning electron microscopy (FESEM)

It has already been demonstrated with the topografiner (section 1.8), as an alternative way of operating this instrument, that the secondary electrons generated by the primary field-emitted electrons can be detected by an electron multiplier close to the sample surface (Young *et al.*, 1972; Young, 1971). By recording the secondary-electron yield while the tip is raster scanned over the sample surface, a scanning electron microscope (SEM)-type image can be obtained. In contrast to conventional SEM, the electron energy can be kept remarkably small while still obtaining high spatial resolution. In particular, it was shown that a lateral resolution of a few nanometers can be achieved by using a primary electron energy of 15 eV and a current of 0.1 nA (Fink, 1988). The overall shape of the field emitter tip was found to be important to guarantee that the secondary electrons can reach the detector at a macroscopic distance from the sample surface despite the presence of the large electric field between tip and sample.

1.20.2.2 Field emission scanning Auger microscopy (FESAM)

Analysis of the energy distribution of the secondary electrons can provide additional important information about the sample surface, particularly about its chemical composition, which is generally impossible to determine by means of STM and tunneling spectroscopy (section 1.13.6). Auger electron spectroscopy (AES) is well known as a surface analysis method which yields chemical information because the Auger process involves the core levels providing a fingerprint for the elemental species. Conventional AES is performed with primary electron energies of typically 3–30 keV. In the scanning mode, spatial resolution at the submicrometer level can be achieved. In analogy, a field emission scanning Auger microscope (FESAM) based on STM technology was developed, using a commercial hemispherical electron analyzer with the entrance lens close to the tip–sample region as illustrated in Fig. 1.101 (Reihl and Gimzewski, 1987). For the excitation of Auger electrons, primary electron energies of about three times the binding energy of the core level involved in the Auger process are needed, requiring a correspondingly high bias voltage of typically at least 1000 V between tip and sample. To exclude field-induced destruction of the tip, the electric field strength has to be kept at reasonably low values despite the high bias voltage, requiring an increase of tip–sample spacing and tip radius. As a consequence, the spatial resolution deteriorates (McCord and Pease, 1985), but submicrometer-diameter primary electron beams can still be achieved. The initial experiments performed with a tip of about 50 nm radius, a tip–sample separation of several hundred submicrometers, a primary beam energy of 1050 eV, and a total emission current of 1.0 μA have proved the feasibility of FESAM (Reihl and Gimzewski, 1987). Electron energy loss spectroscopy (EELS) performed with the same equipment using a primary electron energy of about 550 eV was demonstrated as well.

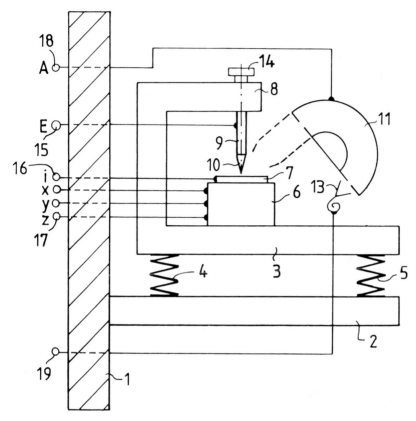

Fig. 1.101. Schematic diagram of the FESAM design (not to scale): (1) Conflat flange, (2) rigid mount, (3) stainless-steel plates with (4,5) vibrational damping, (6) *xyz* sample positioner, (7) sample, (8,14) coarse tip–sample adjustment, (9) tip piezo with (10) field-emission tip, (11) electron-energy analyzer, (13) channeltron or electron multiplier, (15–19) electric feedthroughs (Reihl and Gimzewski, 1987).

1.20.2.3 Field emission scanning electron microscopy with polarization analysis (FESEMPA)

By using a Mott spin detector close to the tip–sample region, the spin polarization of the emitted secondary electrons can be monitored, as illustrated in Fig. 1.102, leading to a field emission scanning electron microscope with polarization analysis (Allenspach and Bischof, 1989; First *et al.*, 1991). The hysteresis loop of an Fe-based metallic glass with up to ±13% polarization was measured by operating the instrument with a bias voltage of 160 V and a current of 0.8 nA (Allenspach and Bischof, 1989). The ability to detect the low-energy (< 20 eV) true

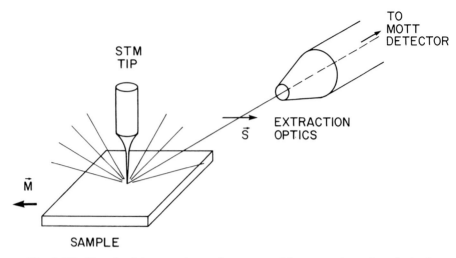

Fig. 1.102. Sketch of the experimental set-up used for extracting spin-polarized secondary electrons from an STM operated in the field emission mode. The extraction optics to the Mott detector is set up at an angle of 65° from the normal of the sample surface (Allenspach and Bischof, 1989).

secondary electrons carrying the desired magnetic information was found to depend strongly on the shape of the tip, because this determines the distribution of the electric field between tip and sample which can strongly affect the low-energy secondary electrons. It is expected that lateral resolution at the nanometer level might be achievable in magnetic imaging by FESEMPA.

1.20.2.4 Field emission inverse photoelectron spectroscopy (FEIPES)

Instead of detecting and analyzing the secondary electrons, one can also measure the yield and energy distribution of the emitted photons while the field emitter tip is raster scanned over the sample surface, providing information similarly to inverse photoelectron spectroscopy (IPES) (Young, 1971; Gimzewski *et al.*, 1988; Coombs *et al.*, 1988b; Reihl *et al.*, 1989). As shown schematically in Fig. 1.103, electrons injected into the sample at an energy considerably higher than its Fermi level can thermalize via a number of energy loss mechanisms; some of these involve photon emission. Since the corresponding transition probability depends upon the final density of states, the spectrum of emitted photons contains detailed information on these unoccupied electronic states. This information can be extracted either in the isochromat mode, where photons of a constant energy are detected as the energy of the injected electrons is changed by varying the applied sample bias voltage,

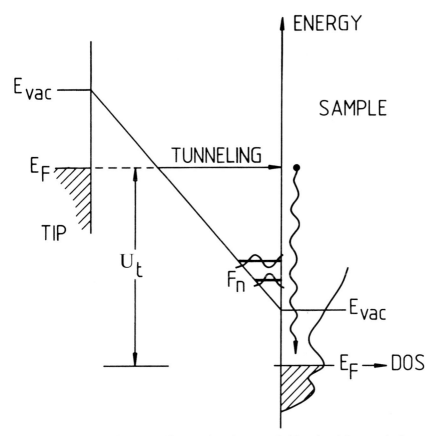

Fig. 1.103. Potential energy diagram for electrons field-emitted from a tip into a sample. The electron may lose energy by photon emission, the emitted light containing information on the density of empty states in the sample. The same process occurs when the electrons are tunnel-injected into states below the vacuum level, except that the energy of the emitted light is lower (Coombs *et al.*, 1988b).

or in the fluorescence mode, where the spectrum of emitted photons is measured for a given electron injection energy. By recording the intensity of the emitted photons at a particular photon energy as a function of surface location, a photon map with a lateral resolution on the nanometer scale can be obtained.

1.20.3 Other field emission experiments

The field emission regime with increased electric field strength and higher electron energies compared with the tunneling regime can also be used to modify the tip and sample surface intentionally. Applying a high bias voltage between tip and sample either for an extremely short time (voltage pulses) or for an extended time while holding the tip position fixed or while scanning the tip, is a well-known method to treat STM tips *in situ*, particularly in UHV. On the other hand, the field emission regime is often used for intentionally modifying the sample surface leading to STM-based nanofabrication methods (chapter 8). The STM writing can be performed either directly or via a resist film.

1.21 Transition to point contact

Apart from considering the transition from the tunneling to the field emission regime (section 1.20) with increasing distance between tip and sample surface, it is also instructive to consider the transition from the tunneling to the point contact regime (section 1.7) with decreasing tip–surface separation (Ciraci, 1990).

1.21.0.1 Nearly independent electrode regime

For tip–surface separations s, defined as distance between the ion cores of the outermost tip atom and the top surface layer, of typically 7 Å and more, the two electrodes can be considered as nearly independent because the proximity of tip and sample causes only a weak perturbation, leaving tip and sample surface wave functions almost unaffected. Electron tunneling in this regime can be well described within Bardeen's transfer Hamiltonian formalism based on a first-order perturbation approach assuming undistorted electrode wave functions (section 1.11.1).

1.21.0.2 Electronic contact or tip-induced localized states (TILS) regime

At tip–surface separations s below 4 Å, the close proximity of the tip causes a strong perturbation and changes in the surface electronic structure accompanied by significant charge rearrangements. In particular, site-specific tip-induced localized states (TILS) appear, associated with a charge accumulation between the tip and the nearest surface atom, providing a net binding interaction between tip and sample surface (Fig. 1.104). Additionally, the ions of the tip and sample might become displaced from their original positions to attain the lowest total energy.

The formation of TILS was studied theoretically at high-symmetry sites by using the empirical tight-binding approximation (Tekman and Ciraci, 1988, 1989) as well as self-consistent field (SCF) pseudopotential calculations within the local-density approximation (LDA) and with periodic boundary conditions (Ciraci and Batra, 1987; Batra and Ciraci,

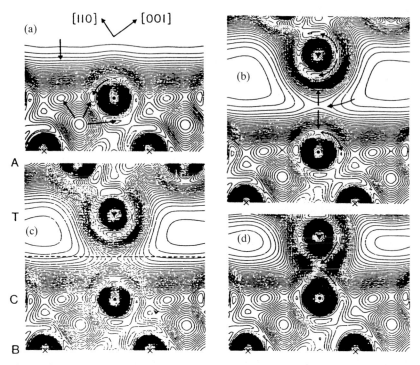

Fig. 1.104. Contours of constant local valence-charge density in the (110) plane of Al. (*a*) Free Al(111) surface. (*b*) Tip atom at a distance of 8 a.u., (*c*) 7 a.u. and (*d*) 5 a.u. The contour spacing is 10^{-3} electron per a.u.; arrows indicate increasing density (Ciraci *et al.*, 1990b).

1988; Ciraci *et al.*, 1990a, 1990b, Tekman and Ciraci, 1990). It was found that it is predominantly the outermost tip atom that interacts with the sample because the character of most TILS was maintained by assuming a multi-atom tip. The type as well as the position of this outermost tip atom, both laterally and vertically with respect to the sample surface, determines the energies, charge distributions and amplitudes of TILS.

The tunneling current can deviate considerably from the proportionality to the local electronic density of states derived within the Tersoff–Hamann theory (section 1.11.1). In particular, the measured corrugation reflects in part site-dependent changes in the electronic surface structure induced by the proximity of the tip. It was found that the STM contrast, determined by the corrugation amplitude, can be significantly enhanced in the presence of TILS. This was demonstrated by calculations based on a generalization of the Tersoff–Hamann approach taking contributions from TILS properly into account. The enhanced contrast can be understood within the framework of symmetry considerations of the tip orbitals (section 1.11.1) by the fact that TILS for metallic samples with sp-derived states at the Fermi level have the character of atomlike p_z orbitals (Ciraci *et al.*, 1990b).

Another consequence of the close proximity of the tip to the sample surface is a significant lowering of the potential barrier (section 1.12). For sufficiently small tip–surface separations, even a channel with local potential lower than the Fermi level is formed, connecting the tip and the sample surface. The formation and enlargement of this channel with decreasing tip–surface separation was studied by Lang (1988) based on a jellium model for the two electrodes, one of which has an adsorbed atom modeling the tip (Fig. 1.105). By taking the atomic structure of the sample surface into account, the formation of this channel can be shown to be site-dependent and thus varies with the lateral position of the tip relative to the sample surface (Ciraci *et al.*, 1990b). However, even if the local potential ϕ in the channel is lower than the Fermi level E_F, the energy E_1 of the lowest propagating state may lie above E_F, leading to an effective barrier $\phi_{eff} = E_1 - E_F$. As a consequence, the current flow between tip and sample may still be due to electron tunneling through this effective barrier ϕ_{eff} even if the local potential barrier ϕ has already collapsed. For electron tunneling within the strong tip–sample interaction regime, the parallel wave vector $k_\parallel$ is no longer conserved as in the independent electrode regime. Therefore, the difference between the effective barrier ϕ_{eff} and the potential barrier ϕ,

Fig. 1.105. Contour maps of the effective potential V_{eff} for the two-electrode case with an adsorbed Na atom (tip). The presence of the atom is represented by a shaded circle with a cross at the position of the nucleus; the positive background regions of both tip and sample electrodes are shaded also. Maps are shown for four values of s (given in bohr, 1 bohr = 0.529 Å), which is the distance between the nucleus of the tip atom and the positive background edge of the sample electrode. The nucleus is at the center of each box, so that the sample electrode in fact lies outside of the box for all but the smallest of the s values shown ($s = 5$ bohr). The contour closest to the atom in each case is that for $\phi_{eff} = E_F$; the contours shown for higher energy values are spaced by 0.25 eV, starting at E_F (Lang, 1988).

though existent, is not as large as expected by assuming strict conservation of $k_{||}$.

1.21.0.3 Point Contact Regime

On further decreasing the tip–sample separation ($s \lesssim 2$ Å), the effective barrier ϕ_{eff} will finally collapse as well. New channels of conduction are opened by the occupation of states quantized in the orifice between tip and sample surface, allowing ballistic transport of electrons if the orifice is large enough. The qualitative change in the character of the conductance from tunneling to ballistic transport can be identified with the onset of point contact. This definition is appropriate for a microscope probing the sample surface via a measurement of current flow.

The transition from tunneling to point contact was studied experimentally by measuring log *I–s* characteristics as illustrated in Fig. 1.106(*a*) (Gimzewski and Möller, 1987; Gimzewski *et al.*, 1987a). For relatively large tip–surface separations *s* of several Ångström units, the current increases exponentially with decreasing *s* and the measured apparent barrier height, as introduced in section 1.12, is as high as 3–4 eV. As the tip–surface separation is reduced, the measured apparent barrier height becomes smaller, which is reflected in a flattening of the log *I–s* curve (Fig. 1.106(*a*)). On further decrease of the tip–sample separation, a discontinuous current jump is observed in the log *I–s* characteristic, which is attributed to mechanical instability caused by significant atomic motion under the influence of adhesive forces (Pethica and Sutton, 1988; Smith *et al.*, 1989b; Taylor *et al.*, 1991), initiating point contact between tip and sample. The plateau in the log *I–s* curve, often observed before discontinuity sets in, corresponds to saturation of the tunneling resistance $R = U/I$ at approximately $R_s \approx 35\,000\ \Omega \approx 1.4R_K$, where

$$R_K = h / e^2 \qquad (1.141)$$

is von Klitzing's constant. Assuming that the radius *a* of the area of contact between tip and sample surface is considerably less than the electron mean free path $l(l/a \gg 1)$, the contact radius *a* might be estimated to a first approximation by using the Sharvin formula (section 1.7, Eq. (1.77))

$$R = \frac{4\rho l}{3\pi a^2}$$

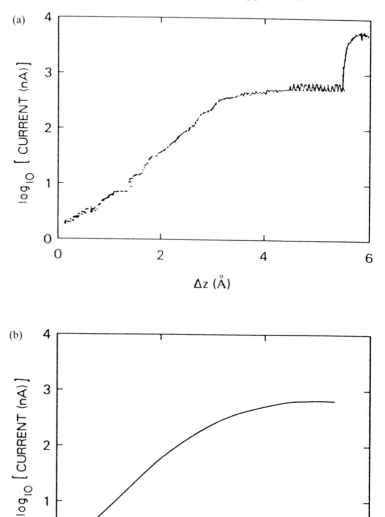

Fig. 1.106. (*a*) Experimental data showing the dependence of the tunnel current ($\log_{10} I$) on distance Δz toward the surface measured from the point at which the resistance is 20 MΩ, for a clean Ir tip and a polycrystalline Ag surface at constant bias voltage of magnitude 20 mV (Gimzewski and Möller, 1987; Lang, 1987b). (*b*) Tunnel current ($\log_{10} I$) versus distance Δz toward the surface measured from the point at which the resistance is 20 MΩ, calculated for a Na tip atom and an applied bias voltage of 20 mV (Lang, 1987b).

In this way, Gimzewski and Möller (1987) concluded that the contact initially formed must be of atomic dimensions.

If the conduction channel associated with the formation of point contact is sufficiently narrow to be regarded as one-dimensional, a 'constriction' resistance of

$$R_c = \frac{h}{2e^2} = \frac{R_K}{2} \approx 12\ 900\ \Omega \qquad (1.142)$$

can theoretically be derived (Lang, 1987b; Landauer, 1987). Based on the adatom-on-jellium model, Lang (1987b) simulated a log *I–s* characteristic for the transition between tunneling and point contact (Fig. 1.106(*b*)). He found that the resistance saturates at a value

$$R_s = A \cdot \frac{h}{2e^2} = \frac{A}{2}\ R_K \qquad (1.143)$$

where the constant A depends on the identity of the tip atom (e.g. $A = 2.5$ for Na and $A = 1.4$ for Ca). By using the tight-binding method, Ferrer *et al.* (1988) argued that the tunneling resistance should saturate at a minimum value of $R_s = h/2e^2 = R_K/2$ if no instability would occur at the onset of point contact.

In summary, by making a very sharp tip approach towards a sample surface, a plateau in the log *I–s* characteristic can be found, corresponding to a saturation resistance close to the fundamental resistance $R_K/2$, indicating formation of a point contact of atomic dimensions.

On the other hand, qualitatively different log *I–s* characteristics without a clearly defined resistance plateau have also been obtained experimentally (Gimzewski and Möller, 1987; Gimzewski *et al.*, 1987a). These log *I–s* curves often exhibit an oscillatory behavior of the current after the onset of point contact. It was argued that such log *I–s* curves are obtained with relatively blunt tips having radii on the order of several Fermi wave lengths λ_F (García-García and García, 1991). The observed oscillatory behavior of the log *I–s* curves was proposed to be due to quantum oscillations caused by the existence of geometry-induced states in the orifice (García-García and García, 1991; García and Escapa, 1989). However, if the length of the orifice formed between tip and sample surface is too small ($<\lambda_F$), then quantization with the associated oscillatory behavior of the resistance is not observable (Ciraci and Tekman, 1989; Ciraci, 1990). In this case, oscillations seen in log *I–s* curves have possibly to be interpreted as an irregular evolution of the point contact, e.g. an irregular enlargement of the contact area.

Imaging in the regime of electronic point contact, where the local potential barrier has already collapsed but a site-dependent effective barrier still exists, can be qualitatively different from imaging in the tunneling regime. For instance, a contrast reversal in constant-current images of the Al (111) surface, with hollow sites appearing as protrusions, was predicted (Tekman and Ciraci, 1990).

Finally, it was demonstrated experimentally that atomic-resolution imaging is possible even beyond the formation of 'mechanical' point contact, where the forces acting between tip and sample (section 1.22) have changed from attractive to repulsive (Smith *et al.*, 1986b). The point-contact microscope (PCM) was operated at low temperatures (4 K) with a resistance down to 400 Ω, while still observing atomic resolution on a graphite surface. The surface deformations induced by the tip were found to be reversible, which was attributed to the exceptional elastic properties of graphite as a layered material (section 4.1.3). The observed atomic resolution in the point-contact mode does not necessarily imply a conductance channel of single-atom width, but rather a fluctuation in total conductivity between tip and sample which varies in registry with the graphite lattice (Pethica, 1986). This can be caused, for instance, by sliding of graphite planes relative to each other induced by the motion of the tip in contact with the sample.

1.22 Forces in STM

Thus far we have only considered the tunneling current flow as a consequence of an overlap of the wave functions of tip and sample surface held at atomic-scale distances. The tunneling current is exploited in STM for vertical tip position control as well as for probing the atomic and electronic structure of the sample surface. However, apart from tunneling, other interaction mechanisms present for closely spaced tip and sample can be important for interpretation of STM results and therefore have to be considered as well.

In this section, we focus on the force interaction which is always present between two bodies such as tip and sample in STM. At relatively large tip–surface separations of 10 Å and more, the force interaction between tip and sample is dominated by long-range van der Waals forces which show a power-law dependence on distance with an exponent dependent on the detailed geometry, i.e. van der Waals forces depend strongly on the tip shape. At small tip–surface separations of only a few Ångström units, the overlap of the wave functions of tip and sample surface is significant and short-range quantum-mechanical exchange-correlation forces become dominant which decay exponentially with increasing distance. For metallic tip and sample, they appear as strongly attractive metallic adhesion forces for tip–surface separations larger than the equilibrium distance. TILS (section 1.21) can be responsible for additional enhancement of the attractive force because the associated accumulation of charge between tip and sample provides a net binding interaction. However, if tip and sample are made to approach one another more closely, the quantum-mechanical forces soon become repulsive as a consequence of the Pauli exclusion principle. The transition between the attractive and repulsive force regime is identified with the onset of mechanical point contact. This definition of point contact is particularly appropriate for a microscope probing the local force interaction between tip and sample (chapter 2).

The importance of considering the force interaction between tip and sample in STM was first realized in conjunction with STM experiments on graphite as a relatively soft sample (Soler *et al.*, 1986). Giant corrugation amplitudes of up to several Ångström units were measured in the constant current mode, which could not be explained by tunneling theory. Instead it was proposed that local elastic deformations of the graphite surface are responsible for significant enhancement of the electronically based corrugation. Direct measurements of the force acting between the tip and the graphite sample while tunneling in air revealed the presence of remarkably high repulsive forces in the range of 10^{-7}–10^{-6} N (Tang *et al.*, 1988; Yamada *et al.*, 1988; Mate *et al.*, 1989a). These strong repulsive forces and the resulting local elastic deformations in STM experiments on the graphite surface performed in air were attributed to the presence of a surface contamination layer (Mamin *et al.*, 1986).

The forces acting between metallic tips and samples during tunneling experiments in UHV were investigated by using a flexible cantilever beam as sample holder (Dürig *et al.*, 1986a). The forces were determined by measurement of the (static) deflection of the cantilever beam from its equilibrium position, whereas force gradients $\partial F/\partial z$ were detected by the induced shift in the resonance properties of the excited beam. Repulsive interaction forces measured in the early experiments were explained by the presence of surface contamination layers. Later, attractive metallic adhesion forces of up to 2×10^{-9} N were observed within a range of 2 Å before making contact between tip and sample (Dürig *et al.*, 1988, 1990; Dürig and Züger, 1990). It was found that the mechanical point of contact coincides with the discontinuity observed in log *I–s* characteristics (section 1.21) to within 0.7 Å. By integrating the measured $(\partial F/\partial z)$–*s* curve, the exponential dependence of the strongly attractive quantum-mechanical adhesion force on the tip–surface sep-

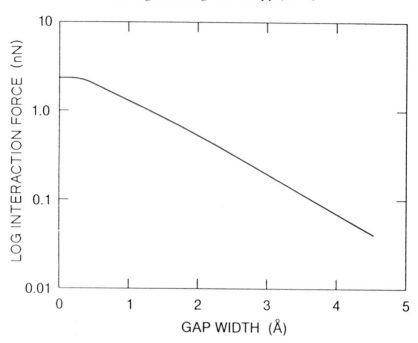

Fig. 1.107. Tip–surface interaction force as a function of tip–surface separation (Dürig *et al.*, 1988).

aration s was verified (Fig. 1.107), with a change of one order of magnitude in the force for a change of about 2 Å in the tip–surface separation s. This means that the force decreases with a rate that is a factor of two smaller than the decay rate $2\varkappa$ of the tunneling conductance σ which decreases by roughly one order of magnitude for 1 Å (section 1.2.1 and 1.11.1) according to

$$\sigma \propto \exp(-2\varkappa s) \qquad (1.144)$$

The observed relationship between the decay rates of the attractive quantum-mechanical force and the tunneling conductance can be explained theoretically by the fundamental equality of Bardeen's tunneling matrix element M (section 1.2.2 and 1.11.1) and the Heisenberg–Pauling resonance energy U, i.e. the energy lowering due to the wavefunction overlap of two atomic systems approaching each other (Chen, 1991b). Since

$$\sigma \propto |M|^2 \qquad (1.145)$$

the energy lowering U is therefore proportional to the square root of the tunneling conductance:

$$U \propto -\sqrt{\sigma} \propto -\exp(-\varkappa s) \tag{1.146}$$

Based on the definition of the force F, we find

$$F = -\frac{\partial U}{\partial s} \propto -\varkappa \exp(-\varkappa s) \propto -\varkappa \sqrt{\sigma} \tag{1.147}$$

From Eq. (1.147) we see directly that the decay rate of the force is equal to $\varkappa$, which is exactly half the value for the decay rate of the tunneling conductance. A more rigorous treatment (Chen, 1991b) shows that the attractive force F and the tunneling conductance σ should conform to the general equation

$$F = -f(\Delta \mathcal{E}_c)\varkappa(\sigma R_K)^{\frac{1}{2}} \tag{1.148}$$

where $(\Delta \mathcal{E}_c)$ is the width of the conductance band of the metal, $R_K = h/e^2$ is von Klitzing's constant (section 1.21), and f is a dimensionless factor on the order of one, which depends on the tip geometry. The relationship (1.148) provides a unified view of tunneling and force measurements. Conceptually, it means that the imaging process in atomic-resolution STM may be regarded as a sequence of bond forming and bond rupturing.

The existence of significant forces in STM at small tip–surface separations also has consequences for the measured apparent barrier height ϕ_A as introduced in section 1.12. Under the influence of an attractive force, the measured barrier height exhibits a slight increase as tip and sample approach one another (Fig. 1.108) before dropping at the onset of (electronic) point contact (Chen and Hamers, 1991). This is in contrast to the predicted continuous slow decrease of ϕ_A over 3–5 Å as tip and sample approach towards the point of contact, neglecting the effect of forces (Lang, 1988).

While the additional influence of forces might not be desired for investigation of truly tunneling-based phenomena, these forces can be exploited to obtain enhanced contrast in STM images. The giant corrugation observed on soft samples such as graphite (Soler *et al.*, 1986) was the first example. If the sample is not soft, one might use a compliant cantilever beam as sample holder to obtain force-enhanced topographic contrast (Dürig *et al.*, 1986a). Alternatively, the usual rigid STM tip can be replaced by a flexible tip, e.g. a thin film tip (Moreland

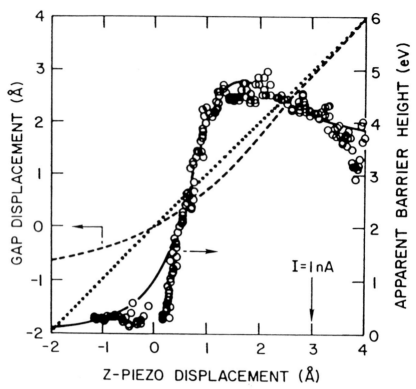

Fig. 1.108. Variation of the measured apparent barrier height with z piezo displacement. Circles are data points. The solid curve is derived by assuming that the apparent barrier height is 3.5 eV by definition over the entire range of operation, but the tip deforms due to the force between the tip and the sample. The dashed curve is the actual gap displacement as a function of the measured z piezo displacement. The dotted curve represents the fictitious gap displacement in the absence of force. The equilibrium distance, where the net force is zero, is taken as the origin of z (Chen and Hamers, 1991).

and Rice, 1990, 1991). In this case, the tip acts as a cantilever (or briefly 'lever') which is deflected according to the force acting between tip and sample. Therefore, the method is often denoted as 'lever-STM'. The tunneling current is still used to regulate the tip–surface separation and serves as a tip position control parameter. However, the contrast in constant-current STM images obtained with a flexible tip having a relatively small spring constant is mainly determined by the force interaction between tip and sample, leading to a tunneling-stabilized scanning force microscope (TSSFM). By using a magnetized flexible tip, magnetic forces can be probed (section 2.7.3), resulting in a tunneling-stabilized magnetic force microscope (TSMFM) (Moreland and Rice, 1990, 1991). In this way, we leave the field of tunneling microscopy and enter the more recently developed field of force microscopy.

2

Scanning force microscopy (SFM)

We have already discussed in section 1.22 that a variety of forces act between the tip and the sample during STM operation with their strength depending on the tip–surface separation. These forces have been exploited to develop another type of scanning probe microscopy, namely scanning force microscopy (SFM), which no longer uses electron tunneling to probe local properties of sample surfaces, but rather the tip–sample force interaction. Since the force interaction does not depend on electrically conducting samples (and tips), SFM can be applied to insulators as well, thereby extending the applicability of local probe studies to an important class of materials which are difficult to investigate by electron microscopical and spectroscopical techniques due to charging problems. Before we focus on SFM, a brief historical review of surface force measurements and surface profilometry, which are closely related to SFM, is given in the next section.

2.1 Historical remarks on surface force measurements and surface profilometry

2.1.1 Surface force apparatus (SFA)

For two electrically neutral and non-magnetic bodies held at a distance of one to several tens of nanometers, the van der Waals (VDW) forces usually dominate the interaction force between them. The VDW forces acting between any two atoms or molecules may be separated into orientation, induction and dispersion forces. Orientation forces result from interaction between two polar molecules having permanent multipole moments, whereas induction forces are due to the interaction of a polar and a neutral molecule where the polar molecule induces polarity in the nearby neutral molecule. Non-polar molecules possess finite fluctuating dipole and higher multipole moments at very short time intervals, which also interact giving rise to dispersion forces between them. Dispersion forces generally dominate over orientation and induction forces except in the case of strongly polar molecules. The dispersion, or generally, VDW forces are usually attractive and rapidly increase as atoms, molecules or bodies approach one another. The force–distance (F–s) dependence is described by

$$F_{VDW}(s) \propto -\frac{1}{s^7} \tag{2.1}$$

For separations s larger than several nanometers, dispersion forces show a retardation effect of relativistic origin leading to faster decay of forces with distance, which can be described by

$$F_{VDW}^{ret}(s) \propto -\frac{1}{s^8} \tag{2.2}$$

These retarded VDW forces are named Casimir forces while the non-retarded VDW forces are named London forces.

The VDW forces acting between two macroscopic bodies can be calculated to a first approximation by neglecting the retarded forces and by assuming the VDW forces to be additive. For a sphere of radius R held at a distance s from a surface, one finds

$$F_{VDW}(s) = -\frac{HR}{6s^2} \tag{2.3}$$

where H is the material-dependent Hamaker constant.

Van der Waals forces had already been studied experimentally for a long time before the invention of scanning probe microscopes by using a surface force apparatus (Tabor and Winterton, 1969; Israelachvili and Tabor, 1972; Israelachvili and Adams, 1976; Israelachvili, 1987; Israelachvili and McGuiggan, 1988, 1990). The VDW forces were usually measured between crossed cylinders of molecularly smooth mica sheets which were made to approach one another by using mechanical reduction mechanisms and piezoelectric transducers implemented in the surface force apparatus (SFA), as shown in Fig. 2.1. The forces can be derived from the deflection Δz of a spring, on which one of the mica sheets is mounted, according to Hooke's law:

$$F = c \cdot \Delta z \tag{2.4}$$

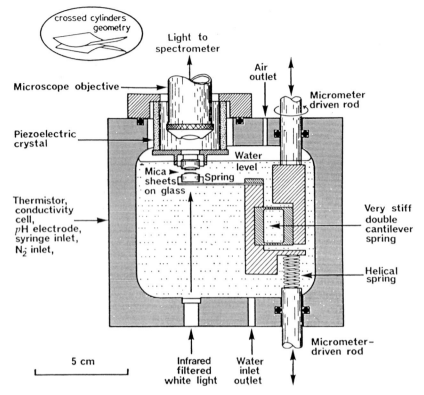

Fig. 2.1. Schematic drawing of a surface force apparatus to measure long-range forces between two crossed cylindrical sheets of mica (thickness about 1 μm and radius of curvature about 1 cm) immersed in liquid. By use of white light and multiple beam interferometry the separation between the two mica surfaces may be measured to about 0.1 nm. The separation between the two mica surfaces may be controlled by use of two micrometer-driven rods and a piezoelectric crystal to better than 0.1 nm (Israelachvili, 1987).

where c is the spring constant. The stiffness of the spring can be varied (from about 10 N m^{-1} to about 10^5 N m^{-1}) by shifting the position of a movable clamp, allowing detection of a wide range (six to seven orders of magnitude) of forces. The separation between the two mica surfaces is measured (to better than 1 Å accuracy) by means of an optical technique using multiple beam interference fringes. A force resolution of typically 10^{-8} N was achieved by using a relatively soft spring.

The SFA allowed study of VDW forces for a wide range of separations (from 1 nm up to more than 100 nm). A gradual transition between non-retarded and retarded forces was found as the separation between the mica sheets was increased from 12 to 50 nm (Israelachvili and Tabor,

1972). The design of the SFA as shown in Fig. 2.1 allowed surface forces to be studied in the presence of vapors and liquids as well. Besides VDW forces, it was possible to investigate double-layer forces, solvation and capillary forces (Israelachvili and McGuiggan, 1988). The surface forces were also studied in the presence of adsorbates on the mica surfaces. The SFA undoubtedly contributed considerably to our knowledge of intermolecular and surface forces (Israelachvili, 1985) before invention of the SFM.

2.1.2 Surface profilometers

Stylus profilometry was developed a long time ago as a valuable means to study the surface roughness of materials, including bulk insulators (Williamson, 1967–1968; Teague *et al.*, 1982). A topographic map is obtained similarly to an STM by raster scanning a sample relative to a stylus tip (Fig. 2.2). However, in contrast to STM, the stylus tip is in

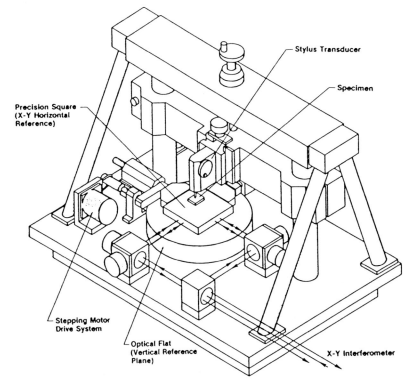

Fig. 2.2. Schematic drawing of a three-dimensional stylus profilometer. The system used for leveling the specimen surface with respect to the optical flat surface is not shown (Teague *et al.*, 1982).

mechanical contact with the sample and the forces on the tip are usually set at 10^{-4} N. This set value of the force corresponds to the set value of the tunneling current in STM. The stylus tip radii are nominally about 1 μm, which is relatively large compared with STM tips. However, at a loading force of 10^{-4} N, the tip radius cannot be chosen much smaller because the resulting huge local pressure would cause considerable surface damage. The effect of the stylus tip radius on the measured roughness values was considered soon after the invention of stylus profilometry (Radhakrishnan, 1970). Clearly, the large size of the stylus tips has been the most serious limitation of the technique of stylus profilometry with respect to lateral resolution (typically about 1000 Å) and reliability of surface roughness values deduced from the topographic maps.

2.2 The birth of SFM

In the previous section we saw that the SFA can offer a relatively high force sensitivity on the order of 10^{-8} N. However, spatially resolved information about the force interaction is not obtained. On the other hand, surface profilometry allows mapping of the surface topography

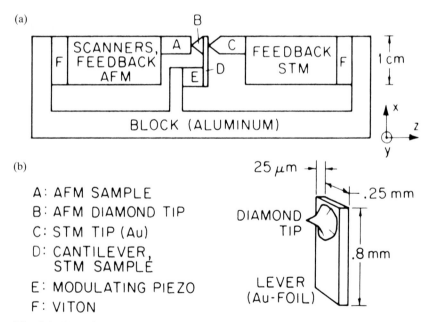

Fig. 2.3. Experimental set-up for the first AFM. The cantilever is not to scale in (*a*). Its dimensions are given in (*b*). The STM and AFM piezoelectric drives face each other, sandwiching the diamond tip that is glued to the cantilever (Binnig *et al.*, 1986c).

of bulk insulators. The lateral resolution is, however, limited by the large stylus tip radius and the relatively large loading force of typically 10^{-4} N.

The scanning force microscope (SFM), invented by Binnig *et al.* (1986c), can be regarded as a hybrid between a SFA and a surface profilometer. As shown in Fig. 2.3, a probe tip is mounted on a cantilever-type spring. The force interaction between the sample and the tip after approaching each other causes the cantilever to deflect according to Hooke's law (Eq. (2.4)). These deflections can be sensed with high sensitivity by a STM tip within tunneling distance of the rear side of the cantilever (Fig. 2.3), or alternatively, by capacitance or optical detection methods (section 2.3). A topographic map of a sample surface is obtained, for instance, by keeping the force constant while scanning the sample relative to the tip. The SFM can be operated either in the contact regime, similar to the stylus profilometer but with considerably lower forces (10^{-6}–10^{-10} N), or in the non-contact regime similar to the SFA where, for instance, VDW forces may be probed.

The most critical component in SFM is certainly the cantilever-type spring. To achieve high sensitivity, a reasonably large deflection for a given force is desired. Therefore, the spring should be as soft as possible. On the other hand, a high resonant frequency is necessary in order to minimize sensitivity to mechanical vibrations, particularly while scanning. Since the resonant frequency of the spring system is given by

$$\omega_0 = \left(\frac{c}{m} \right)^{\frac{1}{2}} \qquad (2.5)$$

where c is the spring constant and m the effective mass loading the spring, it becomes clear that the requirement of a large value of ω_0 for a relatively soft spring (small value of c) can only be fulfilled by keeping the mass m and therefore the geometrical dimension of the force sensor as small as possible. This consideration leads directly to the idea of using microfabrication techniques for the production of cantilever beams. The development of microfabricated force sensors has led to the reliable atomic resolution capability of SFM for both conductors and insulators (Binnig *et al.*, 1987a, 1987b; Albrecht and Quate, 1987, 1988). The applicability of SFM to bulk insulators is particularly important because electron microscopical and spectroscopical studies have proven to be difficult due to charging effects. High-resolution electron microscopy studies of bulk insulators have usually been performed by using the replica technique, preventing direct investigation of the sample surface.

The SFM allows probing of a variety of attractive as well as repulsive forces, including van der Waals forces, ion–ion repulsion forces, electrostatic and magnetic forces, capillary forces, adhesion and frictional forces, to mention only a few. The aim of probing particular types of forces has led to the development of specialized force microscopes which can operate in various environments, including ambient air, liquids and vacuum. The SFM has considerably broadened the applicability of local probe studies with respect to materials as well as types of tip–sample interaction. Therefore, the invention of SFM may be regarded as important as the invention of STM, particularly in view of the fact that most of the materials that surround us are insulators.

2.3 SFM design and instrumentation

Criteria for SFM design and instrumentation follow in many respects those for the STM (section 1.10), particularly concerning vibration isolation, positioning devices, scanning units, electronic feedback system and computer automation. Therefore, we will mainly concentrate now on the force sensors as well as on the experimental techniques to measure the small cantilever deflections.

2.3.1 Force Sensors

The force sensor of a SFM consists of a cantilever beam with a sharp tip at one end. The tip may either be glued onto the cantilever or may be integrated, i.e. the cantilever is directly fabricated with a sharp tip at its end. The force sensors have to fulfill several requirements.

1. The spring constant should be small enough to allow detection of small forces.
2. The resonant frequency ω_0 should be high to minimize sensitivity to mechanical vibrations.
3. For atomic resolution SFM studies – which was originally named atomic force microscopy (AFM) – sharp tips with small effective radius of curvature are required. Ultimately, well-defined monatomic tips are desirable because atomic-resolution SFM images can depend critically on the atomic structure at the outermost end of the probe tip (Abraham *et al.*, 1988d).
4. For large-area scans on relatively rough surfaces, the opening angle of the tips should be as small as possible allowing penetration into deep troughs on the surface.

For atomic resolution SFM studies it is reasonable to consider a typical value of the effective spring constant for interatomic coupling in solids (Binnig *et al.*, 1986c; Rugar and Hansma, 1990), which is on the order of

$$c_{\text{at}} = \omega_{\text{at}}^2 \, m_{\text{at}} \approx 10 \, \text{N m}^{-1} \tag{2.6}$$

if we assume atom vibrational frequencies of typically $\omega_{\text{at}} \approx 10^{13}$ Hz and atom masses of typically $m_{\text{at}} \approx 10^{-25}$ kg. A cantilever of spring constant c less than c_{at} can easily be fabricated from aluminium foil. A piece at 4 mm long, 1 mm wide and 10 μm thick would have a spring constant of about 1 N m^{-1} according to the relation

$$c = \frac{E}{4} \cdot \frac{wt^3}{l^3} \tag{2.7}$$

where E is Young's modulus and l, w and t are the length, width and thickness of the cantilever, respectively. A force as small as 10^{-10} N would deflect this type of cantilever by 1 Å, which can easily be measured (section 2.3.2). Therefore, the first requirement listed above would be fulfilled. However, this type of cantilever has a resonant frequency ω_0 on the order of 1 kHz only, which is far too low for reliable SFM operation.

To increase the resonant frequency of the cantilever, its mass must be considerably reduced. Therefore, most SFM cantilevers used today are microfabricated from silicon oxide, silicon nitride or pure silicon using photolithographic techniques (Albrecht, 1989; Albrecht *et al.*, 1990a). Batch fabrication has the additional advantage that a large number of almost identical cantilevers with similar spring constants is available. For microfabricated cantilevers with typical lateral dimensions on the order of 100 μm and thickness of about 1 μm (Fig. 2.4), spring constants in the range of 0.1–1 N m^{-1} and resonant frequencies of 10–100 kHz are obtained. A high resonant frequency makes the SFM relatively insensitive to low-frequency mechanical vibration because the transmission of external vibration of frequency ω through a mechanical system with resonant frequency ω_0 is reduced by a factor of $(\omega/\omega_0)^2$. In addition, a high resonant frequency of the cantilever allows for reasonably high scan speeds. The V-shaped cantilevers as shown in Fig. 2.4(*b*) have an increased lateral stiffness compared with rectangular cantilevers (Fig. 2.4(*a*)), leading to reduced sensitivity to lateral (frictional) forces which can cause appreciable lateral bending of the cantilever (section 2.5). This lateral bending can result in serious degradation of SFM images, particularly for surface topographies containing very steep height variations (den Boef, 1991).

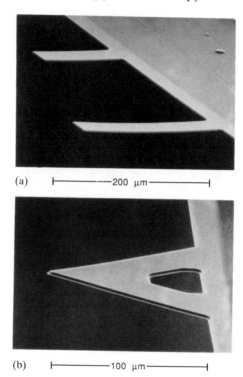

(a) ├────────200 μm────────┤

(b) ├────────100 μm────────┤

Fig. 2.4. SEM micrographs of SiO$_2$ microcantilevers. (*a*) Rectangular cantilevers. The shorter of the two has dimensions 100 μm × 20 μm × 1.5 μm, with
a force constant of 1 N m^{-1} and a resonant frequency of 120 kHz. The V-shaped
cantilever shown in (*b*) has increased lateral rigidity which reduces its sensitivity
to frictional forces (Albrecht *et al.*, 1990a).

Initially, a tip at one end of the cantilever was obtained by glueing a
small piece of diamond onto the cantilever, and a sharp corner of the
diamond fragment was then used for probing the sample surface. The
increasing demand for a large number of cantilevers with reproducible
sharp tips has, however, soon led to the development of microfabricated
cantilevers with integrated tips as shown in Fig. 2.5 (Albrecht, 1989;
Albrecht *et al.*, 1990a). Pyramidal tips (Fig. 2.5 (*a*)–(*d*)) as well as
conical tips (Fig. 2.5 (*e*),(*f*)) integrated on silicon nitride (Si$_3$N$_4$) or
silicon oxide (SiO$_2$) cantilevers with tip radii less than 300 Å have been
achieved, which proved to be sharp enough for atomic-resolution SFM
studies in the contact regime, most likely due to the existence of protruding nanotips.

Improvements with respect to tip radius and particularly tip opening
angle were achieved by microfabrication of pure crystalline silicon canti-

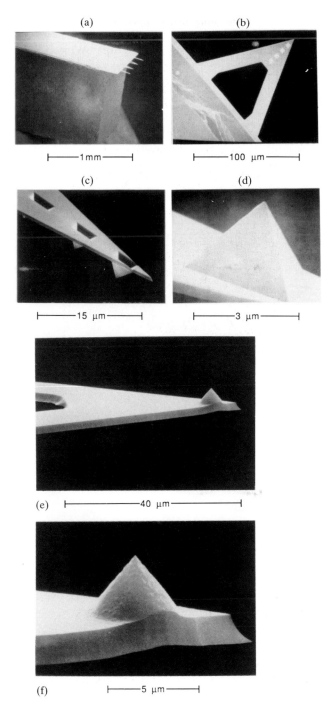

Fig. 2.5. (a)–(d) SEM micrograph of a Si_3N_4 cantilever with integrated pyramidal tip. (e)–(f) SEM micrograph of a SiO_2 cantilever with integrated conical tip (Albrecht et al., 1990a).

levers with integrated single crystal silicon tips (Wolter *et al.*, 1991). Such tips are particularly superior for large-area scans on samples exhibiting considerable surface roughness (Fig. 2.6). The development of SFMs operating under ultra-high vacuum conditions certainly helps to further improve the preparation and *in situ* characterization of these tips, for instance, by combination with field ion microscopy techniques. Well-defined monatomic tips may well be obtained from microfabricated single-crystal silicon force sensors after appropriate *in situ* preparation.

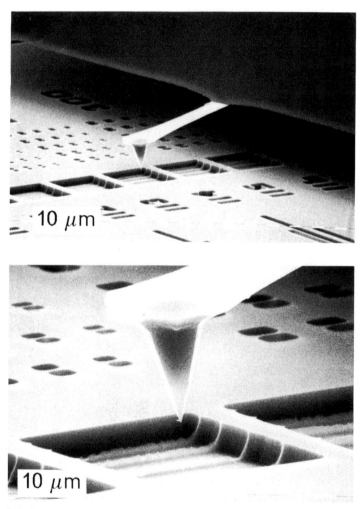

Fig. 2.6. SEM micrograph of a silicon force sensor hovering over a microfabricated test sample (Wolter *et al.*, 1991).

2.3.2 Cantilever deflection measurement techniques

Several different methods for detecting the small cantilever deflections can be used. The different methods may be compared with one another with respect to the following requirements.

1. A high sensitivity at the sub-Ångström unit level is needed.
2. The measurement technique should have negligible influence on the cantilever deflection itself and should not cause imaging artifacts.
3. The deflection measurement system should be easy to implement, particularly for non-ambient environments, e.g. UHV or cryogenic environments.

In the following, we will compare several different techniques for detecting the cantilever deflections and discuss the main fields of application.

2.3.2.1 Tunneling detection

The first method for detecting the small cantilever deflections used by Binnig *et al.*, (1986c) was electron tunneling from the rear side of the cantilever to a STM tip (Fig. 2.3). The exponential dependence of the tunneling current on the separation between the two electrodes (chapter 1) offers a high sensitivity of typically 0.01 Å for detection of cantilever deflections. However, electron tunneling is also extremely sensitive to the surface conditions of the cantilever. When measuring in ambient air or moderate vacuum, the rear side of the cantilever must be coated, for instance, with a freshly prepared gold film to guarantee stable tunneling conditions.

Furthermore, tunneling detection changes the effective spring constant c_{eff} of the cantilever. The interaction of the tunnel tip with the cantilever can be modeled by a spring of spring constant c_{tip} which may have either negative or positive sign, depending on the conditions of the tunnel tip (Meyer, 1990). If the tunneling conditions change as a function of time, e.g. due to an increasing amount of surface contamination, the spring constant c_{tip} and therefore also c_{eff} are changed, which is undesirable.

The presence of thermal drifts can cause considerable problems if tunneling detection is used. The surface roughness of the rear side of the cantilever may then influence the force measurements because the vertical position of the cantilever measured relative to an originally defined zero-point changes as a function of time. Thermal drifts combined with lateral variations in the local tunneling barrier height can additionally influence the force measurements via temporal variation of the spring constant c_{tip}.

The alignment of the tip and cantilever becomes somewhat troublesome in non-ambient conditions such as UHV or a cryogenic environment where other cantilever deflection measurement techniques are usually preferred. SFM measurements with tunneling detection have mainly been performed in the contact mode where atomic resolution is achieved. Some early magnetic force microscope (MFM) studies were based on tunneling detection of the cantilever deflections as well (section 2.7.3).

2.3.2.2 Capacitance detection

The cantilever deflections can alternatively be sensed capacitively by a counterelectrode opposite to the rear side of the cantilever. By using a capacitance transformer bridge for small capacitances, sensitivity at the sub-Ångström unit level can be achieved, which requires capacitance noise levels on the order of 10^{-18} F (Göddenhenrich et al., 1990a; Neubauer et al., 1990). Since the rate of change in capacitance with distance, and therefore the detection sensitivity, increases as the distance between two electrodes decreases, it is desirable to operate with a small electrode separation. On the other hand, the attractive electrostatic force between two electrodes also becomes large at small distances. Therefore, the geometry must be chosen to achieve maximum sensitivity without causing the two electrodes to snap together, which would occur if the gradient of the attractive force exceeds the spring constant of the cantilever.

The capacitance detection method is certainly much less sensitive to the surface conditions of the cantilever and causes less strong detector–cantilever interaction compared with the tunneling-based SFM. However, the convenient and high-sensitivity optical detection methods described below have limited the application of the capacitance detection method which has mainly been used for magnetic force (Göddenhenrich et al., 1990a) and frictional force studies (Neubauer et al., 1990).

2.3.2.3 Laser beam deflection

Most SFM designs used nowadays are based on optical detection methods for sensing the small cantilever displacements. Perhaps the simplest optical method is laser beam deflection (Meyer and Amer, 1988; Alexander et al., 1989), as schematically illustrated in Fig. 2.7. The cantilever displacement is measured by detecting the deflection of the laser beam which is reflected off the rear side of the cantilever. The direction of the reflected laser beam is sensed by a position-sensitive detector (PSD), typically a bicell, which consists of two photoactive (e.g. Si) segments (anodes) that are separated by about 10 μm and have

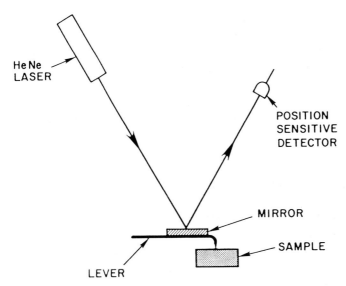

Fig. 2.7. Cantilever deflection detection scheme (Meyer and Amer, 1988).

a common cathode. Sub-Ångström unit sensitivity is routinely achieved with this method.

In contrast to the tunnel tip in the tunneling detection method, the laser beam exerts only negligible forces on the cantilever. Additionally, the laser beam deflection method is much less sensitive to the surface roughness and contamination of the cantilever. However, it still requires a mirror-like surface at the rear side of the cantilever. Furthermore, for optical detection methods the cantilever must be large enough to reflect light without introducing too much diffraction, whereas the size of the cantilever can be almost arbitrarily small if tunneling detection is used.

For SFMs to be operated under UHV conditions, optical detection systems are certainly easier to implement than tunneling detection, which requires a double-junction geometry. On the other hand, for SFMs to be operated in cryogenic environments, the amount of heating of the cantilever by the laser beam has to be considered carefully. The laser beam deflection method has mainly been used for contact force microscopy studies where atomic resolution can routinely be achieved. However, it can also be applied in non-contact force measurements (Ducker *et al.*, 1990).

2.3.2.4 Optical interferometry

Optical interferometry is another important method to sense small cantilever deflections with sub-Ångström unit accuracy. Several different interferometer systems have been developed, based either on homo-dyne (McClelland *et al.*, 1987; Erlandsson *et al.*, 1988b; den Boef, 1989) or heterodyne (Martin *et al.*, 1987) detection methods (Fig. 2.8(*a*) and (*b*)). Advances in the homodyne detection method have mainly come

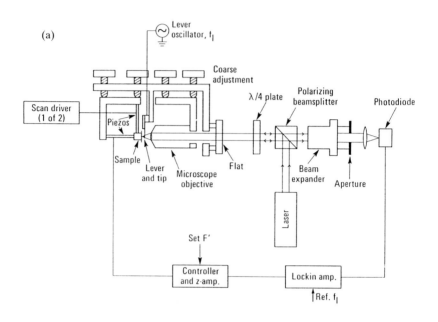

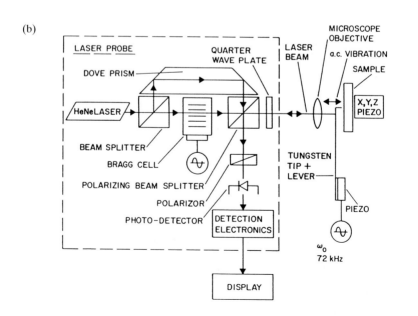

through a fiber-optic technique that places a reference reflector within micrometers of the cantilever (Rugar *et al.*, 1988, 1989; Mulhern *et al.*, 1991). The cantilever deflection measurement is then based on the optical interference occurring in the micrometer-size cavity formed

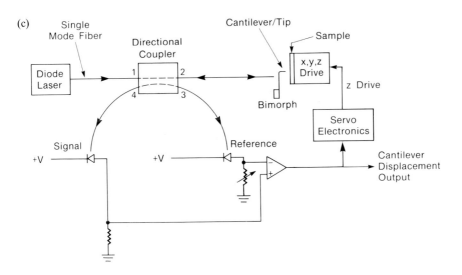

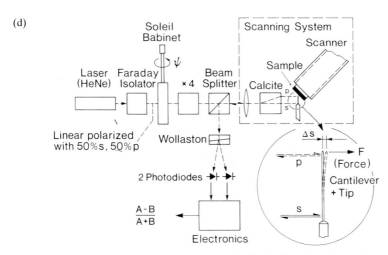

Fig. 2.8. (*a*) Schematic diagram of an AFM design based on homodyne interferometry for cantilever deflection detection (Erlandsson *et al.*, 1988b). (*b*) Experimental set-up for force derivative measurement as a function of tip–sample spacing based on heterodyne interferometry (Martin *et al.*, 1987). (*c*) Schematic diagram of a force microscope based on a fiber-optic interferometer (Rugar *et al.*, 1989). (*d*) Schematic diagram of a force microscope with an optical differential displacement sensor (Schönenberger and Alvarado, 1989).

between the cleaved end of a single-mode optical fiber and the cantilever (Fig. 2.8(c)). Alternatively, a polarizing optical interferometer based on a two-beam differential technique has been developed (Schönenberger and Alvarado, 1989), where the deflections of the cantilever are measured by means of the phase shift of two orthogonally polarized light beams, both reflected off the rear side of the cantilever (Fig. 2.8(d)). This differential interferometer is relatively insensitive to intensity fluctuations of the light source and to perturbations arising from fluctuations of the optical path length.

Compared with the laser beam deflection method, the interferometric detection of the cantilever deflections has the advantage that the cantilever does not need to have a mirror-like reflecting surface. The interferometric detection of the cantilever deflections has initially mainly been applied for non-contact force microscopy using the dynamic (a.c.) mode of SFM operation where the cantilever is vibrated near its resonant frequency (section 2.7). The improved interferometer systems can, however, also be used for contact force microscopy where quasistatic deflections of the cantilever have to be measured.

2.4 Topographic imaging by SFM in the contact mode

In this section we will concentrate on the various modes of SFM operation when tip and sample are in contact. In this case, the interaction force causes the cantilever to deflect quasistatically according to Hooke's law (Eq. (2.4)), and this deflection is directly measured. The first SFM traces were recorded in this contact regime by using electron tunneling detection of the cantilever deflections (Binnig *et al.*, 1986c). More recently, laser beam deflection has become the most widely used technique for sensing quasistatic cantilever deflections in the contact mode. Improved interferometric-based systems, as described in section 2.3.2, are also used for contact force microscopy.

2.4.1 Constant force imaging (CFI)

The most important mode of SFM operation is the constant force imaging (CFI) mode which is analogous to the constant current imaging (CCI) mode in STM (section 1.11.1). The condition of constant force is achieved by keeping the cantilever deflection constant by means of a feedback circuit. For those SFM designs where the sample itself is scanned against the force sensor, this feedback system is connected to the sample z piezodrive. The output signal of the feedback loop U_z, which adjusts the vertical z position of the sample to achieve a constant cantilever deflection (constant force), can be recorded as a function of the (x,y) coordinates which are determined by the corresponding voltages U_x and U_y applied to the x and y piezoelectric drives. The obtained signal $U_z(U_x,U_y)$ can finally be translated into the 'topography' $z(x,y)$, provided that the sensitivities of the three orthogonal piezoelectric drives are known.

Similarly to the CCI mode in STM, the principle of the CFI mode in SFM sounds rather simple. However, we must again discuss the meaning of the 'topography' $z(x,y)$. According to the experimental procedure described above, equiforce surfaces are measured in SFM, ultimately with atomic resolution. Therefore, we have to consider the contributions to the tip–surface force interaction in the contact regime.

2.4.1.1 Interpretation of equiforce surfaces measured by contact force microscopy

The invention and development of SFM with its atomic resolution capability in the contact mode has certainly led to a novel perception about interatomic forces acting in systems which have to be treated quantum-mechanically. Historically, the force concept in quantum physics and chemistry has been relatively neglected. From the viewpoint of quantum theory, the reason might lie in the fact that forces are not quantum-mechanical constants of motion as they do not commute with the corresponding Hamiltonians. Consequently, forces were in most cases looked upon as gradients of appropriate potential energy functions which are considered as 'fundamental' (Deb, 1973). From an experimental point of view, measurement of energy levels has always been found easier for microscopic systems than force measurements. The invention of the force microscope has certainly contributed to increased interest in the force picture of quantum-mechanical systems.

The fundamental relationship between energy and force pictures in quantum theory is expressed by the Hellmann–Feynman theorem (Hellmann, 1937; Feynman, 1939; Slater 1972; Deb, 1973), which states that if ψ is an exact eigenfunction of a Hamiltonian H, and E is the corresponding energy eigenvalue, then

$$\frac{\partial E}{\partial \lambda} \langle \psi \mid \psi \rangle = \left\langle \psi \left| \frac{\partial H}{\partial \lambda} \right| \psi \right\rangle \tag{2.8}$$

for any parameter λ occurring in the Hamiltonian H. This means that, for a normalized wave function, the first derivative of the energy with respect to parameter λ is equal to the expectation value of the corresponding derivative of the Hamiltonian. If λ is taken as a position coordinate of a nucleus, one can derive the electrostatic Hellmann–Feynman theorem (Deb, 1973) which tells us that the force acting on a nucleus in any system of nuclei and electrons can be interpreted solely in terms of classical electrostatics, once the electronic charge density has been obtained by an accurate self-consistent quantum-mechanical electronic structure calculation procedure. However, this theorem is very sensitive to inaccuracies in the determination of the wave function ψ. Nevertheless, it is of central importance for interpretation of SFM data, similarly to Bardeen's transfer Hamiltonian formalism for the STM (section 1.11.1).

Ciraci *et al.* (1990a) have applied the Hellmann–Feynman theorem for derivation of atomic forces in real space by differentiating those terms in the Hamiltonian which explicitly depend on the position of the ions. The wave functions and electron charge densities were obtained from electronic structure calculations performed using the self-consistent-field (SCF) pseudopotential approach in momentum space within the local-density approximation (LDA). The expression obtained for the atomic force was found to have two components. The first, denoted as F_{ion}, originates from the Coulomb repulsion between the ion cores, and the second, denoted as F_{el}, is due to the interaction of valence electrons with the ion cores. At small tip–surface separations, the repulsive force F_{ion} is stronger ($|F_{ion}| > |F_{el}|$) and varies more rapidly with the position of the outermost tip atom than does F_{el}. Therefore, SFM operated in the repulsive contact mode is expected to be mainly sensitive to the repulsive Coulomb interaction between the ion cores of the tip and those of the sample surface. Consequently, SFM is expected to directly probe the position of the ion cores, in contrast to STM where the contrast observed is dominated by the local surface electronic structure near the Fermi level which can significantly differ from the geometric arrangement of the ion cores (section 1.11.1).

As the tip–surface separation is increased, $|F_{el}|$ decays more slowly than $|F_{ion}|$ and $F_{total} = F_{ion} + F_{el}$ changes its sign, leading to a net attractive force. In this regime, the SFM would mainly probe the total charge density distribution of the sample surface rather than ion–ion repulsion. However, most SFM studies are performed in the strongly repulsive regime. In fact, the repulsive force on the outermost tip atom is always underestimated because attractive long-range (VDW) forces felt by tip atoms far from the outermost one can contribute significantly to the total force acting on the tip. In cases of relatively large tip radii of about 1000 Å, the contribution of these long-range attractive forces might be as large as 10^{-8} N or even more (Goodman and García, 1991), so that while the total tip force might be attractive, the outermost tip atoms can still be in the strong repulsive force regime, eventually leading to a local surface deformation, particularly for soft elastic materials (Tománek *et al.*, 1989).

In summary, SFM measurements in the contact regime are expected to probe primarily the ion–ion repulsion forces which decay rapidly with increasing tip–surface separation, very similar to the behavior of the tunneling current in STM. The strong distance-dependence of the ion–ion repulsion forces provides the key for the high spatial resolution achieved by contact force microscopy.

2.4.1.2 Spatial resolution in contact force microscopy

Experimentally, it has been demonstrated that atomic-scale periodicities can be resolved by SFM operated in the contact regime for different classes of materials, including layered materials (e.g. graphite, BN, transition metal dichalcogenides, mica etc.), ionic crystals (e.g. LiF, NaCl, AgBr, KBr etc.) and metals (e.g. Au). For layered materials, such as graphite, atomic scale periodicities were resolved with loading forces of 10^{-8}–10^{-7} N (Binnig *et al.*, 1987a, 1987b; Albrecht and Quate, 1988). Under the assumption of a monatomic tip, it has been shown theoretically that repulsive forces of around 10^{-8} N can lead to a large elastic compressive deformation of the graphite surface and that larger forces should even lead to surface destruction because the monatomic tip then punctures the graphite surface (Abraham and Batra, 1989; Zhong *et al.*, 1991).

To explain the atomically resolved SFM images obtained at loading forces as large as 10^{-7} N or even more, it is assumed that the effective tip is composed of several atoms. In particular, it has been suggested that the SFM tip may pick up a graphite flake which is dragged along the surface while scanning. In this case, the SFM would not image single atoms but repulsion maxima associated with misregistry of the graphite tip unit cell and the graphite surface unit cell. By assuming such a multi-atom tip, anomalous SFM images of the graphite surface were successfully modeled (Abraham and Batra, 1989; Gould *et al.*, 1989). In addition, for a multi-atom tip, the force per atom can be sufficiently small as to cause no surface damage. Such an imaging mechanism could also explain SFM results obtained on other layered materials, e.g. boron nitride (BN) and mica, where only the unit-cell periodicity is resolved, rather than the atomic or molecular structure within the unit cell.

Alternatively, it has been proposed that for BN the SFM is only sensitive to the 'harder' N atoms and does not image the 'softer' B atoms, although they are larger (Albrecht and Quate, 1987; Albrecht, 1989). However, for an experimentally measured loading force as large as 2×10^{-7} N on BN, this force is likely to be distributed over several atoms, and the argument for preferential imaging of N sites cannot be applied.

SFM studies of ionic crystals (e.g. LiF, NaCl and AgBr) proved that atomic-scale periodicities are observable for non-layered materials as well, where the atomic-scale contrast can no longer be explained by the dragging of a flake attached to the tip against the sample surface. However, SFM images of ionic crystal surfaces again show only the unit-cell periodicity. This has been explained in terms of preferential imaging of the sites occupied by the larger negatively charged ions F^-, Cl^- and Br^- (Meyer and Amer, 1990a; Meyer *et al.*, 1990a, 1991a, 1991b).

For a conclusive demonstration of the atomic-resolution capability of contact force microscopy, the observation of surface defects has been particularly important (Binnig, 1992). This means that the force interaction must indeed be spatially highly localized, offering the possibility of probing single atomic sites with the SFM as well, similarly to the STM.

2.4.2 Variable deflection imaging (VDI)

Apart from constant force imaging there exists another important mode of SFM operation which is called variable deflection imaging (VDI). In this mode, the cantilever deflection, and therefore the force, is allowed to change during scanning and these deflections are measured with a suitable detector (section 2.3.2). This mode of SFM operation is analogous to the variable current mode in STM (section 1.11.2), for which the tunneling current instead of the force is monitored while scanning the probe tip. Significantly higher scan rates can be achieved in the VDI mode compared with the constant force mode for which the finite response time of the feedback circuit limits the scan speed. On the other hand, the measured data in the variable deflection mode are more difficult to interpret because the measured contours can generally no longer be considered as equiforce surfaces. However, for SFM images of small surface regions, the relative variations of the force seldomly exceed a few percent and the data obtained in the variable deflection mode can be interpreted similarly to the equiforce mode. In contrast, the force may vary significantly for larger scale images, where the influence of the local force gradient on the recorded signal has to be taken into account (Heinzelmann, 1989).

2.4.3 Differential force microscopy (DFM)

Similar to differential tunneling microscopy (DTM), discussed in section 1.11.3, a lateral modulation mode can also be introduced in contact force microscopy (Maivald *et al.*, 1991). The sample is vibrated parallel to its surface plane (*x* direction) at a frequency well above the feedback response frequency but below the first mechanical resonance frequency. The differential image $\partial F(x,y)/\partial x|_F$ obtained by lock-in detection usually exhibits an improved signal-to-noise ratio and an improved contrast compared with the conventional topographic image.

2.5 Frictional force microscopy (FFM)

Friction occurs if two bodies in contact are in relative motion with respect to one another (Bowden and Tabor, 1950). The theory of friction is governed by Amontons' laws which tell us that the frictional force F_f is proportional to the loading force F_l and independent of the apparent area of contact:

$$F_f = \mu \cdot F_l \qquad (2.9)$$

where μ is the friction coefficient.

To explain the relationship (2.9), we have to consider the structural properties of the contact interface. Since surfaces are usually rough, at least on a microscopic scale, the actual area of contact is limited to discrete spots of the interface between the two bodies. It seems reasonable to assume that the frictional force F_f is proportional to the actual area of contact A:

$$F_f = f \cdot A \qquad (2.10)$$

where the proportionality constant f is the shear strength of the junction. Furthermore, it has been shown that for a rough surface with an exponential distribution of asperity heights, the actual area of contact when such a surface is pressed onto a smooth flat surface must always be proportional to the load F_l, whatever the law of deformation or the shape of the asperities (Greenwood, 1967):

$$A \propto F_l \qquad (2.11)$$

By combining Eq. (2.10) with Eq. (2.11) we finally obtain the basic law of friction (2.9).

To derive the essential relationship (2.11) for a macroscopic contact, we always have to start with a statistical description of the surface roughness for the two bodies in contact. On the other hand, if we would start with a sharp tip in contact with a surface to be studied, we can hope to arrive at an understanding of the microscopic aspects of frictional phenomena. The geometry offered by a SFM is ideally suited for such friction studies on a microscopic scale (McClelland, 1989; McClelland and Cohen, 1990).

Besides measuring the forces normal to the sample surface by detecting the induced bending mode deflections of the cantilever in this direction, it is also possible to study the lateral (frictional) forces with a SFM by detecting the induced torsion mode deflections of the cantilever while scanning (Fig. 2.9). Several different methods can be used for detection of the torsion mode deflections of the cantilever. The first atomic-scale frictional force study was performed using optical interferometry as schematically illustrated in Fig. 2.10 (Mate *et al.*, 1987). More recently developed frictional force microscopes (FFM) allow simultaneous measurement of cantilever deflections in two orthogonal directions. This is achieved, for instance, by a two-dimensional capacitance-based force detector as shown in Fig. 2.11a (Neubauer *et al.*, 1990;

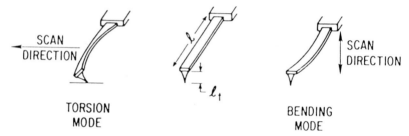

Fig. 2.9. Schematic representation of bending and torsional motions of an AFM cantilever with integrated tip (Meyer and Amer, 1990b).

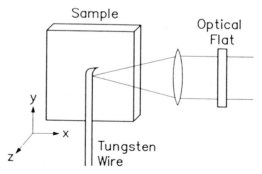

Fig. 2.10. Schematic diagram of a frictional force microscope. The base of the tungsten wire is held fixed, while the sample is moved in the *x*, *y* and *z* directions. Wire deflections parallel to the surface are measured from the intensity change in the interference pattern between light reflected off the wire and light reflected off the optical flat (Mate *et al.*, 1987).

Cohen *et al.*, 1990). A force normal to the surface plane brings the upper plate of this force sensor closer to both of the bottom electrodes and thus increases the capacitance. Lateral (frictional) forces cause twisting of the upper plate, which brings it closer to one of the bottom electrodes and further away from the other one. The frictional force can therefore be monitored by the difference signal from the two bottom electrodes.

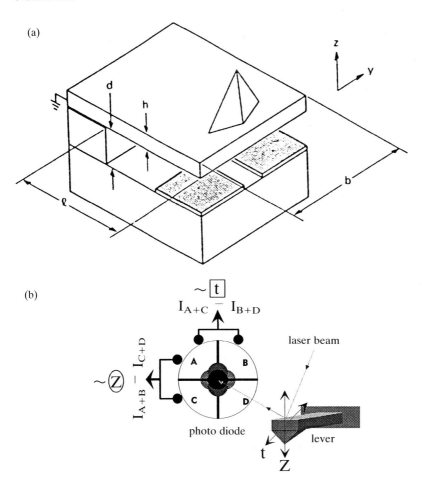

Fig. 2.11. Schematic views of two configurations for a force microscope based on a bidirectional sensor: (*a*) integrated capacitance sensor with tip (Neubauer *et al.*, 1990); (*b*) optical beam deflection with quadrant PSD. The intensity difference of the upper and lower segments of the photo diode is proportional to the z-bending of the cantilever. The intensity difference between the right and left segments is proportional to the torsion *t* of the cantilever (Overney, 1992).

Alternatively, the normal and lateral (frictional) forces can be monitored simultaneously using the laser beam deflection method which measures the orientation of the cantilever and not only its displacement (Meyer and Amer, 1990b; Marti *et al.*, 1990). For conventional force microscopy, bending of the cantilever due to the normal force is detected by a two-segment position-sensitive detector as described in section 2.3.2. Additional acting lateral forces induce a torsion of the cantilever which causes the reflected laser beam to change its direction perpendicular to that due to the bending of the cantilever. Therefore, a combination of two orthogonal bicells, i.e. a quadrant PSD, allows detection of the normal and lateral forces simultaneously (Fig. 2.11b).

Atomic-scale friction was first observed with a tungsten tip scanning a graphite surface (Mate *et al.*, 1987). The tip sliding process was found to be non-uniform, exhibiting a stick–slip behavior. When the restoring force from the bent cantilever becomes high enough to overcome the static friction, the tip begins to slip across the surface in discrete increments. The slips occur instantaneously within the experimentally resolved time scale of 200 μs, while between slips the tip moves with the surface, i.e. tip and sample surface stick together. It was observed that the slips actually exhibit the same spatial periodicity as the graphite surface, leading to the conclusion that the atomic surface structure influences frictional properties of the tip–surface interface. Since the experiment was performed with a loading force on the order of 10^{-5} N, the apparent area of contact is believed not to be of atomic dimensions, but rather on the order of 10^6 Å^2.

Similar experimental observations of atomic-scale friction involving stick–slip behavior were made for the cleavage plane of mica (Fig. 2.12) where the frictional force varies with the periodicity of the hexagonal layer of SiO_4 units (Erlandsson *et al.*, 1988a). Stick–slip phenomena were also revealed in molecular dynamics simulations of a reactive tip–substrate system with tip and substrate consisting of the same material (silicon), and which were scanned against each other under constant force conditions (Landman *et al.*, 1989a, 1989b).

The coefficient of friction μ (Eq. (2.9)) as measured in the microscopic friction studies using a tungsten tip and a graphite sample was found to be on the order of 0.01 and increased slightly with increasing loading force (Mate *et al.*, 1987). This behavior, reflecting a nonlinear F_f versus F_l relationship, has been explained within a first-principles theory of atomic-scale friction (Zhong and Tománek, 1990). A more recent theoretical study of atomic-scale friction led to the conclusion that, for a given loading force, the friction coefficient, and therefore

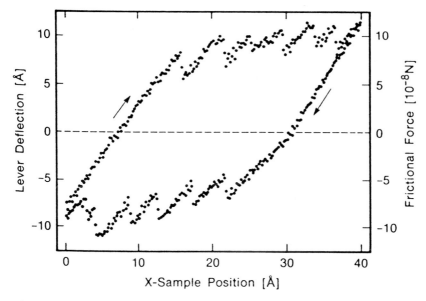

Fig. 2.12. The cantilever deflection and corresponding frictional force in the *x* direction as a function of sample position as a mica sample is scanned back and forth under a tungsten tip (Erlandsson *et al.*, 1988a).

the friction force as well, depend strongly on the interaction potential between the two materials in contact and, even more critically, on the intrinsic spring constant c of the friction force microscope (Tománek, *et al.*, 1991). The effective spring has to be soft enough because no friction can occur if c exceeds a critical value c_{crit} (F_l) for a given loading force F_l. Thus the measured friction force depends on the construction parameters of the FFM. On the other hand, for a given spring constant, the friction force is zero unless a minimum load is exceeded.

2.6 Force spectroscopy (FS)

In this section we will concentrate on the distance-dependence of the measured forces and the information which can be extracted. The investigation of the force–distance relationship has been denoted as force spectroscopy. However, force spectroscopy does not correspond to tunneling spectroscopy in STM, but rather to local tunneling barrier height measurements based on determination of the current–distance relationship.

2.6.1 Local force spectroscopy

In local force spectroscopy (LFS) the force versus distance curve is determined at a particular location on the sample surface. Experimentally, the deflection of the cantilever z_t is measured as a function of the movement of the sample z_s in the z direction perpendicular to the sample surface, yielding z_t (z_s) plots (Meyer *et al.*, 1988; Meyer, 1990). The force F is then obtained by multiplying z_t with the known spring constant of the cantilever c:

$$F = c \cdot z_t \tag{2.12}$$

If we neglect elastic deformations of the sample and the tip, particularly in the contact regime, the interaction distance s between tip and sample is given by

$$s = z_t - z_s \tag{2.13}$$

Therefore, measured z_t (z_s) plots can directly be translated into force–distance curves.

In Fig. 2.13 a simulation of a z_t (z_s) plot for a given $F(s)$ dependence (inset) is shown. As the sample approaches towards the cantilever tip, the cantilever bends towards the sample (z_t negative) due to an attractive force (negative sign). At point 1 the gradient of the attractive force exceeds the spring constant c of the cantilever which leads to an instability resulting in a jump to point 2. The maximum forward deflection of the cantilever at point 1 multiplied by the effective spring constant c of the cantilever yields the maximum attractive force F_{att}. As forward motion of the sample is continued, the sample pushes the cantilever back through its original position at large tip–surface separation, corresponding to zero net interaction force. At this point, the attractive and repulsive forces cancel each other. The sample may further be moved in the forward direction until a predetermined loading force F_l applied by the cantilever tip is achieved. If the direction of sample movement z_s is then reversed, the direction of cantilever movement z_t changes as well. The loading force is continuously decreased and finally one enters the attractive force regime again, where tip and sample adhere to each other. At point 3, a second instability occurs because the force gradient has again become equal to the effective spring constant of the cantilever, and the cantilever tip jumps out of contact. Point 3 therefore corresponds to the position of maximum adhesive force F_{adh}. In general,

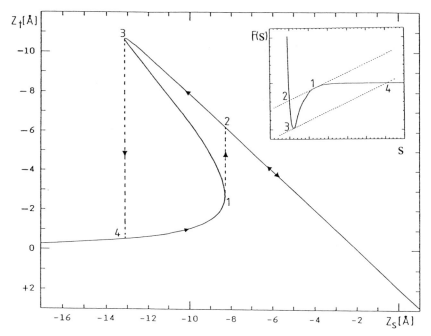

Fig. 2.13. Simulation of a $z_t(z_s)$ characteristic with corresponding force–distance curve. At points of instability 1 and 3, the force gradient becomes equal to the spring constant of the cantilever (Meyer, 1990).

the maximum adhesive force at point 3 is greater than the maximum attractive force at point 2.

In Fig. 2.14 an experimental $z_t(z_s)$ plot is shown which closely resembles the simulated behavior in Fig. 2.13. The significance of the $z_t(z_s)$ plots or, equivalently, of the $F(s)$ curves lies in the fact that the experimentalist can choose to image the sample at a particular point along the $F(s)$ curve, thereby knowing the applied force during imaging within reasonable accuracy. This information is important because force microscope images may depend on the applied loading force (Burnham *et al.*, 1991). A force controller has been developed for automatic re-adjustment of the force to a preset value (Marti *et al.*, 1988).

In addition to establishing the point of contact for imaging purposes, force versus distance curves can provide other valuable kinds of information (Burnham and Colton, 1989; Burnham *et al.*, 1991).

1. The maximum attractive force F_{att} and the maximum adhesive force F_{adh} can be determined locally.

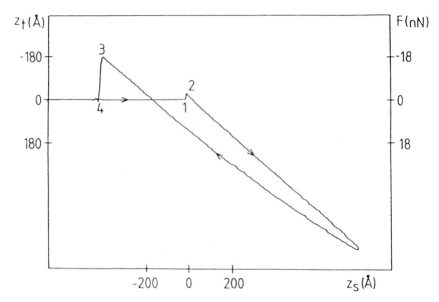

Fig. 2.14. Experimental $z_t(z_s)$ characteristic obtained with a 1T-TaS$_2$ sample and a SiO$_2$ cantilever of spring constant 1 N m^{-1} (Meyer, 1990).

2. The slope of the curve after contact has been established may provide information about local mechanical properties, surface forces, and the local geometry of tip and sample.
3. The amount of hysteresis in the $F(s)$ curve may be indicative for an inelastic response of the tip–sample system.

Unfortunately, a unique interpretation of $F(s)$ curves often cannot be given for the following reasons.

1. If the experiments are performed in ambient air, instability of the cantilever can also arise from formation of a meniscus around the tip as a result of capillary condensation pulling the tip towards the sample. The presence of additional capillary forces complicates quantitative analysis of the tip–sample interaction considerably. The influence of capillary forces can be demonstrated by comparing $z_t(z_s)$ plots obtained in air and in water (Weisenhorn *et al.*, 1989). If the cantilever is fully immersed in water, capillary forces are absent. Under these conditions, the lowest tip–sample interaction force for SFM operation can be reduced typically by two orders of magnitude compared with ambient conditions. Therefore, delicate soft samples, requiring extremely low loading forces for non-destructive imaging, should ideally be studied under liquids.

2. Hysteresis in $F(s)$ curves may not necessarily be due to plastic deformation, but can also result from hysteresis and creep of the piezoelectric ceramics.
3. Lack of knowledge of the detailed tip geometry often prevents quantitative analysis of $F(s)$ curves.

Despite these difficulties in interpretation, $F(s)$ curves measured under an atmosphere of dried and filtered nitrogen have already provided valuable information about surface forces of molecular films (Burnham et al., 1990; Blackman et al., 1990). Analysis of $F(s)$ curves can also yield estimates for the thickness of polymeric liquid films deposited on a solid substrate (Mate et al., 1989b; Mate and Novotny, 1991). Furthermore, force microscopes can be used for highly sensitive nano-indentation measurements by exploiting the repulsive regime of the $F(s)$ curve. Since the applied loads can be chosen to be considerably smaller than those commonly used in conventional indentation hardness testers, force microscopes have a great potential for nanohardness measurements as well (Pethica et al., 1983).

2.6.2 Scanning force spectroscopy (SFS)

Similarly to tunneling barrier height measurements or tunneling spectroscopy in STM (section 1.12 and 1.13), where the determination of local log I–s or I–U characteristics can be extended to each pixel point in a corresponding topographic image, the determination of $F(s)$ curves can be performed at each data point. A particular physical quantity extracted from the local $F(s)$ curves, as discussed in the previous section, may then be plotted as a function of surface location, thereby providing a spatially resolved map.

2.6.2.1 Spatial adhesion maps

A physical quantity of interest, which can easily be extracted from a local $F(s)$ curve, is the maximum adhesion force F_{adh} (section 2.6.1). By using a force microscope with a silicon nitride tip and a polycarbonate film sample, Mizes et al. (1991) have studied the spatial variation of the maximum adhesion force which was found to depend on the surface topography and material inhomogeneities. The spatial resolution in the adhesion maps achieved in these measurements was on the order of a few hundred Ångström units. For each position, no increase in the adhesion was seen with repeated contacts. In Fig. 2.15 the adhesion map for a doped polycarbonate film is compared with the surface topography before and after performing the spatially resolved adhesion measure-

(a) (b) (c)

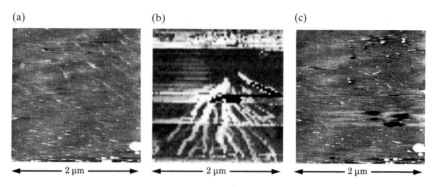

Fig. 2.15. (*a*) Topography of a polycarbonate film. (*b*) Spatial variation in adhesion over the same area. (*c*) Topography of the same area after the adhesion measurement (Mizes *et al.*, 1991).

ment. The increase in surface roughness after the adhesion measurement is small, typically on the order of a few nanometers over a lateral length scale of a few micrometers. High-resolution spatial mapping of adhesion forces by scanning force spectroscopy constitutes significant progress compared with earlier tip–surface contact experiments (Pashley and Pethica, 1985).

Similarly, the maximum attractive force F_{att} (section 2.6.1) may be extracted from the local $F(s)$ curves and mapped as a function of surface location.

2.6.2.2 Spatial maps of surface elasticity

As an alternative to determination of full $F(s)$ curves at each pixel point, a force modulation mode was introduced to probe spatial variations in local surface elasticity (Maivald *et al.*, 1991). This spectroscopic method is based on modulating the sample height (z position) by a fixed amount Δz_s. Motion of the sample causes the cantilever to deflect periodically by an amount Δz_c. For a given applied loading force, a soft surface region deforms more than a hard one. Consequently, the measured cantilever deflection Δz_c is smaller over a soft surface region. By plotting the normalized quantity $\Delta z_c/\Delta z_s$ as a function of surface location, a spatial map of surface elasticity can therefore be obtained.

To achieve reasonable contrast in the images of surface elasticity, the spring constant of the cantilever has to be chosen appropriately. Assuming that the tip material is much harder than the sample, a measured cantilever deflection Δz_c caused by a sample motion Δz_s indicates a sample surface deformation by an amount ($\Delta z_s - \Delta z_c$). The response of the sample surface to a force ΔF can be described by an effective spring constant c_s:

$$c_s = \frac{\Delta F}{(\Delta z_s - \Delta z_c)} \qquad (2.14)$$

Since

$$\Delta F = c \cdot \Delta z_c \qquad (2.15)$$

where c is the effective spring constant of the cantilever, we obtain the following relation:

$$\frac{\Delta z_s}{\Delta z_c} = \frac{c}{c_s} + 1 \qquad (2.16)$$

If the effective spring constant c of the cantilever were chosen considerably smaller than the effective spring constant of the sample surface $c_s (c \ll c_s)$, the measured normalized quantity $\Delta z_c / \Delta z_s$ would always be on the order of one and the spatial variations of c_s would not show up in the $(\Delta z_c / \Delta z_s)$ signal. Therefore, appropriate choice of the effective spring constant of the cantilever depends on the effective spring constant of the sample to be studied.

In Fig. 2.16 a comparison is shown between a topographic SFM image and a simultaneously recorded spatial map of the surface elasticity for a carbon fiber and epoxy composite. To obtain high contrast in the surface elasticity map, an effective cantilever spring constant of $c \simeq 3000$ N m^{-1} was chosen, which is much larger than usually used for conventional contact force microscopy.

2.7 Non-contact force microscopy

In contact force microscopy, discussed in section 2.4, the short-range interatomic forces are probed by measuring the quasistatic deflections of a cantilever beam with known effective spring constant, as the sample is scanned against the cantilever tip. Though long-range van der Waals (VDW) forces are also present in the contact mode of SFM operation, they do not contribute to the image contrast on an atomic scale (section 2.7.1). However, VDW forces can contribute considerably to the total tip force in contact force microscopy, as already mentioned in section 2.4.1.

By increasing the tip–surface separation to 10–100 nm, only the long-

(a)

(b)

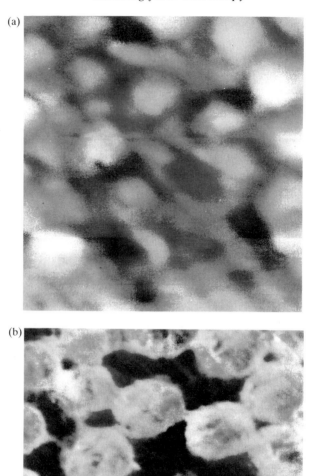

Fig. 2.16. Image of a carbon fiber and epoxy composite in air. Intensity corresponds to (*a*) height in the topographic image, (*b*) stiffness in the force modulation image. The image width is 32 μm (Maivald *et al.*, 1991).

range interaction forces, e.g. VDW, electrostatic, and magnetic dipole forces, remain. These forces can be probed by non-contact force microscopy. Since the magnitude of the long-range forces at relatively large tip–surface separation may be considerably smaller than that of the short-range interatomic forces, the method of interaction force detection in the non-contact mode of SFM operation is usually different from contact force microscopy. Instead of measuring quasistatic cantilever deflections, the cantilever is driven to vibrate near its resonant frequency by means of a piezoelectric element (Binnig *et al.*, 1986c), and changes in the resonant frequency as a result of tip–surface force interaction are measured.

This a.c.-detection method is sensitive to force gradients, rather than to the interaction forces themselves. The presence of a force gradient $F' = \partial F_z/\partial z$ results in a modification of the effective spring constant of the cantilever according to

$$c_{\text{eff}} = c - F' \tag{2.17}$$

where c is the spring constant of the cantilever in the absence of a tip–surface force interaction. An attractive tip–surface force interaction with $F' > 0$ will therefore soften the effective spring ($c_{\text{eff}} < c$), whereas a repulsive tip–surface force interaction ($F' < 0$) will strengthen the effective spring ($c_{\text{eff}} > c$). The change of the effective spring constant causes, in turn, a shift in the resonant frequency ω of the cantilever according to

$$\omega = \left(\frac{c_{\text{eff}}}{m}\right)^{1/2} = \left[\frac{(c - F')}{m}\right]^{1/2} = \left(\frac{c}{m}\right)^{1/2}\left(1 - \frac{F'}{c}\right)^{1/2}$$

$$= \omega_0\left(1 - \frac{F'}{c}\right)^{1/2} \tag{2.18}$$

where m is an effective mass and ω_0 is the resonant frequency of the cantilever in the absence of a force gradient. If F' is small relative to c, then Eq. (2.18) can be approximated by

$$\omega \approx \omega_0\left(1 - \frac{F'}{2c}\right) \tag{2.19}$$

and therefore

$$\Delta\omega = \omega - \omega_0 \approx -\frac{F'}{2c} \tag{2.20}$$

An attractive force with $F' > 0$ will therefore lead to a decrease of the resonant frequency ($\omega < \omega_0$), whereas a repulsive force ($F' < 0$) will lead to an increase ($\omega > \omega_0$).

Two different methods are commonly used to measure the resonance frequency shift.

(a) In the slope detection method the cantilever is driven by means of a piezoelectric element with a typical amplitude at the tip on the order of 1–10 nm and at a fixed frequency ω_d, which is close to the resonant frequency. The amplitude change or phase shift of the vibration as a result of the tip–surface force interaction is measured with the deflection sensor, usually an optical interferometer (section 2.3.2), and a lock-in amplifier (Martin *et al.*, 1987; Erlandsson *et al.*, 1988b). A feedback loop adjusts the tip–surface separation by maintaining a constant force gradient.

(b) In the frequency modulation (FM) detection method oscillation of the cantilever is maintained by a feedback loop using the signal from the deflection sensor. Changes in the oscillation frequency, caused by variations of the force gradient of the tip–sample interaction, are directly measured by a frequency counter or an FM discriminator.

For cantilevers with a small spring constant c, the minimum detectable force gradient is determined by thermal vibration of the cantilever. Its rms amplitude can be derived from the equipartition theorem:

$$\frac{1}{2} c \left\langle (\Delta z)^2 \right\rangle = \frac{1}{2} k_\mathrm{B} T \tag{2.21}$$

$$\left[\left\langle (\Delta z)^2 \right\rangle \right]^{\frac{1}{2}} = \left(\frac{k_\mathrm{B} T}{c} \right)^{\frac{1}{2}} \tag{2.22}$$

where T is the temperature and k_B is Boltzmann's constant. For a typical spring constant of $1\ \mathrm{N\ m^{-1}}$, the thermal vibration amplitude at room temperature is therefore about 0.6Å.

For the FM detection method, the minimum detectable force gradient F'_min is given by

$$F'_\mathrm{min} = \frac{1}{A} \left(\frac{4 c\, k_\mathrm{B}\, TB}{\omega_0\, Q} \right)^{\frac{1}{2}} \tag{2.23}$$

where A is the rms amplitude of the cantilever oscillation, Q is the quality factor of the cantilever and B is the detection bandwidth. To achieve the highest possible detection sensitivity for a given oscillation amplitude A and detection bandwidth B, the following general rules can be derived from Eq. (2.23).

1. The cantilever should have a small spring constant c and a high resonance frequency ω_0. This requires minimization of the mass and therefore size of the cantilever, leading to the need for microfabrication methods (section 2.3.1).
2. A high Q factor of the cantilever is desirable, which is easily achieved for microfabricated cantilevers used in vacuum where Q values of 10^4 and even higher can be attained.
3. Measurements at low temperatures are preferable.

At room temperature, the minimum detectable force gradient is typically on the order of 10^{-4}–10^{-5} N m^{-1}. Assuming a force law (Eq. (2.3))

$$F(s) \propto -\frac{1}{s^2}$$

then

$$|F(s)| = \frac{sF'(s)}{2} \tag{2.24}$$

For a tip–surface separation $s = 10$ nm (100 nm), a force gradient of $F' = 10^{-4}$ N m^{-1} (10^{-5} N m^{-1}) would therefore correspond to a force of 5×10^{-13} N. Using a cantilever with $c = 1$ N m^{-1}, this force would cause a static deflection of only 5×10^{-3} Å which would be difficult to detect with most deflection sensors. Therefore, the a.c. detection method based on vibrating the cantilever offers a significant advantage with respect to detection sensitivity compared with quasistatic cantilever deflection measurements when studying long-range forces.

2.7.1 Van der Waals (VDW) force microscopy

If tip and sample are clean, electrically neutral and non-magnetic, van der Waals (VDW) forces are the sole sources of tip–sample interaction in the non-contact regime. The usually attractive long-range VDW forces can be detected for tip–surface separations up to about 100 nm and can be exploited for non-destructive surface profiling of delicate soft samples, such as biological material that should be imaged with loading forces not exceeding 10^{-11} N (Persson, 1987). The spatial resolution achievable in VDW force microscopy depends critically on the tip geometry and tip–surface separation (Moiseev *et al.*, 1988). As a general rule, in order to resolve structures of lateral size a, both the effective tip radius R and the tip–surface separation s have to be smaller than a. The magnitude of VDW forces is not only determined by R and s, but by the dielectric permittivities of tip, sample and surrounding medium as well (Hartmann, 1990, 1991; Goodman and García, 1991). Polar immersion liquids considerably reduce VDW forces and may even cause transition from attractive to repulsive tip–sample force interaction.

In general, the following rules can be deduced from a rigorous theory of VDW forces (Hartmann, 1990).

1. VDW forces measured in vacuum are always attractive.
2. If tip and sample consist of the same material, VDW forces are always attractive, independent of the surrounding medium. However, they can become repulsive for different tip and sample materials if the intermediate medium exhibits an effective refractive index n_m lying in between that of tip (n_t) and sample (n_s), i.e. $n_t < n_m < n_s$ or $n_s < n_m < n_t$. In particular, by appropriate choice of immersion medium, VDW forces can be reduced by more than two orders of magnitude at small tip–surface separations, which is desirable for imaging delicate soft samples. Furthermore, contamination layers on tip and sample can have significant influence on measured VDW forces, depending on the actual tip–surface separation. Surface contamination of only a few monolayers can dominate VDW-type interactions, even leading to a change in sign of the force.
3. The measured forces remain the same if tip and sample materials are exchanged.
4. The contours of constant force measured by non-contact force microscopy do not necessarily correspond to the surface topography, but rather to profiles of constant dielectric response. Therefore, VDW force microscopy has the potential to provide information on the dielectric response of various media in terms of their Hamaker constants (section 2.1).

The development of non-contact force microscopes with improved force detection sensitivity will further stimulate experimental investigation of long-range VDW-type tip–sample interaction, including retardation effects (section 2.1).

2.7.2 Electrostatic force microscopy (EFM)

Non-contact force microscopy can also be used to non-destructively study the distribution of electrical charge on a surface by probing the long-range electrostatic Coulomb forces (Martin *et al.*, 1988a; Stern *et al.*, 1988). To be able to distinguish charge from other sources of tip–sample interaction forces, and to determine the sign of the charge, a new mode of imaging has been introduced (Martin *et al.*, 1988a; Terris *et al.*, 1989, 1990). As schematically illustrated in Fig. 2.17, an a.c. bias voltage, $U_0 \sin (\omega_2 t)$, is applied between the tip and an electrode behind a dielectric sample. The frequency ω_2 is chosen to be much higher than the feedback response frequency, but much lower than the oscillation frequency ω_1 of the cantilever. The a.c. voltage causes an oscillating charge Q_e on the electrode which itself induces an equal charge of opposite sign on the tip. A local static charge Q_s on the sample surface will additionally induce an equal image charge of opposite sign on the tip, and thus the total tip charge is $Q_t = -(Q_e + Q_s)$. By assuming a simple point charge model, Terris *et al.* (1989) derived an expression for the resulting tip–electrode force, and consequently the force gradient which is measured experimentally. They found that for an uncharged surface ($Q_s = 0$), the force gradient will oscillate at $2\omega_2$, whereas for a charged surface ($Q_s \neq 0$), the force gradient will be modulated at ω_2. Therefore, by detecting the ω_2 signal at the output of the feedback-loop lock-in amplifier with a second lock-in, a 'charge signal' can be extracted. The phase of that signal indicates the sign of the surface charge.

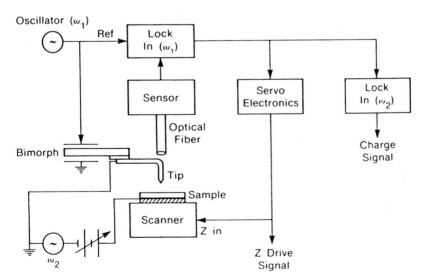

Fig. 2.17. Block diagram of a force microscope used for charge microscopy experiments (Terris *et al.*, 1989).

Insulating samples, such as solvent-cleaned polymethylmethacrylate (PMMA) and single-crystal sapphire, can be charged *in situ*, e.g. by applying a high-voltage pulse (typically 100 V for several milliseconds) to the tip, leading to a localized discharge with a threshold voltage determined by the air breakdown potential for the particular tip–sample geometry (Stern *et al.*, 1988). The sign of the deposited surface charge is determined by the polarity of the voltage pulse. As a result of the attractive interaction between the local surface charge and the induced image charge on the tip, the force gradient increases as the tip scans over the charged surface region. As shown in Fig. 2.18, the contrast is observed to decay with increasing observation time, providing evidence for surface charge mobility. The decay time constant depends critically on the particular insulator to be charged. By using a high-sensitivity EFM, it is possible to monitor the charge decay even with single-electron resolution (Schönenberger and Alvarado, 1990b). After depositing charge on insulating Si_3N_4 films by applying a voltage pulse to the tip, a discontinuous staircase in the force-signal versus time was observed, demonstrating single-charge-carrier resolution (Fig. 2.19).

Charge Decay

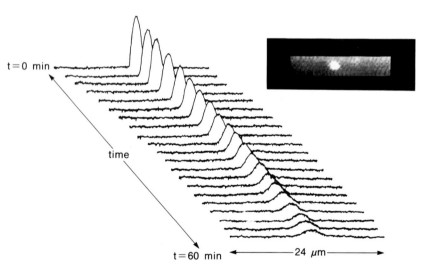

Fig. 2.18. Contours of constant force gradient taken at 3 min intervals over the center of a region of deposited charge on PMMA. The initial peak height corresponds to a 0.5 µm increase in tip-to-sample separation. Inset: gray-scale image (24 µm × 5 µm) of a negatively charged region (Stern *et al.*, 1988).

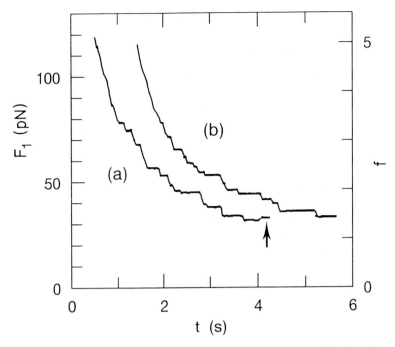

Fig. 2.19. Force versus time curves for a positively charged Si_3N_4 film 20 nm thick. The charge decay, registered by the change in the force signal, shows up as a discontinuous staircase, demonstrating single-carrier resolution (Schönenberger and Alvarado, 1990b).

As an alternative to the pulsing technique for localized surface charging, the tip can be brought into contact with the sample surface while applying a bias voltage (typically several volts) to the tip (Fig. 2.20). The sign of the deposited surface charge is determined by the polarity of the applied tip bias voltage during contact, and the greater the voltage, the more charge is transferred. On the other hand, the amount of charge transferred is nearly independent of the contact time between tip and sample and the number of contacts made.

EFM can also be used to study contact electrification, or triboelectrification, where the surface is tribocharged by bringing the tip in contact without an externally applied bias voltage (Terris *et al.*, 1989). After tribocharging a PMMA surface with a Ni tip, regions of both negative and positive charge were found, though the predominant sign of the deposited charge observed was positive in agreement with macroscopic electrostatic voltmeter measurements. The bipolar charge trans-

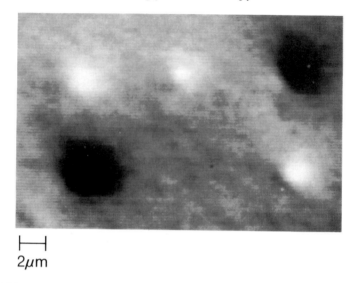

⊢——⊣
2μm

Fig. 2.20. Charge image of five deposited charge regions, three positive (white regions), and two negative (black regions), on a polycarbonate surface (Terris *et al.*, 1989).

fer was unexpected and indicates the potential of EFM for studying microscopic aspects of contact electrification.

Another important field of application for EFM is direct real-space imaging of domains and domain walls in ferroelectrics (Saurenbach and Terris, 1990). The surface polarization charge σ_p associated with the ferroelectric polarization P induces an image charge Q_t on the tip. As the tip passes over the domain wall, the polarization P changes sign and passes through zero. The resulting spatial variation of the electrostatic force interaction between the surface polarization charge and the induced image charge on the tip can be probed by EFM. In Fig. 2.21, a single-line scan of the charge signal across a ferroelectric domain wall in $Gd_2(MoO_4)_3$, measured using the modulation technique described previously (Fig. 2.17), is shown. The charge signal changes its sign as the tip passes over the ferroelectric domain wall, as expected. The transition width of the charge signal across the domain wall is mainly determined by the tip–surface separation and the size of the tip, rather than by the intrinsic width of the ferroelectric domain wall which is expected to be of atomic dimensions. However, with refinements to the EFM technique, it may be possible to investigate the spatial distribution of polarization charges and electric stray fields at ferroelectric surfaces in more detail.

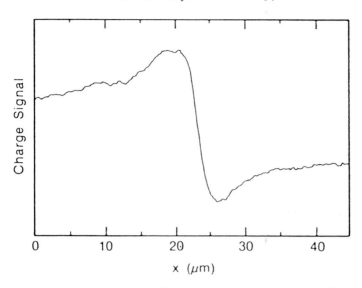

Fig. 2.21. Single-line scan of the charge signal across a ferroelectric domain wall (Saurenbach and Terris, 1990).

2.7.3 Magnetic force microscopy (MFM)

For a magnetic tip and sample, the spatial variation of the magnetic force interaction between them can be probed by using force microscopy (Martin and Wickramasinghe, 1987; Sáenz *et al.*, 1987). If the tip is approached to the sample surface within a distance of typically 10–500 nm, magnetic interaction of the tip with the stray field emanating from the sample becomes noticeable. The magnetic interaction strength determines the amount of cantilever deflection, which can be monitored by one of the deflection sensors described in section 2.3.2. The long-range magnetic dipole interaction is usually probed by using the a.c. detection method introduced at the beginning of section 2.7. Therefore, force gradients rather than magnetic dipole forces themselves are usually measured.

Since magnetic forces F_{mag} can be either attractive or repulsive, problems with feedback loop stability in the non-contact imaging mode are likely to occur. Therefore, an additional attractive 'servoing force' F_{servo} is required that increases in magnitude as the tip approaches the sample surface (Rugar *et al.*, 1990). The van der Waals forces F_{VDW} which are always present can serve as servoing forces. However, it was found useful to have an additional controllable attractive electrostatic force F_{el} by applying a bias voltage of typically 1–10 V between tip and sample

Scanning force microscopy

(Rugar *et al.*, 1990; Schönenberger and Alvarado, 1990a). The total force gradient F' measured by the a.c. detection method is then given by

$$F' = F'_{mag} + F'_{servo} = F'_{mag} + F'_{VDW} + F'_{el} \qquad (2.25)$$

Contours of constant force gradient may therefore not necessarily reflect magnetic contrast only, but can depend on the z-dependence of F'_{servo} as well. Owing to the long-range nature of the magnetic dipole forces, a clear signature for magnetic contrast is its increase with increasing tip–surface separation, as shown in Fig. 2.22 (Martin *et al.*, 1988b). However, this increase in contrast is at the expense of a decrease in spatial resolution and signal-to-noise ratio because the force gradient becomes smaller at increasing tip–surface separation.

Complete separation of magnetic and topographic contrast can be achieved, for instance, by modulating the strength of the magnetic interaction and performing a differential measurement (Martin and Wickramasinghe, 1987; Schönenberger *et al.*, 1990). Alternatively, the applied bias voltage can be modulated, such that $U = U_0 + U_1 \sin(\omega t)$ (Schönenberger *et al.*, 1990; Schönenberger and Alvarado, 1990a). The resulting electrostatic Coulomb forces cause an oscillation of the cantilever at the second harmonic 2ω. The amplitude of this oscillation can

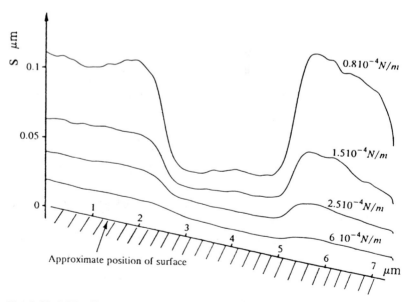

Fig. 2.22. MFM line scans across a single domain for various initial tip–sample spacings (Martin *et al.*, 1988b).

be used to drive the feedback loop adjusting the tip–surface separation. The z piezo movement necessary to keep this oscillation amplitude constant then yields a topographic signal because the contours of constant electrostatic force are almost equivalent to contours of constant tip–surface separation. Simultaneously, the d.c. force can be measured by detecting the quasistatic deflections of the cantilever. Spatial variations of this d.c. force are mainly due to spatial variations in the magnetic interaction because the tip follows the sample surface at approximately constant distance. Successful separation of topography and magnetic structure has been demonstrated by imaging a discrete track magnetic recording sample with written bits (Fig. 2.23).

In the following, we will focus on some of the major issues in MFM, including preparation and characterization of appropriate magnetic force sensors, the theory of MFM response, the spatial resolution achievable in MFM, as well as a comparison with other competing experimental methods.

2.7.3.1 Preparation and characterization of magnetic force sensors
Force sensors for MFM have to fulfill several different requirements concerning their magnetic properties (Grütter, 1989).

1. The spatial extent of the effective magnetic tip volume should be as small as possible, thereby minimizing the spatial extent of the

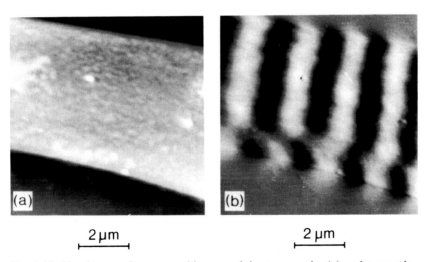

2 μm

2 μm

Fig. 2.23. Simultaneously measured images of the topography (*a*) and magnetic force (*b*) from a discrete magnetic track with recorded bits with longitudinal magnetization (Schönenberger and Alvarado, 1990a).

long-range dipolar magnetic tip–sample interaction and maximizing spatial resolution in MFM.

2. The magnetic stray field of the tip should be small and spatially confined, thereby minimizing undesirable influences on the sample magnetization and maximizing spatial resolution.

3. A large total magnetic moment is desirable, leading to larger measurable forces and therefore better magnetic force sensitivity.

4. A large magnetic anisotropy and coercivity is important to minimize the influence of the sample stray field on the tip magnetization.

5. The force sensors should be magnetically well defined to allow quantitative interpretation of MFM data.

Some of these points cannot be optimized simultaneously. For instance, there is a trade-off between a small effective magnetic tip volume and a high magnetic force sensitivity. For a minimum detectable force of approximately 10^{-12} N at room temperature, the radius of the effective magnetic tip volume must typically be larger than 6 nm (Schönenberger and Alvarado, 1990a).

Several different methods for the preparation of magnetic force sensors exist.

1. A magnetic force sensor consisting of a cantilever with an integrated tip can be obtained, for example, by electrochemical etching of a thin Ni, Fe or Co wire. After etching, the wire must be bent at its foremost end to define the tip, which is then magnetized in a strong d.c. magnetic field. The length of wire acting as the cantilever is typically 400–800 μm, whereas the bent part is typically several tens of micrometers long (Fig. 2.24). The typical spring constant is on the order of 1 N m^{-1}, and the resonant frequency is usually in the range 20–30 kHz. Since most MFM studies are performed under ambient conditions, Ni force sensors are usually preferred because Ni passivates, whereas Fe is continuously oxidized, leading to rapid degradation of MFM data as a function of time. There are at least two serious disadvantages of this preparation method. First, each force sensor must to fabricated individually, thus reproducibility in characteristics and overall shape of the force sensor including its tip is quite low. Second, due to the long-range nature of magnetic dipole forces, magnetic interaction between the sample and the force sensor is not limited to a small volume at the foremost end of the tip, but extends to large distances from the sample surface and therefore includes the shaft of the tip.

2. Alternatively, a sharp non-magnetic tip (e.g. W) may first be prepared by electrochemical etching and subsequently be coated with

100 μm

⊢————⊣

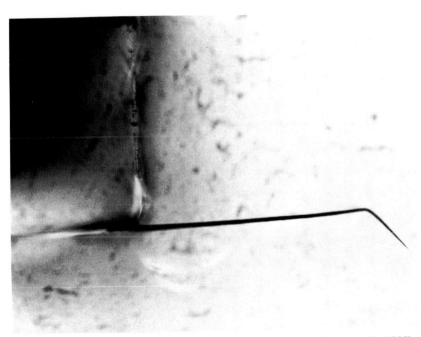

Fig. 2.24. Optical micrograph of a nickel wire cantilever/tip (Rugar *et al.*, 1990).

a thin layer (typically 50 nm) of ferromagnetic material, either by sputtering or by galvanic deposition (den Boef, 1990; Rugar *et al.*, 1990). After deposition, the layer is magnetized along the tip axis in a magnetic field of typically 1–2 T. This preparation method allows minimization of the volume of magnetic tip material that interacts with the sample. However, the force sensors still have to be produced individually.

3. More recently, the magnetic tip coating technique has also been applied to microfabricated silicon cantilevers, opening the possibility of batch fabrication of magnetic force sensors (Grütter *et al.*, 1990b). Their properties can be tailored by choice of coating material and layer thickness. Characterization of magnetic force sensors obtained directly by electrochemical etching of wires made from ferromagnetic material can be performed by Lorentz microscopy (Rugar *et al.*, 1990). It was found that the magnetization is predominantly aligned along the tip axis, in agreement with theoretical expectations

based on the strong tip shape anisotropy. For Ni tips that were not intentionally magnetized it was further found that the last 20 μm of the tip comprise a single magnetic domain. The length of this final domain can be increased by magnetizing the tip in a strong magnetic field. This treatment usually also increases the magnitude of the magnetic stray field at the foremost end of the tip. The length of the final domain depends also on the tip material and is expected to be much less for Co tips than for Ni tips (Schönenberger and Alvarado, 1990a). Alternatively, the geometric and magnetic structure of a magnetic force sensor may be deduced from calibration measurements on well-defined magnetic test structures combined with model calculations. Such magnetic test structures may be given, for example, by magnetic bits written on recording media, or by the current-induced magnetic field of a current-driven thin-film circuit fabricated by nanolithographic methods (Göddenhenrich *et al.*, 1990c).

2.7.3.2 Theory of MFM response

MFM in the non-contact imaging mode is sensitive to magnetostatic dipole–dipole interaction between tip and sample. The total magnetic force acting on the tip in the z direction can be obtained by summing the forces between each dipole in the tip and each dipole in the sample (Sáenz *et al.*, 1987; Wadas, 1988):

$$F(z) = \int_{\text{tip}} d^3 r_1 \int_{\text{sample}} d^3 r_2 \, f_z(r_1 - r_2) \qquad (2.26)$$

where $f_z(r)$ is the interaction between two magnetic dipoles m_1 and m_2 at a distance $r = r_1 - r_2$:

$$f_z(r) = \frac{\mu_0}{4\pi} \frac{\partial}{\partial z} \left(\frac{3(r \cdot m_1)(r \cdot m_2)}{r^5} - \frac{(m_1 \cdot m_2)}{r^3} \right) \qquad (2.27)$$

Alternatively, the magnetic stray field $B_s(r)$ emanating from the sample with a magnetization distribution $M(r)$ can first be determined before calculating the force

$$F = \nabla (m \cdot B_s) \qquad (2.28)$$

summed over the dipoles m in the tip (Rugar *et al.*, 1990).

Model calculations of the MFM response can be performed either analytically or numerically for given magnetization distributions of tip and sample. Unfortunately, neither the magnetization distribution of the tip nor that of the sample are generally known. In addition, the magnetization distributions of tip and sample might change due to the close proximity of the two magnetic bodies, which significantly complicates the theoretical analysis of MFM data (Sáenz *et al.*, 1988; Abraham and McDonald, 1990; Scheinfein *et al.*, 1990). A significant perturbation of the magnetic structure of tip and sample is expected if the magnetic stray fields H_s exceed the anisotropy fields H_K:

$$H_s(\text{sample}) > H_K(\text{tip})$$
$$H_s(\text{tip}) > H_K(\text{sample}) \qquad (2.29)$$

Near a surface the stray field can be a significant fraction of the saturation magnetization M_s. Therefore, the following criterion has been formulated as a limiting constraint for non-destructive MFM operation (Hartmann, 1988):

$$\mu_0 H_K(\text{sample}) / M_s(\text{tip}) \geqslant 1$$
$$\mu_0 H_K(\text{tip}) / M_s(\text{sample}) \geqslant 1 \qquad (2.30)$$

The conditions for non-destructive MFM operation can more easily be achieved for hard magnetic materials with large anisotropy fields, whereas for soft magnetic materials, such as Permalloy (Ni–Fe alloy), changes in the magnetic structure induced by tip–sample interaction can directly be observed during MFM studies with a tip-to-sample spacing of less than about 100 nm, as shown in Fig. 2.25 (Mamin *et al.*, 1989). To reduce the tip stray field at the position of the sample, the tip–surface separation has to be increased which, however, leads to degradation of spatial resolution in MFM. A significant advantage of more recently developed force sensors with thin magnetic film coating on non-magnetic bulk material compared with bulk magnetic wire tips is that the stray field falls off with distance much more rapidly, thereby allowing for non-destructive MFM studies of soft magnetic materials even at relatively small tip–surface separations where high spatial resolution can be obtained (Grütter *et al.*, 1990b). Alternatively, it has been proposed to use paramagnetic tips with a large susceptibility, e.g. Gd close to its Curie temperature, to solve the problems associated with large tip stray fields (Schönenberger and Alvarado, 1990a).

Assuming now that the conditions for non-destructive MFM operation are fulfilled, we have to ask what the MFM actually responds to. It has become clear that the contrast observed in MFM images depends critically on the length of the magnetic domain at the foremost end of

Domains in Permalloy — Tip Effect

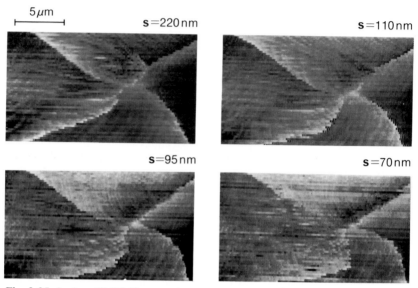

5 μm

s = 220 nm   s = 110 nm

s = 95 nm   s = 70 nm

Fig. 2.25. Series of MFM images showing the effect of decreasing the separation between the tip and sample. Note the increasing distortion of the wall in the lower left part of each image (Mamin *et al.*, 1989).

the sensor tip (Schönenberger and Alvarado, 1990a). If the length of the effective domain at the tip apex is sufficiently small compared with the extension of the sample stray field, the sensor tip can be well approximated as a point dipole with a total magnetic moment m (Mamin *et al.*, 1988; Hartmann, 1989a). In this case, the force acting on the tip in the sample stray field B_s is given by

$$F(\text{dipole}) = \nabla(m \cdot B_s) = (m \cdot \nabla)B_s \qquad (2.31)$$

where the absence of currents ($\nabla \times B = 0$) has explicitly been used. Therefore, in the point-dipole limit, the MFM images are closely related to the spatial distribution of the magnetic stray field gradient, rather than the stray field itself. It is also clear from Eq. (2.31) that the component of the stray field gradient which is sensed depends on the orientation of the tip moment. As discussed previously, the magnetization direction is predominantly aligned along the tip axis due to shape anisotropy. Therefore, the tilt angle of the sensor tip relative to the sample surface determines which component of the stray field gradient is primarily probed (Wadas *et al.*, 1990a, 1990b; Rugar *et al.*, 1990).

We now consider the other limiting case of a long magnetic domain at the tip apex. In this case, only the front portion of the tip domain effectively interacts with the sample stray field, leading to a monopole response due to the magnetic charge q_m at the foremost end of the tip. The measured magnetic force is then directly related to the sample stray field:

$$F\,(\text{monopole}) = q_m B_s \qquad (2.32)$$

and the MFM images reflect the spatial distribution of B_s.

Experimentally, it has been found that some tips (e.g. Co) act as effective point dipole tips and probe the stray field gradient, whereas other tips (e.g. Ni, Fe) act as single-pole tips and directly probe the stray field itself. Therefore, the contrast observed in MFM images can depend critically on the force sensor, as illustrated in Fig. 2.26. Transition from dipole to monopole response can also occur with increasing tip–surface separation (Wadas, 1989).

2.7.3.3 Spatial resolution in MFM

As already discussed, the measured MFM response is due to interaction of a magnetic probe tip with the sample stray field. Therefore, the spatial resolution achieved in MFM depends on the properties of the tip as well as on the properties of the stray field distribution. Even for a point dipole tip and a point dipole object on the sample surface, the width of the measured stray field distribution will be finite and on the order of the tip–surface separation (Rugar *et al.*, 1990). Therefore, the first requirement for high spatial resolution in MFM is to operate the probe tip as close as possible to the sample surface. The improvement of the lateral resolution in MFM with decreasing tip-to-sample spacing is illustrated in Fig. 2.27, showing the sharpening of the domain wall stray field distribution on a single-crystal iron whisker with reduced tip–surface separation (Göddenhenrich *et al.*, 1990b). The minimum practical tip-to-sample spacing is determined by several experimental limitations.

1. For the a.c. detection method usually used in non-contact force microscopy, a minimum oscillation amplitude of the cantilever is required which is typically in the range 1–10 nm. The tip–surface separation obviously has to be chosen larger than this oscillation amplitude.
2. If the derivative of the attractive force becomes too large at small tip–surface separations, exceeding the cantilever spring constant, undesirable instabilities can occur. Stiffer cantilevers allow MFM

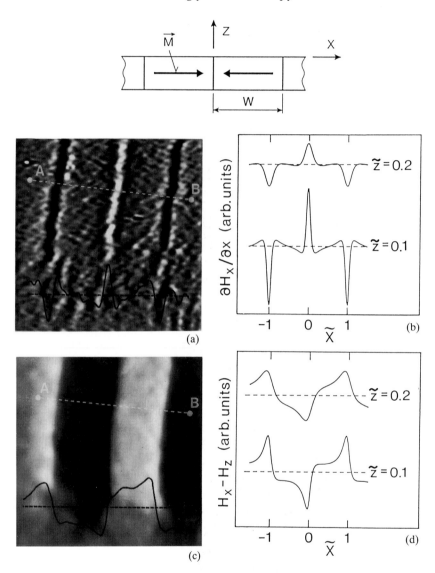

Fig. 2.26. MFM images of bits 2.5 μm wide. Image (*a*), taken with a cobalt tip, is compared to the calculated field derivative in (*b*), and the image (*c*), taken with a nickel tip, to the field component (*d*) (Schönenberger and Alvarado, 1990a).

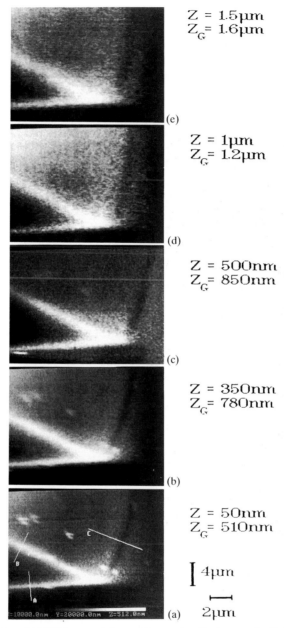

Z = 1.5 µm
Z_G = 1.6 µm

(e)

Z = 1 µm
Z_G = 1.2 µm

(d)

Z = 500 nm
Z_G = 850 nm

(c)

Z = 350 nm
Z_G = 780 nm

(b)

Z = 50 nm
Z_G = 510 nm

4 µm

2 µm

(a)

Fig. 2.27. Series of MFM images taken at increasing tip-to-sample distance from (*a*) z = 50 nm up to (*e*) z = 1.5 µm. z denotes the grey-scale range which corresponds to the maximum sample z deflection in the constant-force mode. Note the different scaling factors in the two lateral directions (Göddenhenrich *et al.*, 1990b).

operation at smaller tip–surface separations at the expense of a reduced sensitivity.

3. The disturbing influence of the tip stray field might become too large at small tip–surface separations, preventing non-destructive MFM operation.

For a given tip-to-sample spacing, the spatial resolution achievable in MFM critically depends on the shape and magnetic structure of the tip (Wadas and Grütter, 1989; Wadas and Güntherodt, 1990). In principle, the effective magnetic tip volume should be as small as possible because the maximum resolution that can be achieved is on the order of the radius of the effective magnetic tip volume, no matter how close the tip is to the sample (Rugar *et al.*, 1990). A multidomain tip can lead to additional fine structure in MFM traces which complicates the interpretation of MFM data (Grütter *et al.*, 1989). Ideally, the probe tip should consist of a small single-domain ferromagnetic particle rigidly attached at the foremost end of a non-magnetic cantilever. However, the minimum size of the effective magnetic tip volume is determined by the force or force gradient detection sensitivity.

At present, the theoretical MFM resolution limit for operation under ambient conditions is around 5–10 nm (Schönenberger and Alvarado, 1990a). There have been some reports on experimental MFM studies approaching this resolution limit (Hobbs *et al.*, 1989; Grütter *et al.*, 1990a). However, standard MFM resolution is typically on the order of 50–100 nm (Rugar *et al.*, 1990; Schönenberger and Alvarado, 1990a).

Significant further improvement in the spatial resolution of MFM towards the atomic level can only be reached by probing the short-range magnetic exchange forces, rather than the long-range magnetic dipole forces. Exchange forces must be probed by contact force microscopy (section 2.4) and the experiments have to be performed under well-defined UHV conditions using clean probe tips and sample surfaces. A first estimate of the magnitude of these exchange forces as a function of the tip-to-sample spacing yields readily measurable values below a tip–surface separation of about 3 Å (Wiesendanger *et al.*, 1991b). The development of UHV-compatible force microscopes will provide the basis for atomic-resolution magnetic force microscopy.

2.7.3.4 Comparison of MFM with competing magnetic imaging techniques

The main features of non-contact MFM may be summarized as follows.

1. Sophisticated sample preparation is not required. In particular, non-contact MFM can be performed even in the presence of non-

magnetic coatings and contamination layers under ambient conditions.

2. Non-contact MFM probes the sample stray field or its gradient, which is the quantity of interest with respect to characterization of magnetic recording media. On the other hand, information about the sample's magnetization distribution, including its magnitude and direction, cannot directly be extracted from MFM data, but rather requires extended MFM response simulations where the input parameters, e.g. shape and magnetic structure of the probe tip, are often not known.

3. The spatial resolution achieved in non-contact MFM is on the order of 10–100 nm, depending on the probe tip and the sample under investigation.

There exist several competing magnetic imaging techniques, some of which are well established whereas others have only recently been developed.

1. The Bitter technique was introduced a long time ago for real-space magnetic domain observations (Bitter, 1931). This method is based on small magnetic particles becoming agglomerated in regions of strong stray field gradients, e.g. at domain walls. The distribution of the magnetic particles as studied by an optical, electron or scanning tunneling microscope (Rice and Moreland, 1991) therefore reflects the magnetic stray field distribution at the sample surface. The spatial resolution is limited by the finite size of the magnetic particles and the resolution of the microscope used for inspection. The magnetic particles used in the Bitter technique and the MFM probe tip respond to the same forces. However, MFM uses only a single 'test particle' at finite distance from the sample surface introducing less perturbation of the sample's magnetic structure. Furthermore, the forces on the 'test particle' are directly measured in MFM which offers additional information about the sample stray field provided that the magnetization state of the tip is known.

2. Kerr microscopy is based on illuminating a magnetic sample with linearly polarized light. As a consequence of magneto-optic effects, the polarization becomes rotated and the amount of rotation is directly related to the sample magnetization. Spatial resolution is limited by the optical wavelength (Williams *et al.*, 1951; Kranz and Hubert, 1963). Kerr microscopy is the ideal technique to combine with MFM. First, Kerr microscopy is a complementary technique because it is directly sensitive to the sample's magnetization distribution. Second, due to its large field-of-view, Kerr microscopy allows

positioning of the MFM tip with respect to the sample's domain wall configuration *in situ* (Göddenhenrich *et al.*, 1990b). Subsequent MFM studies can then cover the length scale beyond the optical resolution limit of Kerr microscopy.

3. Lorentz microscopy is based on the Lorentz force influencing the electron's trajectory in an electron microscope (Hale *et al.*, 1959; Tsuno, 1988). The specimens are usually thinned and then imaged in a phase contrast transmission electron microscope (TEM). Spatial resolution on the order of 10 nm can be achieved. However, it is important to realize that techniques based on TEM studies, such as Lorentz microscopy or electron holography (Tonomura, 1987), yield magnetic information averaged over the entire electron trajectory as it traverses the sample.

4. Scanning electron microscopy with polarization analysis (SEMPA) is based on the fact that the secondary electrons emitted from a ferromagnetic sample surface possess a spin polarization proportional to and oriented with the magnetization of the sample surface. By analyzing the polarization of the secondary electrons while the primary focused electron beam is raster scanned, it is therefore possible to gain spatially resolved information about the magnitude as well as the direction of the surface magnetization. The spatial resolution achieved is on the order of a few tens of nanometers (Celotta and Pierce, 1986; Koike *et al.*, 1987; Hembree *et al.*, 1987). Since the analyzed secondary electrons originate from a surface layer only a few nanometers thick, SEMPA is an extremely surface-sensitive technique which requires UHV conditions and well-prepared sample surfaces free of non-magnetic contaminants. Usually, it takes a long time to obtain a high-contrast image at high spatial resolution due to inefficiencies in collecting and analyzing the secondary electrons.

In summary, while several magnetic imaging techniques can now offer a spatial resolution on the order of 10–100 nm, they can clearly be distinguished according to the sample properties to be probed, e.g. the surface magnetization distribution or the magnetic stray fields. The magnetic imaging techniques listed above also differ significantly with respect to requirements for sample preparation and experimental environment. Certainly, which magnetic imaging technique should be used depends on the particular application.

3

Related scanning probe methods

Scanning probe methods, such as STM and SFM discussed in chapters 1 and 2, represent a novel approach to surface analysis and high-resolution microscopy. Traditional surface analytical and microscopical techniques are based on an experimental set-up schematically illustrated in Fig. 3.1(*a*). The sample in the center of an UHV chamber is probed by electrons, photons, ions or other particles which originate from corresponding sources, typically at a macroscopic distance from the sample surface. As a result of the interaction of the primary electrons, photons, ions etc. with the sample, secondary particles are created and eventually leave the sample together with scattered primary particles. They can be detected by appropriate detectors and analyzers which again are typically a macroscopic distance away from the sample surface. The direction of the emitted particles as well as their energy contain valuable information about the sample under investigation. The spatial resolution achievable is determined by the spatial extent of the primary beam as well as the interaction volume within the sample.

In contrast, scanning probe methods are based on a completely different experimental geometry depicted in Fig. 3.1(*b*). A sharp probe tip is brought into close proximity to the sample surface until interaction between tip and sample sets in. The interaction is spatially localized according to the shape of the probe tip and the finite effective range of the interaction. The distance-dependence of a particular tip–sample interaction can be exploited to guide the tip at a finite tip–surface separation over the sample by keeping the interaction strength at a constant preset value with a feedback circuit. By raster scanning the tip, surface contours of constant interaction strength are obtained. Changing the type of interaction generally leads to changes of the surface contours obtained. Alternatively, the tip can be raster scanned at a fixed height level over the sample surface and variations of the interaction strength, resulting from variations of tip-to-sample spacing, can be recorded.

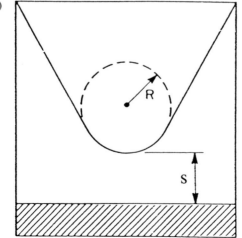

(a)

AES and XPS
hemispherical spectrometer

Slit

Slit plus
retardation
meshes

Channeltron

SIMS,
depth
profiling
ion gun

Transfer lens

AES,
EPMA
electron
gun

EPMA, XRF
X-ray detector

SIMS
mass spectrometer

XPS, XRF
X-ray source

Specimen holder

(b)

R

S

Fig. 3.1. (a) Conventional
surface analysis equipment.
(b) General geometry
for scanning probe methods.

More sophisticated operation modes of scanning probe microscopes use one particular type of tip–sample interaction to control the tip–surface separation while monitoring the spatial variation of a second type of tip–sample interaction of interest.

As a characteristic feature of all scanning probe methods, the tip-to-sample spacing s is kept microscopically small and typically less than the characteristic wavelength λ of the particular interaction to be studied. This means that scanning probe microscopes are generally operated in the near-field (NF) regime ($s \lesssim \lambda$), where the spatial resolution is determined by the tip-to-sample spacing and the effective radius of curvature of the probe tip, rather than by the wavelength λ, which would be typical for diffraction-limited resolution in the far-field (FF) regime ($s > \lambda$). For instance, STM as a typical NF method offers spatial resolution well beyond the wavelength of the electrons that tunnel between tip and sample. Analogously, any type of tip–sample interaction can be exploited to develop a related scanning probe method using the NF regime and therefore offering a spatial resolution beyond the diffraction limit.

In the following, we will present a short overview of some STM-related scanning probe methods which have not yet been discussed in the previous chapters.

3.1 Scanning near-field optical microscopy (SNOM)

It is well known that the spatial resolution achieved in classical optical microscopy is limited by diffraction to about $\lambda/2$ (the Abbé limit), where λ is the optical wavelength (Abbé, 1873). This limit arises because electromagnetic waves interacting with an object to be imaged are always diffracted into two components:

1. propagating waves with low spatial frequencies ($<2/\lambda$), and
2. evanescent waves with high spatial frequencies ($>2/\lambda$).

Classical optics is concerned with the far-field regime where only the propagating waves survive whereas the evanescent waves are confined to sub-wavelength distance from the object corresponding to the near-field regime. Information about the high-spatial-frequency components of the diffracted waves is lost in the far-field regime and therefore sub-wavelength features of the object to be imaged cannot be retrieved. On the other hand, by operating a microscope in the near-field regime, the Abbé resolution limit can easily be surpassed as first discussed by Synge (1928). However, the technology to raster scan a small aperture probe at sub-wavelength distances over a sample surface was not yet developed at that time.

The first experimental realization of a 'super-resolution' scanning microscope was demonstrated by using microwaves with $\lambda \approx 3$ cm (Ash and Nicholls, 1972). After invention of the STM as an extreme example of a super-resolution near-field scanning microscope, STM technology also triggered increasing effort in the development of scanning near-field optical microscopy (SNOM). Several different experimental methods have been introduced to perform SNOM.

3.1.1 SNOM with nanometer-size aperture probes

In the first SNOM experiments a tiny aperture, illuminated by a laser beam from the rear side, was scanned across a sample surface, and the intensity of the light transmitted through the sample was recorded, as schematically illustrated in Fig. 3.2 (Pohl *et al.*, 1984; Dürig *et al.*, 1986b, 1986c; Pohl, 1991). To achieve a high lateral resolution, which was on the order of 25 nm ($\lambda/20$) in the initial experiments, the aperture diameter has to be of nanometer size, and the aperture has to be maintained at a distance of less than 10 nm from the sample surface. The latter requirement arises because at increasing distance s from the aperture, the evanescent waves are damped out rapidly and the field intensity I decreases strongly according to

$$I \propto s^{-4} \tag{3.1}$$

This fourth power dependence of the field intensity on distance in the NF regime is in contrast to the behavior in the FF regime where the field intensity decreases quadratically with distance. Vacuum tunneling between a tiny metallic asperity at the foremost end of the aperture probe and the object to be studied was initially used as a servo-mechanism to maintain the optical probe in close proximity to the sample without actually touching it. Alternatively, the force interaction can be exploited as a servo-mechanism (Betzig *et al.*, 1992; Toledo-Crow *et al.*, 1992).

The optical probes are formed by a sharpened glass or fiber tip coated with a thin metallic layer, leaving a sub-micrometer aperture at the apex. The capability of fabricating nanometer size apertures is essential to obtain high spatial resolution which, however, is gained at the expense of signal intensity. As an alternative to glass or fiber tips, sharpened micropipettes coated with a thin metallic layer can serve as optical probes as well (Harootunian *et al.*, 1986; Betzig *et al.*, 1987; Cline *et al.*, 1991). The micropipette apertures can either be illuminated from the rear side by a laser beam, acting as a light source, or they can

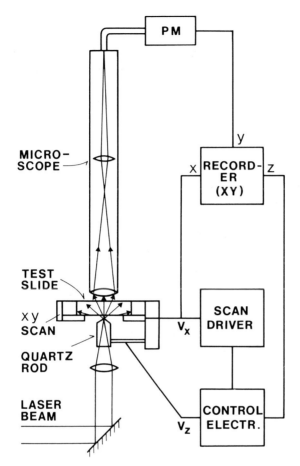

Fig. 3.2. Schematic set-up of a scanning near-field optical microscope (SNOM) (Pohl *et al.*, 1984).

be used as collectors for radiation from a small area of the sample which itself is illuminated as a whole (Betzig *et al.*, 1987). The experimental set-up for this 'collection mode' SNOM is schematically illustrated in Fig. 3.3. To enhance the transmission of electromagnetic energy through the narrow micropipette tube, it has been proposed to fill the micropipette with a fluorescent dye embedded in a plastic matrix (Lieberman *et al.*, 1990; Lewis and Lieberman, 1991).

SNOM can be performed in reflection, as well as in transmission, this being of much larger practical importance (Fischer, 1985; Fischer *et al.*, 1988; Cline *et al.*, 1991). An experimental set-up capable of both

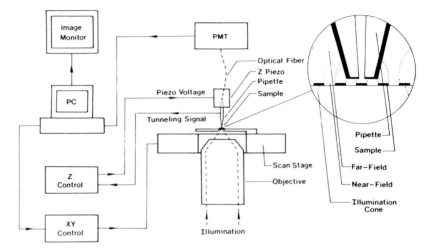

Fig. 3.3. On the left is shown a schematic diagram of an instrument used for collection-mode SNOM. On the right, an expanded view is shown of the light transmitted through a sample being collected in the near-field by an aperture (Betzig *et al.*, 1987).

reflection and transmission SNOM is schematically illustrated in Fig. 3.4. It has also been demonstrated that a small protrusion, e.g. a metallized polystyrene particle on a flat substrate, can replace the probe aperture for reflection SNOM studies (Fischer and Pohl, 1989). Finally, SNOM can be combined with all techniques known in classical optical microscopy including the investigation of luminescence and polarization as well as phase contrast (Trautman *et al.*, 1992).

3.1.2 Photon scanning tunneling microscopy (PSTM)

In close analogy to the STM, a sharpened optical-fiber tip can be used to probe the evanescent field above a dielectric in which total internal reflection (TIR) is made to occur. The sample is placed on or forms the TIR surface and spatially modulates the evanescent field (Fig. 3.5). The tunneling of photons to the tip end of the optical fiber is detected by a photomultiplier tube connected to the other end of the fiber, while the object surface is scanned relative to the tip by means of a piezo-stage. Constant intensity or constant height imaging can be performed similarly to STM. To reduce the noise due to ambient and parasitic light, the primary laser light source is usually modulated, which allows lock-in detection. The lateral resolution achieved with this photon-STM

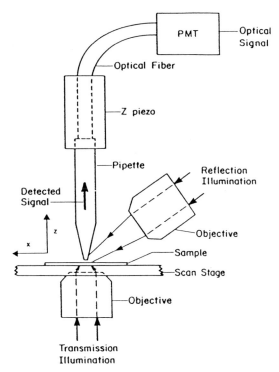

Fig. 3.4. Schematic diagram of a SNOM instrument capable of both reflection and transmission modes (Cline *et al.*, 1991).

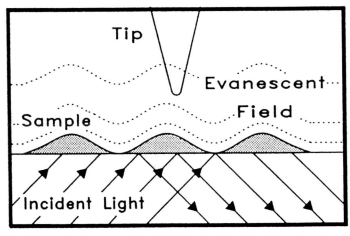

Fig. 3.5. Schematic diagram of the PSTM principle. The tip probes the sample-modulated evanescent field produced by an internally reflected light beam (Reddick *et al.*, 1989).

(PSTM) or evanescent field optical microscope (EFOM) is typically on the order of 50–100 nm (Reddick *et al.*, 1989; Courjon *et al.*, 1989; Courjon, 1990; Guerra, 1990; Tsai *et al.*, 1990; Ferrell *et al.*, 1991; van Hulst *et al.*, 1991). Operation of the PSTM in a reflection geometry has been demonstrated as well (Courjon *et al.*, 1990).

Optical fiber tips can be formed by chemically etching one end of a quartz optical fiber in hydrofluoric acid. The resulting tip radii are typically on the order of 100 nm. The exponential nature of the evanescent field is the primary mechanism for having an effective sharp tip capable of providing sub-wavelength resolution. This is similar to the electron-STM where only the outermost nanotip is effective due to exponential decay of the sample surface wave function. Though the use of sub-micrometer apertures is not required to obtain sub-wavelength resolution in PSTM, it is certainly advantageous to further improve the resolution.

PSTM is particularly suited for investigation of dielectric surfaces which are not directly accessible by electron-based microscopical techniques such as STM or SEM due to charging effects. PSTM images can, in principle, yield information about the sample topography as well as spatial variations in optical properties across the sample, e.g. inhomogeneities in the index of refraction (Tsai *et al.*, 1990). However, for proper interpretation of PSTM images, the sample orientation as well as the shape and position of the probe tip have to be known (Cites *et al.*, 1992).

An analytical PSTM (APSTM) can additionally provide spectroscopic information about the dielectric sample at high spatial resolution (Moyer *et al.*, 1990; Paesler *et al.*, 1990). Two signals have to be measured in APSTM:

1. the signal resulting from detection of the evanescent field, the intensity-dependence of which gives the topographic image of the surface, and
2. the signal resulting from Raman or photoluminescence events, the spatial dependence of which yields information about the chemical nature of the surface.

The spatial resolution achieved in APSTM is considerably better than in micro-Raman spectrometry where a spot of diameter 1 μm is typically probed.

3.1.3 Scanning plasmon near-field microscopy (SPNM)

More recently, another near-field optical microscope has been introduced which achieves a lateral resolution of about 3 nm, corresponding

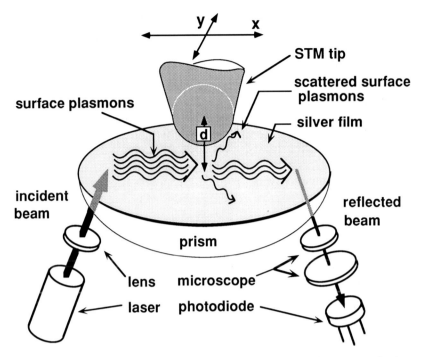

Fig. 3.6. Scheme of the scanning plasmon near-field microscope: Surface plasmons, excited by an incident laser beam, interact with a thin tungsten tip positioned very close to a silver surface (Specht *et al.*, 1992).

to $\lambda/200$ (Specht *et al.*, 1992). The experimental set-up is schematically illustrated in Fig. 3.6. An incident laser beam resonantly excites surface plasmons in a silver film which generate an optical near-field which is locally probed by a sharp tip. The interaction of the surface plasmons with the tip is described by elastic plasmon scattering and radiationless energy transfer from the tip to the sample. Both processes depend strongly on the tip-to-sample spacing. As a consequence, the intensity of the reflected beam, measured with a photodiode, becomes strongly distance-dependent as well, which provides the key for the high spatial resolution achieved in the optical images. It is expected that this scanning plasmon near-field microscope (SPNM) has the potential to detect and identify single adsorbed molecules.

3.2 Scanning near-field acoustic microscopy (SNAM)

Another type of near-field microscopy can be realized by using acoustic waves. A familiar near-field acoustic instrument is the medical doctor's stethoscope which achieves a resolution better than $\lambda/100$.

In analogy to the STM, a scanning near-field acoustic microscope (SNAM) has been developed capable of imaging the topography of non-conducting surfaces (Guethner *et al.*, 1989, 1990). A 32 kHz quartz tuning fork with a high Q factor is used as an oscillator driven at resonance. An edge of one leg of the tuning fork serves as a probe tip, as schematically illustrated in Fig. 3.7. If the tip is made to approach the surface, both the resonance frequency and the oscillation amplitude decrease. These changes of frequency and amplitude as a function of tip-to-sample spacing strongly depend on the pressure of the coupling gas, leading to the conclusion that hydrodynamic damping forces in the air gap predominantly mediate the interaction between tip and sample, rather than the van der Waals forces (section 2.7.1). Interaction sets in at tip–surface separations of typically 200 μm, well below the acoustic wavelength of about 1 cm in air at 32 KHz, and therefore arises in the acoustic near-field regime. To image a surface, the sensor is piezo-

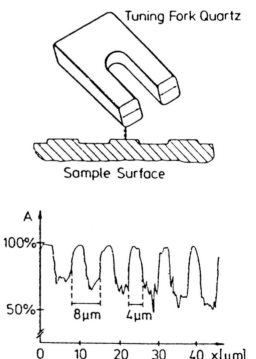

Fig. 3.7. Schematic set-up for scanning near-field acoustic microscopy. The tuning fork is adjusted such that one edge can be used as distance probe. A line scan across an evaporated chromium layer of 30 nm thickness is shown (Guethner *et al.*, 1989).

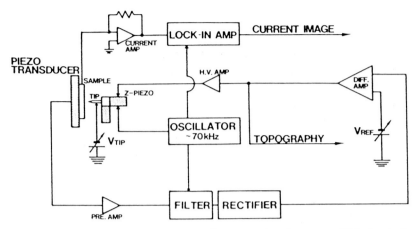

Fig. 3.8. Schematic diagram of a tunneling acoustic microscope (Takata *et al.*, 1989c).

electrically scanned, while the tip-to-sample spacing is controlled by a feedback circuit which maintains a constant oscillator amplitude. A lateral resolution of about 1 µm and a vertical resolution of about 10 nm have been demonstrated with the SNAM.

A different experimental set-up is used for the tunneling acoustic microscope (TAM), as schematically illustrated in Fig. 3.8 (Takata *et al.*, 1989a, 1989c, 1991, 1992). An electrochemically etched STM tip is vibrated at typically 70 kHz by applying a sinusoidal voltage to the *z* piezoelectric actuator. As the tip is made to approach the sample, the vibration of the tip is transmitted to the sample via a force interaction, generating acoustic waves which are detected by a piezotransducer coupled to the rear side of the sample. The transducer output voltage is sensitive to the distance between tip and sample and can therefore be used to control the vertical tip position. The maximum tip-to-sample spacing is limited by the force sensitivity of the TAM which is on the order of 10^{-11} N. By scanning the tip over the sample surface, an acoustic imaging topography can be obtained. For conducting samples a tunneling current image may be recorded simultaneously.

3.3 Scanning near-field thermal microscopy (SNTM)

Thermal interaction between a probe tip and a sample to be studied can also be used for surface profiling of both electrical conductors and insulators. Depending on the experimental environment and the tip-to-sample spacing, different types of heat transfer mechanisms are relevant, and different methods of temperature measurement might be applied.

3.3.1 Scanning thermal profiler (STHP)

The scanning thermal profiler (STHP) is a non-contacting near-field thermal probe consisting of a miniaturized thermocouple sensor at the end of a sharp tip, as schematically illustrated in Fig. 3.9 (Williams and Wickramasinghe, 1986a, 1986b; Martin *et al.*, 1988c). The thermal probe is heated in order to produce a temperature difference between the tip and the sample. As the heated thermal probe is made to approach the sample, the tip temperature becomes reduced as a result of the thermal coupling between tip and sample. The strong dependence of the thermal transport on the tip-to-sample spacing can then be exploited to perform non-contacting surface profiling by scanning the thermal probe, for instance, at constant thermal interaction strength, i.e. at constant tip temperature, over the sample surface. To avoid imaging disturbances due to ambient temperature variations, the tip-to-sample spacing is usually modulated, and the resultant a.c. thermal signal is detected and rectified before being sent to the feedback loop.

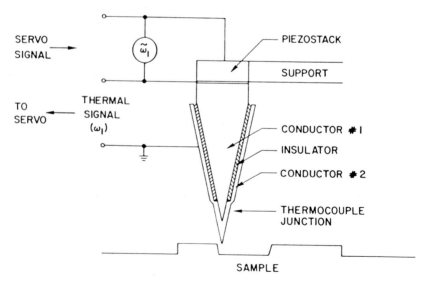

Fig. 3.9. Schematic diagram of a thermocouple probe supported on a piezo-electric element for modulation of tip-to-sample distance as well as to provide average servo positioning. The a.c. thermal signal is detected, rectified, and sent to a servo loop, which supplies a voltage to the piezostack to maintain the average tip-to-sample spacing constant (Williams and Wickramasinghe, 1986a).

The lateral resolution achieved with the STHP is on the order of 30–100 nm and depends on the tip-to-sample spacing as well as on the size of the thermocouple sensor. Thermal probe tips with dimensions on the order of 100 nm have been fabricated. The vertical resolution was found to be on the order of 3 nm. The minimum detectable change in tip temperature is less than 0.1 mK, which corresponds to a heat flow of only a few nanowatts.

The mechanism responsible for heat conduction between tip and sample depends critically on the experimental environment and on the tip-to-sample spacing. For experiments performed in ambient air and at tip–surface separations s well above the mean free path l of the air molecules ($s \gg l \approx 66$ nm), the observed thermal coupling between tip and sample might be attributed to classical mechanisms for heat conduction. However, as the tip–surface separation becomes on the order of the mean free path l or even less ($s \lesssim l$), an alternative mechanism for heat transport, for example via thermally excited infrared electric fields, must be considered (Dransfeld and Xu, 1988). For ionic crystals, near-field coupling to the optical phonon field becomes important, whereas for electrical conductors near-field coupling to the plasmon field has to be taken into account. Similarly to the case of total internal reflection (section 3.1.2), the thermally excited electric fields associated with charge fluctuations with short wavelengths λ do not propagate but decay exponentially with distance s from the surface as $\exp(-s/\lambda)$. Energy can be extracted from these localized electric surface fields only if tip and sample are made to approach to within the near-field regime. The heat flow $\dot{Q}$ via thermally excited electric fields increases strongly with decreasing tip–surface separation s (Dransfeld and Xu, 1988):

$$\dot{Q} \propto s^{-3} \tag{3.2}$$

Therefore, for a tip-to-sample spacing of less than about 100 nm, this near-field coupling mechanism is expected to dominate the heat transfer even in the presence of air and may explain the high spatial resolution achieved by the STHP. Any thermal interaction via thermally excited electric fields would obviously be present under vacuum conditions as well as in ambient air, and would depend strongly on the dielectric properties of the surface to be studied. In contrast, classical mechanisms for heat conduction would lead to surface profiles being nearly independent of material properties (Williams and Wickramasinghe, 1986a, 1986b).

3.3.2 Tunneling thermometer (TT)

As an alternative experimental set-up, a thermocouple can be formed by making a conducting STM tip approach a conducting sample of a different material until a tunneling current is detected. To measure the temperature, the tunneling feedback loop is periodically switched off by means of a sample-and-hold circuit (section 1.10.4), and the external bias voltage is removed. The open-circuit potential developed across tip and sample then provides a measure of the local thermoelectric potential (Weaver *et al.*, 1989; Williams and Wickramasinghe, 1990). The spatial variations of the thermoelectric signal obtained by scanning the tip over the surface are determined by the spatial variations of the product of the local gradient of the sample chemical potential with respect to temperature $\partial \mu_s(x,y)/\partial T$ times the temperature differential ΔT normal to the surface layer being imaged. It is therefore possible to develop different types of microscopes.

3.3.2.1 Scanning optical absorption microscope (SOAM)

A scanning optical absorption microscope (SOAM), as schematically illustrated in Fig. 3.10, measures the local variations in temperature ΔT of a thin film as a result of spatial variations in the absorption of laser radiation (Weaver *et al.*, 1989). To be able to perform sensitive lock-in detection measurements, the junction region is illuminated with a chopped laser beam tuned to a wavelength of spectroscopic interest. As a result, the tunnel junction is periodically heated, leading to a periodic thermoelectric potential difference developed across an external circuit which can be lock-in detected at the laser modulation frequency. Spatial variations in the amount of light absorbed by the thin film sample lead to spatial variations in the temperature of the junction and therefore to spatial variations in the recorded thermoelectric potential signal. To be able to distinguish between spatial variations in the absorption due to spatially varying film thickness and other sources of absorption variations, the sample is illuminated by two laser beams of different wavelength. To first approximation, the ratio of absorption at the two different wavelengths is then independent of film thickness. The minimum detectable temperature difference by SOAM is on the order of 0.01 K.

3.3.2.2 Scanning chemical-potential microscope (SCPM)

In scanning chemical-potential microscopy (SCPM), a constant temperature gradient ΔT is generated normal to the sample surface by heating the sample from the rear side up to about 50 K above ambient temper-

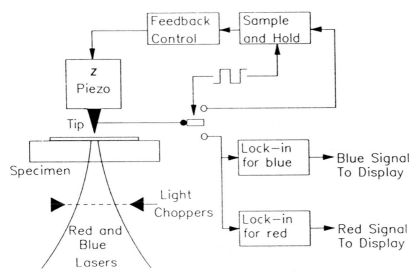

Fig. 3.10. Schematic diagram of an absorption microscope. Beams from a 'reference' He–Ne laser at 633 nm and from a tunable argon-ion laser are independently chopped at frequencies of 1 and 1.5 kHz respectively. The beams are then combined and focused to a 2 μm spot on the sample directly below the tip of the STM. With the tunneling feedback loop closed, the gap resistance is regulated and the surface expansion measured. With the z piezo position held constant the junction is isolated and the a.c. 'thermoelectric' voltage developed across the gap is synchronously detected (Weaver *et al.*, 1989).

ature. Spatial variations in the thermoelectric voltage signal recorded are then attributed to spatial variations in the chemical potential gradient $\partial\mu_s/\partial T$ (Williams and Wickramasinghe, 1990, 1991). By using a MoS_2 and a graphite test sample, local variations in the thermoelectric voltage signal were found even on an atomic scale. The maxima for the chemical-potential gradient signal did not overlap strongly with the tunneling-current maxima. This is not expected either because, to a first approximation, the tunneling current probes the local electronic density of states (section 1.11.1), whereas the chemical-potential gradient should depend on the logarithmic derivative of the local density of states (Stovneng and Lipavsky, 1990).

3.4 Scanning capacitance microscopy (SCAM)

In scanning capacitance microscopy (SCAM), a small probe electrode is brought within a distance of a few nanometers from the sample surface to be studied. Assuming a simple parallel plate geometry to a first approximation, the capacitance C is inversely proportional to the probe-to-sample spacing s:

$$C \propto s^{-1} \qquad (3.3)$$

This $C(s)$-dependence can be exploited to control the height of the probe above the sample surface, leading to another technique for non-contact surface profiling of both electrically conducting and semiconducting samples. In general, the capacitance depends in a complicated way on the geometry of the probe electrode and the sample. It further depends on the dielectric constants of the sample and the medium between the probe electrode and the sample. Therefore, SCAM can be applied to map spatial variations in dielectric properties in or through insulating layers as well.

In a first version of a SCAM, the probe electrode was attached to the edge of a diamond stylus in mechanical contact with the sample, as schematically illustrated in Fig. 3.11(*a*) (Matey and Blanc, 1985). The probe electrode was about 5 μm long, 2.5 μm wide and 0.15 μm thick, and its bottom was at a distance of about 20 nm from the bottom of the stylus. As the stylus is scanned across the sample, the electrode–sample capacitance is monitored. This is achieved by incorporating the electrode–sample capacitance C_{ES} into a resonant circuit which is excited by an ultra-high-frequency (UHF) oscillator. Changes in C_{ES} lead to changes in the resonant frequency and consequently to changes

(a)

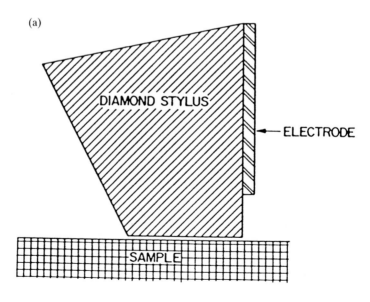

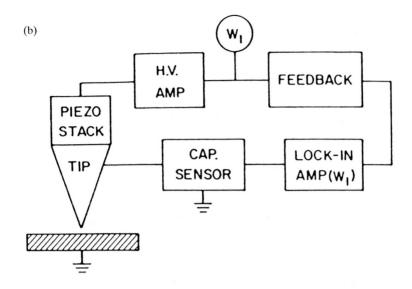

Fig. 3.11. (*a*) A capacitance probe for scanning capacitance microscopy (Matey and Blanc, 1985). (*b*) Schematic illustration of the capacitance servo loop (Williams *et al.*, 1989a). (*c*) Schematic diagram of a capacitance microscope added to an atomic force microscope. This instrument can simultaneously and independently measure surface topography and sample capacitance variations (Barrett and Quate, 1991b).

in the oscillation amplitude of the resonant circuit. Detection of the oscillation amplitude therefore provides a measure of C_{ES}. The thin-film electrode SCAM achieved lateral resolution of about 100 nm and vertical resolution of about 0.3 nm.

In a scanning near-field capacitance microscope, an STM tip is used as capacitance probe, as schematically illustrated in Fig. 3.11(*b*) (Williams *et al.*, 1989). To avoid large, low-frequency drifts in the capacitance output caused by stray capacitance variations, the tip-to-sample spacing and therefore the capacitance between tip and sample is modulated at a high frequency (typically 30 kHz), providing the means to apply lock-in detection for the $\partial C/\partial s$ signal. This type of SCAM therefore probes the capacitance gradient which depends on the inverse square of the tip-to-sample spacing:

$$\frac{\partial C}{\partial s} \propto s^{-2} \tag{3.4}$$

By using tips with an effective radius of curvature below 100 nm, the SCAM achieved a lateral resolution of about 25 nm. Imaging below 10 nm might be possible with tips having even smaller effective radii. It has been estimated that the capacitive sensitivity is as small as $C_{min} \approx 2 \times 10^{-22}$ F in a 1 Hz bandwidth. Capacitive variations ΔC of this size at a voltage $U = 10$ V correspond to charge variations $\Delta Q = U \cdot \Delta C \approx 10^{-2} e$.

The scanning near-field capacitance microscope has been applied for lateral dopant profiling in semiconductor structures with 100–200 nm lateral resolution (Williams *et al.*, 1989, 1990; Abraham *et al.*, 1991). The technique is based on measurements of the voltage-dependent capacitance between tip and sample. When the tip is biased, the semiconductor surface becomes depleted, and the measured depletion capacitance or capacitive gradient as a function of bias voltage provides the information necessary to determine the local activated dopant density. An additional a.c. and/or d.c. bias voltage between tip and sample allows one to obtain C–U curves at any position of the sample surface, or to record $\partial C/\partial U$ images at constant d.c. bias voltage, as the tip is scanned over the surface, simultaneously with the acquisition of topographic $\partial C/\partial s$ images. The largest $\partial C/\partial U$ signal comes from the lightly doped regions where the capacitive variation with bias voltage is strongest. Measurements of dopant density were demonstrated over a dopant range of 10^{15}–10^{20} cm^{-3}.

Another type of SCAM was developed by adding a capacitance sensor to an atomic force microscope (section 2.4) and measuring the capacitance between a metallized cantilever and the sample, as schematically illustrated in Fig. 3.11(*c*) (Barrett and Quate, 1991b). As the sample is scanned relative to the tip, simultaneous and independent measurements of the electrode–sample capacitance and the sample topography can be performed. To eliminate the strong low-frequency noise of the capacitance sensor, the derivative of the capacitance signal with respect to the sample bias $\partial C/\partial U$ is usually recorded by lock-in detection, rather than the capacitance signal itself. The cantilevers are fabricated either from electrochemically etched tungsten wires or by coating silicon dioxide cantilevers with metal layers. The probe tip can be in direct contact with the electrically conducting or semiconducting sample as long as the sample is covered by a sufficiently thick insulating dielectric film.

3.5 Scanning electrochemical microscopy (SECM)

The scanning electrochemical microscope (SECM) probes the Faradaic current that flows through a small electrode probe in close proximity to an electrically conducting or semiconducting sample, both being immersed in an electrolytic solution (Liu *et al.*, 1986; Kwak and Bard, 1989; Bard *et al.*, 1989, 1990, 1992). The Faradaic current results from redox processes at tip and sample and is controlled by electron transfer kinetics at the interfaces as well as mass transfer processes in solution (Fig. 3.12). Tip and sample are part of an electrochemical cell con-

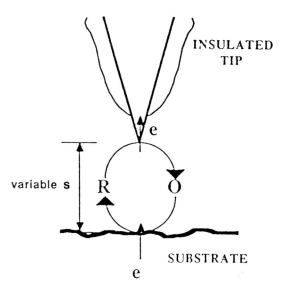

Fig. 3.12. Schematic illustration of scanning electrochemical microscopy (Bard *et al.*, 1989).

taining other auxiliary and reference electrodes to control the tip and sample potentials. SECM measurements can be performed over a wide range of tip-to-sample spacings, typically from 1 nm to 10 μm. This is in contrast to STM where the tip-to-sample spacing has to be on the order of 1 nm or below to detect a tunneling current.

The dependence of the Faradaic current I_F on the tip-to-sample spacing s has been evaluated for different probe geometries (Bard *et al.*, 1989). For a planar probe, assuming the absence of edge effects, the Faradaic current is inversely proportional to the gap spacing s:

$$I_F \propto s^{-1} \tag{3.5}$$

This distance-dependence of the Faradaic current can be exploited for non-contact surface profiling in electrolytical solutions. In analogy to STM, the SECM can be operated either in the constant current mode, yielding 'topographic images' of the sample surface, or in the constant height mode, where spatial variation of the Faradaic current is probed. In this second mode, the SECM can be used for 'chemical imaging', where a material contrast is obtained through a corresponding change in the measured Faradaic current.

The lateral resolution achieved by SECM is mainly determined by the size and shape of the probe tip. To minimize the effective probe surface area exposed to the electrolyte, tips with an insulating coating, made from glass, epoxy or Apiezon wax are used, leaving only a small exposed metal area (Heben *et al.*, 1988; Nagahara *et al.*, 1989). It has been pointed out that conical or tapered cylindrical tips are less useful for obtaining high spatial resolution than inlaid disks or hemispheres, for which the Faradaic currents depend more strongly on the probe-to-sample spacing (Bard *et al.*, 1989). The ultimate resolution achieved by SECM is on the order of 10 nm, both laterally and vertically.

3.6 Scanning micropipette microscopy (SMM)

3.6.1 Scanning ion conductance microscopy (SICM)

The scanning ion conductance microscope (SICM) is based on an electrolyte-filled micropipette used as a local probe for insulating samples immersed in an electrolytic solution, as schematically illustrated in Fig. 3.13 (Hansma *et al.*, 1989). As the micropipette is made to approach the sample, the flow of ions through the opening of the pipette is blocked at small probe-to-sample separations, resulting in a decrease of the ion conductance between an electrode inside the pipette and an electrode

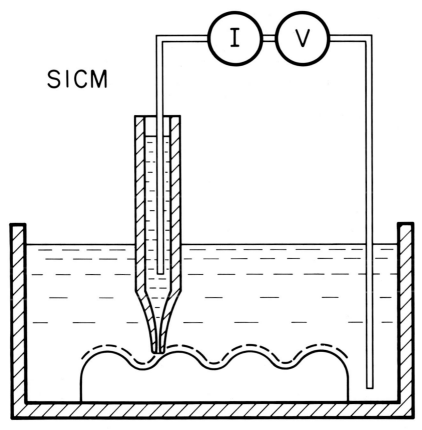

Fig. 3.13. In scanning ion-conductance microscopy a micropipette is scanned over the contours of a surface by keeping the electrical conductance through the tip of the micropipette constant by adjusting the vertical height of the probe (Hansma *et al.*, 1989).

in the electrolyte reservoir. The distance-dependence of the ion conductance provides the key to the ability to perform non-contact surface profiling, similarly to other scanning probe microscopies.

The SICM can be operated in two different modes. In the first mode, a feedback circuit is used to maintain a given conductance while the tip is scanned over the sample surface. The constant-conductance contours obtained can be interpreted as topographic images to a first approximation. The set-value for the conductance is typically 0.9 to 0.98 as great as the conductance when the micropipette is retracted from the sample surface. Alternatively, the micropipette can be scanned at a constant height over the sample surface while monitoring the local ion current coming out through pores in a surface, e.g. of a membrane. In this second mode of operation, spatial variations in the local electrical properties of the sample are probed.

The spatial resolution achieved by SICM is primarily determined by the inner diameter of the micropipette probe. In the initial experiments, micropipettes with inner diameters of 50–100 nm and outer diameters of 100–200 nm were used. The lateral resolution obtained with these pipettes is on the order of 200 nm, i.e. a few times the inner diameter of the micropipettes.

Improved versions of the SICM make use of silicon microfabricated probe tips with a small aperture at the tip apex (Prater *et al.*, 1991a). These probe tips are compact and mechanically robust, thereby allowing for high scan speeds and reliable long-term performance of the SICM.

3.6.2 Scanning micropipette molecule microscopy (SMMM)

In scanning micropipette molecule microscopy (SMMM), the opening of a micropipette probe is sealed with a small plug of a permeable material. The shank of the micropipette is connected by means of a flexible vacuum coupling to a vacuum system containing a quadrupole mass spectrometer (QMS), as schematically illustrated in Fig. 3.14 (Jarrell *et al.*, 1981). As the micropipette is approached towards a sample immersed in solution, molecules coming from the sample may permeate the plug and can be detected by the QMS. By scanning the pipette over the sample surface, spatial variations in the concentrations of the permeating molecules can be mapped out and correlated with the sample morphology as observed with a light microscope.

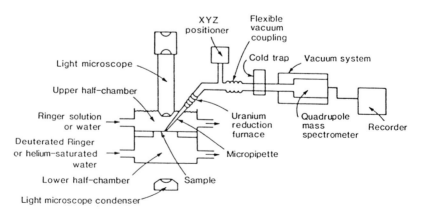

Fig. 3.14. Schematic of a scanning micropipette molecule microscope (Jarrell *et al.*, 1981).

The sensitivity of the SMMM is determined by the inner diameter of the micropipette, the permeability of the plug, and the detection efficiency of the QMS. For a given pipette-to-sample spacing, the spatial resolution achieved by SMMM depends on the inner diameter of the micropipette, as for SICM.

Part two
Applications of scanning probe microscopy and spectroscopy

Scanning tunneling microscopy (STM) and related scanning probe microscopies (SPM) have found numerous applications in various scientific disciplines and are already documented by thousands of publications devoted to this research field. It is certainly impossible to present a review of all that work which would in any sense be complete. Therefore, the second part of this book aims at a description and discussion of representative and important applications of STM and related SPM, rather than a comprehensive review of published work in this field. The selection of topics as well as the emphasis put on particular topics is mainly influenced by the number and significance of publications within a particular field of application. However, it naturally reflects the particular interests of the author of this book as well.

4

Condensed matter physics

Condensed matter physics deals with systems containing a large number of atoms (typically 10^{23} cm^{-3}) that form a dense aggregate. Theoretical treatment of such complex systems has traditionally been based on simplifying concepts.

1. The concept of totally neglecting the microscopic structure is used in phenomenological theories for macroscopic systems, such as thermodynamical or continuum mechanical descriptions.
2. The concept of a statistical theoretical treatment of a large number of atoms appears as a link between a macroscopic and microscopic description, where macroscopic observables are attributed to the average properties of the constituents of the whole aggregate, whereas fluctuations resulting from the individual behavior of single atoms are assumed to be negligible.
3. The concept of idealizing the microscopic structure of macroscopic systems has led to the foundation of modern condensed matter theory. Most notably, the concept of the perfect crystal, whose atoms are arranged in space with strict periodic order, has been used as a starting point for a profound description of real crystalline solids. The strict periodicity present in a perfect crystal allows rigorous theoretical treatment, leading to exact solutions of the Schrödinger equation. However, these solutions emphasize the collective properties of the many atoms chemically bound in the crystal, whereas the individuality of each atom is again lost.

The experimental techniques traditionally applied to investigation of the condensed state of matter, such as diffraction, specific heat, electrical transport or magnetization measurements, have also focused on average and collective properties of the many atoms in a macroscopic system. More recently, however, both theorists and experimentalists have increasingly aimed at a microscopic understanding of condensed matter. Progress has been achieved based mainly on advances in computational science and development of appropriate experimental methods. Most notably, scanning probe methods have enabled us to probe physical properties of condensed matter down to the scale of individual atoms and to correlate these properties with the microscopic real-space structure. Unfortunately, the SPM experimental set-up restricts these investigations to surface or near-surface regions.

4.1 Surface science

In the first ten years following its invention, STM certainly had its greatest impact in the field of surface science. For instance, numerous structures of clean and adsorbate-covered surfaces have been solved. However, in most cases, it was not solely the STM technique that successfully led to the determination of these surface structures, but rather the combination of STM with other (conventional) surface analytical techniques. STM is particularly powerful for revealing the local aspects of the atomic and electronic surface structure and its dynamics. On the other hand, STM probes only the vacuum tails of the surface wave functions (section 1.11.1) and generally lacks chemical specifity (section 1.13). Therefore, the combination of STM with other surface analytical techniques, capable of providing subsurface and chemically specific information, has proven to be important (Demuth *et al.*, 1988). Furthermore, insight into the local nature of surface structures as revealed by STM is complemented by the statistical information averaged over large surface areas as provided by conventional surface analytical techniques.

Initially, the application of STM to semiconductor surfaces has been most fruitful. Since atomic resolution is routinely obtained on metal surfaces as well, STM has successfully been applied to solve surface structures for all kinds of electrically conducting substrates. Finally, the invention of the scanning force microscope (SFM) has opened new opportunities for addressing unsolved issues in surface science of bulk insulators, a particular challenging field where most conventional surface analytical techniques involving charged probe particles fail.

4.1.1 Semiconductors

STM investigations of semiconductor surfaces generally reveal a pronounced bias voltage dependence of the observed images. The 'topographic' maxima are usually associated with surface dangling bonds (DB). According to their energy, certain bonds may be seen at some particular sample bias voltage, but not at others. To gain insight into the surface structure, it is therefore mandatory to perform voltage-dependent STM/STS studies. Other important issues in STM investigations of semiconductors are possible surface band bending and Fermi-level pinning. These effects critically depend on the particular semiconductor to be studied (Ihm, 1988; Feenstra, 1990).

4.1.1.1 Clean semiconductor surfaces: Si(111)

When starting to describe and discuss applications of STM and related SPM, one certainly has to begin with the Si(111) 7×7 surface structure because the first atomically resolved STM image of this surface in 1982 marked the breakthrough for the STM technique as discussed in section 1.9 (Binnig *et al.*, 1983a).

The (7×7) reconstructed Si(111) surface can be obtained by thermally annealing a silicon single crystal typically above 1000 °C under ultra-high-vacuum (UHV) conditions in the 10^{-10} mbar range, and subsequent slow cooling to ambient temperature. The thermal treatment leads to a sublimation of the surface oxide layer which is present on silicon samples exposed to ambient air. By slowly cooling the silicon crystal, the high-temperature (1×1) surface structure is transformed at about 860 °C into the energetically more favorable (7×7) surface structure. The driving force for this (1×1) → (7×7) transformation is the reduction of dangling bond density. Dangling bonds generally contribute significantly to the total surface energy and are therefore unfavorable. For the (7×7) reconstructed Si(111) surface, the number of dangling bonds per (7×7) unit cell is reduced to 19, compared with 49 for the (1×1) surface structure.

STM images of the Si(111) 7×7 surface reveal 12 'topographic' maxima per unit cell as shown in Fig. 4.1. These topographic maxima can be attributed to the dangling bonds on the adatoms of the (7×7) surface structure (Fig. 4.2). There are twelve adatoms per (7×7) unit cell. Each adatom ties up three dangling bonds from the underlying atomic layer, leading to a single dangling bond on each adatom due to the fourfold coordination of silicon. The dangling bonds on the adatoms are partially filled and therefore contribute to both empty and filled states. The positions of the observed maxima do not depend on the polarity of the applied bias voltage, i.e. the maxima of the empty and filled states are spatially coincident. Consequently, Si(111) 7×7 is a special example of a semiconductor surface for which STM images directly provide geometric information about the positions of surface atoms. For instance, the shifted topographic maximum in Fig. 4.1, marked by an arrow, can directly be interpreted as a point defect in the arrangement of the adatoms on the Si(111) 7×7 surface.

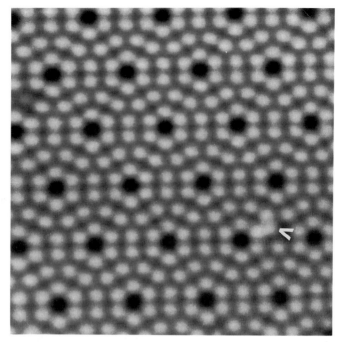

Fig. 4.1. STM top-view image of the clean Si(111) (7×7) surface. The side
length of the rhombohedral unit cell connecting the dark corner holes is 26.9
Å. Two point defects (a shifted adatom in the lower right part of the image
and a reacted adatom or surface impurity in the upper left part of the image)
are observed (Wiesendanger *et al.*, 1990g).

   The topographic STM images of the Si(111) 7×7 surface were found
to be in excellent agreement with the structure model proposed by
Takayanagi *et al.*, (1985a, 1985b) on the basis of transmission electron
diffraction (TED) data (Fig. 4.2). Takayanagi's dimer–adatom–stack-
ing-fault (DAS) model consists of 12 adatoms arranged locally in a
(2×2) structure, a stacking fault layer, as well as a layer with a vacancy
at the corner ('corner hole') and nine dimers on the sides of each of
the two triangular subcells of the (7×7) unit cell. The silicon layers in
one subcell (right-hand triangle in Fig. 4.2) are stacked regularly, while
those in the other subcell (left-hand triangle in Fig. 4.2) are stacked
with a faulted sequence. The 19 dangling bonds per unit cell are located
on the 12 adatoms, the six 'rest atoms' (triply coordinated silicon
atoms), and on the atom at the bottom of the corner hole. Other structure models proposed for the Si(111) 7×7 surface fail to fit the experimental STM data satisfactorily, as shown in Fig. 4.3 (Tromp *et al.*,
1986). Si(111) 7×7 is therefore a typical example of a surface structure
that has been solved not solely by STM. However, STM has certainly

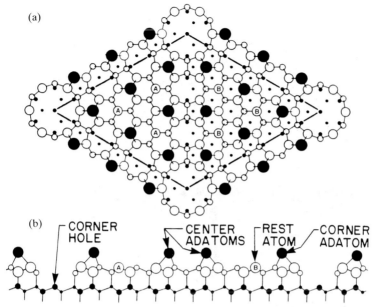

(a)

(b)

CORNER HOLE CENTER ADATOMS REST ATOM CORNER ADATOM

Fig. 4.2. DAS model of the Si(111) (7×7) surface. (a) Top view. Atoms on (111) layers at decreasing heights are indicated by circles of decreasing sizes. Heavily outlined circles represent 12 adatoms. Larger open circles represent atoms in the stacking fault layer. Smaller open circles represent atoms in the dimer layer. Solid circles and dots represent atoms in the unreconstructed layers beneath the reconstructed surface. (b) Side view. Larger open and solid circles indicate atoms on the ($\bar{1}01$) plane parallel to the long diagonal across the corner vacancies of the (7×7) unit cell. Smaller open and solid circles indicate atoms on the next ($\bar{1}01$) plane (Takayanagi et al., 1985b).

contributed significantly to discrimination between different structure models.

Though STM images of the Si(111) 7×7 surface directly reveal the positions of the adatoms, as already discussed, there are still significant electronic contributions to the observed image contrast. For instance, STM images obtained with positive sample bias voltage usually reveal 12 adatoms of equal height in each unit cell (Fig. 4.1 and Fig. 4.4(a)). In contrast, in STM images obtained with negative sample bias voltage the adatoms in the faulted half of the unit cell appear 'higher' than those in the unfaulted half (Fig. 4.4(b)). Furthermore, the adatoms located next to a corner hole ('corner adatoms') appear slightly higher than the central adatoms. By increasing the sample bias to −3 V, the rest atoms become visible as well, in addition to the adatoms (Fig. 4.4(c)).

These voltage-dependent STM studies of the Si(111) 7×7 surface have been complemented by local tunneling spectroscopy measurements as shown in Fig. 4.5. The conductance curves (I/U versus U), measured at different specific locations within the Si(111) 7×7 unit cell

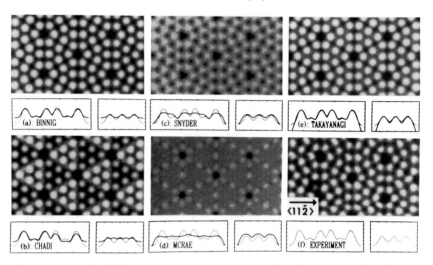

Fig. 4.3. (*a*)–(*e*) STM images calculated for (*a*) Binnig's model, (*b*) Chadi's model, (*c*) Snyder's model, (*d*) McRae's model, and (*e*) Takayanagi's model. (*f*) Measured STM image. The line scans run from corner hole to corner hole along the long (left) and short (right) diagonal of the (7×7) unit cell. The vertical range in these line scans is 4 Å. Solid lines are calculations, dashed lines represent the experimental results (Tromp *et al.*, 1986).

(Fig. 4.5(*a*)), show maxima at particular bias voltages corresponding to the energies of the surface states (section 1.13). Good agreement between ultraviolet photoemission spectroscopy (UPS) data and STM area-averaged data is found as shown in Fig. 4.5(*b*) and Fig. 4.5(*c*). At the energies of surface states, symmetry changes are observed in STM current images (CIs). By performing atomically resolved current imaging tunneling spectroscopy (CITS) measurements, the atomic origins of the various electronic states can directly be determined (Fig. 4.6). The electronic states near -0.35 eV (Fig. 4.6(*a*)) and $+0.5$ eV arise from the 12 adatoms within each unit cell, whereas the state near -0.8 eV (Fig. 4.6(*b*)) arises from the six rest atoms. Finally, the state near -1.7 eV (Fig. 4.6(*c*)) is ascribed to a backbond state. The state near -0.35 eV (Fig. 4.6(*a*)) appears to have some contributions from underlying layers because the corresponding CITS images show an asymmetry between the faulted and unfaulted halves of the (7×7) unit cell. The ability to map out the electronic states of a surface with a lateral resolution of about 3 Å, as demonstrated for the Si(111) 7×7 surface, is a unique feature of the STM technique and can successfully be applied to study atom-resolved surface chemistry (section 5.1).

STM is also a powerful method for characterization of non-periodic surface structures, such as defects, and their relation to the local surface electronic structure which can be probed via the atomically resolved

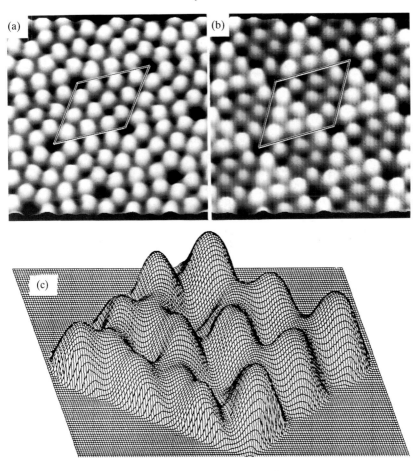

UNFAULTED HALF   FAULTED HALF

Fig. 4.4. STM topographs of the clean Si(111) 7×7 surface. (*a*) Topograph of the unoccupied states obtained with the sample biased at +1.5 V. The (7×7) unit cell is outlined and the 12 adatoms are clearly visible. (*b*) Topograph of the occupied states obtained with the sample biased at −1.5 V. The stacking fault and the differences between corner and center adatoms are visible. (*c*) A three-dimensional rendering of a topograph of the occupied states obtained with the sample biased at −3 V. The rest atoms are now visible (Avouris and Wolkow, 1989a).

tunneling current distribution. The influence of such defects on surface processes, e.g. chemical reactions, epitaxial growth, sputtering etc., can be studied in detail on a local scale. Different kinds of defects have been observed on the Si(111) 7×7 surface. Point defects appear, for instance, as shifted adatoms (Fig. 4.1), surface vacancies, or impurity sites. Line defects have been observed either within the surface plane

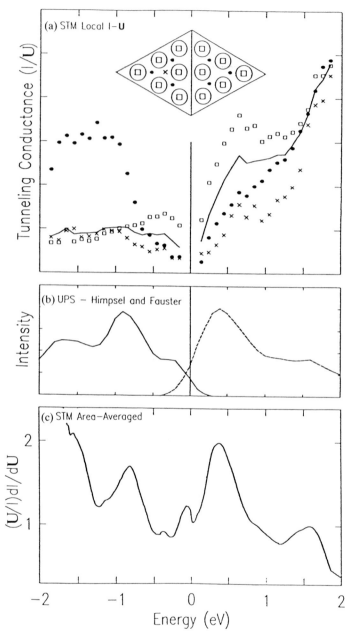

Fig. 4.5. Plots of the conductivity (I/U) measured at various locations within the Si(111) (7×7) unit cell (*a*), compared with the structure observed in photoemission and inverse photoemission spectroscopies (*b*). Also included is the normalized tunneling spectrum averaged over an area encompassing many unit cells (*c*) (Hamers, 1989a).

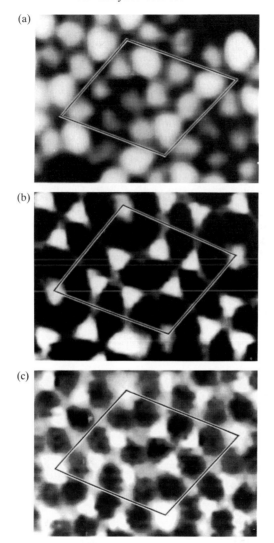

Fig. 4.6. CITS images of occupied Si(111) (7×7) surface states. (*a*) Adatom state at −0.35 V, (*b*) dangling-bond state at −0.8 V, (*c*) backbond state at −1.7 V (Hamers *et al.*, 1986a).

or as steps separating (111) terraces. The in-plane line defects appear as boundaries between translational domains of the (7×7) structure (Demuth *et al.*, 1986b; Berghaus *et al.*, 1988c; Köhler *et al.*, 1989; Sumita *et al.*, 1990; Hadley and Tear, 1991). These domain boundaries were found to play an essential role in the nucleation and growth behavior of the Si(111) 7×7 surface.

Condensed matter physics

STM studies of single and multiple steps on the Si(111) 7×7 surface have shown that the (7×7) reconstruction persists essentially undistorted right up to the steps (Becker *et al.*, 1985b; Wiesendanger *et al.*, 1990d, 1990f). This is illustrated in Fig. 4.7 and Fig. 4.8 for a sequence of multiple steps, separating narrow (7×7) reconstructed (111) terraces. The persistence of the (7×7) reconstruction right up to the step edges implies that the reconstruction itself depends on very short-range local interactions. The positions of the monatomic steps usually coincide with the (7×7) unit-mesh edges in both the upper and lower terraces (Becker *et al.*, 1985b). This leads to a quantization of terrace widths on vicinal Si(111) 7×7 surfaces (Wang *et al.*, 1990b; Goldberg *et al.*, 1991). As a direct consequence, kinks along step edges are quantized as well and follow the corner holes of the (7×7) structure. Though a quantization of terrace widths is also observed for surface areas with multiple steps, the lower step edges do not necessarily coincide with the edges of the (7×7) unit mesh, in contrast to the upper step edges (Fig. 4.8). In addition, point defects are likely to be observed in the vicinity of the lower step edges, as shown in Fig. 4.9 (Wiesendanger *et al.*, 1990d, 1990f).

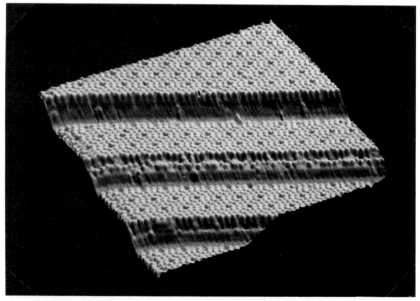

Fig. 4.7. Perspective STM image of a 32 nm × 36 nm area on the Si(111) 7×7 surface. Three steps as high as four times a double-layer step separate narrow (7×7) reconstructed terraces (Wiesendanger *et al.*, 1990d).

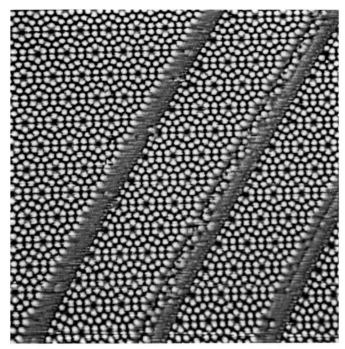

Fig. 4.8. Top-view image corresponding to Fig. 4.7 after using statistical differencing. Note the same termination of all (7×7) reconstructed terraces toward both sides of the multiple-step edges as well as the presence of defects in the vicinity of the multiple-step edges (Wiesendanger *et al.*, 1990d).

In general, the step structure of the Si(111) 7×7 surface depends critically on the rate of cooling through the $(1 \times 1) \rightarrow (7 \times 7)$ phase transition at about 860 °C during surface preparation, as well as on the presence of pinning centers such as surface impurities. A slow rate of cooling $(< 1 \text{ K s}^{-1})$ through the $(1 \times 1) \rightarrow (7 \times 7)$ transition is required to reach the equilibrium step and atomic surface structure. By increasing the cooling rate above 1 K s^{-1}, the high-temperature (1×1) surface is unable to transform into the energetically favorable (7×7) structure and non-equilibrium (2×1) and (2×2) structures can be quenched in (Pashley *et al.*, 1988a; Becker *et al.*, 1988). The (2×2) structure can be regarded as the basic building unit of the (7×7) reconstruction and is necessarily formed as an intermediate structure in the $(1 \times 1) \rightarrow (7 \times 7)$ transition. Other non-equilibrium Si(111) surface structures can be obtained by a combination of laser and thermal annealing (Becker *et al.*, 1986). Directly after laser annealing, which is associated with a high cooling rate,

(a)

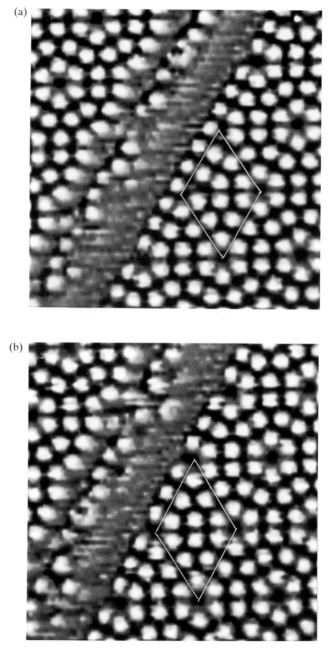

(b)

Fig. 4.9. (*a*) Selected part of the STM image shown in Fig. 4.8. A typical defect present on the lower side adjacent to multiple steps is marked by an arrow. (*b*) Another selected part of the STM image shown in Fig. 4.8 without the defect visible in (*a*) (Wiesendanger *et al.*, 1990d).

the non-DAS structures c-(4×2) and (2×2) are observed, whereas subsequent partial thermal anneals lead to several different DAS structures, including (5×5), (7×7) and (9×9). The (5×5) and (9×9) structures can be constructed by extension of Takayanagi's DAS model for the (7×7) structure (Fig. 4.10). Laser and thermally annealed Si(111) surfaces have also been prepared to study spatial transitions between ordered and disordered surface structures with atomic resolution (Wiesendanger *et al.*, 1990f, 1990g).

Cleaving silicon single crystals in UHV is another preparation method for obtaining clean Si(111) surfaces. In contrast to the annealing treatment, the resulting surface structure is not the equilibrium (7×7), but rather a metastable (2×1) structure which forms because of kinetic limitations at room temperature. STM images of this Si(111) 2×1 surface are dominated by the spatial dependence of the surface electronic structure and not by the geometrical positions of the adatoms, as for the Si(111) 7×7 surface. This is reflected by the observation of a lateral shift of the 'topographic' maxima between positive- and negative-bias STM images (Fig. 4.11), indicating a different spatial location of unoccupied and occupied electronic states (Stroscio *et al.*, 1986, 1987a; Feenstra and Stroscio, 1987a). The lateral shift between empty and filled surface states can be explained by charge transfer from one dangling

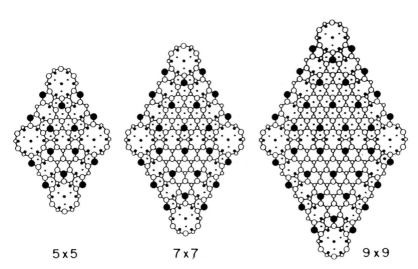

5 x 5   7 x 7   9 x 9

Fig. 4.10. Top view of (5×5), (7×7) and (9×9) structures according to the DAS model. Shaded large circles are adatoms. Open circles are atoms in the partially faulted double layers directly below. Filled circles represent the bulk unreconstructed double layers of the bulk (Becker *et al.*, 1986).

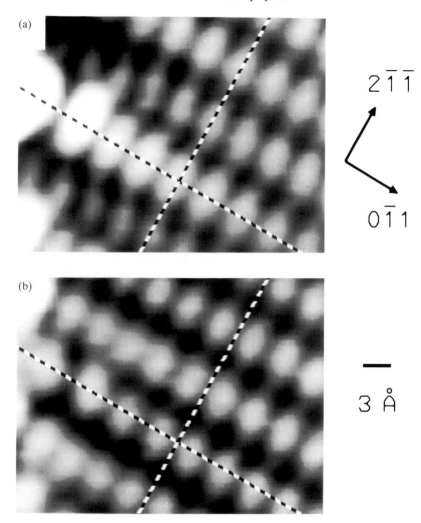

bond to a neighboring one, leading to a preferred appearance of one bond in the image of the unoccupied states, whereas the neighboring bond predominantly appears in the image of the occupied states. The charge transfer between dangling bonds can eventually lead to surface buckling as well (Feenstra, 1990). The observed difference in lateral position between empty and filled states provides a valuable simple measure of the surface structure which can be used to discriminate between different structure models. Only the π-bonded chain model of the Si(111) 2×1 surface (Pandey, 1981) was found to be consistent with the experimental STM data.

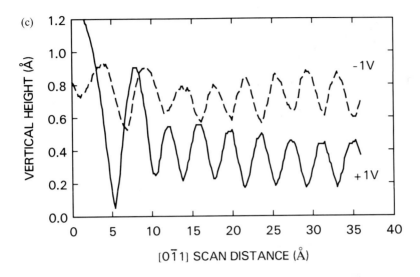

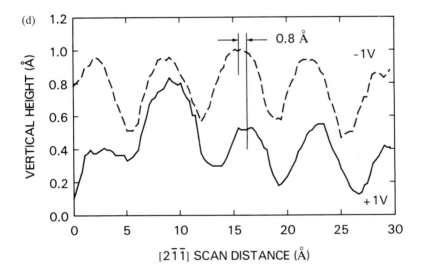

Fig. 4.11. Two STM images of the Si(111) 2×1 surface, acquired simultaneously at sample bias voltages of (*a*) +1 V and (*b*) −1 V. Also shown are cross-sections of the topography, (*c*) along the [0$\bar{1}$1] direction of the (2×1) unit cell and (*d*) along the [2$\bar{1}\bar{1}$] direction of the unit cell, as shown in (*a*). For the [0$\bar{1}$1] direction, the corrugation shifts by half a unit cell (1.92 Å) when the voltage changes from −1 V to +1 V, and for the [2$\bar{1}\bar{1}$] direction the corrugation shifts by 0.8 Å (Feenstra and Stroscio, 1987a).

The electronic structure of the Si(111) 2×1 surface was studied in more detail by local tunneling spectroscopy (Stroscio *et al.*, 1986). The normalized differential conductance curve $(dI/dU)/(I/U)$ versus U, as shown in Fig. 4.12, exhibits several peaks which can be identified with known electronic features of the Si(111) 2×1 surface. The peaks at -0.3 eV and $+0.2$ eV separate the occupied (π-bonding) and unoccupied (π^*-antibonding) bands of surface states. The peaks at -1.1 eV and $+1.2$ eV correspond to the bottom of the occupied band and the top of the unoccupied band, respectively. Therefore, the width of the occupied band is found to be 0.8 eV, which is less than the width of the unoccupied band, 1.0 eV. The band gap is determined to be 0.50 ± 0.05 eV (Feenstra, 1991). This value has to be compared with the bulk band gap of 1.12 eV at room temperature.

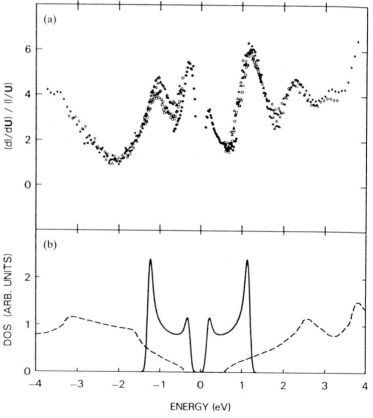

Fig. 4.12. (*a*) Ratio of differential to total conductivity for the Si(111) 2×1 surface. The different symbols refer to different tip–sample separations. (*b*) Theoretical DOS for the bulk valence band and conduction band of silicon (dashed line), and the DOS from a one-dimensional tight-binding model of the π-bonded chains (solid line) (Stroscio *et al.*, 1986).

STM has also been applied to study the step structure on cleaved Si(111) 2×1 surfaces (Feenstra and Stroscio, 1987b; Tokumoto *et al.*, 1990b). The (2×1) reconstruction was found to extend right up to the steps (Fig. 4.13). For detailed interpretation of the structure along the step edge, voltage-dependent STM studies are again required.

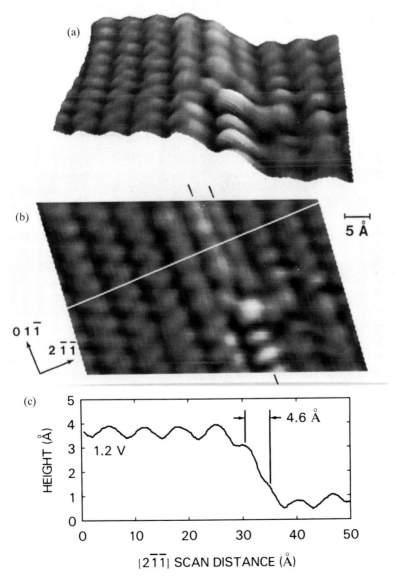

Fig. 4.13. STM image of a step on Si(111), acquired at a sample bias voltage of +1.2 V; (*a*) perspective view, (*b*) top view of the same data, and (*c*) cross-sectional cut along the line indicated in (*b*). The step edge is identified by tick marks at the border of the image in (*b*) (Feenstra and Stroscio, 1987b).

Condensed matter physics

Starting with the metastable (2×1) structure, a transformation into the equilibrium (7×7) structure can be achieved by thermal annealing of the cleaved Si(111) sample. Two intermediate surface structures are observed during this transition: first, a disordered adatom arrangement which nucleates at steps or (2×1) domain boundaries, and second, a well-ordered (5×5) structure which grows from small domains of locally ordered adatom arrangements (Feenstra and Lutz, 1990, 1991a, 1991b). The (5×5) structure, as shown in Fig. 4.14, is theoretically predicted to have a total energy only slightly above that of the (7×7) structure. The formation of an intermediate (5×5) structure is believed to be favored

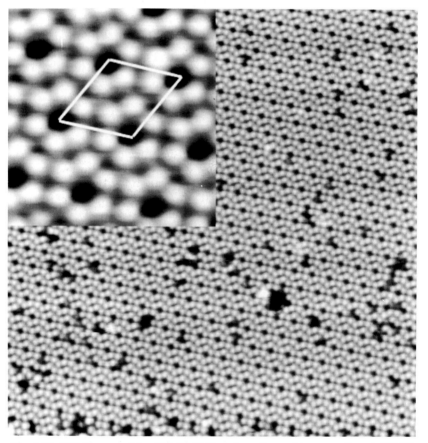

Fig. 4.14. STM image of the Si(111) 5×5 surface, formed by annealing at 535 °C for 10 s. The large image extends over an area of 50 nm × 50 nm. The inset shows the surface on a four times expanded scale. The 5×5 unit cell is indicated by white lines in the inset (Feenstra and Lutz, 1991b).

over a direct $(2 \times 1) \rightarrow (7 \times 7)$ transformation because the (5×5) structure contains the same number of atoms as the (2×1) structure.

Si(111) as the semiconductor surface that has most often been studied by STM certainly provides an outstanding example for the variety of information which can be gained by detailed STM/STS investigations. We now focus on STM studies of other silicon surfaces.

4.1.1.2 Clean semiconductor surfaces: Si(100)

Clean Si(100) surfaces can be obtained by the thermal annealing treatment already described for the Si(111) surface, leading to a Si(100) 2×1 reconstruction. STM images of the Si(100) 2×1 surface are dominated by the surface electronic structure and show a different spatial distribution of the filled and empty states (Fig. 4.15). At negative sample bias, the STM images show a single bean-shaped structure per (2×1) unit cell, whereas at positive sample bias, two clearly resolved 'topographic' maxima appear in the (2×1) unit cell with a deep trough between them and less pronounced 'topographic' minima towards the neighboring unit cells (Tromp *et al.*, 1985; Hamers *et al.*, 1986b, 1987). The experimental STM results are consistent with a dimer-type reconstruction of the Si(100) 2×1 surface (Levine, 1973). Pairs of silicon atoms, each of which is bonded to two silicon atoms in the underlying atomic layer, dimerize (Fig. 4.16), leaving a single dangling bond per surface adatom. The resulting (2×1) unit cell therefore contains two dangling bonds.

The electronic structure of the dimers has been described in terms of a π-bonding state below the Fermi level E_F and a π^*-antibonding state above E_F. The antisymmetric nature of the π^*-antibonding state is associated with a node in the wave function at the center of the dimer bond, explaining the deep trough there in STM images obtained with a positive sample bias. In contrast, the symmetric nature of the π-bonding state leads only to a single maximum per dimer in negative-bias STM images. The Si(100) 2×1 surface therefore provides an instructive example for cases where STM images directly reflect the spatial symmetry of the surface-state wave functions. Experimental investigation of the local electronic structure of the Si(100) 2×1 surface has been complemented by local tunneling spectroscopy measurements (Hamers *et al.*, 1987). The normalized differential conductance curve $(\mathrm{d}I/\mathrm{d}U)/(I/U)$, as shown in Fig. 4.17, exhibits two pronounced peaks at -0.8 eV and $+0.35$ eV which have been assigned to the π-bonding and π^*-antibonding states, respectively. The surface state band gap is nearly 1 eV.

An important issue with respect to the Si(100) 2×1 surface has been the question of under which circumstances the dimers, as imaged by

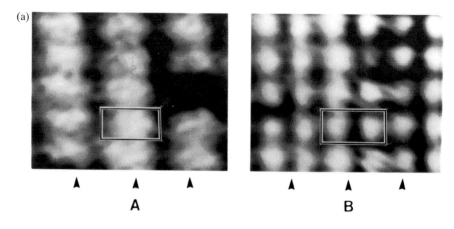

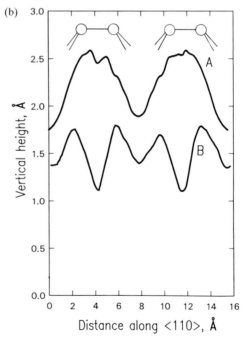

Fig. 4.15. (*a*) STM topographic images probing the (A) occupied (−2 V bias) and (B) unoccupied (+1.2 V bias) surface states of the clean Si(001) 2×1 surface. The arrows point in the [1$\bar{1}$0] direction and point directly at the center of the dimer rows. Each box outlines a (2×1) unit cell with a dimer centered in the box. (*b*) STM corrugation profiles along the [110] direction on clean Si(001) 2×1 at −2 V bias (curve A) and +1.2 V bias (curve B). Each profile starts between two dimer rows, passes over the centers of two dimers, and ends between two rows (Hamers *et al.*, 1987).

Si (100)

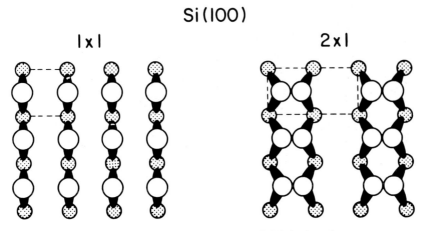

Fig. 4.16. Structural model for the Si(001) 2×1 surface.

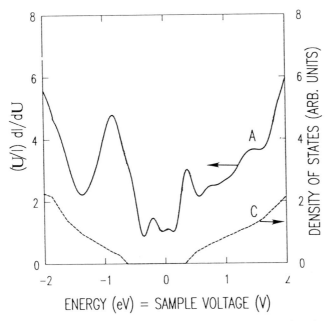

Fig. 4.17. Tunneling spectroscopy measurements on ordered (2×1) regions of a clean Si(001) 2×1 surface (curve A). For comparison, curve C shows the calculated silicon bulk density of states (Hamers *et al.*, 1987).

STM, appear symmetric or asymmetric, non-buckled or buckled. The first atomic-resolution STM study of the Si(100) 2×1 surface revealed the presence of both non-buckled and buckled dimers (Tromp *et al.*, 1985; Hamers *et al.*, 1986b). The regions containing non-buckled dimers exhibit (2×1) symmetry, whereas ordered arrangements of buckled dimers generate various higher-order surface reconstructions, e.g. p(2×2) or c(4×2). It was proposed that at room temperature the time-averaged configuration for the dimers as seen by STM is symmetric and non-buckled, although the dimers may be dynamically buckling about the equilibrium configuration on a time scale which is short relative to the STM measurement time (Kochanski and Griffith, 1991). Buckled dimers were often observed near surface defects such as steps or dimer defects on terraces. It was proposed that the dimer buckling is stabilized by these defects by pinning the dimers in particular buckling orientations. However, it can depend on the particular type of dimer defect whether static buckling in adjacent regions is observed or not (Hamers and Köhler, 1989). In addition, indications for a temperature-dependence of the number of buckled dimers have been found more recently by means of variable-temperature UHV STM (Wolkow, 1992). Other STM studies of the Si(100) 2×1 surface have revealed the presence of a strong dimer asymmetry over extended surface areas without defects in both occupied and unoccupied state images even at room temperature (Sugihara *et al.*, 1991; Wiesendanger *et al.*, 1992a). The observed dimer asymmetry, as shown in Fig. 4.18, was found to be independent of the scanning direction. It is likely that the ratio of symmetric to asymmetric, non-buckled to buckled dimers, as imaged by STM at room temperature, depends on the exact preparation conditions of the Si(100) 2×1 surface (Tromp *et al.*, 1985).

The density of defects on the Si(100) 2×1 surface is typically higher than on the Si(111) 7×7 surface for the same preparation conditions. Different types of surface defects have been distinguished on the Si(100) 2×1 surface: single and double dimer vacancy-type defects (Figs. 4.19(*a*) and 4.19(*b*)) which show semiconductor-like local *I–U* characteristics, and another type of characteristic defect (Fig. 4.19(*c*)) which gives rise to metal-like local *I–U* characteristics (Hamers and Köhler, 1989). The latter type of defect (a 'C-defect') is believed to contribute considerably to the relatively high density of states at the Fermi level and is likely active in Fermi-level pinning on the Si(100) 2×1 surface.

STM has also been applied extensively to study the step structure of vicinal Si(100) surfaces (Griffith and Kochanski, 1990b). If the miscut angle is 2° or less, the steps were found to be predominantly of mon-

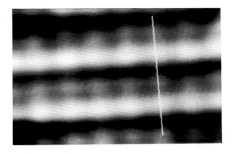

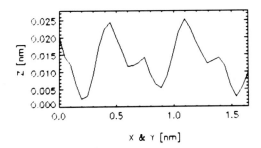

Fig. 4.18. STM image of the Si(001) 2×1 surface showing highly asymmetric dimers, as can also be seen from a line section perpendicular to the dimer rows (Wiesendanger *et al.*, 1992a).

atomic height. The dimer rows of the (2×1) reconstruction are orientated perpendicular to each other by going from one terrace to the next, which is a consequence of the diamond-type crystal structure of silicon. Two different types of steps can be distinguished, according to the orientation of the dimer rows on the upper terrace relative to the step edges, being either parallel (A-type step) or perpendicular (B-type step). If the equilibrium step structure had been reached during surface preparation, the A-type steps were found to be rather smooth whereas B-type steps appeared rough with many kink sites, as shown in Fig. 4.20 (Swartzentruber *et al.*, 1990; Kariotis and Lagally, 1991). According to theoretical calculations, A-type steps require less energy to form than B-type steps (Chadi, 1987).

The STM image of Fig. 4.20 reveals nearly equal surface areas of (2×1) and (1×2) domains. This is typical for an externally unstrained

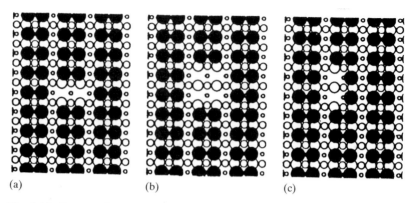

(a)   (b)   (c)

Fig. 4.19. Structural models illustrating the positions of the atoms observed in the STM (black) relative to those in the bulk crystal (open circles). Atoms with larger diameters are closer to the viewer; four atomic layers are visible. (*a*) Type-A single dimer vacancy defect; (*b*) type-B double dimer vacancy defect; (*c*) type-C defect (Hamers and Köhler, 1989).

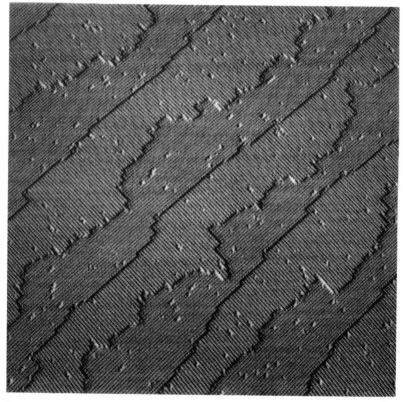

Fig. 4.20. STM image of a vicinal Si(001) 2×1 surface (Kariotis and Lagally, 1991).

Si(100) sample. By applying a uniaxial strain, an asymmetry in the relative population of the (2×1) and (1×2) domains is induced, as shown in Fig. 4.21 (Webb *et al.*, 1990, 1991; Packard *et al.*, 1990). To produce this asymmetry, the motion of monatomic steps, forming the domain boundaries, is required. The population asymmetry can be reversed by reversing the strain. It is found that the domains compressed along the direction of the dimer bond are favored. The observations can be explained by a reduction of the energy associated with a

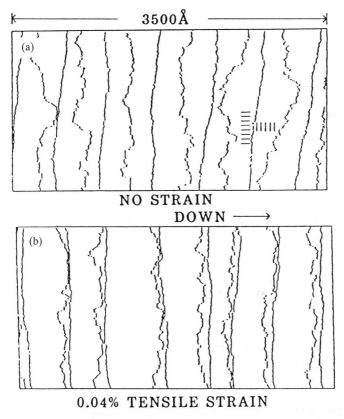

Fig. 4.21. Large-scale STM images of the Si(001) 2×1 surface before and after tensile strain. To show the steps most easily, the image has been processed so that the positions of steps along each scan line are displayed as a black pixel. In the top panel, the sample was miscut by 0.3° in the [110] direction. The average terrace width is 270 Å. The two domain populations are equal to within 1%. The direction of the dimer bonds is indicated. In the bottom panel, after applying a tensile strain of 0.04% in the horizontal direction, 80% of the area is occupied by the majority domain. The step density is unchanged after tension. For both panels the highest terrace is at the left (Webb *et al.*, 1991).

long-range strain field due to the anisotropy of the intrinsic surface stress for the two types of reconstructed domains.

For vicinal Si(100) surfaces with a miscut angle of about 4°, STM images reveal the presence of straight, evenly spaced double steps of type-B (Fig. 4.22), whereas A-type double steps or monatomic steps are not observed (Wierenga *et al.*, 1987; Griffith *et al.*, 1989; Swartzentruber *et al.*, 1989). Theoretical calculations show that a B-type double step is less energetic than the two different types of monatomic steps combined (Chadi, 1987). Therefore, the surface can lower its energy by pairing the single steps into B-type double steps. In contrast, A-type double steps would require considerably more energy to form and consequently are not observed. The tendency of vicinal Si(100) surfaces, with a miscut angle as large as 4°, to form double steps has important implications for semiconductor heteroepitaxy, e.g. optimized growth of GaAs on Si(100) substrates (Bringans *et al.*, 1991). For vicinal Si(100) surfaces with a small miscut angle (2° or below), the formation of double steps is not favored, presumably due to a long-range repulsive interaction between steps.

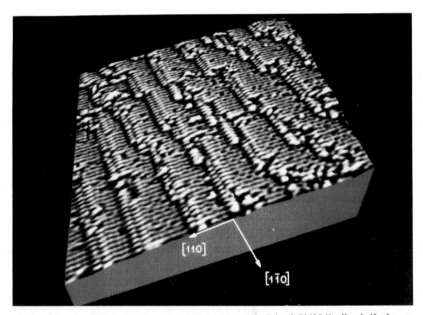

Fig. 4.22. Perspective view of an STM image of vicinal Si(001) tilted 4° about [1$\bar{1}$0]. The dimensions of the region are 425 Å along [1$\bar{1}$0] and 225 Å along [110]. Double steps running along [1$\bar{1}$0] separate the terraces. At kinks the double steps usually split into single steps. Curvature-keyed shading is used (Griffith *et al.*, 1989).

4.1.1.3   Clean semiconductor surfaces: Si(110)

The Si(110) surface has less often been studied by STM than the Si(111) and Si(100) surfaces. The clean Si(110) surface was found to exhibit a '16×2' reconstruction (van Loenen *et al.*, 1988; Hoeven *et al.*, 1989a). A STM image showing a '16×2' reconstructed area together with a partly faceted region is presented in Fig. 4.23.

4.1.1.4 Clean semiconductor surfaces: High-index silicon surfaces

High-index silicon surfaces, such as Si(112) or Si(223), generally tend to be unstable and prefer to facet into lower-index planes, e.g. (111) planes, upon annealing (Berghaus *et al.*, 1987, 1988a). An exception is given by the Si(113) surface which is expected to have an energy comparable to that of the low-index surfaces. STM studies of the Si(113) surface revealed a (3×2) reconstruction and a high density of domain boundaries which introduce energy states that pin the Fermi level (Knall *et al.*, 1991).

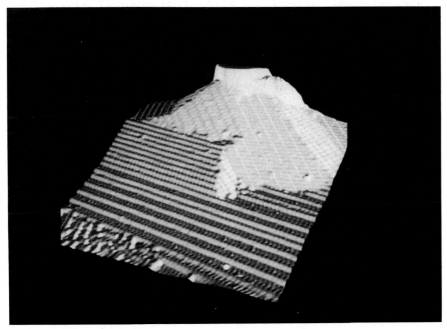

Fig. 4.23. Three-dimensional representation of a topographic STM image (80 nm × 91 nm) of the Si(110) surface (Hoeven *et al.*, 1989a).

4.1.1.5 Clean semiconductor surfaces: Ge(111)

The clean equilibrium Ge(111) surface obtained by ion sputtering and thermal annealing exhibits a c(2×8) reconstruction (Fig. 4.24). STM studies have shown that this c(2×8) structure coexists with local (2×2) and c(4×2) arrangements (Becker *et al.*, 1985d, 1989; Feenstra and Slavin, 1991). Evidence for a full or partial stacking fault on the Ge(111) c(2×8) surface was not found. Therefore, the c(2×8) reconstruction cannot be described by the DAS model. A lateral shift is observed between STM images of the empty and filled states, indicating the dominance of electronic structure effects (Feenstra and Slavin, 1991). Local

Fig. 4.24. STM image of the Ge(111) c(2×8) surface, acquired at a sample bias voltage of 2.5 V. A rectangular unit cell is shown in the center of the image. Arrows indicate an extra row of adatoms, forming a local (2×2) structure (Feenstra and Slavin, 1991).

tunneling spectroscopy data for the Ge(111) c(2×8) surface are presented in Fig. 4.25. The peaks at −0.7 eV and +0.5 eV are ascribed to
the 'rest atom' and adatom dangling bond states, respectively. Consequently, the 'rest atoms' are the dominant features in STM images
taken with a negative sample bias voltage, whereas the adatoms become

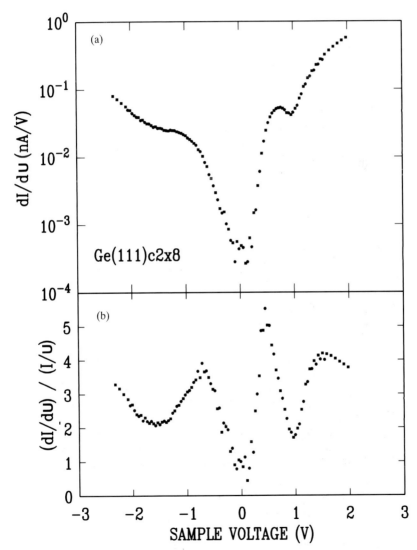

Fig. 4.25. Spectroscopic results for the Ge(111) c(2×1) surface, showing (*a*)
d*I*/d*U* on a logarithmic scale, and (*b*) the normalized conductivity (d*I*/d*U*)/(*I*/*U*)
(Feenstra and Slavin, 1991).

the dominant features in positive-bias images (Hirschorn *et al.*, 1991). The surface state band gap is found to be about 0.6 eV.

A clean Ge(111) surface can alternatively be obtained by cleaving a germanium single crystal in UHV. In this case, a metastable (2×1) reconstruction is found, similar to cleaved Si(111). STM studies of the Ge(111) 2×1 surface are in agreement with the π-bonded chain model, which had already described the Si(111) 2×1 surface structure (Feenstra, 1991). The surface state band gap was determined to be 0.65 ± 0.09 eV. This value should be compared with the bulk band gap of 0.67 eV at room temperature.

By annealing a cleaved Ge(111) crystal at about 100 °C, the metastable (2×1) structure transforms into the equilibrium c(2×8) structure (Feenstra and Slavin, 1991).

4.1.1.6 Clean semiconductor surfaces: Ge(100)

STM has also been applied to study the clean Ge(100) surface obtained by ion sputtering and thermal annealing. A dimer-type reconstruction is observed, similar to the case for the Si(100) surface (Kubby *et al.*, 1987). The dimers on Ge(100) are found to be preferentially buckled. The different arrangements of the asymmetric buckled dimers lead to local p(2×2) or c(4×2) structures besides the basic (2×1) structure. The concentration of dimer defects on Ge(100) is much less than on Si(100), and defects do not appear to be necessary to induce dimer buckling on the Ge(100) surface.

4.1.1.7 Clean semiconductor surfaces: GaAs(110)

Among compound semiconductors, GaAs has most often been studied by STM. By cleaving a GaAs single crystal in UHV, a clean unreconstructed (110) surface can be obtained which consists of equal numbers of Ga and As atoms. The (1×1) surface unit cell contains one Ga and one As atom, each of which has one dangling bond. The Ga and As sites define separate atomic rows along the [1$\bar{1}$0] direction. STM images of the GaAs(110) surface reveal these rows, as shown in Fig. 4.26 (Feenstra *et al.*, 1987b). However, at each polarity of the sample bias voltage, only one 'topographic' maximum per surface unit cell is observed. In addition, a lateral shift of the maxima between positive- and negative-bias polarity images is found (Fig. 4.26(*a*) and (*b*)). These experimental STM results can be explained by the charge transfer which occurs from the Ga to the As atoms, leaving the Ga dangling bond somewhat more empty, and the As dangling bond somewhat more full. Therefore, the unoccupied state density is concentrated around the Ga sites, and the occupied state density around the As sites. Consequently,

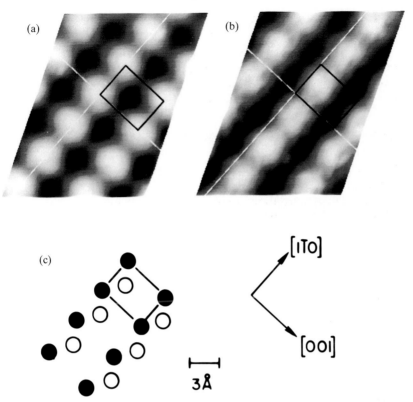

Fig. 4.26. Constant-current STM images of the GaAs(110) surface acquired at sample bias voltages of (*a*) +1.9 V and (*b*) −1.9 V. (*c*) Top view of the surface atoms. The As atoms are represented by open circles and the Ga atoms by closed circles. The rectangle indicates a unit cell, whose position is the same in all three figures (Feenstra *et al.*, 1987b).

the maxima observed at positive sample bias voltage correspond to Ga sites, whereas the maxima at negative bias reveal As sites. This way, atom-selective imaging of the GaAs(110) surface is feasible by STM. However, the distinction between Ga and As as chemically different species relies on external knowledge, such as the direction of charge transfer and identification of electronic states with atomic species.

Charge transfer from Ga to As atoms is additionally accompanied by vertical displacement (buckling) of As atoms with respect to Ga atoms. The precise value of the lateral shift between the maxima in the unoccupied and occupied states depends on that difference in vertical position of the As and Ga atoms because the bond lengths are nearly invariant. Therefore, the tilt or buckling angle ω, as defined in Fig. 4.27, can

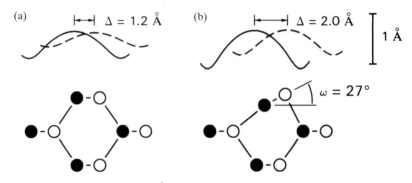

Fig. 4.27. [1$\bar{1}$0] cross-sectional cuts of the state-density contours for occupied (dashed line) and unoccupied (solid line) states. (*a*), (*b*) Theoretical results for buckling angles of (*a*) 0° and (*b*) 27°. A side view of the atomic structure is also shown, with As atoms represented by open circles and Ga atoms by closed circles. (Feenstra *et al.*, 1987b).

be quantitatively determined by comparison between experiment and theory (Feenstra *et al.*, 1987b).

Investigations of the local surface electronic structure of the GaAs(110) surface were complemented by local tunneling spectroscopy measurements (Feenstra *et al.*, 1987b; Feenstra and Stroscio, 1987c). The normalized differential conductance curve $(dI/dU)/(I/U)$ versus U, as shown in Fig. 4.28, reveals three peaks labeled V, D and C. Peaks V and C are attributed to tunneling out of GaAs valence-band states and tunneling into conduction-band states, respectively. The separation between the edges of the C and V peaks yields roughly the band gap of 1.4 eV. Peak D is expected to be associated with tunneling through dopant-induced states in the band gap region, whereas surface states are not believed to exist in the band gap of the clean GaAs(110) surface.

STM has also been applied to study defects on the cleaved GaAs(110) surface. Point defects are observed as depressions along the atomic rows running in the [1$\bar{1}$0] direction for both atomic species (Cox *et al.*, 1990a). In contrast to the As point defects, the Ga point defects were found to be negatively charged, which was deduced from the dependence of their apparent size on the applied bias voltage. The observed point defects were interpreted as Ga or As vacancies.

STM images of vicinal GaAs(110) surfaces reveal terraces of triangular shape which are predominantly separated by monatomic steps (Möller *et al.*, 1989b; Cox *et al.*, 1990a). On cleaved (110) surfaces of plastically deformed GaAs crystals, dislocations have been imaged with

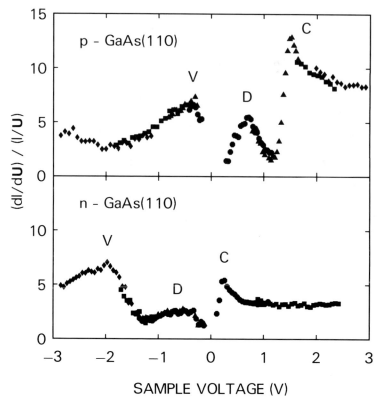

Fig. 4.28. Ratio of differential to total conductivity for n- and p-type GaAs. The peaks are labeled as conduction band (C), valence band (V), and dopant-induced (D) (Feenstra *et al.*, 1987b).

atomic resolution, as shown in Fig. 4.29 (Cox *et al.*, 1990b, 1991). During observation the dislocations were found to be mobile over nanometer distances. The motion of the dislocations may be a result of local stresses induced by the close proximity of the scanned tip or may be stimulated by charge-carrier injection and subsequent recombination.

4.1.1.8 Adsorbates and metal overlayers on semiconductor surfaces
Starting with clean semiconductor substrates, different kinds of species can be adsorbed, resulting in new surface structures. Thus, the early stages of oxidation or metal thin film growth, for instance, can be studied. Depending on the substrate temperature and the degree of coverage, many different phases may appear for these adsorbate or overlayer systems. Even for a given preparation procedure, the whole

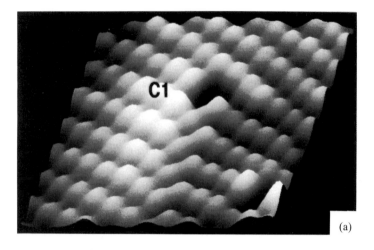

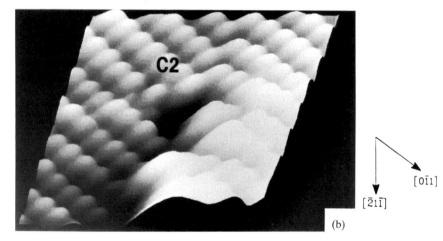

Fig. 4.29. (*a*) STM image of the GaAs(110) surface showing the first partial dislocation of a dissociated dislocation. The core can be seen at C1. From it a stacking fault runs along $[\bar{2}1\bar{1}]$ which ends at the core of the second partial shown in (*b*). From the core seen at C2, a monoatomic step extends along $[\bar{2}1\bar{1}]$. (Cox *et al.*, 1991).

surface usually exhibits several different phases simultaneously in different surface regions, rather than being spatially homogeneous on a macroscopic length scale. Spatial inhomogeneities in the adsorption process may arise, for instance, from the presence of substrate defects, inequivalent surface sites (even for an idealized defect-free substrate), or from spatial variations in substrate temperature or adsorbate coverage. STM as a local probe is ideally suited to study the local aspects of adsorption and early stages of overlayer growth.

4.1.1.9 Oxygen adsorption

Investigation of the initial stages of oxidation of semiconductor surfaces is important for optimizing the growth of thin, insulating oxide layers in semiconductor devices. We will first discuss the adsorption of oxygen on the Si(111) 7×7 surface because this system has been most widely studied by STM. Oxidation of the Si(111) 7×7 surface is a rather complex and non-uniform reaction process which does not lead to an ordered surface reconstruction. STM studies have shown that defect sites on the clean Si(111) 7×7 surface, such as vacancies or (7×7) domain boundaries, act as nucleation centers for the oxidation process. The oxidation rate is significantly enhanced if the initially prepared surface has a substantial number of such defects. On the other hand, straight atomic steps were found to be relatively insensitive to oxygen exposure (Leibsle, *et al.*, 1988). On defect-free surface regions, it was found that inequivalent surface sites exhibit a different degree of local reactivity, which has been explained in terms of a site-dependent sticking coefficient. For instance, the adatoms in the faulted half of the (7×7) unit cell appear to be more reactive than those in the unfaulted half (Tokumoto *et al.*, 1990a; Pelz and Koch, 1990b, 1991; Avouris *et al.*, 1991). In addition, the adatoms near the corner holes seem to be more reactive than the center adatoms (Pelz and Koch, 1990b; Avouris *et al.*, 1991). The higher degree of local reactivity of the faulted half and the corner-adatom sites is directly related to the higher LDOS of occupied states near E_F for these sites (section 4.1.1.1).

Furthermore, two distinct reaction species have been identified in STM topographs of the unoccupied states of the sample which appear as either bright or dark sites (Fig. 4.30). Based on a comparison of STM spectroscopic data with photoemission spectroscopy data and theoretical electronic structure calculations, the bright sites are attributed to an oxygen atom inserted in one of the adatom's back bonds, whereas the dark sites are attributed to an oxygen atom saturating the adatom's dangling bond while another oxygen atom is inserted in one of the adatom's back bonds (Lyo *et al.*, 1990; Avouris *et al.*, 1991). Without doubt, STM has contributed significantly to our present understanding of the early stages of oxidation of the Si(111) 7×7 surface. However, the assignment of modified surface sites to particular reaction products remains a central problem in the interpretation of STM data in general.

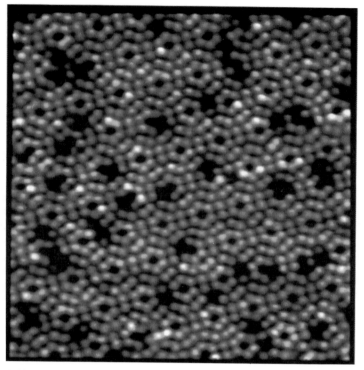

Fig. 4.30. STM topograph of the unoccupied states of a Si(111) 7×7 surface
exposed to 0.2 L of O_2 at 300 K. Sample bias voltage is +2 V (Avouris *et al.*,
1991).

More recently, STM has also been applied to study the initial stages
of oxidation of the Si(100) 2×1 surface, which is technologically more
important than the Si(111) surface (Cahill and Avouris, 1992). The
initial reaction appears to be essentially localized at 'C-type' defects
which show a high LDOS at E_F (section 4.1.1.2). The oxidation of
Ge(111) c(2×8) surfaces has been investigated as well, though the oxide
of germanium is technologically less useful because it reacts with water.
Similarly to the Si(111) 7×7 surface, defect sites such as domain boundaries were identified as primary nucleation sites, whereas step edges
appeared to be surprisingly inert. At elevated temperature, oxidation
was found to be facilitated, primarily due to degradation of the c(2×8)
order (Klitsner *et al.*, 1991).

STM studies of oxygen adsorption on n-type GaAs(110) surfaces have
revealed a pronounced voltage dependence of the images obtained
(Stroscio *et al.*, 1987b). The adsorbed oxygen appears as surface protrusion for negative sample bias, and as surface depression for positive

sample bias (Fig. 4.31). This behavior can be explained by assuming the electronegative oxygen adsorbate to be negatively charged. Consequently, electrons in the semiconductor's conduction band will experience repulsion at positive bias, leading to a decrease in the tunnel current, whereas attraction will be experienced by holes in the semiconductor's valence band at negative bias, leading to an increase in the tunnel current. The electric field from the charged adsorbate was

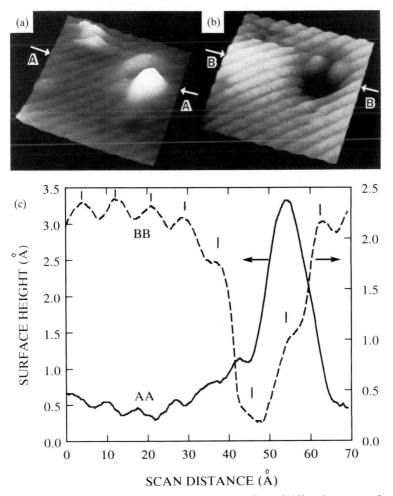

Fig. 4.31. STM images of an oxygen defect on a GaAs(110) substrate surface acquired simultaneously at sample bias voltages of (*a*) -2.6 V and (*b*) 1.5 V. (*c*) Surface-height contours along the lines AA (solid line) and BB (dashed line) shown in (*a*) and (*b*), respectively. Oxygen exposure is 120 L. The tick marks indicate the position of the Ga lattice (Stroscio *et al.*, 1987b).

found to be screened by the conduction electrons with a characteristic screening length of about 5 nm, which can be viewed as microscopic local band bending. This local band bending is directly seen in real-space STM contours around the defects, as well as in spectroscopic I–U curves (Stroscio *et al.*, 1987b). In contrast to the experimental results obtained for n-type GaAs samples, no evidence for local band bending was found for oxygen adsorbates on p-type GaAs(110) surfaces, indicating that these adsorbates are electrically neutral. Consequently, the oxygen sites always appeared as surface protrusions, irrespective of the applied bias voltage polarity (Stroscio *et al.*, 1987c).

4.1.1.10 Hydrogen adsorption

The formation and breaking of silicon–hydrogen bonds are involved in many technologically important processes, motivating STM studies of atomic hydrogen adsorption on silicon surfaces. For the Si(111) 7×7 surface, direct binding of the hydrogen atoms to the existing surface dangling bonds is observed at low coverage (Mortensen *et al.*, 1991; Boland, 1991a, 1991b). At increasing hydrogen exposure, the uppermost silicon adatom layer is removed, accompanied by binding of hydrogen atoms to the created dangling bonds of the next atomic layer containing the stacking fault. The STM images then reveal a (1×1) lattice periodicity, while the boundaries of the (7×7) unit cells are still visible.

STM has also been used to study the adsorption of hydrogen on Si(100) 2×1 surfaces (Boland, 1990, 1992). At low exposure, hydrogen reacts with the surface dangling bonds, while the dimer bonds remain intact, resulting in a monohydride phase which preserves the (2×1) surface structure. On the other hand, saturation exposures were found to lead to formation of a dihydride phase with a (1×1) surface structure.

4.1.1.11 Metal overlayers

The initial stages of growth of thin metal overlayers on semiconductor substrates have been studied extensively by STM, aiming at a better understanding of metal–semiconductor interfaces including Schottky barrier formation. In the following, we mainly focus on Si(111) 7×7 and GaAs(110) substrates which have been used most widely.

STM as a local probe has been found to be particularly useful in cases where several different surface phases coexist. An example is presented in Fig. 4.32 for half a monolayer (ML) of tin on Si(111) 7×7, where three different surface structures were observed simultaneously (Nogami *et al.*, 1989). Evidently, surface analytical techniques averaging over large surface areas must fail to provide detailed information

Fig. 4.32. STM image (30 nm × 30 nm) of a Si(111) surface after deposition of half a monolayer of Sn. Areas covered by three different surface phases (7×7, $\sqrt{3}\times\sqrt{3}$, and $2\sqrt{3}\times2\sqrt{3}$) are clearly visible. Two different orientations of the $2\sqrt{3}\times2\sqrt{3}$ structure are seen (Nogami *et al.*, 1989).

about the atomic and electronic structure of such inhomogeneous surfaces, where particular surface phases might be confined to small regions only a few nanometers across.

Even for a given surface phase, inhomogeneities arising from point defects may be present in the metal overlayer which can be characterized on a local scale by STM. An example is given by the Si(111) $(\sqrt{3}\times\sqrt{3})$Al system (Hamers and Demuth, 1988; Hamers, 1988, 1989b). STM images obtained at positive sample bias reveal the $(\sqrt{3}\times\sqrt{3})$ lattice together with defect sites which appear less bright (Fig. 4.33(*a*)). At negative sample bias, the contrast reverses as shown in Fig. 4.33(*b*). Based on the observed dependence of the defect density on Al coverage, it was concluded that the bright protrusions seen at positive bias must correspond to Al adatoms whereas the defect sites were attributed to Si adatoms substituting for Al in the $(\sqrt{3}\times\sqrt{3})$ structure below an Al coverage of one third of a ML. The Si adatoms give rise to an extra dangling-bond defect state near -0.4 eV, as seen by local tunneling spectroscopy (Fig. 4.34). This defect state causes the Si adatoms to appear brighter than the Al adatoms for negative sample

(a)

(b)

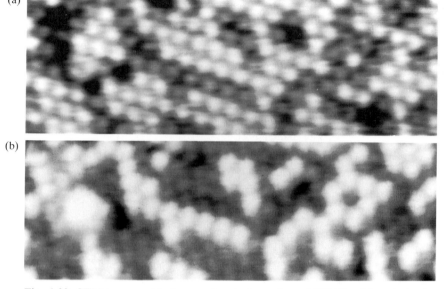

Fig. 4.33. STM topographic image of a single region of Si(111) with a sixth of a monolayer Al coverage. The local nature of the $\sqrt{3}\times\sqrt{3}$ bonding permits selective imaging of (*a*) Al adatoms at positive sample bias and (*b*) Si adatoms at negative sample bias (Hamers, 1989b).

bias. Based on the polarity-dependent contrast, atom-selective imaging can be achieved for the Si(111) ($\sqrt{3}\times\sqrt{3}$)Al system, as already found for the GaAs(110) surface. However, assignment of the imaged surface protrusions to specific atomic species again depends on external knowledge and cannot be inferred directly from the STM data.

In the vicinity of the Si defect sites, no evidence of microscopic band bending was found on either n-type or p-type substrate material, leading to the conclusion that these defects are electrically neutral. It was argued that the absence of local band bending and the tunneling spectroscopy data presented in Fig. 4.34 are both inconsistent with a one-electron band picture of the highly localized defect state, requiring that many-electron effects be taken into account for an appropriate description.

We now come back to the problem of assigning observed surface protrusions to specific atomic species. For the Si(111) ($\sqrt{3}\times\sqrt{3}$)Ag system, STM images have been interpreted in two different ways by attributing the topographic protrusions to either Ag or Si atoms in the top surface layer (Wilson and Chiang, 1987a; van Loenen *et al.*, 1987). Assignment of surface protrusions to specific atomic species is often facilitated by imaging several different structural phases simultaneously,

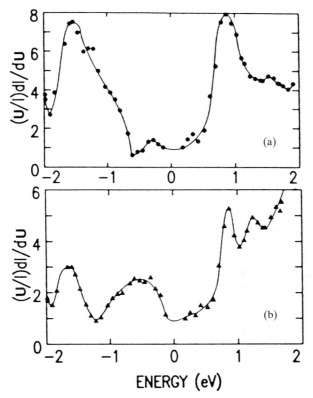

Fig. 4.34. Tunneling density-of-states measurements taken above (*a*) normal and (*b*) defected unit cells. The quantity $(dI/dU)/(I/U)$ is normalized to unity at $U = 0$ (Hamers and Demuth, 1988).

and thereby revealing the lateral registration of one particular surface structure relative to another. STM data analysis can then be performed by extending a grid or a mesh, fitted to a known surface structure in one part of an STM image, over another part exhibiting an unknown surface structure. This method is demonstrated in Fig. 4.35 for the Si(111) ($\sqrt{3}\times\sqrt{3}$)Ag system, where the lateral registration of the ($\sqrt{3}\times\sqrt{3}$) overlayer structure relative to the Si(111) 7×7 substrate structure has been determined (Wilson and Chiang, 1987b, 1988). In this way, some structure models for the Si(111) ($\sqrt{3}\times\sqrt{3}$)Ag surface have been ruled out. However, a conclusive surface structure determination, based solely on STM data, is still not possible. In particular, the observed protrusions may not necessarily represent Ag or Si surface atomic sites, but may correspond, for instance, to the center of a surface Ag trimer (Watanabe *et al.*, 1991). This particular example clearly demonstrates the need for a combination of STM with other surface analyt-

(a)

(b)

Fig. 4.35. STM image, with contrast enhancement, of a Si(111) 7×7 to Ag/Si(111) ($\sqrt{3}\times\sqrt{3}$)R30° domain boundary. (*a*) Three-dimensional view with (7×7) and ($\sqrt{3}\times\sqrt{3}$) cell boundaries marked in white. Negative sample-bias images are shown. Positive-bias images of the same area have confirmed that the right half of the (7×7) cell is faulted. (*b*) Top view with dark Si(111) mesh superimposed to show registration. White represents elevated features and small squares on the mesh mark adatom sites of the DAS model (Wilson and Chiang, 1987b).

ical techniques, providing additional, complementary surface structure information.

On the other hand, STM provides a unique experimental tool for studying spatially selective nucleation phenomena. For instance, if Ag condenses on a low-temperature (<<500 °C) Si(111) 7×7 substrate, it forms triangular-shaped two-dimensional islands which are preferentially located on the faulted halves of the (7×7) unit cells (Tosch and Neddermeyer, 1988a, 1988b). Similar spatially localized nucleation has been observed for Cu on Si(111) 7×7 (Tosch and Neddermeyer, 1989), Pd on Si(111) 7×7 (Köhler *et al.*, 1988), and Pb on Si(111) 7×7 (Ganz *et al.*, 1991a, 1991b). In the case of Pd deposition on Si(111) 7×7, 95% of the nucleated islands were found to be located on the faulted halves of the (7×7) structure (Fig. 4.36). These experimental results directly

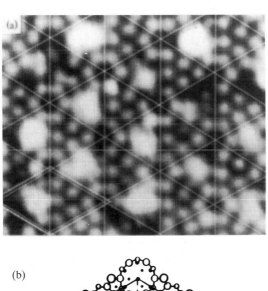

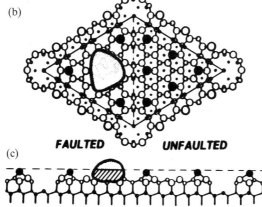

Fig. 4.36. (*a*) STM topograph of an 11.5 nm × 9.5 nm area of Si(111) 7×7 covered with a quarter monolayer of Pd. A grid is overlaid to indicate the substrate (7×7) lattice. The size of one (7×7) unit cell is highlighted in the lower left corner. (*b*) Top view of the DAS model for the Si(111) (7×7) reconstruction indicating the location of the Pd–silicide cluster within the faulted half of the unit cell. (*c*) Cut through the long diagonal of the (7×7) unit cell showing the vertical extension of the Pd–silicide nucleus. Only the part above the dashed horizontal line is visible to STM (Köhler *et al.*, 1988).

reveal a significant influence of the Si(111) 7×7 subsurface stacking fault on adsorption which is caused by the difference in electronic and atomic structures between the faulted and unfaulted halves of the (7×7) unit cell. In contrast, steps and other surface imperfections were found to play a minor role in the nucleation process (Köhler *et al.*, 1988).

Site-specific adsorption has also been observed for alkali metals (K, Cs) on Si(111) 7×7 substrates, where K or Cs atoms prefer the on-top sites of the central adatoms (Hashizume *et al.*, 1991a, 1991b).

STM has also been applied extensively to study metal growth on GaAs(110) substrates, including highly reactive (e.g. Ti, Sm), disruptive but weakly reactive (e.g. Cr, Fe, Au), and non-disruptive (e.g. Cs, Ag, Sn, Sb, Bi) overlayer systems. In many cases, formation of three-dimensional clusters (Volmer–Weber growth) is observed, e.g. for Ti (Yang *et al.*, 1991), Cr (Trafas *et al.*, 1991a), Fe (Dragoset *et al.*, 1989; First *et al.*, 1989b; Stroscio *et al.*, 1990), Ag (Trafas *et al.*, 1991b), Au (Feenstra, 1989a, 1989b), and Al (Suzuki and Fukuda, 1991; Patrin *et al.*, 1992). For some metals (e.g. Cr), preferential clustering along step edges occurs, whereas for other metals (e.g. Al), no evidence of pre-ferred decoration of step edges is found. For Fe on GaAs(110) (Fig. 4.37(*a*)), the onset of metallicity has been studied as a function of cluster size by using local tunneling spectroscopy (First *et al.*, 1989b). Fe clus-ters with volumes of about 150 $Å^3$, corresponding to about 13 atoms, are found to be non-metallic with an energy gap at the Fermi level. Larger Fe clusters consisting of more than about 35 atoms begin to show metallic characteristics, whereas Fe clusters with volumes of about 1000 $Å^3$, corresponding to about 85 atoms, exhibit fully metallic charac-teristics (Fig. 4.37(*b*)). However, the onset of metallicity may depend on the particular overlayer system under study (Suzuki and Fukuda, 1991).

Non-reactive overlayer systems, such as Cs on GaAs(110), are likely to form well-defined ordered surface structures. At low coverage, one-dimensional Cs chains have been observed by STM (Fig. 4.38), which sometimes extend over several hundred Ångström units (First *et al.*, 1989a; Whitman *et al.*, 1991a). The formation of these extended chains can be explained by an attractive interaction between Cs atoms along the [1$\bar{1}$0] direction of the GaAs(110) substrate. Local tunneling spectro-scopy data (Fig. 4.39) indicates that the one-dimensional Cs chains are non-metallic, but show an energy gap of about 1.10 eV at the Fermi level. Additional Cs adsorption on GaAs(110) results in formation of a two-dimensional overlayer, consisting of five-atom Cs polygons arranged in a c(4×4) superlattice (Whitman *et al.*, 1991a). This two-dimensional Cs overlayer still exhibits non-metallic characteristics,

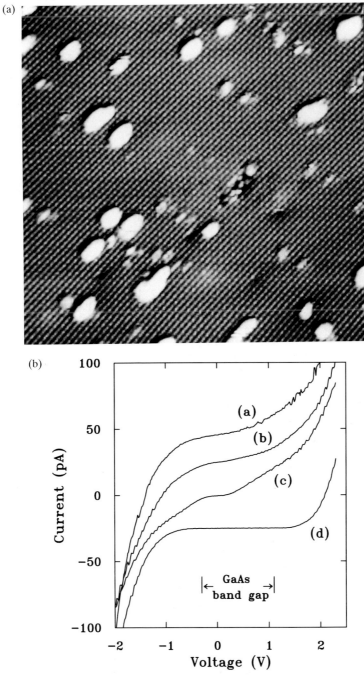

Fig. 4.37. (a) STM image (40 nm × 40 nm) of 0.1 Å Fe/p-GaAs(110) obtained with a sample bias voltage of −2.5 V. The image is displayed using a gray scale keyed to the gradient of the surface height to increase the dynamic range (Stroscio *et al.*, 1990). (b) *I–U* characteristics of Fe/p-GaAs(110). The four curves correspond to a 17 Å Fe film (a), a 1150 Å cluster (b), a 150 Å cluster (c), and the GaAs(110) substrate surface, 40 Å from the nearest cluster (d). The curves have been shifted vertically for display (First *et al.*, 1989b).

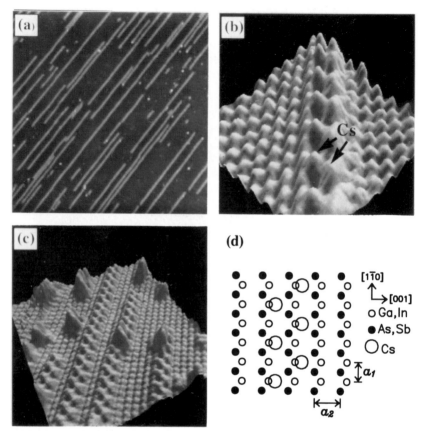

Fig. 4.38. STM images recorded with negative sample bias: (*a*) one-dimensional Cs zig-zag chains on GaAs(110) at about 0.03 monolayer coverage (image size 137 nm × 137 nm); (*b*) a single Cs zig-zag chain on GaAs(110) (image size 7 nm × 7 nm); (*c*) Cs zig-zag chains on InSb(110) (image size 20 nm × 20 nm). (*d*) Schematic drawing of the Cs zig-zag chains (Whitman *et al.*, 1991a).

though the energy gap is narrowed to about 0.65 eV (Fig. 4.39). Finally, upon growth of a second Cs layer, metallic characteristics are found by local tunneling spectroscopy (Fig. 4.39), in agreement with experimental results from electron energy loss spectroscopy (DiNardo *et al.*, 1990).

A similar observation is made for Cs on InSb(110) (Fig. 4.40), where one- and two-dimensional Cs overlayers also exhibit a band gap at the Fermi level whereas a three-dimensional Cs overlayer shows metallic behavior (Whitman *et al.*, 1991b). Interestingly, the band gap of about 0.6 eV for the two-dimensional Cs overlayer on InSb(110) is nearly

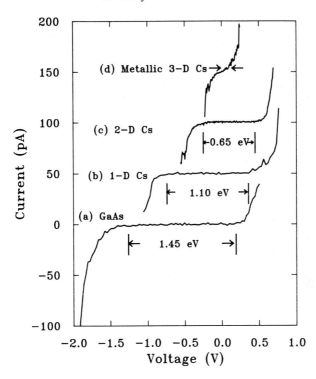

Fig. 4.39. *I–U* characteristics over various Cs structures on room-temperature GaAs(110): curve (*a*), region of clean n-GaAs(110); curve (*b*), one-dimensional zig-zag chain on n-GaAs(110); curve (*c*), quasi-two-dimensional triple chain on p-GaAs(110); curve (*d*), saturation three-dimensional bilayer on p-GaAs(110). The indicated band gaps were determined on a more sensitive logarithmic scale. The curves (*b*)–(*d*) are offset from zero current for clarity (Whitman *et al.*, 1991a).

identical to that observed for a two-dimensional Cs overlayer on GaAs(110), suggesting that the measured energy gap may be character-istic of the two-dimensional Cs overlayer. Since the bulk band gap of InSb is only on the order of 0.16 eV, the Cs on InSb(110) overlayer system provides an outstanding example where a single layer of adsorbed 'metal atoms' opens up a band gap larger than that of the underlying semiconductor substrate. The insulating behavior of the two-dimensional Cs overlayer on III–V compounds may be due to electron correlation effects, as described by a Mott–Hubbard model (Whitman *et al.*, 1991b).

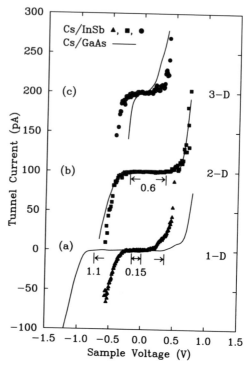

Fig. 4.40. A comparison of *I–U* characteristics recorded on various Cs structures on InSb(110) (symbols) and GaAs(110) (lines): (*a*) one-dimensional zig-zag chain; (*b*) two-dimensional overlayer; (*c*) three-dimensional saturation bilayer. The band gaps indicate the regions of zero conductance (determined on a more sensitive scale). Note that the spectra in (*b*) and (*c*) are offset from zero current, and that the GaAs data have been slightly smoothed (Whitman *et al.*, 1991b).

4.1.1.12 Interfaces and heterostructures

The electronic properties of buried interfaces of metal–semiconductor junctions have been studied by scanning probe methods which can provide subsurface information, such as BEEM and BEES (section 1.19), or electron interferometry (section 1.20.1). These techniques are based on either transmission or reflection of injected electrons at the buried interface, with the tip aligned perpendicular to the interface plane. For semiconductor junctions and heterostructures, interfaces can alternatively be studied with STM and related methods by positioning the probe tip above a cross-sectional cleavage plane, as shown in Fig. 4.41. By using this cross-sectional STM geometry and the current imaging method, it has been possible to distinguish between n-type and p-type areas of silicon p–n junctions (Hosaka *et al.*, 1988; Kordic *et al.*, 1990).

(a)

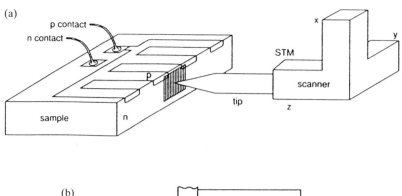

(b)

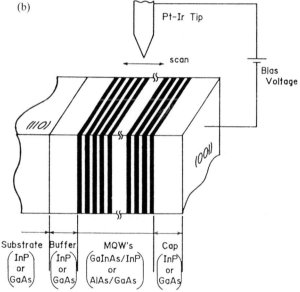

Fig. 4.41. (*a*) An array of p–n junctions is exposed by sample cleavage in UHV (Kordic *et al.*, 1991). (*b*) Schematic diagram of a multiple quantum well sample and the experimental set-up for STM measurements (Tanaka *et al.*, 1990a).

Even the depletion layer between the n-type and p-type regions has been made visible (Kordic *et al.*, 1991). The potential distribution across the cleavage plane of a GaAs p–n junction has been studied by scanning tunneling potentiometry (section 1.18), which allowed location of the space-charge region along the interface with nanometer resolution (Muralt, 1986). By using scanning capacitance microscopy (section 3.4), lateral dopant profiling across semiconductor interfaces has been demonstrated with a lateral resolution on the order of 200 nm (Williams *et al.*, 1989b, 1990; Abraham *et al.*, 1991).

Semiconductor heterostructures, such as MBE-grown GaAs–AlGaAs superlattices, have been studied by cross-sectional STM and scanning tunneling potentiometry (Muralt *et al.*, 1987; Salemink *et al.*, 1989; Tanaka *et al.*, 1990a; Tanimoto and Nakano, 1990; Dagata *et al.*, 1991b). For GaAs–AlGaAs heterostructures cleaved *in situ* in UHV, the atomic registry in the epitaxial GaAs and AlGaAs layers as well as at their interfaces has been observed directly by topographic STM studies, as shown in Fig. 4.42 (Albrektsen *et al.*, 1990; Salemink and Albrektsen, 1991). The GaAs layers appear as atomically flat planes,

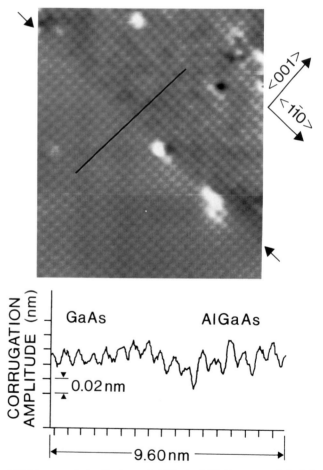

Fig. 4.42. STM image of the GaAs and AlGaAs MBE layers and their interface obtained at a sample bias voltage of −1.8 V. The electronic interface extends over 1–2 unit cells. Note the texture in AlGaAs which has been attributed to composition fluctuations (Salemink and Albrektsen, 1991).

whereas the AlGaAs layers exhibit a patchy texture which has been attributed to inhomogeneous distribution of the local electronic density of states in the AlGaAs layers resulting from composition fluctuations on the nanometer scale. In addition, preferential adsorption of oxygen on the Al-containing layers is observed, which reflects the enhanced oxygen affinity of Al compounds. The spatial transition between the homogeneous GaAs and the inhomogeneous AlGaAs layers occurs over an interfacial width of one to two unit cells. Local tunneling spectroscopy has been used to characterize the band structure of the different layers on the nanometer scale, including the semiconductor band gaps and the valence-band offsets relative to the Fermi level (Fig. 4.43). From the observed local variations in the AlGaAs band gap of about ± 50 meV, fluctuations in the Al concentration of about 4% have been deduced (Albrektsen *et al.*, 1990; Salemink *et al.*, 1992).

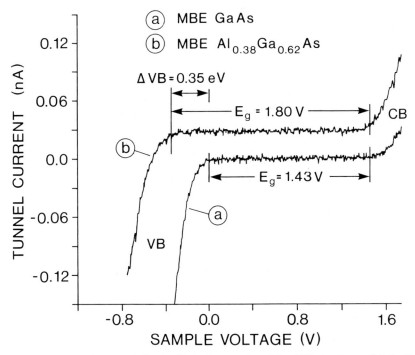

Fig. 4.43. *I–U* characteristics in GaAs (curve a) and in AlGaAs (curve b), both taken approximately 5 nm from the interface. Curve b is displaced vertically by 0.03 nA for clarity. The GaAs band gap is 1.43 eV and the valence-band edge is very near the Fermi level. The measured AlGaAs band gap is 1.80 eV. The measured valence-band offset from the Fermi level is 0.35 eV. The valence band (VB) and conduction band (CB) are indicated (Albrektsen *et al.*, 1990).

GaAs–AlGaAs multilayers have also been studied by tunneling-induced luminescence microscopy and spectroscopy (section 1.16), based on the recombination of holes with electrons tunneling into the semiconductor heterostructure (Abraham *et al.*, 1990; Alvarado *et al.*, 1991). The spatially resolved luminescence image, as presented in Fig. 4.44, shows a sequence of bright and dark regions corresponding to the GaAs and AlGaAs layers, respectively.

Scanning probe methods certainly have great potential also for characterization of one-dimensional quantum wires and zero-dimensional quantum dots (Weiner *et al.*, 1991).

4.1.1.13 *Epitaxial growth of semiconductors on semiconductors*

Besides investigation of static surface structures, STM can be applied to study the temporal evolution of surfaces during material deposition (e.g. by MBE growth) or removal (e.g. via sputtering by ion bombardment; see section 4.1.1.5). Interest in epitaxially grown semiconductor heterostructures is driven by the desire to optimize particular band structure properties, leading to improved or completely new electronic

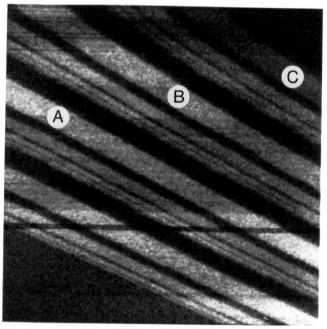

Fig. 4.44. Luminescence image (1.73 μm × 1.73 μm) of a multiple quantum well structure of four GaAs–AlGaAs series with different doping levels (Alvarado *et al.*, 1991).

or optoelectronic devices. STM has been found particularly useful for characterization of initial stages of growth including the influence of the preceding substrate preparation procedure and deposition conditions (e.g. substrate temperature and deposition rate).

Epitaxial growth of Si on Si(111) 7×7 surfaces, for instance, has been studied in detail by STM as a function of substrate temperature (Köhler *et al.*, 1989; Hamers *et al.*, 1989). At room temperature, formation of small clusters with random shapes is observed (Fig. 4.45(*a*)), suggesting that they consist of amorphous silicon, which exhibits a similar surface morphology (Wiesendanger *et al.*, 1988b). The Si clusters cover neither the corner holes of the underlying (7×7) substrate structure, nor the lines connecting them. On the other hand, the Si clusters show no preference for any particular triangular subunit of the (7×7) unit cell. This behavior is unlike the adsorption behavior of Ag, Cu, Pd and Pb on Si(111) 7×7 surfaces at room temperature, where preferred nucleation on the faulted half of the unit cell was observed (section 4.1.1.11).

Deposition of Si at elevated substrate temperatures leads to formation of triangular islands of epitaxial silicon, as shown in Fig. 4.45. This epitaxial growth of Si on Si(111) 7×7 involves a sequence of complicated rearrangements, including breaking of dimers and propagation of the stacking fault in one half of each (7×7) unit cell. The Si islands were found to nucleate preferentially along boundaries between (7×7) translational domains of the substrate, which can be attributed to the higher number of broken bonds along such domain boundaries. A preference for nucleation at superstructure domain boundaries was also found for the growth of metal overlayers (section 4.1.1.11), and therefore seems to be a phenomenon of general importance.

At early stages in the epitaxy of Si on Si(100) 2×1, held at a temperature below 350 °C, anisotropic islands have been observed by STM, as shown in Fig. 4.46 (Tsao *et al.*, 1989; Hamers *et al.*, 1990). The long dimension of each island is along the dimer-bonding direction of the underlying layer. Si growth on Si(100) 2×1 requires breaking of surface dimers, as schematically illustrated in Fig. 4.47. Upon annealing, the shapes of the islands become more isotropic, indicating that the initially formed anisotropic islands represent, to a large degree, non-equilibrium growth structures (Mo *et al.*, 1989; Lagally *et al.*, 1989). For an elevated substrate temperature of about 500 °C, the diffusion length of Si adatoms becomes larger than the average terrace width on vicinal substrates with a misorientation angle of 0.5° or more, and hence it is favourable for adatoms to be incorporated into steps. In this case, initial growth was found to occur almost exclusively at one of the two inequivalent types of step edges of the Si(100) 2×1 surface (section 4.1.1.2).

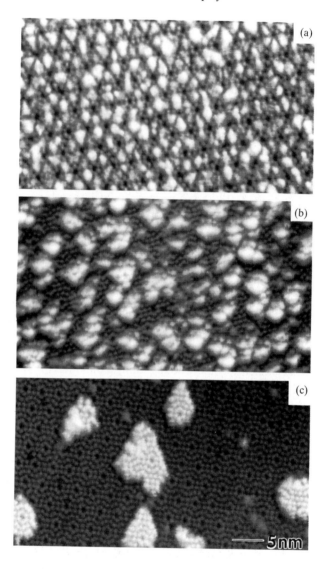

The deposited material is preferentially incorporated at type-B step
edges, at the end of the dimer rows, which has been ascribed to a
different reactivity at the two inequivalent step edges, caused by the
different bonding geometries (Hoeven *et al.*, 1989b, 1990). Consequently, a single-domain surface is formed with an array of evenly
spaced straight steps of biatomic height, vhile, later on, both island
growth and step flow occur.

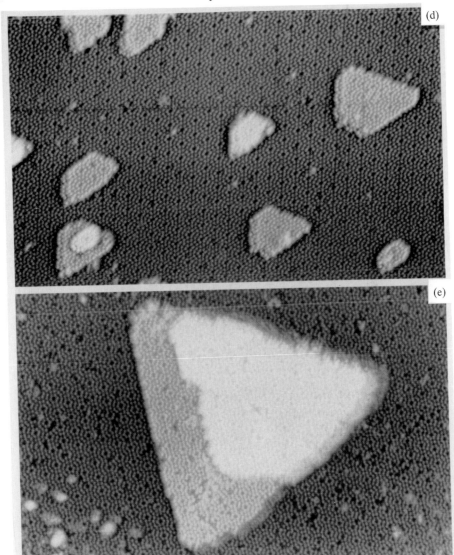

Fig. 4.45. Sequence of STM topographic images showing nucleation of silicon islands on a Si(111) 7×7 surface at different substrate temperature. The area for each is about 20 nm × 40 nm. The amount of silicon deposited is roughly the same for all images: 0.25 ML. The size of a (7×7) unit cell is indicated in (*a*). Substrate temperature during deposition: (*a*) 20 °C, (*b*) 250 °C, (*c*) 350 °C, (*d*) 450 °C and (*e*) 550 °C. On the perfectly triangular island in (*e*) nucleation of the second epitaxial layer is visible (Köhler *et al.*, 1989).

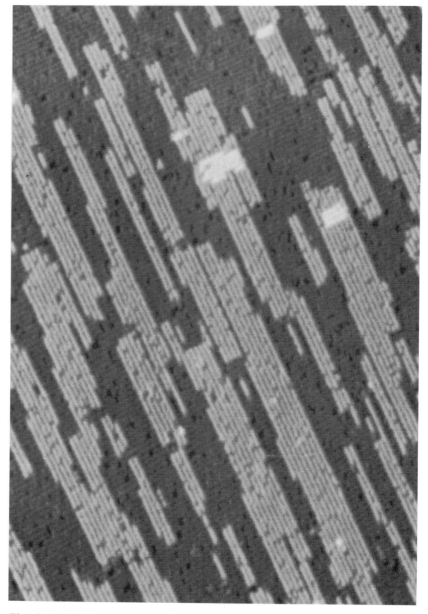

Fig. 4.46. STM image obtained after deposition of 0.1 ML Si onto a Si(001) (2×1) substrate held at a temperature of 580 K. Most of the silicon atoms have nucleated into one-dimensional 'dimer strings' whose orientation changes in alternating layers (Hamers *et al.*, 1990).

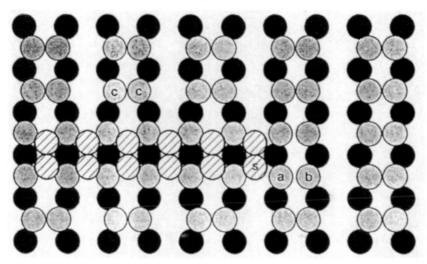

Fig. 4.47. Schematic plan view of the top two layers of a Si(001) surface, on which a single (cross-hatched) dimer string is growing horizontally (Tsao *et al.*, 1989).

Epitaxial growth of Ge on Si has been studied for several different substrate orientations, including (111), (100) and (113) (Becker *et al.*, 1985c; Köhler *et al.*, 1991; Mo *et al.*, 1990; Iwawaki *et al.*, 1991; Knall and Pethica, 1992). Since there is a lattice mismatch between Ge and Si of about 4%, Ge grows on Si in a layer-by-layer mode for only several layers, after which three-dimensional islands form (Stranski–Krastanov growth). In the transition regime from two- to three-dimensional growth of Ge on Si(100) 2×1, held at a temperature below 500 °C, the existence of an intermediate phase between two-dimensional layers and macroscopic three-dimensional clusters has been established (Mo *et al.*, 1990). This intermediate phase, observed for a Ge coverage beyond three ML, shows small clusters with well-defined shapes, crystal structure, and orientations with respect to the substrate (Fig. 4.48). The clusters consist of rectangular pyramids with four equivalent (105)facets. Upon annealing at about 600 °C, the small clusters initially formed at lower growth temperatures disappear and macroscopic clusters form, with mostly (113) facets. Only these macroscopic clusters can be observed by scanning electron microscopy (SEM), whereas the small clusters of the intermediate phase have remained invisible prior to STM studies.

The Stranski–Krastanov growth mode has also been studied in detail for Ge on Si(111) 7×7 substrates (Köhler *et al.*, 1991). It was found

Condensed matter physics

(a)

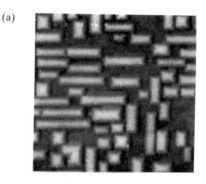

(b)

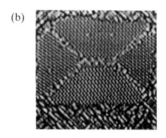

Fig. 4.48. (*a*) STM image (250 nm × 250 nm) of Ge clusters on Si(001). The clusters have rectangular or square bases, in two orthogonal orientations, corresponding to [100] directions of the substrate. The clusters are <100 nm long and 2–4 nm high. (*b*) Curvature-mode grey-scale plot of a single Ge cluster. The crystal structure on all four facets as well as the dimer rows in the two-dimensional Ge layer around the cluster are visible. The two-dimensional layer dimer rows are 45° to the axis of the cluster (Mo *et al.*, 1990).

that initially Ge nucleates preferentially at (7×7) domain boundaries. A (5×5) reconstruction appears for Ge coverages exceeding two ML. Finally, above four ML the formation of three-dimensional islands is observed, which exhibit mainly (113) and (111) facets. On top of these three-dimensional islands, the typical Ge reconstructions (c(2×8), c(2×4), and (2×2)) appear (section 4.1.1.5). In addition, bulk defects in the three-dimensional islands can be identified, such as the stacking fault tetrahedron shown in Fig. 4.49.

Epitaxial growth of CaF_2 on Si(111) substrates has also been studied by STM (Wolkow and Avouris, 1988a; Avouris and Wolkow, 1989b). At monolayer coverage, (2×3) and ($\sqrt{3} \times \sqrt{3}$) surface structures were

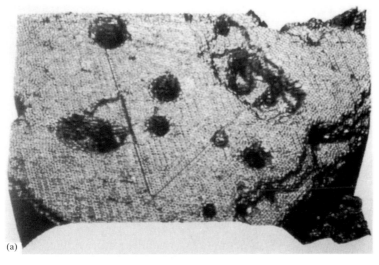

(a)

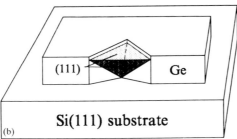

(b)

Fig. 4.49. (*a*) Perspective view (120 nm × 75 nm) of a fraction of a three-dimensional island of Ge on Si(111), containing a stacking fault tetrahedron. The hole at the upper right corner of the triangle allows a view of lower layers of the defect. (*b*) Schematic diagram of a stacking fault tetrahedron (Köhler *et al.*, 1991).

observed. Despite the large band gap of about 12 eV, it was possible to image even CaF_2 films of 30 Å thickness. This experiment directly showed that the ability of STM to image thick insulating films does not depend primarily on the width of the band gap, but rather on the relative positions of the valence and conduction band edges of the insulating film with respect to the corresponding edges of the conducting substrate. Since the conduction band of CaF_2 films is located about 3 eV above the Fermi level, it is accessible to STM.

Diamond films homoepitaxially grown on the (100) surface of synthesized diamond single crystals by means of chemical vapor deposition (CVD) show an intrinsic resistivity that is sufficiently low to allow STM investigations. Atomic resolution images of such (100)-orientated diamond films have revealed the presence of a (2×1) dimer-type reconstruction, similar to the Si(100) surface (Tsuno *et al.*, 1991). However, the C(100) 2×1 surface was most likely covered by hydrogen adsorbates, originating from the CH_4–H_2 plasma during film growth. Hydrogen is known to preserve the (2×1) surface reconstruction, while rendering the surface relatively inert by saturating the surface dangling bonds. Consequently, atomic-resolution STM studies can be performed even under ambient conditions.

STM has also been applied extensively to the investigation of MBE-grown GaAs(100) surfaces (Pashley *et al.*, 1988b; 1988c, 1991a, 1991b, 1992; Biegelsen *et al.*, 1990a). Depending on the relative surface concentration of As and Ga, a variety of surface reconstructions with different symmetries have been observed, including c(4×4), c(2×8), 2×4, 2×6, 4×2 and c(8×2) structures. The complex nature of the GaAs(100) surface has proven difficult for structure determination based on diffraction techniques, but rather requires local probe studies. For MBE-grown GaAs($\overline{111}$) surfaces, As-rich (2×2) and Ga-rich ($\sqrt{19}\times\sqrt{19}$) reconstructions are found (Biegelsen *et al.*, 1990b).

4.1.1.14 Sputtering by ion bombardment

Layer-by-layer removal by sputtering with low-energy (<400 eV) noble gas ions can be considered as the inverse process to epitaxial growth. A close analogy between deposition and sputtering has been found for Si(111) and Si(100) surfaces (Bedrossian and Klitsner, 1991, 1992). Epitaxial growth, as discussed in the previous section 4.1.1.4, involves mobile adatoms which nucleate islands or migrate to steps, whereas layer-by-layer removal by low-energy sputtering involves mobile vacancies which nucleate 'vacancy islands', i.e. depressions, or annihilate at step edges. Therefore, a close correspondence between mobile adatoms and mobile vacancies exists.

For a low-temperature (<300 °C) Si(111) 7×7 substrate bombarded with Xe ions of energy 225 eV, STM studies have revealed formation of monolayer-deep depressions which grow with increasing sputtering time. The (7×7) reconstruction remains only on the top non-eroded layer, while the newly exposed layer, though atomically flat, shows a disordered adatom arrangement without long-range reconstruction (Fig. 4.50). In contrast, at higher substrate temperatures (>500 °C), sputtering leads to newly exposed layers which are predominantly (7×7) reconstructed, with (7×7) domain widths exceeding 100 Å. The observed absence of vacancy islands on the narrower terraces at elevated substrate temperatures indicates that a vacancy created on such a

Fig. 4.50. STM image (15 nm × 15 nm) showing the initial stages of sputtering of a Si(111) 7×7 surface by 225 eV Xe at $T = 250$ °C (Bedrossian and Klitsner, 1991).

terrace can reach a step edge before nucleating a depression. Consequently, the layer-by-layer removal by sputtering becomes dominated by step retraction.

STM studies of the layer-by-layer removal of low-temperature (400 °C) Si(100) 2×1 substrates under 225 eV Xe ion bombardment reveal the nucleation of monolayer-deep depressions, or vacancy islands on the terraces. These vacancy islands exhibit a highly anisotropic shape, being elongated parallel to the dimer rows in the outermost atomic layer (Fig. 4.51). In contrast, sputtering at higher substrate temperatures (450 °C) does not lead to formation of isolated depressions within the terraces, but rather to surface erosion by step retraction. Preferential annihilation of mobile vacancies at the ends, rather than the sides, of dimer rows results in more rapid retraction of B-type steps compared with A-type steps. Consequently, a new type of non-equilibrium, single-domain Si(100) surface is obtained that is not accessible by epitaxial growth because epitaxial growth in the step-flow regime contrarily leads to an advance of B-type steps relative to A-type steps (section 1.1.1.13).

Besides investigations of layer-by-layer removal by low-energy noble-gas ion sputtering, STM can, of course, also be applied to characterize the morphology of semiconductor surfaces after high-energy ion bombardment (Wilson *et al.*, 1988, 1989).

4.1.1.15 Surface diffusion

The surface diffusion of adatoms is fundamentally important in epitaxial growth, sputtering, surface phase transitions and chemical reactions. Direct atomic-scale observations of surface diffusion allow correlation of adatom migration processes with specific atomic sites, as well as determination of diffusion coefficients, activation energies and atomic migration frequencies. Before the invention of STM, atomic-scale studies of surface diffusion were performed primarily by field ion microscopy (FIM). However, the number of materials suitable for FIM, as well as the size of the terraces at the foremost end of the tips ($\ll 100$ Å) gave rise to severe limitations. In contrast, STM allows one to probe considerable larger surface areas of any electrically conducting substrate.

Atomic-scale observations of surface migration by STM have been performed, for instance, for Si and Ge adatoms on top of Si(100) 2×1 substrates (Mo and Lagally, 1991; Mo *et al.*, 1991, 1992). It was found that the dimer-type (2×1) reconstruction of the Si(100) surface leads to a strong anisotropy in the surface diffusion process, which is about 1000 times faster along the surface dimer rows than perpendicular to them. The activation energy for adatom diffusion in the fast direction along the dimer rows was determined to be 0.67 ± 0.08 eV.

Atomic-scale studies of surface diffusion have also been performed for adatoms which lie within the plane of the reconstructed surface layer, rather than on top of it. The system studied was Pb on Ge(111) c(2×8) where the isolated, randomly distributed Pb adatoms occupy substitutional sites on the reconstructed Ge surface (Ganz *et al.*, 1992). The Pb and Ge adatoms are distinguishable over a range of biases, which allows monitoring of the changes in position of Pb adatoms on the surface (Fig. 4.52). The interval between STM data acquisition is selected so that only a few atoms move between successive images. By analyzing the STM data from 171 images, it was found that the number

Fig. 4.51. STM image (36 nm × 36 nm) of a Si(001) 2×1 surface obtained upon removal of about 0.5 ML by Xe sputtering with the substrate held at 400 °C. Anisotropic, monolayer-deep depressions elongated parallel to the dimer rows of the top layer can be seen (Bedrossian and Klitsner, 1992).

of Pb adatoms which move per image follows a binomial distribution, which is consistent with the assumption that each Pb adatom moves independently. The STM studies revealed adatom interchanges between nearest-neighbor sites, as well as over longer distances. For the nearest-neighbor interchanges, the diffusion process appears to be anisotropic within a single c(2×8) domain. It is expected that the exchanging Pb and Ge atoms move in concert, resulting in a rather small effective attempt frequency. Variable-temperature STM studies between 24 °C and 79 °C allowed verification that the adatom diffusion process obeys an Arrhénius law with an activation energy of 0.54±0.03 eV (Fig. 4.53). The STM scanning process was found not to affect the measured activation energy.

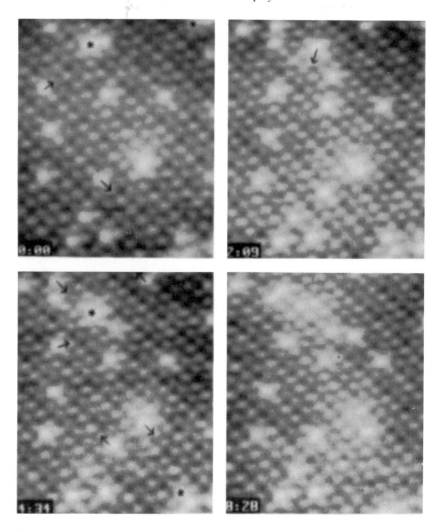

Fig. 4.52. Four successive STM images (9 nm × 10 nm) of Pb adatoms in substitutional sites of a Ge(111) c(2×8) reconstructed surface. Elapsed time (minutes:seconds) is indicated at the bottom of each image. The black arrows indicate the direction and length of eight single interchanges. Four atoms which do not appear in the same near-neighbor positions in the subsequent frame are marked with black stars (Ganz *et al.*, 1992).

4.1.1.16 Surface phase transitions

The development of variable-temperature UHV STM systems has made possible investigation of local aspects of surface phase transitions, including the influence of surface defects such as steps or domain bound-

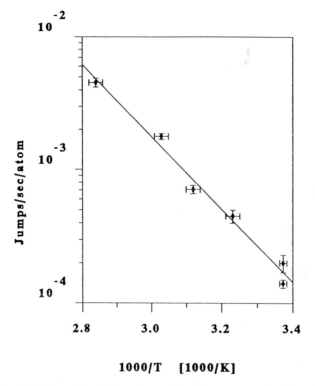

Fig. 4.53. Arrhénius plot of the observed movement rate for Pb adatoms at temperatures from 24 °C to 79 °C (Ganz *et al.*, 1992).

aries. Different kinds of phase transitions may be observed, depending on the particular surface to be studied.

(a) Reconstructive phase transitions. A high-temperature STM has been used, for instance, to study the $(1\times1) \to (7\times7)$ phase transition of the Si(111) surface at about 860 °C (Kitamura *et al.*, 1991). Step formation and migration occurring near the phase transition could be observed with atomic resolution. Starting with the Si(111) 2×1 surface, obtained by cleaving a silicon single crystal at room temperature (section 4.1.1.1), variable-temperature STM has also been applied successfully to study the $(2\times1) \to (5\times5) \to (7\times7)$ transformation in the temperature range between 280 °C and 425 °C (Feenstra and Lutz, 1990, 1991a, 1991b). It was found that the transition rate is dominated by the availability of adatoms on the Si(111) surface and that steps and (2×1) domain boundaries play an important role in nucleation of the new surface phases.

(b) Melting transitions. A surface order–disorder (melting) transition has been observed directly on Ge(111) as a function of temperature (Feenstra *et al.*, 1991). The formation of disordered regions was found to start exclusively at domain boundaries of the c(2×8) reconstructed Ge(111) surface (Fig. 4.54). The individual adatoms cannot be distinguished in the disordered surface regions, but appear to move rapidly, whereas those in the ordered regions appear to be stationary. This observation can be explained by the thermally activated hopping motion of the atoms in the disordered surface regions above about 220 °C, which is expected to occur on

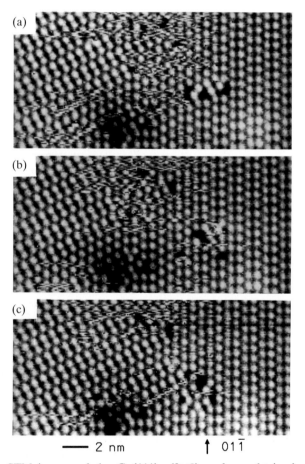

—— 2 nm   ↑ 01$\bar{1}$

Fig. 4.54. STM images of the Ge(111) c(2×8) surface, obtained at 215 °C. Three consecutive images are shown, separated in time by 85 s each. A domain boundary extends through the center of each image, and motion of the atoms near the boundary is evident (Feenstra *et al.*, 1991).

a time scale much shorter than the time scale of the STM measurement (typically 10–100 ms), leading to a fuzzy appearance of the disordered surface areas in the STM images. With increasing temperature the disordered parts of the surface grow, until the entire surface becomes disordered at about 300 °C. The presence of disordered surface regions already at temperatures well below the order–disorder transition temperature indicates that a premelting phenomenon in two dimensions, occurring at domain boundaries, has been observed. This premelting of two-dimensional systems ('edge melting') is analogous to the well-known surface melting phenomenon of three-dimensional systems (Zangwill, 1988).

4.1.2 Metals

On semiconductor surfaces, as discussed in section 4.1.1, the spatial distribution and energies of dangling bonds with s or p electron character have determined the STM images and their voltage-dependence. In contrast, electronic states in metals can have s, p, d, f or hybrid character. Generally, s- and p-type states dominate the STM image contrast if such states exist in the energy window probed, whereas d or f states are more difficult to observe (Lang, 1987a). This can be explained by a stronger localization of the valence d states around the atomic cores compared with the valence s or p states in transition metals, leading to a smaller amplitude of the d states at the tip position. The effect of the stronger localization of d states can be described by an additional 'centrifugal barrier' for tunneling.

The delocalized character of s- and p-type states, usually probed by STM, leads to smooth electron-density distributions, resulting in small measured corrugation amplitudes for non-reconstructed metal surfaces, which are typically on the order of 0.1 Å or below. However, the presence of surface states or tunneling through non-spherical tip states (section 1.11.1) may enhance the corrugation. At small tip–surface separations, the possible influence of forces on the measured corrugation amplitudes have additionally to be taken into account (section 1.22).

In contrast to semiconductor surfaces, a pronounced bias-voltage dependence is usually not observed in STM studies of clean metal surfaces. However, for adsorbate-covered metal surfaces, the measured STM contours again become strongly bias-dependent. The applied bias voltage can be chosen in the millivolt regime for metals due to the lack of a bulk energy gap at the Fermi level. For a given tunnel current, a small bias is preferable for obtaining high spatial resolution, as a consequence of small tip–surface separation (section 1.11.1).

4.1.2.1 Clean metal surfaces: Au(111)

The first observation of atomic resolution on the close-packed Au(111) surface (Hallmark *et al.*, 1987) had a strong impact on STM investigations of metal surfaces, similarly to the role of the Si(111) 7×7 surface for STM investigations of semiconductors (section 4.1.1.1). Since that time, Au(111) has been the metal surface most widely studied by STM.

A clean Au(111) surface can be prepared under UHV conditions by repeated cycles of ion sputtering and annealing of a polished Au(111) single crystal surface, as generally done for obtaining clean and well-defined metal surfaces. Alternatively, a Au(111) film can be grown epitaxially on a suitable substrate, e.g. mica. Since Au surfaces are relatively inert, atomic-resolution STM studies can be performed in air as well as under UHV conditions. However, this is no longer true for other metal surfaces which are more reactive.

The measured atomic corrugation in constant current STM images of the clean Au(111) surface is typically on the order of 0.3 Å or below, depending on the tunnel gap resistance, i.e. the tip-to-sample spacing, and the sharpness of the probe tip. Since the atomic corrugation on the Au(111) surface, as measured by helium-atom diffraction (Harten *et al.*, 1985), is only 0.07 Å, it has been speculated that a surface state near the Fermi level on Au(111) may contribute strongly to the measured STM corrugation. Such a surface state with a peak at about -0.4 eV appears, in fact, in both tunneling spectroscopy and photoemission data, as shown in Fig. 4.55 (Kaiser and Jaklevic, 1986). However, bias-dependent atomic-resolution STM studies of another close-packed metal surface, Al(111), have shown that the enhanced STM corrugation is unlikely to be due to a surface state, though such a state also exists for Al(111) (Wintterlin *et al.*, 1988b, 1989).

On the other hand, the surface state on Au(111) was found to depend on structural surface features such as steps. The surface state peak intensity is substantially reduced at step edges compared to the value obtained over extended terraces, as shown in Fig. 4.56 (Everson *et al.*, 1990, 1991). This has been explained by an increase of the local barrier potential across surface steps, which in turn affects the local density of states, leading to a suppression of the dI/dU surface state peak intensity (Davis *et al.*, 1991).

STM has also been applied to study the $(23 \times \sqrt{3})$ reconstruction of the Au(111) surface. Gold is the only face-centered-cubic (FCC) metal whose (111) surface reconstructs, as has been verified, for instance, by LEED (van Hove *et al.*, 1981), TEM (Takayanagi and Yagi, 1983), and helium-atom diffraction (Harten *et al.*, 1985). STM images of the

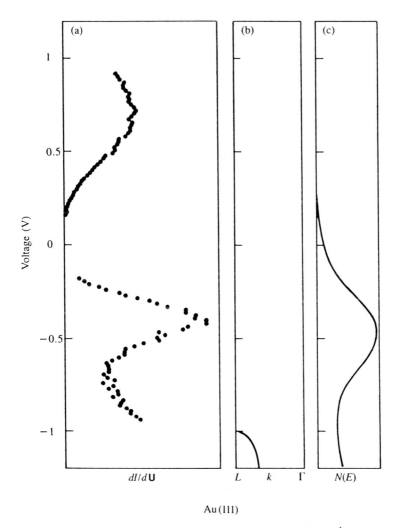

Au(111)

Fig. 4.55. (*a*) d*I*/d*U*–U characteristic measured at constant resistance on a Au(111) substrate held at room temperature. (*b*) Experimentally determined portion of the Au(111) band structure; there is a wide forbidden gap extending upward from the saddle point at L at −0.9 V. The UPS results for the surface state are shown in (*c*) (Kaiser and Jaklevic, 1986).

Au(111) surface reveal two bright stripes within the reconstructed unit cell, in addition to the underlying atomic lattice, as shown in Fig. 4.57 (Wöll *et al.*, 1989; Barth *et al.*, 1990). The observations are consistent with a stacking-fault-domain model involving periodic transitions between surface regions with FCC-type stacking (ABC . . .) and hexa-

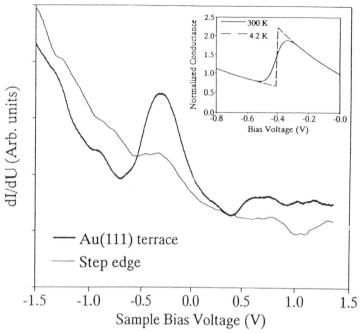

Fig. 4.56. Experimental dI/dU–U characteristics obtained on a flat Au(111) plane (heavy line) and at a monatomic step (light line). The inset shows a model calculation for surface-state tunneling from a flat (111) plane at 4 K (dashed line) and 300 K (solid line) (Everson et al., 1991).

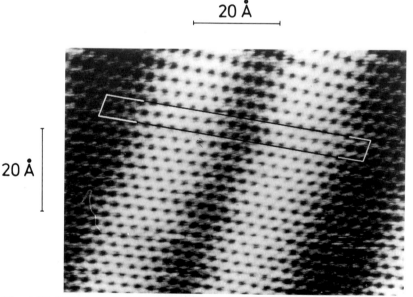

Fig. 4.57. STM image of the reconstructed Au(111) surface showing a pair of corrugation lines (white); the unit cell has been outlined for clarity (Barth et al., 1990).

gonal-close-packed (HCP) stacking (ABA . . .) of top-layer atoms
induced by surface elastic strain. The parallel corrugation lines along
the [11$\bar{2}$] direction (Fig. 4.57) can be identified as transition regions
between FCC and HCP stacking domains. The FCC domains are found
to be more than twice as wide as the HCP domains, indicating that
FCC stacking is energetically more favorable, which is consistent with
the FCC bulk structure of gold. The observation of 23 gold surface
atoms per unit cell on 22 bulk lattice sites implies an average uniaxial
contraction of 4.55%. The vertical displacement in the FCC–HCP transition regions is found to be 0.20±0.05 Å, independent of the tunneling
parameters, and in agreement with helium-atom diffraction data
(Harten *et al.*, 1985).

The local twofold symmetry of the (23× √3) reconstruction of the
Au(111) surface breaks the higher threefold symmetry of a FCC (111)
surface. On a macroscopic scale, threefold symmetry is regained by the
existence of three rotational domains of the (23× √3) reconstruction
which can simultaneously be observed in large-area STM images (Barth
et al., 1990). Interestingly, the rotational domains often form a long-
range superstructure consisting of a correlated periodic bending of the
parallel corrugation lines by 120°, as shown in Fig. 4.58 (Barth *et al.*,
1990). This superstructure, often called 'herringbone structure', is
caused by long-range elastic strain favoring an effective isotropic contraction on a larger scale against the locally favorable uniaxial contraction. The non-parallel arrangement of corrugation lines observed at the
bending points is related to the simultaneous change of the contraction
direction.

STM has also been used to study the interaction between the
(23× √3) reconstruction and surface defects. In many cases, the reconstruction lines are found to continue across surface steps with a strict
correlation both in phase and orientation on the different terraces. This
indicates the existence of a sufficiently strong interaction between the
topmost reconstructed layers on both sides of the step, which is presumably mediated by the second atomic layer of the upper terrace. In some
cases, the line-pairs are terminated by U-connections towards a step
edge, or they are arranged in a complicated non-periodic pattern (Barth
et al., 1990). On the other hand, interaction between the (23× √3)
reconstruction and surface steps does not appear to lead to significant
stabilization of certain step directions.

The influence of the (23× √3) reconstruction on the Au(111) surface
state has been studied by local and scanning tunneling spectroscopy
(Everson *et al.*, 1990). A change in the surface state peak intensity by
a factor of two across the (23× √3) reconstruction unit cell was found.

300 Å

300 Å

Fig. 4.58. STM image showing periodic rotational domains on the reconstructed Au(111) surface. Note the non-parallel arrangement of the corrugation lines at the bending points (Barth *et al.*, 1990).

It is expected that the variation in the crystal surface ordering between FCC and HCP stacking affects the surface potential, which in turn determines the local density of states being probed by tunneling spectroscopy.

Further STM studies of the Au(111) surface have shown that the $(23 \times \sqrt{3})$ reconstruction remains stable in air and even under aqueous solutions (Tao and Lindsay, 1991). However, on smaller terraces of Au(111) surfaces, or on (111) facets of grains in polycrystalline gold samples, the reconstruction may totally be absent. In the latter case, the elastic surface strain can be released by the overall shape of the individual grains (Wiesendanger *et al.*, 1990e).

4.1.2.2 Other clean noble metal surfaces

STM has also been applied to study the other low-index surfaces of gold. Clean Au(110) surfaces exhibit a (1×2) reconstruction of the 'missing-row' type (Binnig *et al.*, 1983b; Kuk *et al.*, 1988a; Gritsch *et al.*, 1991). The (1×2) reconstruction was found to interact with structural defects such as steps, for instance, by a stabilization of (111) micro-facets at step edges. A STM image of two screw dislocations on the (1×2) reconstructed Au(110) surface is shown in Fig. 4.59(*a*) (Gritsch *et al.*, 1991). The reconstruction remains stable in the strain field surrounding the dislocation line. The steps emerging from the screw dislocations exhibit many kink sites, a phenomenon which is observed for other surface steps as well. A similar (1×2) missing-row reconstruction can be found on clean Pt(110) surfaces. The Au(100) surface exhibits (1×5) and (20×5) reconstructions which have also been studied by STM and local tunneling spectroscopy (Binnig *et al.*, 1984b; Kuk *et al.*, 1988a; Kuk and Silverman, 1990).

In contrast to the clean low-index surfaces of Au, the corresponding surfaces of Ag and Cu do not reconstruct. The primary interest in STM investigations of low-index Ag and Cu surfaces has been the character-ization of surface defects such as steps or dislocations. An STM image showing the emergence point of a dislocation on a close-packed Cu(111) surface with atomic resolution is shown in Fig. 4.59(*b*) (Samsavar *et al.*, 1990). On Cu(100), evidence for a repulsive short-range step–step interaction, but an attractive interaction at intermediate distances was found by STM (Frohn *et al.*, 1991).

4.1.2.3 Clean metal surfaces: alloys

Clean surfaces of metal alloys can also be obtained by repeated cycles of ion sputtering and annealing as already described for Au(111). How-ever, preferential sputtering or surface segregation of one particular component can cause problems. Only a few STM investigations of ordered metal alloys have been reported, for instance, on NiAl(111) (Niehus *et al.*, 1990). The (111) planes of NiAl consist of alternating Ni and Al layers stacked in an ABAB . . . sequence. Therefore, trunca-tion of the bulk structure can result in either a Ni- or an Al-terminated surface. STM studies of the NiAl(111) surface have revealed the exist-ence of a terrace-and-step structure with the terraces covered by small triangular-shaped islands. Different terraces were found to be separated by biatomic steps, whereas the islands were only of monatomic height. Since STM is unable to identify the different elemental species (section 1.13), additional information from other surface analytical techniques

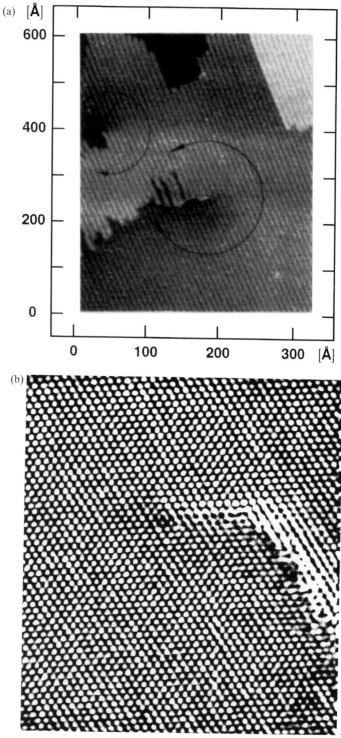

Fig. 4.59. (*a*) Top-view STM image (33 nm × 60 nm) showing two screw dislocations of opposite orientation on a (1×2) reconstructed Au(110) surface (Gritsch
et al., 1991). (*b*) STM image of an area encompassing the emergence point of
a dislocation on a clean Cu(111) surface (Samsavar *et al.*, 1990).

was required to show that the terraces are terminated by Ni, whereas the small islands are built up from Al which might be segregated over the top surface Ni layer under the influence of oxygen.

A clean, ordered surface alloy can alternatively be prepared by vacuum deposition of one metal onto a substrate of another metal, provided that intermixing of the two different elemental species results in surface alloy formation. For instance, Au deposited on Cu(100) at room temperature leads to a well-ordered CuAu surface alloy layer which resembles a (100) surface of Cu_3Au. STM studies have shown that alloy formation arises from replacement of surface Cu by Au atoms, constrained by the condition that Au atoms tend not to be nearest neighbors. The Cu atoms replaced by Au atoms diffuse over the surface and finally aggregate into islands (Chambliss and Chiang, 1992). The CuAu surface alloy layer exhibits a c(2×2) structure. Two inequivalent atoms per unit cell with a slightly different apparent height are observed in STM images (Fig. 4.60). The 'higher' sites were tentatively assigned to the Au atoms because of the larger atomic radius for Au compared with Cu, and the 0.1 Å outward displacement of the Au atoms as derived from an analysis of LEED data. This STM observation of an ordered metal alloy surface is in contrast to the selective STM imaging of different atomic species for compound semiconductor surfaces such as GaAs(110) (section 4.1.1.7).

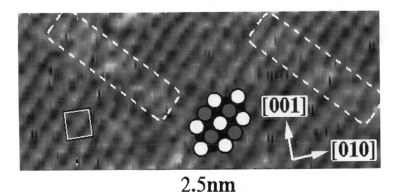

Fig. 4.60. STM image of a Cu(100) + Au c(2×2) alloy structure. The atomic lattice with both Cu and Au atoms is seen in STM with a sample bias voltage of +0.5 V. The small square marks a single c(2×2) unit cell with one 'higher' atom (presumably Au) at the center. The alloy structure is indicated by a superimposed array of white circles (Au atoms) and gray circles (Cu atoms). Dashed white rectangles contain regions of disrupted atomic positions associated with faults along the [01$\bar{1}$] direction (Chambliss and Chiang, 1992).

4.1.2.4 Clean metal surfaces: quasicrystals

In 1984, a new phase of condensed matter was discovered in an Al_6Mn alloy which exhibits icosahedral point group symmetry (Shechtman *et al.*, 1984). This symmetry is inconsistent with the presence of long-range translational periodicity, but still allows for long-range orientational order. Several models have been developed to explain the occurrence of this unusual symmetry in condensed matter. For instance, a random packing of icosahedral subunits leads to the correct symmetry and accounts for many experimental results obtained for icosahedral alloys. The existence of structural subunits on a nanometer scale in icosahedral Al_6Mn quasicrystals was confirmed by high-resolution microscopies, such as FIM (Melmed *et al.*, 1986) and STM (Wiesendanger, 1987; Wiesendanger *et al.*, 1988a). At the surface, the structural subunits appear as patches having a surface cross-section size of about 2–7 nm. Sometimes they exhibit pentagonal, although not necessarily regular, shapes. SEM studies of the same samples reveal pentagonal dodecahedra as larger units on a micrometer scale which themselves build up even larger pentagonal-shaped structures. Therefore, a hierarchy of structural units appears to exist in icosahedral Al_6Mn quasicrystals.

The presence of fivefold symmetry on the atomic scale was directly revealed by STM studies of icosahedral $Al_{65}Cu_{20}Fe_{15}$ quasicrystals (Becker *et al.*, 1991). The terraces imaged by STM were found to be remarkably flat, with a measured corrugation on the order of 0.1 Å. This extremely small corrugation is difficult to reconcile with the three-dimensional quasiperiodicity which should be present in the icosahedral phase.

Decagonal quasicrystals exhibit quasiperiodic order in two dimensions and periodic order in the third dimension and can therefore be regarded as intermediate between icosahedral quasicrystals and crystalline materials (Bendersky, 1985). STM studies of decagonal $Al_{65}Co_{20}Cu_{15}$ quasicrystals revealed terraces exhibiting decagonal symmetry on the atomic scale, as shown in Fig. 4.61 (Kortan *et al.*, 1990; Becker *et al.*, 1991). The local decagonal structures, as seen by STM, resemble typical decagon patterns. A stacking structure for the decagonal phase has been proposed consisting of four layers, each 4.13 Å high. The step-height values as measured by STM are typically half this fundamental length of 4.13 Å. The STM results for the decagonal $Al_{65}Co_{20}Cu_{15}$ quasicrystals can be well described by a tiling model (Levine and Steinhardt, 1984) and rule out the proposed multiple twinning of small crystallites (Pauling, 1987) as an explanation for the fivefold symmetry.

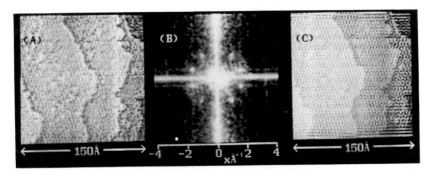

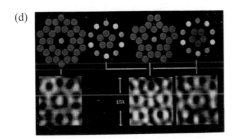

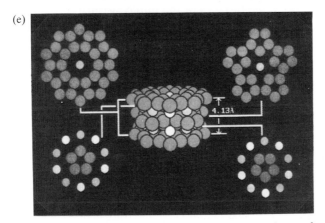

Fig. 4.61. (*a*) STM image of the quasiperiodic surface of decagonal $Al_{65}Co_{20}Cu_{15}$. (*b*) Normalized power spectrum of the image. (*c*) The same STM image after small-scale feature enhancement. (*d*) Local decagonal structure of atoms in the two-dimensional quasilattice and their proposed stacking along the periodic direction. These are compared to small-scale decagonal features in the tunneling images. (*e*) Space-filling model of the proposed decagonal stacking sequence. The four layers are shown individually, along with their positions in the stack (Kortan *et al.*, 1990; Becker *et al.*, 1991).

4.1.2.5 Adsorbate-covered metal surfaces

STM images of adsorbate-covered metal surfaces can depend strongly
on the applied bias voltage, the tip–surface separation, and the
adsorbate species (Lang, 1986a). Therefore, it is not *a priori* clear
whether a particular adsorbate will appear as protrusion or depression,
or will not be seen in the STM contours for given tunneling parameters.
The invisibility of adsorbates need not necessarily be due to the surface
electronic structure of the adsorbate-on-metal system, but may also
arise from a high surface mobility of the adsorbates at the temperature
of the experiment.

4.1.2.6 Oxygen adsorption on metal surfaces

Adsorption of oxygen has been studied for a variety of metal substrates,
motivated by the desire to understand the initial stages of metal oxide
formation. Most STM studies of oxygen adsorption have concentrated
on Cu substrates. For oxygen on Cu(110), a (2×1) 'added-row' type
surface reconstruction has been established based on dynamic STM
observations (Coulman *et al.*, 1990b; Jensen *et al.*, 1990a; Kuk *et al.*,
1990b; Besenbacher *et al.*, 1991). At room temperature, the highly
mobile oxygen adatoms, originating from dissociative chemisorption of
O_2, can diffuse on the terraces of the unreconstructed Cu(110) surface.
Cu adatoms, evaporating from surface step edges, are highly mobile as
well, and also diffuse across the terraces of the substrate surface. As
oxygen and copper adatoms approach each other, the strongly attractive
Cu–O interaction along the [001] direction can induce formation of
nuclei of the (2×1) phase, consisting of single strings of Cu atoms running in the [001] direction, which are glued together by oxygen atoms
(Fig. 4.62). According to this formation process, the resulting (2×1)
phase has been denoted as an 'added-row' rather than a 'missing-row'
structure, although both structures are identical with respect to the O
and Cu positions within the (2×1) phase. The apparent 'topographic'
height of the Cu–O strings with respect to the surrounding Cu(110)
substrate was found to depend strongly on the experimental parameters,
i.e. the tip–surface separation and the applied bias voltage, as shown
in Fig. 4.63 (Coulman *et al.*, 1990b). The Cu–O strings appear as eleva-
tions for sufficiently high negative bias, otherwise they are imaged as
depressions. These STM observations can be explained by local tunneling spectroscopy data (Fig. 4.64), indicating an increase in the local
density of states above the Cu–O strings relative to the Cu(110) substrate only for sufficiently large negative bias (Kuk *et al.*, 1990b).

STM studies of oxygen adsorption on Cu(110), held at a temperature

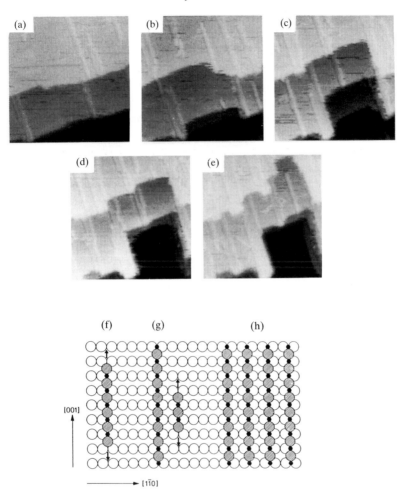

Fig. 4.62. (*a*)–(*e*) Five snapshots from an STM movie showing how Cu atoms are removed from step edges between terraces during exposure at room temperature to oxygen, and how 'added rows' are simultaneously formed on the terraces. The area imaged is 30 nm × 30 nm and the height of a single step is 1.28 Å. In the first image the surface has been exposed to 4 L of oxygen. Each of the following images corresponds to an additional 2 L exposure (Besenbacher *et al.*, 1991). (*f*)–(*h*) Atomistic model of the different stages of (2×1)O–Cu(110) formation. (*f*) Single string of Cu–O adatoms along [001] ('added row'); arrows indicate the preferential growth direction. (*g*) Growth of a single-row (2×1)O island along [110]; nucleation of a neighbored added row. (*h*) Two-dimensional island of (2×1)O added-row phase, the structure being equivalent to the 'missing-row' structure. Filled circles: O atoms; shaded circles: added-row Cu atoms on top of the substrate atoms (open circles) (Coulman *et al.*, 1990b).

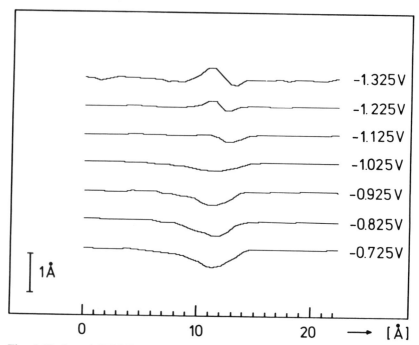

Fig. 4.63. Set of STM line scans taken at different bias voltages, in the [1$\bar{1}$0] direction, over a (2×1)O island surrounded by a clean Cu(110) (1×1) substrate area (Coulman *et al.*, 1990b).

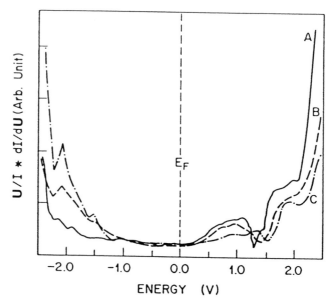

Fig. 4.64. (dI/dU)/(U/I)–U characteristics taken above the Cu(110) substrate (A) and above the Cu–O strings (B,C) (Kuk *et al.*, 1990b).

above 175 °C during exposure, revealed a long-range spatial self-organization of the two-dimensional anisotropic Cu–O islands which arrange themselves in a striped periodic supergrating, with the stripes running along the [001] direction (Kern *et al.*, 1991). The periodicity of the supergrating was found to depend on the oxygen coverage and the substrate temperature.

STM has also been applied to study the formation of the c(6×2) structure of oxygen on Cu(110), which starts after completion of the (2×1) structure (Coulman *et al.*, 1990a; Feidenhans'l *et al.*, 1990). The c(6×2) phase was found to nucleate preferentially at step edges and to grow nearly isotropically, in contrast to the (2×1) phase which nucleates on flat terraces and grows anisotropically in Cu–O added rows. Oxygen-induced reconstructions on Cu(100) and Cu(111) substrates have been studied by STM as well (Wöll *et al.*, 1990; Jensen *et al.*, 1990b, 1991).

STM studies of oxygen on Ni(100), ordered in p(2×2) and c(2×2) structures, have shown that the oxygen atoms adsorb on the fourfold hollow sites of the Ni(100) substrate (Kopatzki and Behm, 1991). At these positions, the adatoms appear as local 'topographic' minima 0.3 Å in depth. Experimental conditions for which the oxygen sites would appear as local maxima were not found, in contrast to an earlier theoretical prediction (Doyen *et al.*, 1988). For higher oxygen exposures, the formation of NiO clusters on top of the ordered oxygen adlayer was observed (Bäumer *et al.*, 1991). The step edges of the Ni(100) substrate were found to constitute preferential sites for growth of NiO clusters.

For oxygen on Al(111), the formation of a (1×1) structure is observed with the O atoms located in the threefold hollow sites. The adsorbed O atoms are practically immobile at room temperature (Brune *et al.*, 1992). A tendency for island formation of the adlayer phase has been revealed by STM, indicating a net attractive force between O atoms adsorbed on neighboring sites (Wintterlin *et al.*, 1988a; Brune *et al.*, 1992). These oxygen islands usually appear as local depressions in STM images. However, for small tip–sample distances, the oxygen sites can be seen as protrusions, which would better agree with theoretical predictions (Kopatzki *et al.*, 1988).

On Ru(0001) substrates, the oxygen sites again appear as local 'topographic' minima (Günther and Behm, 1992).

4.1.2.7 Adsorption of other species on metal surfaces

(a) *Sulfur adsorption.* The adsorption of a monolayer of sulfur can render a reactive metal surface completely inert towards adsorption of atmospheric reactants, allowing for atomic-resolution STM studies of sulfur-on-metal systems even in air, as has been demonstrated, for instance, for S on Mo(100) (Marchon *et al.*, 1988a, 1988b), S on Re(0001) (Ogletree *et al.*, 1990), and S on Pt(100) (Maurice and Marcus, 1992). STM investigations of S on Re(0001) have revealed aggregate clustering of sulfur above a critical coverage which has been ascribed to substrate-mediated three- and more-adatom interactions (Hwang *et al.*, 1991b).

(b) Noble gas adsorption. At room temperature, noble gas adatoms on metal substrates are generally highly mobile because the adsorption energy is small relative to the thermal energy. Therefore, STM studies of noble gas adsorbates on metal surfaces require low temperatures. Xe atoms adsorbed on Ni(110) or Pt(111) substrates at 4 K have been imaged as surface protrusions with an apparent 'topographic' height of about 1.5 Å, as shown in Fig. 4.65 (Eigler and Schweizer, 1990; Eigler *et al.*, 1991a). This might be surprising at first sight because the Xe 6s atomic resonance lies far above the Fermi level, outside the energy window being probed by STM (Fig. 4.66(*a*)). However, the 6s resonance of Xe nevertheless makes the dominant contribution to the STM image contrast because the size

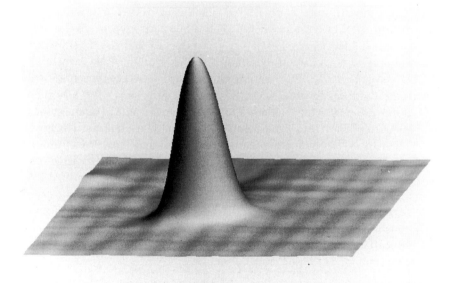

Fig. 4.65. Constant current STM image (4 nm × 4 nm) of a single Xe atom adsorbed on a Ni(110) substrate. The Xe atom appears as a protrusion 1.53 Å high at a sample bias voltage of −0.02 V and with a tunnel current of 1 nA. The corrugation of the unreconstructed Ni(110) surface is just visible (Eigler *et al.*, 1991a).

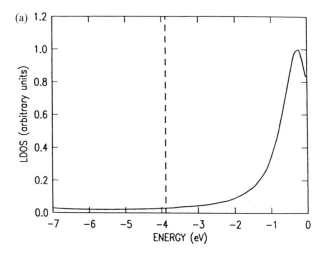

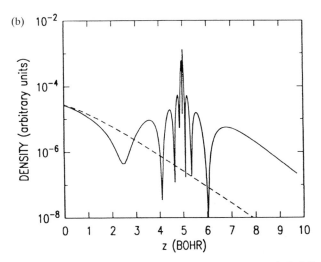

Fig. 4.66. (*a*) The calculated additional local density of states (LDOS) due to an adsorbed Xe atom versus energy measured from the vacuum level. The LDOS at the Fermi level (dashed line) is due to the tail of the almost completely unfilled Xe 6s atomic resonance, which is seen to peak just below the vacuum level. The residual density of the 6s resonance at the Fermi level renders the Xe atom visible to the STM. (*b*) The Fermi-level conduction-electron density along a normal to the surface through the nucleus of a Xe atom adsorbed on a metal substrate. The bare metal substrate density (dashed curve) is shown in order to emphasize the form and extent of the conduction-electron density redistribution. The conduction electrons extend further out into the vacuum at the Xe atom site (Eigler *et al.*, 1991b).

of the orbital associated with that resonance is such that it extends considerably further out into the vacuum than the bare-surface wave functions (Fig. 4.66(*b*)).

(c) Adsorption of group IV elements. STM has also been used for a comparative study of the adsorption of Si atoms and Si_{10} clusters on Au(100) substrates (Kuk *et al.*, 1989, 1988b). Deposition of Si atoms was found to lead to the formation of flat islands one atom high. These islands exhibit metallic behavior, as deduced from local tunneling spectroscopy data (Fig. 4.67), which has been ascribed to silicide formation. In contrast, the deposited Si_{10} clusters show up as three-dimensional structures in STM images. However, they appear as surface protrusions only if the applied bias voltage is sufficiently high. In the bias range between -0.8 and $+0.5$ V, the clusters appear dark. This STM observation is explained by local tunneling spectroscopy measurements (Fig. 4.67), showing a band gap of about 1 eV between the empty and filled states for the tip positioned above a Si_{10} cluster. This measured band gap width for a Si_{10} cluster is close to the value for bulk Si (1.12 eV) and indicates

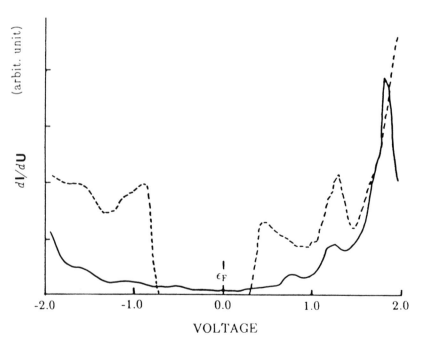

Fig. 4.67. dI/dU–U characteristics obtained on a clean Au(110) surface (solid line) and above a Si_{10} cluster (dashed line). (Kuk *et al.*, 1989).

that the semiconductor characteristics already appear for aggregates of only ten silicon atoms. The Fermi level of the Au substrate lies closer to the Si_{10} conduction band edge, indicating charge transfer from the substrate to the cluster. The amount of this charge transfer was found to depend on the structure and orientation of the cluster on the substrate.

4.1.2.8 Epitaxial growth of metals on metals

Detailed investigation of epitaxial growth of metals on metal substrates is important for optimizing the growth of ultrathin metal films and metallic superlattices which reveal a variety of interesting physical phenomena, most remarkably in the field of magnetism. The nucleation and growth process is influenced both by the surface structure of the substrate and the deposition conditions (e.g. substrate temperature and deposition rate).

Most STM studies of growth of metals on metals have been performed using Au(111) substrates which were already well characterized by numerous STM investigations of the clean Au(111) surface (section 4.1.2.1). The nucleation of metal overlayers was found to be strongly influenced by the $(23 \times \sqrt{3})$ reconstruction of the Au(111) substrate, particularly for deposition of Ni, Co and Fe (Chambliss *et al.*, 1991a, 1991b; Voigtländer *et al.*, 1991a, 1991b). Ordered two-dimensional arrays of small polygonal metal islands are formed at submonolayer coverage by preferential decoration of the 'elbows' of the long-range 'herringbone' reconstruction of Au(111), as shown in Fig. 4.68(*a*) and (*b*). The formation of these island arrays is explained by a model in which individual deposited atoms diffuse on the Au(111) surface until they become bound at the dislocations located at the 'elbow' sites, and subsequently act as nuclei for further aggregation. Three equivalent orientations ('hyperdomains') of the island arrays can be found, related by 120° rotations, reflecting the three possible orientations of the underlying 'herringbone' reconstruction. The observed nucleation phenomenon raises the possibility of well-controlled lateral texturing or patterning of metal films on a 10 nm length scale.

For Ni and Fe deposition on Au(111), the nucleated islands, being pseudomorphic with the Au(111) substrate, are of one monolayer height, whereas for Co they are two atomic layers high. With increasing coverage, the islands grow laterally until they start to coalesce. Film growth usually proceeds in a non-ideal layer-by-layer growth mode in which new layers start to form before preceding layers have been completed.

The STM investigations of the structure of the thin metal films as a

(a)

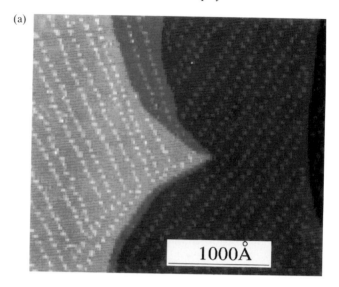

1000Å

(b)

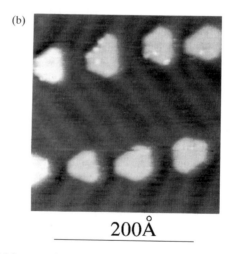

200Å

Fig. 4.68. (*a*) STM image of 0.11 ML Ni on Au(111). Several atomically flat Au terraces are seen, separated by steps of single-atom height. Small light dots on each terrace are monolayer Ni islands, in rows along [1$\bar{2}$1]. (*b*) Correlation of Ni island nucleation with the Au(111) surface reconstruction. About 99% of the Ni islands nucleate at the elbow sites of the Au(111) reconstruction lines (Chambliss *et al.*, 1991a).

function of coverage can be correlated with their magnetic properties. For Co layers on Au(111), a contiguous Co film is formed at coverage exceeding two ML, which is a consequence of the biatomic height of the initially nucleated islands (Voigtländer *et al.*, 1991a). Only a connected Co layer is expected to lead to ferromagnetic coupling between the islands and the formation of magnetic domains in the thin film. Such magnetic domains with perpendicular magnetization anisotropy are indeed observed by SEMPA for a Co coverage exceeding two ML (Allenspach *et al.*, 1990). No apparent structural change of the Co thin film is observed by STM in the Co coverage range between four and seven ML, where the magnetization axis changes from perpendicular to parallel relative to the surface. In contrast, for Fe films on Au(111), this switching of the easy magnetization axis from out-of-plane to in-plane at a Fe layer thickness of about three ML can be correlated with a structural transition from coherent FCC Fe layer growth to BCC iron crystal structure (Voigtländer *et al.*, 1991b). The combination of microscopic structural investigations of thin magnetic films by STM with characterization of their magnetic properties is of great importance for improved understanding of magnetism in reduced dimensions, such as in thin magnetic films or magnetic metal multilayers.

In contrast to Ni, Co and Fe on Au(111), nucleation of Au and Ag on Au(111) substrates preferentially occurs at step edges (Dovek *et al.*, 1989; Chambliss and Wilson, 1991; Chambliss *et al.*, 1991b). This observation can be explained by the reduced probability of Au and Ag atoms sticking to an 'elbow' site of the Au(111) 'herringbone' reconstruction, compared with Ni, Co and Fe. However, this sticking probability can strongly depend on the presence of surface contaminants, decorating the 'elbow' sites and acting as nucleation centers (Lang *et al.*, 1989; Chambliss *et al.*, 1991b). Though the Ag atoms diffuse freely on the terraces of the Au(111) substrate, they appear to be immobile along or across step edges separating the terraces. As a result, the monolayer-high Ag deposits nucleated at the steps grow outward with increasing deposition, forming finger-like Ag islands (Dovek *et al.*, 1989; Chambliss and Wilson, 1991). These growth structures can be explained within a diffusion-limited aggregation model.

Several other metal-on-metal systems have been studied by STM. For instance, deposition of Au on Ru(0001) substrates, held at room temperature, was found to lead to formation of dendritic islands in the submonolayer regime, as shown in Fig. 4.69 (Hwang *et al.*, 1991a). These dendritic islands exhibit a fractal character with a fractal dimension of 1.72 ± 0.07. The growth of the dendritic islands can again be described by a two-dimensional diffusion-limited aggregation model

(a)

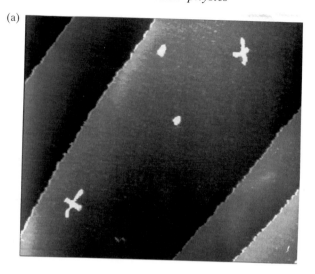

(b)

based on a high mobility of individual Au atoms on extended terraces of the Ru(0001) substrate combined with limited diffusion along island edges. The nucleation of these islands was found to be influenced by substrate defects (Pötschke *et al.*, 1991). The observation of extended dendritic-shaped islands therefore requires sufficiently large defect-free substrate regions leading to a low density of Au island nuclei. Upon annealing to about 375 °C, Au atoms can diffuse along island edges,

(c)

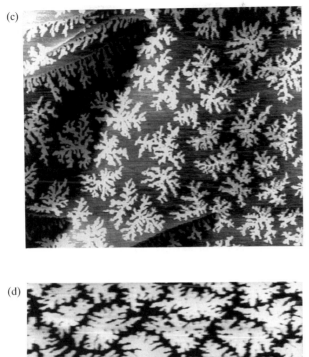

(d)

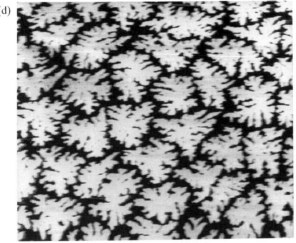

Fig. 4.69. STM images of Au films on a Ru(0001) substrate at varying coverages: (*a*) 0.03 ML (0.70 μm × 0.60 μm), (*b*) 0.15 ML (1.31 μm × 1.20 μm), (*c*) 0.37 ML (1.13 μm × 0.96 μm), (*d*) 0.69 ML (1.32 μm × 1.13 μm) (Hwang *et al.*, 1991a).

leading to more compact structures. At 825 °C, the Au islands dissolve and can recondense into extended connected structures during cooling.

Another interesting STM observation was made for Cu on Ru(0001) substrates which had been slightly annealed at 250 °C (Pötschke and Behm, 1992; Pötschke *et al.*, 1991). The first Cu layer was found to grow pseudomorphically, whereas for the second layer, a one-dimensional corrugation appears, similar to the $(23 \times \sqrt{3})$ reconstruction of the Au(111) surface (Fig. 4.70). In both cases, the one-dimensional corrugation arises from uniaxial contraction of the top hexagonal surface layer, which is more closely packed than the underlying hexagonal substrate lattice (section 4.1.2.1). Therefore, the one-dimensional reconstruction appears to be a universal phenomenon which can occur in strained hexagonal metal overlayers on hexagonal substrates. The shape of the bilayer Cu islands is affected by the one-dimensional reconstruction because the island edges tend to follow one of the three possible orientations of the one-dimensional corrugation lines (Fig. 4.70). At increasing Cu coverage (above five ML), formation of multilayer islands is observed with a characteristic shape of slightly irregular hexagonal pyramids (Fig. 4.71). Their edges are aligned parallel to the close-packed directions of the Ru(0001) substrate.

4.1.2.9 Sputtering by ion bombardment

STM is also ideally suited to study the microscopic morphological changes of metal surfaces as a result of material removal by sputtering

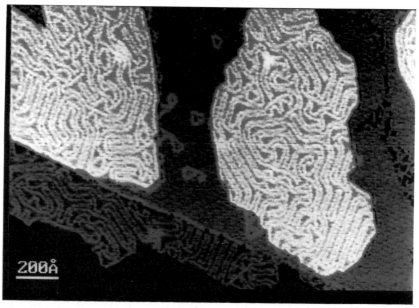

Fig. 4.70. STM image (200 nm × 127.5 nm) of a slightly annealed (520 K), medium-coverage (1.5 ML) Cu film. Second-layer islands are reconstructed while the first layer exhibits no corrugation except for structural defects (Pötschke *et al.*, 1991).

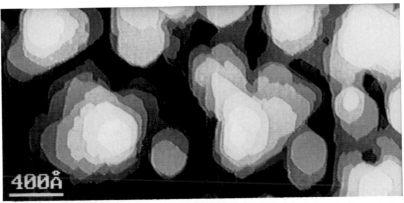

Fig. 4.71. STM image (300 nm × 150 nm) of a high-coverage (>5 ML) Cu film after room-temperature deposition. Multilayered islands of characteristic pyramidal shape are formed, whose edges are oriented along the close-packed directions of the substrate (Pötschke *et al.*, 1991).

with noble gas ions (Michely *et al.*, 1990; Michely and Comsa, 1991a, 1991b, 1991c; Lang *et al.*, 1991). The impinging ions create individual vacancies which are more or less mobile, depending on the substrate temperature. These vacancies can cluster into 'vacancy islands', i.e. depressions, one atomic layer deep. Further material removal results either in creation of new vacancy islands at the next lower level, leading to three-dimensional 'pit' structures, or in layer-by-layer removal which is observed at higher substrate temperatures.

The transition from pit formation at relatively low sputtering temperature (350 °C) to the layer-by-layer removal regime at high temperature (650 °C) has been directly observed by STM for a Pt(111) surface sputtered with 600 eV Ar$^+$ ions, as shown in Fig. 4.72(*a*)–(*c*) (Michely and Comsa, 1991a). The pits consist of vacancy islands of similar shape stacked one on the other. These vacancy islands exhibit nearly regular hexagonal shapes, reflecting the symmetry of the Pt(111) surface. The monatomic steps separating the vacancy islands follow the energetically favorable close-packed [110] directions. In contrast, only two layers are exposed in the layer-by-layer removal regime. The formation of three-dimensional pit structures does not occur because monatomic vacancies produced within a vacancy island are rapidly annealed by mobile atoms originating from a higher layer. It was found that the shape and size of the vacancy islands are independent of whether or not the surface had been exposed to the highest annealing temperature during or after sputtering.

In general, the surface morphologies after sputtering appear to be

(a)

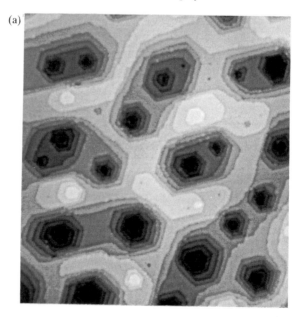

(b)

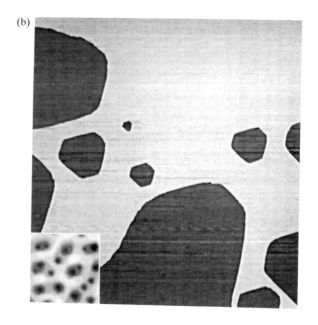

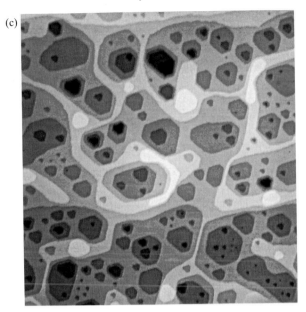

Fig. 4.72. Three STM topographs of Pt(111) after removal of 5 ML at different sputtering temperatures: (*a*) image size 82.5 nm × 82.5 nm, substrate temperature during sputtering 625 K, pit formation temperature regime; (*b*) 330 nm × 330 nm, sputtering temperature 910 K, layer-by-layer removal temperature regime (inset shows topography of (*a*) drawn to the same scale); (*c*) 330 nm × 330 nm, sputtering temperature 735 K, transition temperature regime (Michely and Comsa, 1991a).

similar and complementary to the morphologies resulting from adatom deposition during thin-film growth (section 4.1.2.8): the formation of vacancy islands corresponds to the formation of adatom islands, three-dimensional pit formation to three-dimensional island growth, and layer-by-layer removal to layer-by-layer growth. However, sputtering by ion bombardment can lead to generation of adatom islands as well as vacancy islands. This has been directly observed by STM for Pt(111) surfaces sputtered with 600 eV Ar$^+$ ions at temperatures below 300 °C (Michely and Comsa, 1991b). The formation of adatom islands can be explained by ion-induced atom-replacement sequences in a near-surface region, leading to a large number of individual atoms being pushed onto the surface layer, typically on regular lattice sites. Alternatively, adatom island formation can result from sputtering with He$^+$ ions (Michely and Comsa, 1991c). The implanted He agglomerates into subsurface bubbles which can 'push' entire adatom islands out of the surface.

4.1.2.10 Surface diffusion

Surface diffusion plays an important role for the transformation of surfaces, e.g. during reconstruction, epitaxial growth or sputtering. Being a dynamical phenomenon, surface diffusion can be studied by STM in two different ways.

1. A detailed analysis of the time-dependence of the tunneling current, with the tip held at a fixed position, can provide information about diffusion processes occurring on a short time scale (Binnig *et al.*, 1986b).
2. Alternatively, direct real-space observation of the surface dynamics by time-dependent STM studies can be performed, provided that the structural changes on the surface occur on a time scale accessible to STM.

At room temperature, significant surface diffusion is revealed in STM studies of noble metal surfaces. The step structure of Au, Ag and Cu surfaces is found to change rapidly with time. Surface depressions tend to fill up and mounds tend to smooth out, leading to a decrease of surface roughness as a function of time (Jaklevic and Elie, 1988; Lin and Chung, 1989; Trevor and Chidsey, 1991). In addition, high mobility of kink sites, due to rapid emission and capture of adatoms, diffusing along step edges, is observed in STM studies of stepped Ag and Cu surfaces (Wolf *et al.*, 1991; Frohn *et al.*, 1991; Wintterlin *et al.*, 1991). The rapid kink diffusion leads to a fuzzy appearance of the step edges in STM images. By reducing the temperature, the diffusion process is stopped and the step edges appear sharp.

Structural changes of steps are observed on reconstructed as well as on non-reconstructed noble metal surfaces. An example for the temporal evolution of the step structure on a Au(110) (2×1) reconstructed surface is presented in Fig. 4.73. It was found that the kink sites on the Au(110) (2×1) surface are highly mobile, as on non-reconstructed noble metal surfaces, which appears to be important for nucleation and growth of the (2×1) phase (Gimzewski *et al.*, 1991a, 1991b, 1992).

Surface diffusion can also be studied on adsorbate-covered metal surfaces, often yielding an improved understanding of the origin of the adsorbate-induced surface structures. This has clearly been demonstrated for the O on Cu(110) system, where STM studies of the surface dynamics finally established the formation of a (2×1) 'added-row'-type reconstruction (Coulman *et al.*, 1990b; Wintterlin *et al.*, 1991; Besenbacher *et al.*, 1991). Two STM images showing the 'condensation' of two isolated mobile Cu–O strings to a nucleus of the (2×1) 'added-row' phase are presented in Fig. 4.74. These images clearly demonstrate

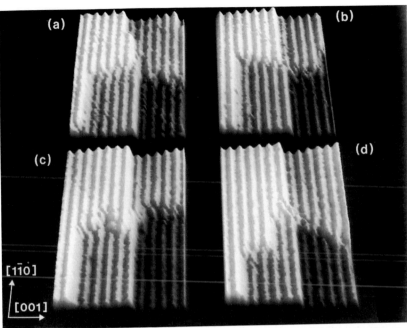

Fig. 4.73. Time series of STM topographs recorded every 60 min showing the development of a (2×1) reconstructed area on a Au(110) surface which exhibits three terraces separated by monoatomic steps. The middle and top terrace expose unusually wide steps along the unfavorable [001] direction. These steps meet in all the pictures, creating a convenient reference point (Gimzewski *et al.*, 1991b).

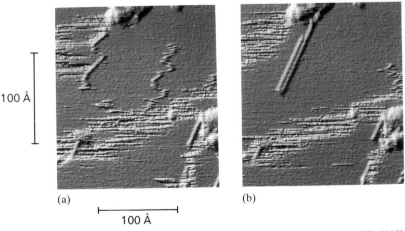

Fig. 4.74. STM images recorded consecutively on the same area of an O/Cu(110) (2×1) surface. Two isolated, mobile (2×1) strings in (*a*) are seen to be stabilized by condensation in (*b*) (Wintterlin *et al.*, 1991).

the power of STM and related scanning probe techniques for revealing the dynamics of surface processes ultimately with atomic resolution, thereby visualizing the birth of new structures of matter.

4.1.2.11 Surface phase transitions

A variety of phase transitions occurring on metal surfaces can be studied by variable-temperature STM. Local probe studies are particularly useful for investigation of the onset of phase transitions and their relation to structural defects present on the surface.

(a) Reconstructive phase transitions. Reconstructive phase transitions are less frequently observed on clean metal surfaces than on clean semiconductor surfaces (section 4.1.1.16), but are often induced by adsorption, e.g. of oxygen or alkali metals. Particularly well-studied systems are O on Cu(110), as discussed in section 4.1.2.6, and K on Cu(110) (Schuster *et al.*, 1991a, 1991b).

(b) Roughening transitions. Some metal surfaces exhibit a roughening transition below the melting temperature, involving spontaneous creation of vacancies and steps for low-index surfaces or spontaneous creation of kinks for vicinal surfaces. This phenomenon is caused by the vanishing of the free energy for formation of steps or kinks at the roughening transition temperature (Zangwill, 1988). STM has been used to directly observe the thermal roughening transition at a vicinal Ag(115) surface (Frenken *et al.*, 1990). The roughening transition was found to occur between 58 °C and 98 °C. An additional statistical analysis of the STM images, involving determination of height–height, kink–kink, and step–step correlation functions can yield more detailed information about the nature of the roughening transition.

(c) Melting transitions. It is now well established that some metal surfaces exhibit a melting (order–disorder) transition at a temperature considerably below the bulk melting temperature (Zangwill, 1988). However, there also exist examples of surfaces which do not show any sign of structural changes up to the bulk melting temperature. For instance, STM studies of the Ga(001) surface proved the absence of premelting phenomena below the bulk melting temperature at 29.78 °C (Züger and Dürig, 1992a). At the onset of bulk melting, large droplets of Ga are observed on the otherwise undistorted surface, indicating that macroscopic amounts of bulk material must have been molten, while the (001) surface remains in its crystalline state and does not show two-dimensional disordering. Step motion or creation of new steps (roughening) do not occur either, indicating that the binding energy within the surface plane is high enough to prevent even thermal disordering at step edges up to the bulk melting temperature. The absence of premelting phenomena on the Ga(001) surface has been explained by stabilization of this surface via slight reconstruction. On the other hand, similar STM studies of the more loosely packed Ga(110) surface have revealed a tendency to disorder below the bulk melting temperature (Züger and Dürig, 1992b).

4.1.3 Layered materials

Layered materials exhibit a variety of properties which make them exceptional with respect to the application of scanning probe microscopies (SPM). True layer-type or quasi-two-dimensional materials are built up by layers or neutral sandwiches held together only by the weak van der Waals (VDW) forces. Consequently, layered materials can easily be cleaved, providing atomically flat terraces of up to several thousand square Ångström units. The flat surfaces of cleaved layered materials are ideally suited for atomic-resolution SPM studies. In addition, they often serve as substrates for deposition of material to be studied by SPM. Perhaps most importantly, cleavage of true layer-type compounds does not create dangling bonds and therefore a freshly cleaved surface may stay clean for a long time. This allows atomic-resolution SPM studies of layered materials even under ambient conditions. The elastic properties of layered materials usually differ significantly from materials of three-dimensional structure. Layer-type materials generally tend to be much softer, increasing the influence of tip–sample interaction forces (section 1.22). Finally, the electronic properties of quasi-low-dimensional materials tend to be exceptional as well, which can have a strong influence on the observed contrast in STM images.

4.1.3.1 Clean surfaces of layered materials: graphite

Graphite is certainly the layered material that has most often been studied by scanning probe microscopies. Since the graphite surface is relatively inert, its atomic lattice is usually used as a reference for atomic-scale calibration of STM and SFM instruments working under ambient conditions.

The most common form of graphite in nature is hexagonal graphite, of which the crystal structure is shown in Fig. 4.75(*a*). Graphite is built up by layers with a honeycomb arrangement of carbon atoms being strongly covalently bonded to one another. The nearest-neighbor dis-

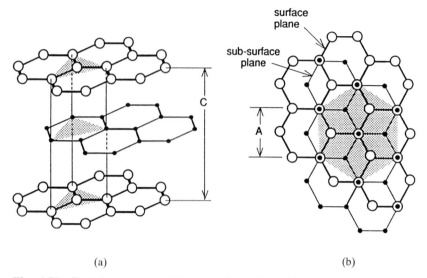

(a)   (b)

Fig. 4.75. Crystal structure of hexagonal graphite. (*a*) A side view showing three planes and the stacking between layers. (*b*) Top view. 'A'-sites in the top layer have an atom directly below them in the middle layer, 'B'-sites do not. The lattice parameters are A = 2.456 Å and C = 6.67 Å (Albrecht, 1989).

tance is 1.42 Å, whereas the in-plane lattice constant is 2.46 Å. The layers are spaced 3.35 Å apart and are held together by VDW forces. For hexagonal graphite, neighboring layers are shifted relative to each other, resulting in an ABAB . . . stacking sequence and a c axis lattice constant of 6.70 Å perpendicular to the layers. This stacking sequence gives rise to two non-equivalent carbon atom sites within the surface unit cell (Fig. 4.75(*b*)): one carbon atom (A-site) has a neighboring carbon atom directly below in the next layer, whereas the other carbon atom (B-site) is located above the center (hollow site) of the sixfold carbon ring in the second layer.

STM studies of the graphite (0001) surface (basal plane) have revealed a dominance of images showing a triangular, rather than a honeycomb lattice (Binnig *et al.*, 1986a). The spacing between the 'topographic' maxima as observed in most STM images is 2.46 Å (Fig. 4.76), which suggests that only every other surface carbon atom appears as a protrusion. The sensitivity of STM to the 'carbon-site asymmetry' indicates the importance of the electronic structure for understanding the observed image contrast, because a surface reconstruction is not known to occur. Electronic structure calculations for graphite have shown that the local electronic density of states (LDOS) near the Fermi

Fig. 4.76. STM image (6 nm × 6 nm) of graphite obtained at 100 K (Anselmetti, 1990).

level, E_F, being probed by STM to a first approximation (section 1.11.1), is different for the two inequivalent carbon atoms within the surface unit cell (Batra *et al.*, 1987; Tománek *et al.*, 1987; Tománek and Louie, 1988). The B-sites exhibit a higher LDOS near E_F than the A-sites and are therefore expected to appear as protrusions in STM images, whereas the A-sites should only appear as saddle points. The hollow sites should give rise to depressions because they exhibit the lowest LDOS near E_F.

For a single layer of graphite, a distinction between the two carbon atoms within the unit cell can no longer be made, and therefore one would expect to see the honeycomb lattice in STM images. In contrast, STM studies of a single layer of graphite on a Pt(111) substrate reveal the triangular lattice with 2.46 Å spacing between 'topographic' maxima, although a unique environment of one particular carbon species with respect to the Pt substrate does not exist, which could explain a preferential imaging (Land *et al.*, 1992). An explanation for this experimental observation can be given by considering the effect of the two-dimensionality of a single graphite layer on its electronic structure

(Tersoff, 1986): the Fermi surface collapses to a point at the corner of the Brillouin zone and, consequently, the STM image corresponds to an individual electronic state. The nodes in the wave function of this state then give rise to the measured STM corrugation with the period-icity of the unit cell (2.46 Å). Whether the hollow sites appear as depressions or protrusions depends critically on the kind of atom at the front of the tip (Tersoff and Lang, 1990). In bulk graphite, the weak interlayer interactions lead to a finite, though narrow, Fermi surface which lifts the strict node in the electronic state wave function. However, the observed STM image contrast is still dominated by this individual state (Tersoff, 1986).

Besides the triangular lattice shown in Fig. 4.76, STM images of graphite surfaces can also reveal a variety of different structures, including honeycomb arrays, triangular arrays of triangles, triangular arrays of ellipses, or linear row-like structures. Several explanations have been given for these STM images, which are considered to be 'anomalous'. One of these explanations is based on the assumption of imaging with a multiple atomic tip (Mizes *et al.*, 1987; Tsukada *et al.*, 1990; Isshiki *et al.*, 1990). It has been argued that STM images of graphite are dominated by only three independent Fourier components. A non-ideal, multiple tip can change the relative amplitudes and shift the relative phases of these components, leading to changes of shape and amplitude of the observed protrusions. Alternatively, anomalous STM images can be explained by assuming special tip electronic levels to be active in tunneling (Kobayashi and Tsukada, 1990).

Another anomaly which can appear in STM studies of graphite surfaces is a large measured corrugation, significantly higher than the value of 0.2 Å determined by helium atom diffraction. It has been argued that the nodal structure of the individual electronic state, being probed by STM measurements on graphite, can give rise to an enhanced corrugation on the order of 1 Å which is nearly independent of the tip–surface separation (Tersoff, 1986). On the other hand, experimentally observed giant corrugations of up to 8 Å and even more in STM studies of graphite surfaces performed in air, in the constant current mode of operation, can only be understood on the basis of a considerable force interaction between tip and sample mediated by a surface contamination layer (section 1.22).

Bias-dependent STM studies of the graphite surface have revealed a strong decrease of the measured corrugation with increasing magnitude of applied bias voltage (Binnig *et al.*, 1986a), in qualitative agreement with theoretical predictions (Selloni *et al.*, 1985; Tománek and Louie, 1988). The carbon-site asymmetry also decreases with increasing bias, and is found to be independent of the bias polarity (Tománek *et al.*, 1987). Bias-dependent STM imaging of graphite has been complemented by local tunneling spectroscopy measurements (Reihl *et al.*, 1986; Fuchs and Tosatti, 1987). The local conductance curve d*I*/d*U* versus *U*, as shown in Fig. 4.77, exhibits two and sometimes three peaks within the energy range between E_F and $E_F + 5$ eV. These are ascribed to a π^*-antibonding state at $+1.5$ eV, an interlayer state at $+3.7$ eV, and a possible defect-related state at $+2.7$ eV, which is only occasionally visible. The spectroscopic STM data agree reasonably well with experimental results from inverse photoemission and with theoretical predictions (Selloni *et al.*, 1985).

The graphite surface has also been studied by contact force microscopy, or atomic force microscopy (AFM), with atomic resolution (Binnig *et al.*, 1987a, 1987b; Albrecht, 1989; Meyer, 1990). A variety of different images have been obtained showing, for instance, honeycomb

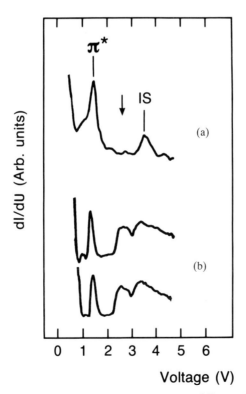

Fig. 4.77. d*I*/d*U*–*U* characteristics obtained at two different locations on a graphite surface (Fuchs and Tosatti, 1987).

lattices, triangular lattices, or linear row-like structures. According to theoretical calculations, a monatomic tip should experience very similar forces above an A- or a B-site of the graphite lattice, whereas the force above a hollow site should be somewhat lower (Abraham and Batra, 1989). Therefore, a monatomic tip is expected to image every carbon atom on the graphite surface, leading to a honeycomb lattice. On the other hand, repulsive forces on the order of 10^{-8} N, as typically applied in AFM experiments, are likely to lead to a large elastic compressive deformation of the graphite surface, eventually leading to its destruction (Abraham and Batra, 1989; Zhong *et al.*, 1991). In addition, a single-atomic tip is believed to be unstable during scanning with a tip load on the order of 10^{-8} N (Paik *et al.*, 1991). Therefore, it is reasonable to assume that the AFM images of the graphite surface are usually obtained with a multi-atom tip. In particular, it has been proposed that the as-prepared AFM tip might pick up a flake of graphite which is then dragged along the surface as the tip is scanned. In this case, the AFM would image repulsion maxima associated with misregistry of the graphite tip unit cell and the graphite surface unit cell (Abraham and Batra, 1989). This way, 'every other atom' images as well as anomalous images obtained by AFM can be explained.

STM and AFM have also been applied to study defects on graphite surfaces, including point defects, grain and tilt boundaries. For instance, point defects on gold-sputtered graphite samples have been investigated by STM (Bryant *et al.*, 1986b). The defect sites appeared as protrusions as long as the tip–surface separation exceeded a critical value. At smaller distances, however, the protrusions disappeared in the STM images, which were obtained in the variable current mode of operation (section 1.11.2). To explain these experimental observations, the defects were modeled as gold atoms which lie just below the surface layer. It has been argued that the STM image contrast at small tip–surface separations is dominated by the graphite electronic states. As the tip is moved away from the surface, only the slowly decaying component of the gold s-like electronic state is left. Subsequent theoretical calculations have shown that the protrusions, as seen by STM, do not directly represent images of the impurities themselves, but rather their effect on the graphite wave functions (Mizes and Harrison, 1988; Soto, 1988, 1990). Since the distance-dependences of the apparent 'topographic' height of the protrusions at the defect sites were found to depend on the type of impurity, it has been speculated whether an assignment of chemical identity to the impurity species could be based on distance-dependent STM studies.

A different kind of point defect on graphite surfaces has been created by C^{+} ion bombardment (Coratger *et al.*, 1990). In this case, a distance-dependence in the appearance of the defect sites was not found by STM.

An interesting STM observation of a local $(\sqrt{3} \times \sqrt{3})$ superlattice structure can often be made in the vicinity of defects on graphite surfaces, arising from adsorbates (Rabe *et al.*, 1988; Mizes and Foster, 1989), ion impacts (Shedd and Russell, 1991, 1992), or intrinsic structural imperfections. An example of a STM image showing the $(\sqrt{3} \times \sqrt{3})$ superlattice in the vicinity of an isolated structural defect, which can occasionally be found on a cleaved graphite surface, is presented in Fig. 4.78. The origin of this superlattice is ascribed to long-range electronic perturbations caused by defect sites (Mizes and Foster, 1989). It has been argued that local perturbation of the Fermi level charge density by the presence of a defect can lead to periodic oscillations similar to Friedel oscillations. These oscillations have a wavelength $\sqrt{3}$ times the periodicity of the graphite lattice and decay away from the defect over some ten lattice constants according to a power law. The symmetry of the oscillations was shown to reflect the nature of the defect (Mizes and Foster, 1989).

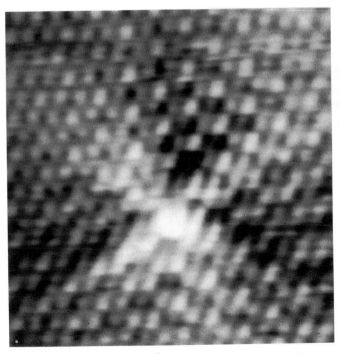

Fig. 4.78. STM image (42 Å × 42 Å) of a graphite surface with a structural defect. A $(\sqrt{3} \times \sqrt{3})$ superlattice of electronic origin is observed in the vicinity of the defect (Anselmetti, 1990).

Since AFM primarily probes ion–ion repulsion forces or forces associated with the total charge density distribution (section 2.4.1), rather than the Fermi level charge density distribution, AFM is not expected to be sensitive to the $(\sqrt{3} \times \sqrt{3})$ superlattice to the same extent as STM. This is indeed confirmed experimentally (Albrecht, 1989), although in one case, a $(\sqrt{3} \times \sqrt{3})$ superlattice associated with the presence of adsorbates on a graphite surface was seen by AFM as well (Albrecht *et al.*, 1988a).

Naturally occurring graphite single crystals are relatively small and difficult to obtain. Therefore, the most widely studied form of graphite by SPM has been highly oriented pyrolytic graphite (HOPG). This polycrystalline material with a hexagonal structure has a relatively large grain size (about 3–10 μm) and a good c axis orientation (misorientation angle less than 2°). Since the scan sizes typically used in SPM studies are small relative to the grain size of HOPG, grain boundaries are seldom encountered and thus only a few observations have been reported. STM studies of the grain boundary region between crystallites have revealed a disordered appearance with a width varying between 10 and 100 Å (Albrecht *et al.*, 1988b). No preferential orientation of grains within the basal plane of graphite has been found. If multiple tips scanning over different grains contribute to the STM image simultaneously, Moiré patterns can be observed near the grain boundaries, as shown in Fig. 4.79 (Albrecht *et al.*, 1988b).

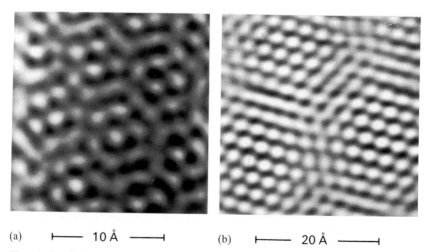

(a) ⊢——— 10 Å ———⊣   (b) ⊢——— 20 Å ———⊣

Fig. 4.79. Moiré patterns observed in the vicinity of grain boundaries of a graphite sample due to multiple-tip imaging. Analysis of the patterns reveals the orientation mismatch of the grains (Albrecht *et al.*, 1988b).

Anomalous superperiodicities, sometimes observed in STM images of graphite (Fig. 4.80), have been attributed to Moiré patterns as well (Kuwabara *et al.*, 1990; Buckley *et al.*, 1991; Oden *et al.*, 1991). They can arise from misorientation of a surface layer of graphite with respect to the underlying crystal. The superperiodicity S of a rotational Moiré pattern between two lattices with unit cell spacing a, misoriented by an angle θ, is given by the expression $S = a/[2 \sin (\theta/2)]$ (Kuwabara *et al.*, 1990). By using this relationship, the determination of S from STM images directly yields the misorientation angle θ. Subsurface defects can in this way be characterized by STM.

4.1.3.2 Clean surfaces of layered materials: boron nitride (BN)

Boron nitride (BN) was the first electrically insulating material to be imaged with atomic resolution by using AFM (Albrecht and Quate, 1987; Albrecht, 1989). It has a layered-type crystal structure with an in-plane lattice constant of 2.50 Å and a nearest neighbor spacing of 1.45 Å (Fig. 4.81). The basal planes are stacked 3.33 Å apart and each atom has a neighbor of the opposite atomic species located above and below in the next layer. AFM images of BN reveal a triangular lattice, rather than a honeycomb lattice, with only one 'topographic' maximum per unit cell, as shown in Fig. 4.82. In contrast to graphite, the inequivalence of the two sites within the surface unit cell is given by the two different atomic species present in BN. However, determining which

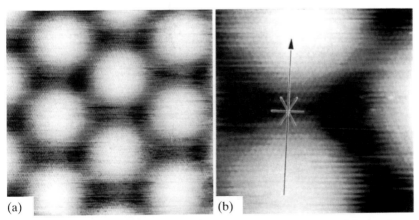

(a)   (b)

Fig. 4.80. STM images of an anomalous contrast area on a graphite surface. The graphite atomic fringes are seen in addition to the superperiodicity. (*a*) An image of 22.5 nm × 25.0 nm area. (*b*) An image of an 8.5 nm × 10 nm area. Short bars indicate the graphite fringe directions and a long arrow indicates one of the superperiodicity directions (Kuwabara *et al.*, 1990).

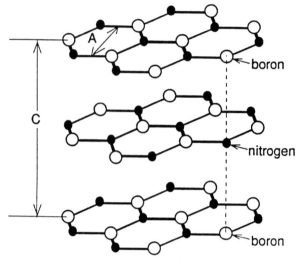

Fig. 4.81. The crystal structure of boron nitride. The lattice parameters are $A =$ 2.504 Å and $C = 6.66$ Å (Albrecht, 1989).

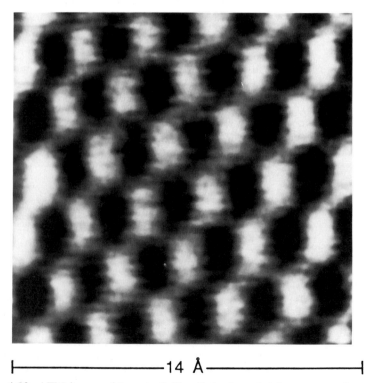

Fig. 4.82. AFM image of boron nitride. Only three bright spots per hexagon appear with a spacing of 2.5 Å (Albrecht, 1989).

site within the surface unit cell appears as protrusion is complicated. It has been argued that the B atom is a larger 'soft' atom while the N atom is a smaller 'hard' atom. Assuming a monatomic probe tip, the applied force would compress the B atom making it appear smaller than the N atom, and therefore the protrusions were tentatively attributed to N sites (Albrecht, 1989). However, this conclusion depends critically on the tip shape and the details of the image contrast mechanism in AFM which are still poorly understood. In general, atom-selective imaging of different elemental species in a compound insulator, such as BN, by using AFM is as difficult as in the case of STM (e.g. applied to GaAs(110)). Some preknowledge is required to attribute the protrusions imaged by AFM to specific surface sites.

4.1.3.3 Clean surfaces of layered materials: transition metal dichalcogenides (TMD)

Transition metal dichalcogenides (TMD) are built up by repetition of a three-layer sandwich consisting of a top chalcogenide layer (S, Se or Te), a middle transition metal layer (e.g. Ti, V, Nb, Mo, Ta, W etc.), and a bottom chalcogenide layer (Fig. 4.83). The bonding within each

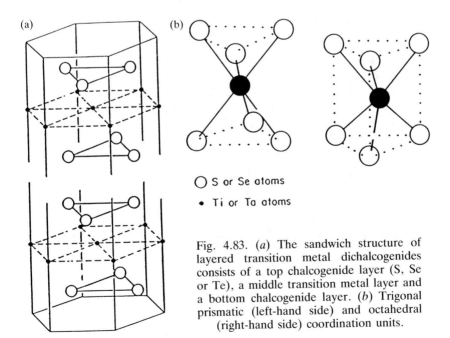

○ S or Se atoms

● Ti or Ta atoms

Fig. 4.83. (*a*) The sandwich structure of layered transition metal dichalcogenides consists of a top chalcogenide layer (S, Se or Te), a middle transition metal layer and a bottom chalcogenide layer. (*b*) Trigonal prismatic (left-hand side) and octahedral (right-hand side) coordination units.

sandwich is covalent, while neighboring sandwiches are held together mainly by weak VDW forces. The layers usually form a triangular lattice. The coordination around the transition metal atoms can be either trigonal prismatic or octahedral. Accordingly, one can distinguish between the 1T phase of TMD (pure octahedral coordination), the 2H phase (pure trigonal prismatic coordination), and the 4Hb phase (mixed octahedral and trigonal prismatic coordination). The surfaces of TMD, consisting of a chalcogenide layer, are relatively inert. Therefore, atomic-resolution SPM studies can be performed in an ambient environment. Electrically, the TMD cover a wide spectrum from true metals (e.g. NbS_2), through semimetals (e.g. WTe_2) and semiconductors (e.g. MoS_2, $MoTe_2$, WSe_2, SnS_2), to insulators (e.g. HfS_2). Depending on the degree of conductivity, either STM or AFM have been applied to study the surfaces of TMD.

STM images of $NbSe_2$ (Tokumoto *et al.*, 1986; Kajimura *et al.*, 1987; Bando *et al.*, 1987; Coleman *et al.*, 1988c; Dahn *et al.*, 1988; Watanabe *et al.*, 1988), MoS_2 (Stupian and Leung, 1987; Weimer *et al.*, 1988; Sarid *et al.*, 1988a; Henson *et al.*, 1988; Takata *et al.*, 1989b; Ichinokawa *et al.*, 1990), WSe_2 (Henson *et al.*, 1988; Akari *et al.*, 1988), and $MoTe_2$ (Tang *et al.*, 1989) all show a triangular lattice (Fig. 4.84). Since both the top chalcogenide layer and the second transition metal layer of these materials are built up from a triangular lattice with the same in-plane lattice constant, it cannot be decided *a priori* whether the top chalcogenide or the second transition metal layer is imaged by STM. On the other hand, some TMD exist for which the transition metal layer does not exhibit the same hexagonal arrangement as the chalcogenide layers. In WTe_2, for instance, every two rows of W atoms pair to form zig-zag and slightly buckled chains, thereby making the chalcogenide and transition metal layers inequivalent. STM images of WTe_2, showing zig-zag patterns, were initially interpreted as direct images of the subsurface W atoms (Tang *et al.*, 1989). However, possible multiple-tip imaging (Albrecht, 1989) and surface electronic structure effects (Tang *et al.*, 1990a, 1990b) have to be taken into account, leading to the final conclusion that the STM image contrast obtained on WTe_2 is more likely due to the surface Te layer, rather than the subsurface W layer.

STS studies of $NbSe_2$ have shown that the sensitivity to either the chalcogenide or transition metal layer in TMD might depend on the bias voltage polarity (Bando *et al.*, 1988). More recently, force microscopy has been used to characterize the tribological properties of TMD (Kim *et al.*, 1991). It was found that wear proceeds at defects and is, for instance, at least five times more slowly for MoS_2 than for $NbSe_2$.

Fig. 4.84. STM image of MoS$_2$ showing a triangular lattice with an in-plane lattice constant of 3.2 Å (Takata *et al.*, 1989b).

4.1.3.4 Clean surfaces of layered materials: mica

Mica is an insulating layered material of complex structure. Layers with a honeycomb arrangement of SiO$_4$ tetrahedra form the most prominent structural feature. AFM studies of mica surfaces reveal the unit cell periodicity of 5.2 Å (Drake *et al.*, 1989). However, individual atoms within the unit cell are not resolved. Defects on mica surfaces, created by irradiation with highly energetic Kr ions, have been studied by AFM as well. The tracks induced by the Kr ions are imaged as local depressions which are not believed to be of topographic origin, but rather result from changes in the local elastic properties: the cores of the tracks consist of disordered mica leading to softer regions than the surrounding crystalline mica.

4.1.3.5 Clean surfaces of layered copper oxides

The discovery of high-T_c superconductivity in layered copper oxides has led to an increasing interest in layered materials of complex structure (Bednorz and Müller, 1986, 1988). SPM is a promising technique to investigate the relationship between the local atomic and electronic structure and properties of the superconducting state of high-T_c superconductors (HTSC). In the following, we will concentrate on room temperature SPM studies of layered copper oxides, whereas low-temperature SPM investigations of the superconducting state in general will be discussed in section 4.3.

Among the different families of HTSC, the Bi- and Tl-based copper oxides have been most widely studied by STM because clean surfaces are easily obtained by cleaving the samples, as for other layered materials of less complex structure. The crystal structure of a Bi-based copper oxide is shown in Fig. 4.85. The cleavage process most likely leaves a BiO plane (Bi compounds) or TlO plane (Tl compounds) on top of the surface. STM images of the Bi-based copper oxide $Bi_2Sr_2CaCu_2O_{8+x}$ ('2212' compounds) reveal the atomic lattice and a superimposed incommensurate superstructure consisting of a sinusoidal modulation with a periodicity of 9–10 unit cells along the [010] direction

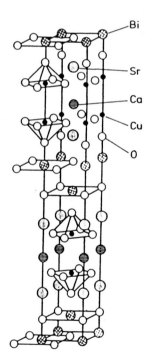

Fig. 4.85. Crystal structure of $Bi_2Sr_2CaCu_2O_8$.

and a periodicity of 4.5–5 units along [1$\bar{1}$0], as shown in Fig. 4.86 (Kirk *et al.*, 1988; Shih *et al.*, 1989; Anselmetti, 1990). The superstructure is also observed, for instance, by LEED, TEM, x-ray and neutron diffraction. Its origin can be explained by a periodic insertion of additional oxygen rows into the BiO planes, which in turn causes large displacive modulation in all layers (Le Page *et al.*, 1989; Beskrovnyi *et al.*, 1990). The atomic lattice periodicity, as determined from STM images, suggests that only one atomic species, either Bi or O, appear as 'topographic' protrusions. The assignment of these protrusions to specific atomic sites is, however, controversial (Kirk *et al.*, 1988; Shih *et al.*, 1989).

(a)

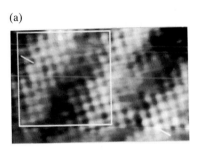

(b)

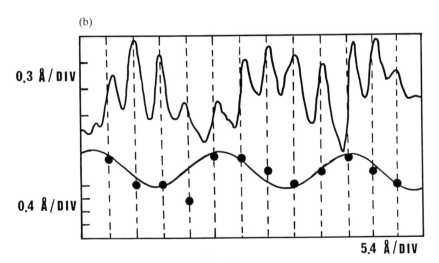

0.3 Å / DIV

0.4 Å / DIV

5.4 Å / DIV

Fig. 4.86. (*a*) STM image (6 nm × 4 nm) of $Bi_2Sr_2CaCu_2O_8$. The line section (AB) is shown in (*b*) together with the corresponding sinusoidal displacement of atomic positions with a period of 4.95 lattice constants (Anselmetti, 1990).

Local tunneling spectroscopy measurements on Bi-based '2212' compounds have revealed a lack of density of states at the Fermi level (Fig. 4.87), indicating that the surface BiO layer is not metallic (Shih *et al.*, 1989; Tanaka *et al.*, 1989b, 1990b; Shih *et al.*, 1991). Consequently, a model for the '2212' compounds has been proposed, consisting of an alternating stacking of non-metallic BiO and metallic CuO_2 planes. Photoemission and inverse photoemission experiments, with a probing depth of about 20 Å, represent more the density of states (DOS) averaged over the whole unit cell, and therefore yield a finite DOS at the Fermi level, in contrast to local tunneling spectroscopy (Fig. 4.87).

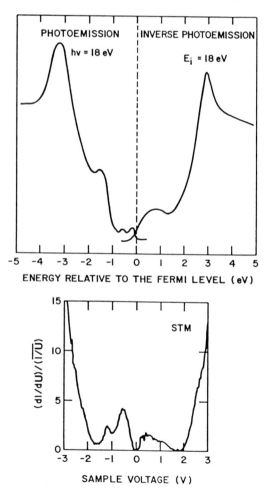

Fig. 4.87. Comparison of the STM spectrum of $Bi_2Sr_2CaCu_2O_8$ and the spectra obtained by using photoelectron spectroscopy and inverse photoelectron spectroscopy (Shih *et al.*, 1989).

Local tunneling spectra obtained on non-superconducting $Bi_2Sr_2CuO_{8+x}$ ('2201' compounds) show electronic gap structures similar to those observed for the BiO surface of superconducting '2212' compounds (Ikeda *et al.*, 1992). This suggests that superconductivity is not directly related to the electronic structure of the BiO plane. STM has also been applied to study the role of oxygen (Wu *et al.*, 1991; Awana *et al.*, 1991), as well as the effect of Pb substitution in the Bi-based copper oxides (Wu *et al.*, 1990; Zhang *et al.*, 1990).

STM studies of the Tl-based copper oxide $Tl_2Ba_2CaCu_2O_8$ ('2212' compound) reveal the atomic lattice together with a superimposed one-dimensional superstructure with a period of about 10 Å, as shown in Fig. 4.88. In contrast to the Bi-based copper oxides, both Tl and O

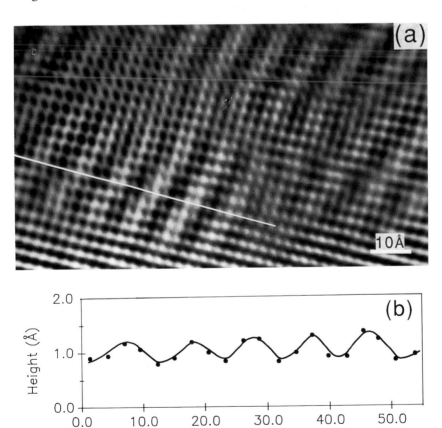

Fig. 4.88. (*a*) STM image of $Tl_2Ba_2Ca_1Cu_2O_8$ showing a one-dimensional super-lattice modulation of period 10 Å. (*b*) Profile of the surface corrugation measured along the line marked in (*a*) (Wu *et al.*, 1989).

sites of the top TlO surface plane are visible in STM images, indicating that Tl and O make similar contributions to the density of states near the Fermi level (Wu *et al.*, 1989; Zhang *et al.*, 1991b). For the $Tl_2Ba_2Ca_2Cu_3O_{10}$ ('2223' compounds), STM and local tunneling spectroscopy have been found to be useful techniques to study the relationship between crystal chemistry and the local structure and electronic properties (Zhang *et al.*, 1992).

Other families of HTSC, such as the '123' compounds, are more difficult to study on the atomic scale by STM because the samples do not cleave as easily as the Bi- or Tl-based copper oxides. However, a few atomic resolution STM studies of '123' compounds, e.g. $YBa_2Cu_3O_{7-x}$ single crystals or thin films, have been reported (van de Leemput *et al.*, 1988a, 1988b; Lang *et al.*, 1991, 1992a). Simultaneously recorded topographic and local conductance images of polycrystalline $YBa_2Cu_3O_{7-x}$ samples, fractured in UHV, revealed an alternation of metallic and semiconductor characteristics on a nanometer scale (Chen and Tsuei, 1989). The sequence of metallic and semiconducting strips was attributed to alternating Cu–O and Ba–O exposed surface regions, respectively.

STM and SFM have also been widely applied to study the surface morphology of sputtered and laser ablated $YBa_2Cu_3O_{7-x}$ thin films. Spiral-shaped growth terraces emanating from screw dislocations have been revealed as a typical structural feature on a nanometer scale, as shown in Fig. 4.89 (Hawley *et al.*, 1991; Gerber *et al.*, 1991; Norton *et al.*, 1991; Moreland *et al.*, 1991; Schlom *et al.*, 1992). Similar growth spirals on a larger scale were also observed by scanning electron and optical microscopy (Taylor *et al.*, 1988; Sun *et al.*, 1989, 1991). The density of screw dislocations was found to decrease with increasing substrate temperature during film growth (Norton *et al.*, 1991; Schlom *et al.*, 1992). It has been proposed that a correlation between the number of screw dislocations and the critical current density of the superconducting thin films exists (Gerber *et al.*, 1991; Mannhart *et al.*, 1992; McElfresh *et al.*, 1992), which is, however, controversial (Norton *et al.*, 1991; Moreland *et al.*, 1991).

4.1.3.6 Adsorbates and metal overlayers on surfaces of layered materials

Freshly cleaved layered materials provide ideal atomically flat substrates for the investigation of adsorbates, clusters and overlayers. Their static structure and dynamic behavior can be studied by STM down to the atomic scale. This was done, for instance, for Au, Ag, Cu, Al and Pt clusters and islands on graphite substrates. The adsorption sites of

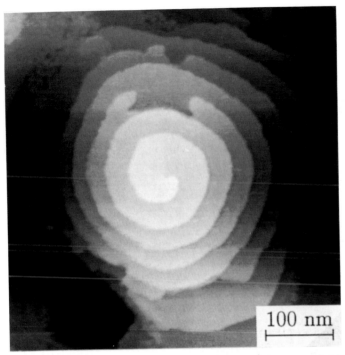

Fig. 4.89. STM image of an $YBa_2Cu_3O_7$ thin film showing a growth hill with a screw dislocation (Lang *et al.*, 1992a).

monomers, dimers and trimers, and the internal structure of extended monolayer islands have been determined (Abraham *et al.*, 1986; Ganz *et al.*, 1988a, 1988b, 1989). The two-dimensional metal islands often contain ordered regions separated by grain boundaries, while the atomic arrangement at the periphery is disordered. The ordered parts exhibit rectangular lattices which are not close-packed as in the bulk FCC structure and are incommensurate with the graphite substrate lattice (Fig. 4.90).

STM has also been used to study three-dimensional gold clusters on graphite substrates (Baró *et al.*, 1987; Humbert *et al.*, 1989; Nishitani *et al.*, 1991), as well as the initial stages of gold film growth as a function of substrate temperature (Nishitani *et al.*, 1991). Small nanometer-size gold particles usually nucleate at defect sites of the graphite substrate. With increasing coverage, coalescence of particles leads to the formation of large dendritic islands. Their detailed shape depends strongly on the substrate temperature, as shown in Fig. 4.91.

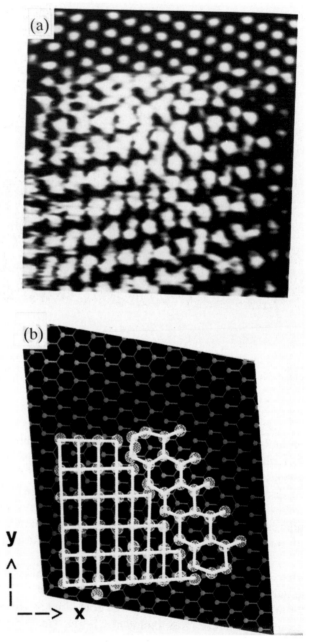

Fig. 4.90. (*a*) STM image (35 Å × 35 Å) of a monolayer Au island on graphite. The graphite substrate lattice is visible at the top. (*b*) Computer model of the Au overlayer showing a rectangular lattice on the left and a honeycomb lattice on the right (Ganz *et al.*, 1988a).

(a)   (b)   (c)

(d)   (e)   (f)

Fig. 4.91. STM micrographs (1800 nm × 1800 nm) of gold films deposited on graphite surfaces. Substrate temperatures were (a) 50 °C, (b) 60 °C, (c) 70 °C, (d) 80 °C, (e) 90 °C and (f) 100 °C (Nishitani *et al.*, 1991).

STM studies of small clusters on graphite substrates have additionally revealed the presence of superstructures in the vicinity of the clusters, very similar to the $(\sqrt{3} \times \sqrt{3})$ superlattices which are observed around graphite defect sites (section 4.1.3.1). The superstructures have been ascribed to a periodic modulation of the surface charge density in the surrounding regions of the clusters, decaying within a distance of 2–5 nm from the cluster sites (Xhie *et al.*, 1991a, 1991b).

4.1.3.7 Intercalated layered materials

Intercalated layered materials provide interesting model systems for investigation of two-dimensional physics. The most widely studied class of materials have been the graphite intercalation compounds (GIC), which are obtained by intercalating atomic or molecular species into the graphite galleries (Fig. 4.92). Intercalation can be achieved by a number of different preparation techniques, such as the two-temperature gas phase method, reactions with the liquid phase intercalant, or electrochemical preparation methods (Dresselhaus and Dressel-

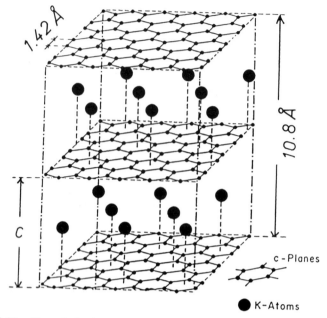

1.42 Å

10.8 Å

C

c-Planes

● K-Atoms

Fig. 4.92. Crystal lattice of a stage-1 potassium graphite intercalation compound.

haus, 1981; Pfluger and Güntherodt, 1981). Depending on the direction of charge transfer between the intercalant and the graphite host, donor GIC and acceptor GIC can be distinguished. Different 'stages' of GIC are obtained by appropriate choice of preparation conditions. The stage number n denotes the number of carbon (C) layers between two neighboring intercalant (I) layers. The overall intercalant concentration of GIC is primarily given by the relative amount of I layers to C layers, which is represented by the reciprocal stage number $1/n$ and, for a given stage number n, by the in-plane density of the I layer. The intercalant sublattice can exhibit in-plane order or disorder, depending on the intercalant species, its concentration and the investigated temperature range. Ordered intercalant sublattices can be either commensurate or incommensurate with the carbon sublattice.

The in-plane lattice constant of the graphite host lattice is not affected much by intercalation, whereas the distance between the carbon layers of 3.35 Å in pure graphite can be drastically increased in GIC up to 10 Å for some acceptor compounds. The ABAB . . . stacking sequence of hexagonal graphite (section 4.1.3.1) is generally not preserved in GIC. The same type of C layer is usually found on each side of an I layer, leading to an AIAI . . . sequence for stage-1 GIC and, for instance, an AIABABIBABAIA . . . sequence for stage-4 GIC. The I layers may or may not be regularly stacked in the *c* direction. To understand the phenomenon of stage transformations, a domain model for GIC has been proposed (Daumas and Hérold, 1969). In this model, all graphite galleries are equally filled, but not continuously in space, thereby exhibiting islands of the intercalated material.

The application of STM to GIC is of interest for several reasons. Firstly, the surface structure of GIC is almost unknown, whereas the microscopic bulk structure has been studied extensively by high-resolution TEM. STM can reveal the local atomic-scale surface structure of GIC, study its relation to defects in the graphitic host lattice, and investigate possible domain formation. Secondly, charge transfer between the intercalant and the graphite host leads to changes in the electronic structure of GIC compared with pure graphite. In particular, the local density of states at the Fermi level and the Fermi surfaces are significantly modified in GIC (Pfluger and Güntherodt, 1981); in fact, the semimetal graphite is turned into a synthetic metal. Therefore, STM studies of GIC can complement the experimental results obtained on pure graphite (section 4.1.3.1). Thirdly, as another consequence of intercalation, the elastic response of GIC generally differs from that of pure graphite, which might lead to a different STM response at low tunnel gap resistances.

4.1.3.8 Donor graphite intercalation compounds

Among the donor compounds, alkali metal GIC provide the simplest prototype systems. However, in contrast to pure graphite, the surfaces of alkali metal GIC are highly reactive and immediately oxidize when exposed to air. On the other hand, the low vapor pressure of the alkali metals prevents successful STM studies in UHV. Therefore, STM measurements on alkali metal GIC have been performed in an inert gas environment, provided by a stainless steel glove box containing a high-purity argon atmosphere. The samples are freshly cleaved *in situ* on the microscope's stage just before STM measurements are started. Atomic-resolution studies can be performed for time periods between half an hour and several hours, depending on the particular system to be studied. After that time, the STM image quality deteriorates, most likely due to the onset of surface oxidation.

Large-scale STM images of GIC generally show atomically flat terraces together with highly defective surface regions. The mean terrace size is significantly smaller and the density of defects higher compared with pure graphite. This observation is most likely explained by the creation of faults in the graphite host lattice during intercalation. The atomically flat terraces exhibit a variety of surface structures, even for a given alkali metal GIC.

Atomic-resolution STM studies of the surface of stage-1 Li GIC (C_6Li) have revealed three different types of triangular superlattices with in-plane lattice constants of 3.5, 4.2 and 4.9 Å, as shown in Fig. 4.93 (Anselmetti *et al.*, 1988, 1989; Wiesendanger *et al.*, 1989; Lang *et al.*, 1992b). The lattice constant of 4.2 Å corresponds to a commensurate ($\sqrt{3} \times \sqrt{3}$) superlattice (Fig. 4.94(a)) which can also be found in the bulk of C_6Li, whereas the lattice constant of 4.9 Å indicates the presence of a commensurate (2×2) superlattice (Fig. 4.94(b)), which is usually found in the bulk of stage-1 heavy alkali metal GIC. The third observed lattice constant of about 3.5 Å has tentatively been ascribed to an incommensurate superlattice of nearly close-packed lithium. For

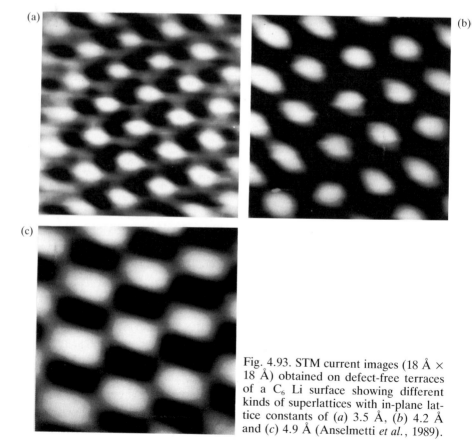

Fig. 4.93. STM current images (18 Å × 18 Å) obtained on defect-free terraces of a C_6 Li surface showing different kinds of superlattices with in-plane lattice constants of (*a*) 3.5 Å, (*b*) 4.2 Å and (*c*) 4.9 Å (Anselmetti *et al.*, 1989).

(a)   (b)

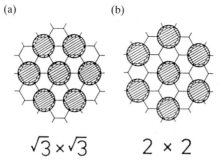

$$\sqrt{3} \times \sqrt{3} \qquad 2 \times 2$$

Fig. 4.94. Bulk structures of (a) C_6M (M denotes an alkali metal) and (b) C_8M graphite intercalation compounds at room temperature.

comparison, pure hexagonal close-packed lithium has a lattice constant of 3.1 Å. However, electrostatic repulsion, originating from almost complete charge transfer from lithium to the graphite layers in C_6Li, is likely to increase this spacing. On the other hand, the assignment of the superlattice with the smallest in-plane lattice constant remains uncertain as long as the underlying graphitic host lattice cannot be imaged simultaneously. The non-observation of the underlying host lattice in STM images of C_6Li seems to be typical, in contrast to STM images of the heavier alkali metal GIC described below.

Detection of the superlattices on C_6Li was found to be independent of the applied bias voltage, whereas the measured STM corrugation showed a pronounced bias-dependence. At small bias voltage (<200 mV), the corrugation appears to be similar to that of pure graphite (about 1 Å), whereas at larger bias voltage, a strong decrease of the corrugation is found (Anselmetti et al., 1989). The latter observation agrees with theoretical predictions (Selloni et al., 1988), while the corrugation at small bias voltage (low tunnel gap resistance) might be enhanced by elastic deformations of the GIC surface.

Apart from the atomic-scale superlattices, a remarkable superstructure on a scale of several hundred Ångström units can occasionally be seen by STM on the surface of C_6Li (Anselmetti, 1990). This superstructure, as shown in Fig. 4.95, is built up by individual clusters with a 'topographic' height of about 20 Å and an average apparent diameter of about 60 Å. The origin of this superstructure is not clear yet, but nevertheless its observation emphasizes the potential of STM for discovering previously unknown surface structures on a mesoscopic scale as well. STM studies of C_6Li on a sub-micrometer scale have also

Fig. 4.95. STM image (200 nm × 200 nm) of a superstructure with a period of about 30 nm observed on the surface of a C_6Li compound (Anselmetti, 1990).

revealed island structures of typical dimensions 500–2000 Å (Fig. 4.96), indicating inhomogeneous distribution of the intercalated material in a near-surface region (Lang *et al.*, 1992b).

Surfaces of stage-1 K GIC (C_8K) are most difficult to study because of their extremely high surface reactivity. Atomic-scale STM images predominantly show a commensurate (2×2) superlattice, as present in the bulk of C_8K, together with the underlying graphitic host lattice (Kelty and Lieber, 1989a; Lang *et al.*, 1992b). In addition, a variety of other superlattices can occasionally be found, including ($\sqrt{3} \times \sqrt{3}$), non-hexagonal one-dimensional, and non-hexagonal orthorhombic superstructures (Lang *et al.*, 1992b).

STM studies of surfaces of the heavier alkali metal stage-1 GIC, C_8Rb and C_8Cs, again reveal a variety of superlattices superimposed on the underlying graphitic host lattice (Kelty and Lieber, 1989b; Anselmetti *et al.*, 1990a, 1990b). Apart from a commensurate (2×2) superlattice (Fig. 4.97), which is also present in the bulk of these compounds, several different non-hexagonal one-dimensional superstructures can be

(a)

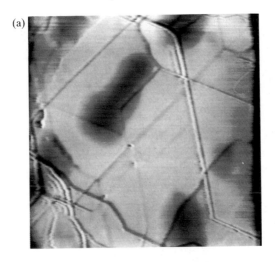

(b)

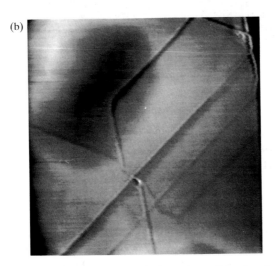

Fig. 4.96. STM images of a stage-1 Li-GIC surface. Islands of apparently reduced topographic height can be seen. Image size: (*a*) 414 nm × 414 nm and (*b*) 200 nm × 200 nm (Lang *et al.*, 1992b).

found, for instance, a ($\sqrt{3}\times4$) superlattice as shown in Fig. 4.98. The occasional observation of local superlattice defects indicates that the one-dimensional superstructures are not caused by imaging artifacts, e.g. Moiré patterns (section 4.1.3.1). Similar non-hexagonal superlat-

Fig. 4.97. STM image (5 nm × 6 nm) of a C_8Rb surface showing a (2×2) superlattice together with the underlying graphite host lattice (Anselmetti *et al.*, 1990b).

tices, ($\sqrt{3}\times2$) and ($\sqrt{3}\times\sqrt{13}$), have been found in the bulk of unsaturated Cs GIC by using scanning transmission electron microscopy (Hwang *et al.*, 1983). By analogy, dilution of the intercalant at the surface of stage-1 Rb and Cs GIC might be responsible for the observed one-dimensional superstructures. A model for the ($\sqrt{3}\times4$) and ($\sqrt{3}\times\sqrt{13}$) superlattices is shown in Fig. 4.99. The one-dimensional superstructures obviously break the threefold symmetry of the underlying graphitic host lattice locally. On a macroscopic scale, the threefold symmetry is regained by the occurrence of three rotational domains of the one-dimensional superstructures. A STM image showing all three possible rotational domains of a ($\sqrt{3}\times4$) superlattice simultaneously is presented in Fig. 4.100.

Interestingly, STM images of the heavy alkali metal GIC, showing the underlying graphitic host lattice in addition to the superlattices induced by the intercalant, typically reveal the same carbon site asymmetry as for pure graphite (section 4.1.3.1). This experimental observation clearly contradicts theoretical predictions based on the AlAl . . .

Fig. 4.98. STM image (8.5 nm × 8.5 nm) of a C_8Cs GIC surface showing a one-dimensional ($\sqrt{3}\times4$) superlattice structure with the simultaneously imaged graphite host lattice. The regularity of the superstructure is disturbed by a link between two neighboring linear chains (Anselmetti *et al.*, 1990a).

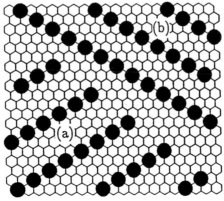

Fig. 4.99. Schematic illustration of (*a*) ($\sqrt{3}\times4$) and (*b*) ($\sqrt{3}\times\sqrt{13}$) superlattices with the underlying graphite honeycomb lattice (Anselmetti *et al.*, 1990b).

Fig. 4.100. This STM image of C_8Rb shows simultaneously the three possible rotational domains of the ($\sqrt{3}\times4$) superlattice (Anselmetti *et al.*, 1990b).

stacking sequence of stage-1 GIC (Tománek and Louie, 1988; Qin and Kirczenow, 1989, 1990).

STM studies of stage-2 K, Rb and Cs GIC show the absence of superlattice structures at the surface (Kelty and Lieber, 1989b). Instead, the images appear similar to those obtained on pure graphite with very pronounced carbon-site asymmetry. The non-observation of a superlattice can be explained by the liquid-like behavior of the intercalant for the stage-2 compounds at room temperature. The measured corrugation is found to be significantly lower than on pure graphite, which is consistent with expansion of the Fermi surface for the intercalation compounds.

More recently, ternary GIC have also been studied by STM (Kelty and Lieber, 1991; Kelty *et al.*, 1991; Lang *et al.*, 1992b). Hexagonal as well as non-hexagonal one-dimensional and orthorhombic superlattices have been found. On the surface of a stage-1 K–Cs GIC, a hexagonal (2×2) and a long-period orthorhombic superlattice have simultaneously been observed, in addition to the underlying graphitic host lattice. It is most likely that these superstructures, exhibiting incompatible symmetries, have a different origin. Apart from topographic and surface electronic structure effects, it has been argued that charge density wave (CDW) states (section 4.4) could probably explain some of the superlattice structures observed at the surfaces of donor GIC. An indication for the existence of surface-driven CDW in stage-1 alkali metal GIC had previously been found by means of angle-resolved photoemission spectroscopy (Laguës *et al.*, 1988). However, more experimental and theoretical work is needed to clarify this issue.

4.1.3.9 Acceptor graphite intercalation compounds

Only a few STM studies have concentrated on acceptor GIC. A graphitic surface structure was found in most cases, for instance, on stage-1 $FeCl_3$ GIC (Gauthier *et al.*, 1988) and stage-1 $CuCl_2$ GIC (Olk *et al.*, 1990, 1991; Pappas *et al.*, 1992). The carbon site asymmetry appears to be absent in most STM images of these acceptor GIC, and the measured corrugation is found to be reduced considerably compared with pure graphite. For stage-1 $CuCl_2$ GIC, a bias-dependence of the STM images has been reported (Olk *et al.*, 1990, 1991). At positive sample bias voltage ($+10$ mV), the honeycomb lattice of graphite was observed, whereas at negative bias voltage (-10 mV), a monoclinic superlattice was found, which was ascribed to the intercalated $CuCl_2$ layer lying just below the top graphitic surface layer.

Local tunneling spectroscopy measurements have been performed on a stage-1 $CoCl_2$ GIC (Tanaka *et al.*, 1988). A shift in the Fermi level position of about 0.4 eV towards lower energy compared with pure graphite was observed, in agreement with theoretical expectations for acceptor-type compounds (Fig. 4.101).

Application of SFM to GIC is expected to provide additional complementary information about these compounds. For instance, it has been shown theoretically that contact force microscopy should be a unique tool to determine the local surface rigidity and to measure the local surface layer distortion of graphitic planes in the vicinity of intercalated species (Tománek *et al.*, 1989). In addition, SFM is a promising technique to study the interesting frictional properties of GIC on a local scale.

4.1.4 Insulators

The invention of contact force microscopy or atomic force microscopy (AFM) has made possible surface science studies of bulk insulators. This field has been relatively neglected in the past due to the problem of surface charging when using conventional surface analytical techniques

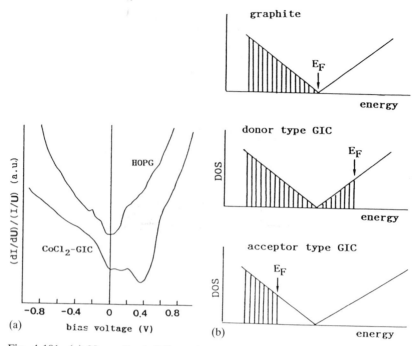

Fig. 4.101. (*a*) Normalized differential conductance versus bias voltage for a CoCl$_2$-GIC and graphite measured at room temperature in ambient air. (*b*) Comparison of the Fermi level position in idealized pure graphite, donor GIC and acceptor GIC. (Tanaka *et al.*, 1988).

involving charged particles. AFM has initially mainly been applied to study the clean surfaces of insulators, such as ionic crystals as discussed below. However, UHV AFM will also become a powerful technique to study adsorption processes and epitaxial growth on insulator substrates.

(a) LiF. Atomic resolution AFM studies of the LiF(100) surface, performed in an ambient environment, have revealed a square lattice with a periodicity of 2.8 Å, as shown in Fig. 4.102(*a*) (Meyer, 1990; Meyer *et al.*, 1990a, 1991b). The observed periodicity corresponds to the nearest-neighbor distance between ions of the same kind on a non-reconstructed LiF(100) surface (Fig. 4.102(*b*)). Since the ionic radius for F$^-$ (1.33 Å) is considerably larger than for Li$^+$ (0.68 Å), the 'topographic' protrusions as seen in the AFM images have been attributed to F$^-$ ions, whereas Li$^+$ is believed to remain 'invisible' due to its small size. This interpretation is based on the 'contact' hard-sphere model (Fig. 4.103), which is widely used in the context of helium-atom scattering. The measured AFM corrugations along

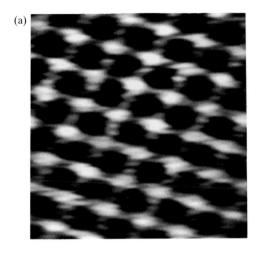

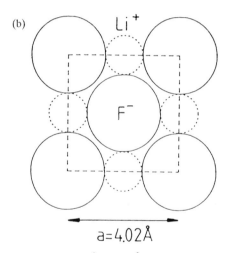

Fig. 4.102. (*a*) AFM image (15 Å × 21 Å) of a LiF(100) surface. The bright spots are attributed to fluorine ions (Meyer *et al.*, 1991b). (*b*) Schematic diagram of the arrangement of Li$^+$ (small dotted circles) and F$^-$ ions (large circles). The nearest-neighbor distance of the fluorine ions is 2.84 Å (Meyer *et al.*, 1990a).

the [100] and [110] directions were found to agree reasonably well with the corrugations derived from helium-atom scattering experiments.

(b) NaCl. The NaCl(100) surface has been studied by AFM under UHV conditions (Meyer and Amer, 1990a). Atomic-resolution images again show only one ionic species on the non-reconstructed surface. The 'topographic' protrusions in the AFM images have

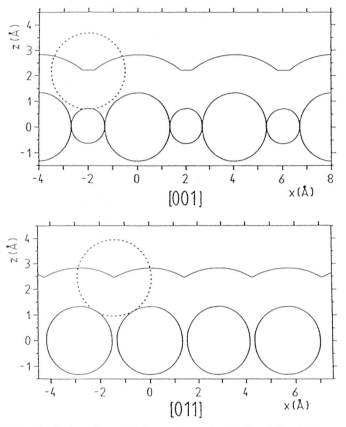

Fig. 4.103. Simulation of an AFM scan along the [001] and [011] directions on the LiF(001) surface using a contact hard sphere model. The dotted sphere represents the tip atom (Meyer *et al.*, 1991b).

been attributed to Cl⁻, since it is the larger of the two constituent ions of NaCl.

(c) AgBr. Surfaces of AgBr(100) films, epitaxially grown on NaCl(100) substrates, have also been atomically resolved using AFM (Meyer *et al.*, 1991a). A square lattice is observed with a periodicity of 4.1 Å, corresponding to the nearest-neighbor distance between ions of the same species (Fig. 4.104(*a*)). Based on the contact hard-sphere model, the protrusions have been attributed to Br⁻ ions, whereas the smaller Ag⁺ ions are not expected to be 'visible' (Fig. 4.104(*b*)). Larger scale AFM images have revealed terraced growth hills and pits, as well as screw dislocations. The surface steps were found to

(a)

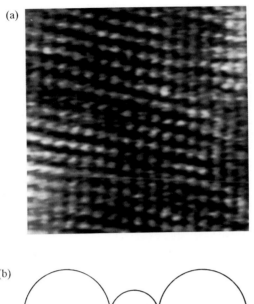

(b)

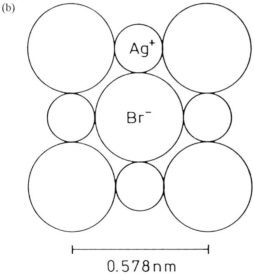

Ag$^+$

Br$^-$

0.578nm

Fig. 4.104. (*a*) AFM image (70 Å × 70 Å) of a AgBr(001) surface. The protrusions are separated by 4.1 Å and are attributed to bromine ions (Meyer *et al.*, 1991a). (*b*) Schematic diagram of the atomic arrangement on the AgBr(001) surface. The small circles and the large circles represent silver and bromine ions, respectively. The nearest-neighbor distance of the bromine ions is 4.08 Å (Haefke *et al.*, 1991).

Fig. 4.105. Force micrograph (4.2 μm × 4.2 μm) of an individual plate-like AgBr microcrystal of height 200 nm displayed in a three-dimensional perspective projection (Schwarz *et al.*, 1992).

be typically of monatomic height. No preferential orientation of the steps was observed. In contrast, AFM studies of AgBr single-crystal surfaces have revealed preferential alignment of surface steps along particular crystallographic directions (Heinzelmann *et al.*, 1989; Meyer *et al.*, 1989b; Meyer, 1990). A striking difference in the step structure between (100) and (111) surfaces has been found which is probably related to the different adsorption and growth behavior of J-aggregating dye molecules on these surfaces, as is important in photographic emulsion technology. Large-scale AFM studies have also been performed on AgBr tabular grains (T-grains). AFM allows resolution of surface fine structure on top of the terraced emulsion grains (Fig. 4.105), which is of interest for understanding the adsorption behavior of site-directing dye molecules (Schwarz *et al.*, 1992).

4.2 Magnetism

Magnetism is a well-known collective phenomenon in condensed matter, discovered several thousand years ago. It is of quantum mechanical origin and results from the existence of electron spin. Magnetic structures are determined by long-range dipolar and short-range exchange interactions. Magnetic-sensitive SPM techniques contribute in two ways to the field of magnetism and magnetic materials.

1. They provide valuable information about micromagnetic configurations, such as magnetic domains and domain wall structures on a micrometer down to a sub-micrometer scale. Even surface spin structures at the atomic level are experimentally accessible. Most important, magnetic-sensitive SPM techniques allow correlation of structural and magnetic properties from sub-micrometer down to atomic scale, which is of general interest for improved understanding of magnetism as well as for tailoring the properties of new magnetic materials.
2. Techniques such as MFM allow characterization of the surface stray field distribution of magnetic storage media on a sub-micrometer scale under ambient conditions and without the need for special sample preparation. Based on the magnetic imaging capability on a refined length scale, it becomes possible to improve magnetic storage media or even develop new ones. Thus, magnetic-sensitive SPM techniques play an important role in the technology as well as in the science of magnetism and magnetic materials.

4.2.1 Micromagnetic configurations

4.2.1.1 Magnetic domains and domain wall structures

Magnetic domains and domain walls have been studied on a sub-micrometer scale by means of magnetic force microscopy (MFM), as introduced in section 2.7.3. By using single-crystal iron whiskers as samples, it was possible to investigate well-defined micromagnetic objects, such as 90° and 180° Bloch walls (Göddenhenrich *et al.*, 1988, 1990b; Hartmann and Heiden, 1988; Hartmann, 1989b). Exact positioning of the MFM probe tip with respect to the domain wall configuration was achieved by simultaneous magneto-optic Kerr microscope observations during MFM operation. A MFM image showing the intersection of two 90° Bloch walls with a 180° Bloch wall is presented in Fig. 4.106(*a*). The observed contrast is primarily determined by the vertical component of the surface magnetic stray field distribution (section 2.7.3). Differences in wall contrast, including stray field wall width and shape, between the 90° and 180° domain walls are clearly revealed by MFM (Fig. 4.106(*b*)). MFM even allows resolution of fine

(a)

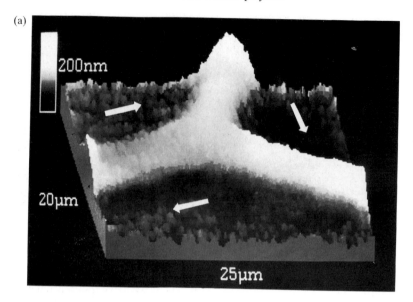

(b)

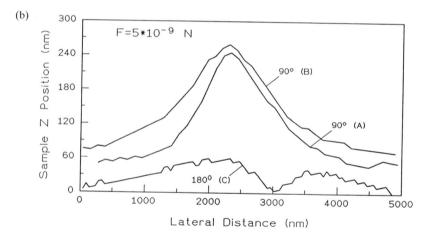

Fig. 4.106. (*a*) MFM image of the intersection of two 90 ° Bloch walls with a 180 ° wall. The arrows represent the polarizations of the adjacent domains (Hartmann *et al.*, 1990). (*b*) Averaged MFM profiles over 90 ° and 180 ° domain walls taken at a tip-to-sample distance of about 50 nm (Göddenhenrich *et al.*, 1990b).

structures within domain boundaries. This is demonstrated by Fig. 4.107, showing a bipolar 180° Bloch wall with a Bloch line separating two adjacent antiparallel wall segments, leading to a reversal in MFM contrast.

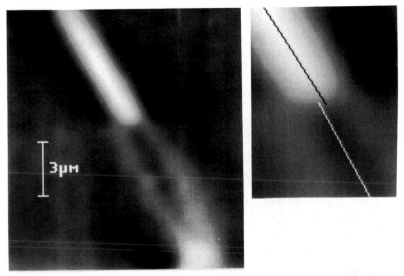

Fig. 4.107. MFM image of the transition zone between two oppositely polarized segments of a 180 ° Bloch wall in an iron whisker (Göddenhenrich *et al.*, 1990b).

Domain wall configurations have also been studied on permalloy ($Ni_{80}Fe_{20}$) thin films (Mamin *et al.*, 1989). Fig. 4.108(*a*) shows a MFM image together with the corresponding Kerr microscope image and a schematic outline of the direction of magnetization in each domain inferred from these images. The typical closure structure of the domain walls is apparent. In the MFM image, the walls show up as either bright or dark lines, depending on the sense of the rotation of the magnetization across the wall (Fig. 4.108(*b*)), leading to either repulsive or attractive tip–sample force interaction. Contrast reversal along the domain wall, as marked by an arrow in Fig. 4.108(*a*), indicates the presence of a Bloch line.

4.2.1.2 Surface spin structures

By using STM-based techniques under well-defined UHV conditions, it has been possible to probe surface spin structures even down to the atomic level. For instance, individual paramagnetic spin centers on oxidized silicon surfaces have been detected by the spin precession in a constant magnetic field which induces a modulation in the tunnel current at the Larmor frequency (Manassen *et al.*, 1989). The corresponding radio-frequency signal was shown to be localized over distances less than 10 Å.

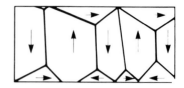

(a)

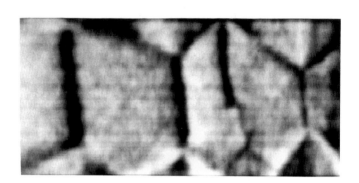

(b)

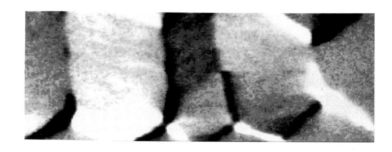

(c)

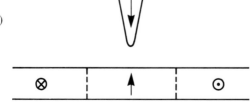

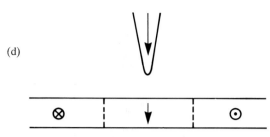

(d)

Fig. 4.108. (*a*) MFM image and (*b*) Kerr image of domains in a Permalloy film. The imaged region is 50 μm × 30 μm; magnifications differ in horizontal and vertical directions. The inset shows inferred magnetization directions. The arrow marks the location of a Bloch line. (*c*) Highly simplified picture of a MFM tip interacting with a 180° Bloch wall. The overall interaction is repulsive because the wall and tip magnetizations are antiparallel; in (*d*) the magnetizations are parallel and the interaction is attractive (Mamin *et al.*, 1989).

Spin-polarized scanning tunneling microscopy (SPSTM), based on vacuum tunneling of spin-polarized electrons between a magnetic tip and sample (section 1.14), has allowed probing of the nanometer-scale magnetic order of the Cr(100) surface with terraces separated by monatomic steps (Fig. 4.109(*a*)). The alternating monatomic step-height values, as measured with a ferromagnetic CrO_2 tip (Fig. 4.109(*b*)), have directly proved the existence of alternately magnetized (100) terraces (Wiesendanger *et al.*, 1990c, 1991b). This 'topological antiferromagnetic order' of the Cr(100) surface was theoretically predicted earlier, based on self-consistent total-energy calculations (Blügel, *et al.*, 1989).

More recently, SPSTM has been applied to study the Fe_3O_4 (100) surface down to the atomic scale (Wiesendanger *et al.*, 1992b, 1992c, 1992d). Clear contrast between the different magnetic ions Fe^{3+} and Fe^{2+} on the octahedrally coordinated B-sites of magnetite was obtained by using *in situ* prepared ferromagnetic Fe probe tips (Fig. 4.110), which has been attributed to the different spin configurations of Fe^{3+} and Fe^{2+} being $3d^{5\uparrow}$ and $3d^{5\uparrow} 3d^{\downarrow}$, respectively. The static arrangement of Fe^{3+} and Fe^{2+}, observed at the (100) surface of magnetite at room temperature has indicated the presence of a surface Wigner glass state for which, in contrast to the bulk, the mobility of the $3d^{\downarrow}$ electrons among the octahedrally coordinated B-sites has ceased due to a localization of these electrons. As a consequence, the surface is in an insulating ionic state, rather than in a metallic state, which has been verified by detection of a finite energy gap by means of local tunneling spectroscopy. The ability of SPSTM to directly probe surface spin configurations at the atomic level complements other magnetic-sensitive SPM

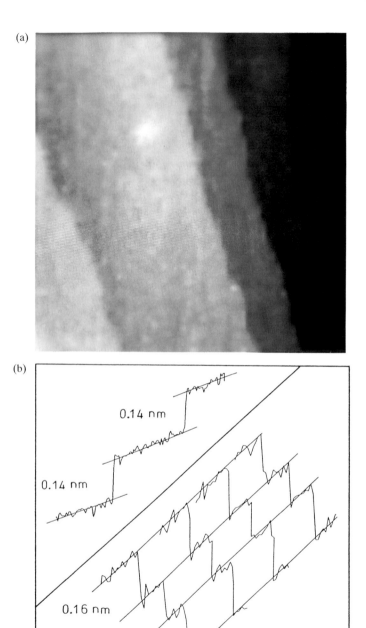

(a)

(b)

0.14 nm

0.14 nm

0.16 nm

0.12 nm

0.16 nm

Fig. 4.109. (*a*) Constant current STM image (32 nm × 32 nm) of a Cr(001) surface obtained with a CrO_2 tip, showing terraces separated by monatomic steps with different gray levels for each terrace (Wiesendanger *et al.*, 1990c). (*b*) Single-line scans over the same three monatomic steps taken from the STM image shown in (*a*). The same alternation of step-height values (0.16 nm, 0.12 nm and again 0.16 nm) is evident in all single-line scans. The line scans are 22 nm long. Inset: for comparison, a single-line scan over two monatomic steps as measured with a tungsten tip is presented. In this case, the measured step-height value is constant and corresponds to the topographic monatomic step height. This line scan is 70 nm long (Wiesendanger *et al.*, 1990c).

(a)

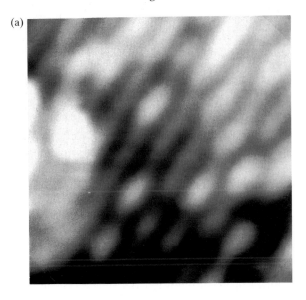

(b)

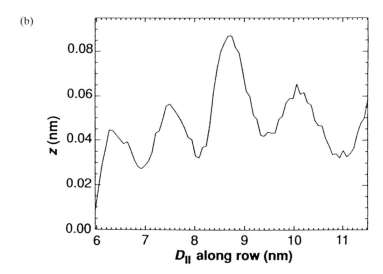

Fig. 4.110. STM image of the Fe–O (001) plane of magnetite obtained with an Fe tip. The rows of Fe B-sites of spacing 6 Å and with modulation along these rows can clearly be seen. A period of 12 Å is preferentially observed along the Fe rows. This is particularly evident from single-line sections along the Fe rows as shown in (b). The period 12 Å corresponds to the repeat pattern of Fe^{2+} and Fe^{3+} in the Fe–O (001) plane of magnetite (Wiesendanger *et al.*, 1992b).

techniques, providing surface magnetic structure information on a mesoscopic length scale.

4.2.2 Magnetic storage media

Magnetic force microscopy (MFM) and tunneling-stabilized MFM (section 1.22) have found important applications in characterization of magnetic storage media on a sub-micrometer scale. Information about the magnetic writing process can be obtained, including the size and shape of the written bits, the media noise, the overwrite characteristics, as well as the ability of the recording media to sustain high data densities.

For instance, thermomagnetically written bits on magneto-optical recording media, such as Fe–Tb thin films, have been characterized by MFM (Martin, *et al.*, 1988b). Based on the MFM images, it can easily be checked under which conditions subdomain-free bits are written. For Co/Pt multilayers, it was found that the structure of the written bits is strongly influenced by multilayer composition, preparation conditions and the structure of the Pt base layer (van Kesteren *et al.*, 1991). Regularly shaped and subdomain-free bits were obtained for multilayers consisting of an alternating sequence of Co layers about 4 Å thick and Pt layers thinner than 15 Å (Fig. 4.111).

Fig. 4.111. MFM image of marks written with laser pulses on a $10 \times (4.1$ Å Co+14.7 Å Pt) multilayer substrate (van Kesteren *et al.*, 1991).

Bit tracks written on various magnetic storage media have been characterized by MFM (Abraham *et al.*, 1988c; Mamin *et al.*, 1988; Rugar *et al.*, 1990; Schönenberger and Alvarado, 1990a) as well as by tunneling-stabilized MFM (Moreland and Rice, 1991; Gomez *et al.*, 1992). A bit track written on a Co–Pt–Cr longitudinal recording medium is shown in the MFM image presented in Fig. 4.112(a). The sample was first d.c.-erased (magnetized) using a recording head of width 25 μm. The erased region generates little or no magnetic stray field and therefore appears smooth in the MFM image. After the erasure, a magnetic bit track was written using a recording head of width 8 μm. The recorded transitions, which are spaced with a periodicity of 2 μm along the track, generate a significant magnetic stray field which alternates in sign (Fig. 4.112(b)). The spatially varying interaction of this stray field with the MFM probe tip generates the strong magnetic contrast seen in Fig. 4.112(a). Outside the erased band, the MFM image shows apparently rough surface regions. This roughness is of magnetic rather than topographic origin, and reflects the complex magnetic stray field distribution arising from the demagnetized state of the as-deposited recording medium.

Apart from characterization of written bit tracks on magnetic storage media, MFM was also found to be a useful technique for studying the magnetic stray field distribution of recording heads (Martin and Wickramasinghe, 1987). The ability to operate magnetic-sensitive SPM techniques under ambient conditions has certainly proved to be of great importance for their various applications in magnetic storage technology.

4.3 Superconductivity

Superconductivity is another collective phenomenon in condensed matter that has been known since the beginning of the 20th century (Onnes, 1911). The recent discovery of high-T_c superconductivity in copper oxides with transition temperatures T_c above 30 K has led to an unprecedented effort in the experimental and theoretical investigation of the superconducting state (Bednorz and Müller, 1986, 1988).

Since the early experiments with planar metal–oxide–metal junctions (section 1.3.1 and 1.3.2), tunneling has been recognized as one of the most important experimental techniques for studying the superconducting state because tunneling directly probes the quasiparticle density of states of the superconductor (Buckel, 1984). STM as a local probe is particularly powerful because it allows to determine various quantities characteristic of the superconducting state for different locations on the

2 μm Bits on Co−alloy Disk

10μm

(a)

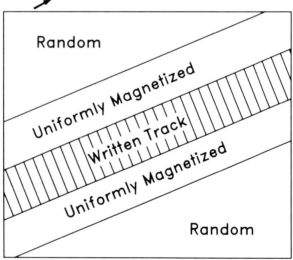

Random

Uniformly Magnetized

Written Track

Uniformly Magnetized

Random

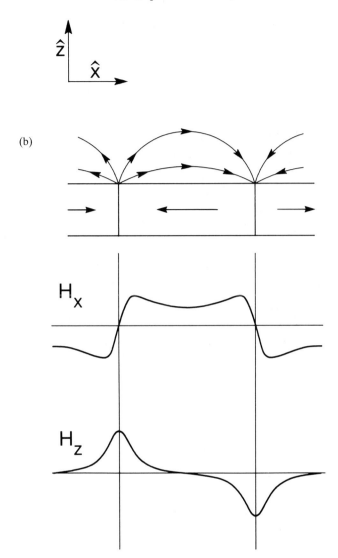

Fig. 4.112. (*a*) MFM image of a track 8 μm wide. The bit transitions are spaced every 2 μm along the track. Arrows point to the edges of the d.c.-erased region. The image was taken with a straight tip oriented at an oblique angle with respect to the surface. (*b*) Transition geometry and stray field configuration in longitudinal magnetic media (Rugar *et al.*, 1990).

Fermi surface of single crystals and for different sample locations for inhomogeneous superconductors, with a spatial resolution beyond the coherence length. Specifically, STM has found important applications for studying spatial variations of the local superconducting properties of polycrystalline materials, the influence of defects and impurities on the superconducting state, and the investigation of the Abrikosov flux lattice as well as flux motion in type-II superconductors.

4.3.1 Local energy gap

Local tunneling spectroscopy (section 1.13.3) allows measurement of a local I–U characteristic at a preselected location (x_0, y_0), from which a local energy gap $\Delta(x_0, y_0)$ can be extracted by fitting the experimental curve according to the theoretical I–U relationship given by Eq. (1.49) in section 1.3.1 (Fig. 4.113). For an appropriate interpretation of the derived energy gap value, it is important to consider the following points.

1. Tunneling probes the electronic density of states within a probing depth on the order of the coherence length ξ of the superconductor. For superconductors exhibiting an extremely small coherence length on the order of one unit cell, such as for the high-T_c superconductors (HTSC), an atomic-level perfect surface is required. The presence of insulating surface contamination layers can lead to energy gap values which are not characteristic of the superconducting state. In addition, even for perfect surface layers, possible surface effects in superconductivity have to be considered (Giamarchi *et al.*, 1990).
2. Contamination layers on the tip can affect the measured energy gap values as well (Wilkins *et al.*, 1990b). Since tungsten tips easily become oxidized when exposed to air or moderate vacuum, platinum or gold tips are usually preferred for spectroscopic tunneling studies of superconductors.
3. In the presence of insulating surface layers, the tip and sample are likely to get in contact to reach the demanded tunnel current flow. In this case, point contact rather than vacuum tunneling experiments are performed. For point contact junctions, the possible influence of the local applied pressure on the measured energy gap value has to be considered (van Bentum *et al.*, 1987).
4. For polycrystalline superconductors, such as sintered or thin-film HTSC samples, capacitive charging effects caused by Coulomb blockade of tunneling (section 1.3.3) can determine the measured energy gap (van Bentum *et al.*, 1989; Gallagher and Adler, 1990; Wilkins *et al.*, 1990b, 1991; Wan *et al.*, 1990).

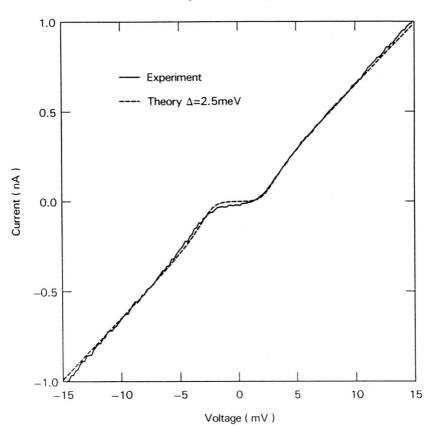

Fig. 4.113. *I–U* characteristic obtained with a Pt–Rh tip and a NbN sputtered film at a temperature of 5 K, with no externally applied field. The solid curve represents experimental data; the dashed curve is a best fit to the theoretical prediction using the BCS density of states, with an energy gap value of 2.5 meV (Kirtley *et al.*, 1987b).

5. If the tunnel gap resistance is chosen to be small (<10 MΩ), 'leakage' currents for bias voltages $U < \Delta/e$ can exist even at zero temperature (García *et al.*, 1988; Ferrer *et al.*, 1989).

In general, it is preferable to apply surface science techniques for preparation and characterization of well-defined surfaces of the superconductors to be investigated. This requires development of STM instruments which can simultaneously be operated at low temperatures and under ultra-high-vacuum conditions. Most of the initial STM studies of superconductors have been performed only under high-vacuum conditions on surfaces that have not additionally been characterized by other surface-sensitive experimental techniques.

4.3.1.1 'Conventional' superconductors

Local tunneling spectroscopy has been applied to measure local I–U characteristics and to determine local energy gaps for polycrystalline samples of several 'conventional' superconductors, such as Nb_3Sn (Elrod *et al.*, 1984), NbN (Kirtley *et al.*, 1987b; LeDuc *et al.*, 1987, 1989), and Pb (LeDuc *et al.*, 1987; Ramos *et al.*, 1988; Wilkins *et al.*, 1990b). The local energy gap values were found to be in reasonable agreement with results of earlier tunneling experiments using planar tunnel junctions.

4.3.1.2 Copper oxide HTSC

Soon after the discovery of high-T_c superconductivity in copper oxides, local tunneling spectroscopy was also applied to the different families of HTSC, including the La-based compounds (Pan *et al.*, 1987; Kirtley *et al.*, 1987c; Hawley *et al.*, 1987; Naito *et al.*, 1987; Fein *et al.*, 1988; Vieira *et al.*, 1988a), the '123' compounds (Kirtley *et al.*, 1987a; Kirk *et al.*, 1987; Crommie *et al.*, 1987; van Bentum *et al.*, 1987; Gallagher *et al.*, 1988; Gallagher and Adler, 1990; Wilkins *et al.*, 1990a), and the Bi-based copper oxides (Vieira *et al.*, 1988b; Briceno and Zettl, 1989; Hasegawa *et al.*, 1989, 1990; Escudero *et al.*, 1990; Jiang *et al.*, 1991; Koltun *et al.*, 1991; Chen and Ng, 1992). Generally, the dI/dU–U characteristics obtained appear to be broadened by an order of magnitude in excess of what would be expected from thermal processes. A satisfactory fit of the experimental data can only be achieved by introducing a broadening parameter Γ in the relationship (1.51) derived in section 1.3.1:

$$\frac{dI}{dU} \propto \mathrm{Re}\left(\frac{eU - i\Gamma}{\left[(eU - i\Gamma)^2 - \Delta^2\right]^{\frac{1}{2}}}\right) \tag{4.1}$$

where i is the imaginary unit. The parameter Γ accounts for finite quasiparticle lifetime effects associated with mechanisms, such as inelastic scattering caused by defects or inhomogeneities, that introduce tunneling states below the superconductor energy gap (Dynes *et al.*, 1978).

A large variation of the derived energy gap values has typically been found even for a given sample, which is most likely explained by the ill-defined state of the surfaces being investigated. A possible anisotropy of the energy gap for the directions perpendicular and parallel to the basal plane of the layered copper oxides could, in principle, also account for the observed variations, particularly for polycrystalline samples. The amount of gap anisotropy as deduced from local tunneling spectroscopy on single crystals of '123' and Bi-based copper oxides is, however, controversial (Kirtley *et al.*, 1987a; Hasegawa *et al.*, 1989, 1990; Briceno and Zettl, 1989; Chen and Ng, 1992).

Unfortunately, only a few STM investigations have been reported where the atomic-scale surface structure was checked by 'topographic' measurements at low temperatures in addition to the local tunneling spectroscopic studies. Atomic-resolution STM images at low temperatures have been obtained, for instance, for the Bi-based copper oxides $(Bi_2Sr_2CaCu_2O_{8+x})$, which are less susceptible to surface contamination than are other families of HTSC (Hasegawa *et al.*, 1990; Wang *et al.*, 1990b; Liu *et al.*, 1991). Local tunneling spectroscopy measurements on these well-characterized surfaces have revealed energy gap values in the range 30–35 meV, as shown in Fig. 4.114, yielding values of $2\Delta/k_B T_c \approx 8$, well above the BCS prediction of 3.53, which indicates extremely

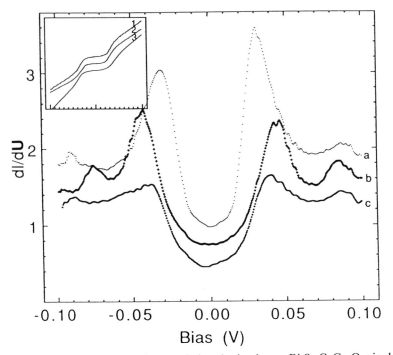

Fig. 4.114. Three dI/dU–U characteristics obtained on a $Bi_2Sr_2CaCu_2O_8$ single-crystal surface. A conductance peak resembling the BCS energy-gap feature is apparent in the three curves. The inset displays three I–U characteristics. They differ in tip-to-surface distance (Liu *et al.*, 1991).

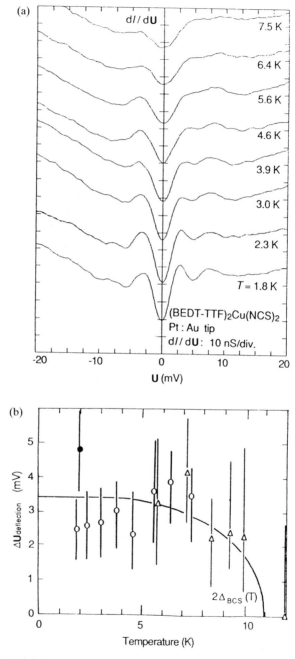

Fig. 4.115. (*a*) Temperature-dependence of the differential conductance measured on the surface of an organic superconductor $\varkappa$-(BEDT-TTF)$_2$Cu(NCS)$_2$. (*b*) Temperature-dependence of the spacing between the deflection points of the dI/dU–U curves, compared with the temperature-dependence of the superconductor energy gap according to the BCS theory (solid curve) (Bando *et al.*, 1990).

strong coupling superconductivity (Wang *et al.*, 1990b; Liu *et al.*, 1991). However, it has been argued that the observation of such large energy gap values may correspond to the semiconductor characteristics of the BiO layer on top of the surface, whereas the measured energy gap value for the superconducting CuO layer beneath the surface BiO layer is consistent with BCS theory (Hasegawa *et al.*, 1990).

4.3.1.3 Organic superconductors

Local tunneling spectroscopy has also been applied to study the energy gap of single-crystalline organic superconductors, such as $\varkappa$-(BEDT-TTF)$_2$Cu(NCS)$_2$ with a transition temperature T_c as high as 11 K (Bando *et al.*, 1990). The measured temperature-dependence of the differential conductance curve (Fig. 4.115(*a*)) has been used to derive the variation of the superconductor energy gap as a function of temperature (Fig. 4.115(*b*)), which was found to be in agreement with BCS theory. However, the shape of the dI/dU–U curves tended to deviate from the BCS prediction.

4.3.1.4 Doped fullerenes

Following the discovery of high-T_c superconductivity in alkali-metal-doped C$_{60}$ (Hebard *et al.*, 1991), local tunneling spectroscopy has been applied to these exciting materials as well (Zhang *et al.*, 1991a). For single-phase superconducting Rb$_3$C$_{60}$ with $T_c \approx 29$ K, an energy gap of $\Delta \approx 6.6$ meV has been derived (Fig. 4.116), corresponding to a value of $2\Delta/k_B T_c \approx 5.3$. This experimental result is indicative for strong coupling superconductivity.

4.3.2 Spatially resolved energy gap

The spectroscopic capabilities can be combined with the high lateral resolution of STM in order to study spatial variations in the superconductor energy gap which might be caused, for instance, by spatial inhomogeneities in the samples. Several different experimental procedures have been proposed.

1. As a first approach, a single parameter derived from the local I–U relationship can be recorded as a measure of local superconductivity. This parameter should preferably be insensitive to the spatially varying surface topography. It has been found that, by recording the spatial variations in the ratio of the zero-bias conductance to the large-bias conductance, a clear contrast between superconducting and normal regions is obtained. This has been demonstrated by using

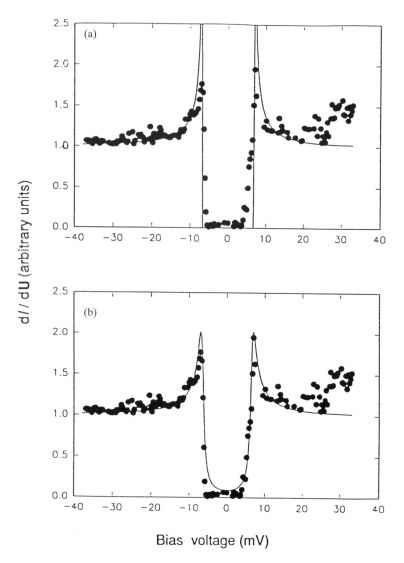

Fig. 4.116. Plots of dI/dU versus U for Rb$_3$C$_{60}$ at 4.2 K. The experimental data for conductance (solid circles) were calculated numerically from measured I–U characteristics. The data are fitted (*a*) without and (*b*) with a broadening parameter. The values of the energy gap and the broadening parameter for the fit in (*b*) are 6.6 meV and 0.6 meV, respectively (Zhang *et al.*, 1991a).

a superconducting Nb$_3$Sn thin film sample for which strong spatial variations, with reproducible transitions between fully superconducting and fully normal behavior over distances as small as 13 nm, have directly been observed (de Lozanne *et al.*, 1985).

2. Recording only a single parameter derived from the local *I–U* relationship certainly provides an incomplete measure of local superconductivity. Ultimately, a complete *I–U* characteristic should be measured at every topographic data point, allowing detailed quantitative analysis of the spatial variations in superconductivity. The local *I–U* characteristics can be fitted to yield, for instance, the local energy gap at each spatial point. Consequently, spatial maps of the energy gap value can directly be compared with simultaneously recorded topographic images, as has been demonstrated for a granular superconducting NbN film exposed to a magnetic field of varying strength (Kirtley *et al.*, 1987b, 1988c). The energy gap images revealed significant spatial variations on a nanometer scale, including areas with small or zero gap when a magnetic field of 2–6 T was applied (Fig. 4.117). In some cases, a clear correlation between features in the energy gap image and the corresponding topographic image can be found.

Similar studies on HTSC would be of great interest. Initial experiments have been performed using '123' compounds with ill-defined surface layers (Volodin and Khaikin, 1987). It was necessary to form point contacts at each data point before reliable local *I–U* characteristics were obtained. Consequently, spatial variations of the local energy gap, derived from the *I–U* characteristics, could be studied only on a micrometer scale.

4.3.3 Phonon density of states

The superconductor energy gap is by far not the sole parameter that can be derived from local tunneling spectroscopy data. For strongly coupled superconductors, for instance, conductance versus voltage characteristics additionally yield information about the density of states of the phonons effective in mediating electron–electron coupling in the superconducting state, as discussed in section 1.3.1. This has been verified by low-temperature spectroscopic studies of Pb thin films as shown in Fig. 4.118 (LeDuc *et al.*, 1989). In principle, these measurements could be extended to study spatial variations of phonon effects in strongly coupled superconductors by means of STM/STS.

25 nm

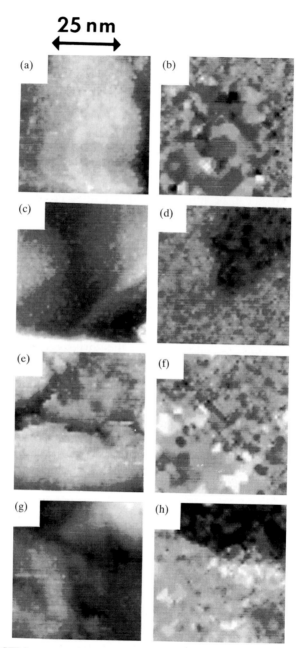

Fig. 4.117. STM topographic images (left column) and simultaneously acquired energy gap images (right column) of a NbN film taken at a temperature of 5 K. Image size is 49 nm × 50 nm. The applied magnetic fields were 0 T (*a*), (*b*), 2 T (*c*), (*d*), 4 T (*e*), (*f*) and 6 T (*g*), (*h*), respectively. The gray scales for the topographic images correspond to 137 Å (*a*), 80 Å (*c*), 170 Å (*e*) and 101 Å (*g*) vertical height from black to white. The gray scales for the gap images correspond to 0.0–4.0 meV from black (low) to white (high) (Kirtley *et al.*, 1987b).

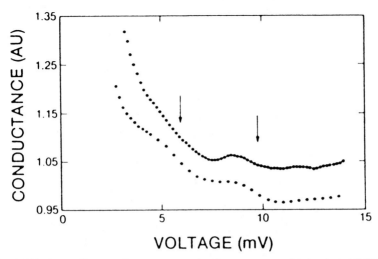

Fig. 4.118. Tunneling conductance versus voltage spectra obtained at 4.2 K for a Pb thin film. The upper curve has been measured by STM and the lower curve by using a macroscopic Pb/AlO$_x$/Al tunnel junction. The arrows indicate features associated with the transverse (left-hand arrow) and longitudinal (right-hand arrow) peaks in the phonon density of states (LeDuc *et al.*, 1989).

4.3.4 Abrikosov flux lattice

Soon after the development of the spectroscopic capabilities of STM, it was suggested that one of the most powerful applications would be investigation of the Abrikosov flux lattice in type-II superconductors exposed to an external magnetic field (d'Ambrumenil and White, 1984). However, initial experiments in this direction failed, most likely due to the poor surface quality of the superconductors that were studied. By choosing an appropriate model system, the layered transition metal dichalcogenide 2H-NbSe$_2$ (section 4.1.3.3) with its excellent surface quality after sample cleavage, Hess *et al.*, (1989, 1990a) finally achieved direct STM observation of the Abrikosov flux lattice and the density of states near and inside a single vortex core.

Upon cooling, NbSe$_2$ first undergoes a charge density wave (CDW) transition at 33 K (section 4.4) before becoming superconducting at $T = 7.2$ K with an average in-plane coherence length of 77 Å and an upper critical magnetic field of 3.2 T, applied perpendicular to the basal plane. Local tunneling spectroscopy data for superconducting NbSe$_2$, obtained in zero applied magnetic field, revealed a superconducting energy gap of 1.1 meV, as shown in Fig. 4.119. The temperature-dependence of the energy gap was found to agree well with the BCS prediction. If a

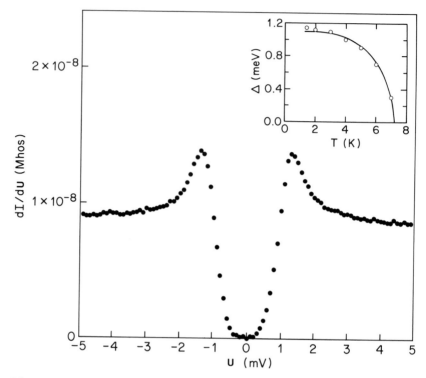

Fig. 4.119. Plot of dI/dU versus U for $NbSe_2$ with zero applied magnetic field used to determine the gap at 1.45 K. Inset: The energy gap versus temperature and the corresponding BCS fit (Hess *et al.*, 1989).

magnetic field is applied, the density of states of $NbSe_2$ becomes spatially modulated by the Abrikosov flux lattice, which can be revealed by spatially resolved conductance images (Fig. 4.120). To get high contrast, the differential conductance dI/dU is monitored at a bias voltage of 1.3 mV, which corresponds to the peak position in the (dI/dU)–U characteristic of superconducting $NbSe_2$ (Fig. 4.119). High values of the differential conductance therefore correspond to superconducting regions, whereas low values correspond to vortex cores in Fig. 4.120. The measured vortex spacing in the triangular flux lattice depends on the strength of the applied magnetic field and was found to be in agreement with theoretical predictions.

Subsequently, local tunneling spectroscopy was applied to study the density of states inside a single vortex core. Surprisingly, a zero-bias conductance peak was found with the tip positioned directly above the

1 Tesla

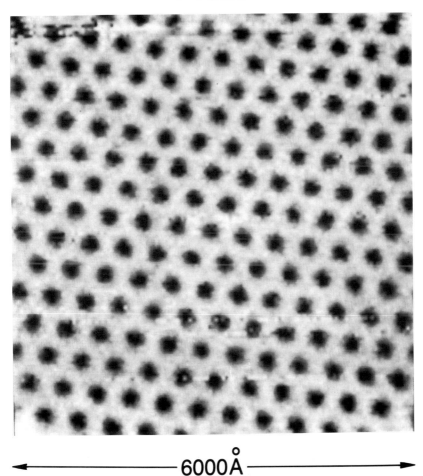

◄─────────── 6000 Å ───────────►

Fig. 4.120. Abrikosov flux lattice produced by 1 T magnetic field in NbSe$_2$ at 1.8 K. The gray scale corresponds to dI/dU (Hess *et al.*, 1989).

center of a vortex core, as shown in Fig. 4.121. If the density of states in the core had the characteristic appearance of a normal metal, a featureless (dI/dU)–U curve would have been expected. By moving the tip away from the center of the vortex core, the zero-bias conductance peak decreases and finally vanishes above the superconducting regions (Fig. 4.121). The spatial extent of the zero-bias peak was found to be on the order of the coherence length of superconducting NbSe$_2$ ($\xi = 77$ Å).

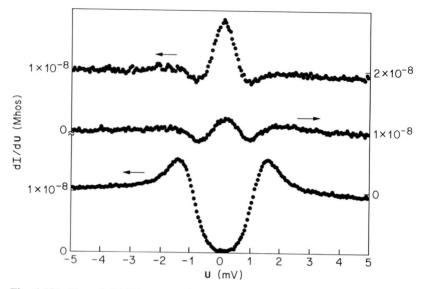

Fig. 4.121. Plot of dI/dU versus U for NbSe$_2$ at 1.85 K and 0.02 T field, taken at three different positions: on a vortex, about 75 Å from a vortex, and 2000 Å from a vortex (Hess *et al.*, 1989).

Theoretical studies, motivated by experimental STM results, have shown that the zero-bias conductance peak can be attributed to low-angular-momentum quasiparticle bound states in the vortex core, whose localized wave functions are maximum near the vortex center (Shore *et al.*, 1989; Klein, 1989, 1990; Gygi and Schluter, 1990a; Ullah *et al.*, 1990). In addition, it has been predicted that larger angular-momentum bound states exist, having the maximum of their wave functions at some distance from the center of the vortex core (Fig. 4.122). These states have higher energies E, $0 < E < \Delta$, that are symmetric about the Fermi energy, causing a pair of symmetric maxima in the (dI/dU)–U characteristic.

Upon improving the experimental voltage (energy) resolution to 0.1 mV at lower temperatures, the splitting of the zero-bias conductance peak into a double peak, corresponding to the higher-energy quasiparticle bound states, has indeed been observed as the tip was moved away from the vortex core center (Hess *et al.*, 1990b, 1991). This can be illustrated by a voltage–distance plot as presented in Fig. 4.123. Sufficiently far away from the vortex center, the energy of the quasiparticle excitations become of the order of the superconductor energy gap, and merge with the quasiparticle excitations at the gap edge.

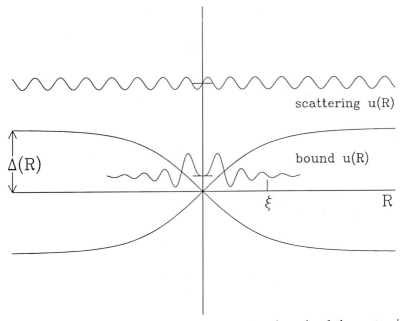

Fig. 4.122. The energy gap collapses to zero as the axis of the vortex is approached. This forms a radial potential well for the quasiparticle excitations. Bound states with energies less than the energy gap value exist in this spatially confined region. A bound state wave function is drawn symbolically, centered on its energy eigenvalue. Likewise a scattering state has been sketched with its energy greater than the energy gap value (Hess *et al.*, 1990a).

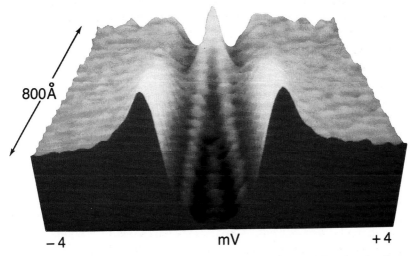

Fig. 4.123. Perspective view of dI/dU (U, r, $\theta = 26°$) for NbSe$_2$ showing how the superconductor energy gap evolves into a zero-bias peak as spectra are taken along a radial line that goes through the core. The zero-bias peak not only diminishes in amplitude but actually splits as spectra are taken away from the core (Hess *et al.*, 1991).

Condensed matter physics

Moreover, the spatial distribution of vortex-core bound states has directly been determined in real-space (Hess *et al.*, 1990b; Hess, 1991). Surprisingly, sixfold star-shaped, rather than circular structures were found in the normalized conductance images (Fig. 4.124). Subsequent theoretical studies have shown that breaking of the rotational invariance

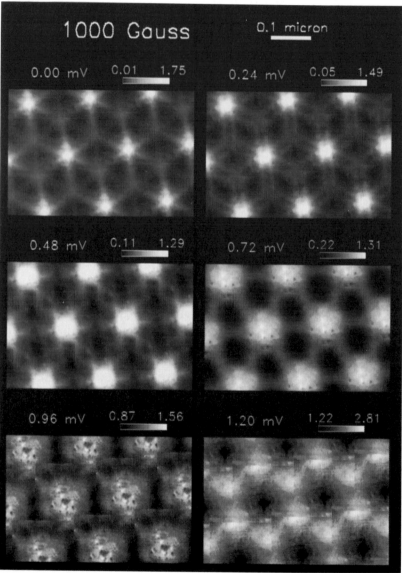

Fig. 4.124. Normalized real-space conductance patterns of vortices in 2H-NbSe$_2$ at different energies (Hess, 1991).

of a vortex line by an anisotropic crystal potential and band structure can induce an angular band structure in the low-energy part of the quasiparticle spectrum, leading to the observed star-shaped structures whose orientation is energy-dependent (Gygi and Schluter, 1990b, 1991). The ground-state wave function at 0 mV exhibits a star-like distribution with the streamers rotated by 30° relative to the crystal and Abrikosov lattice vectors, while the streamers of the '0.5 mV star', corresponding to antibonding states, are aligned with the crystal and Abrikosov lattice vectors (Fig. 4.124).

The energy-dependence of the star orientation excludes interaction between flux lines as a primary source for the observed anisotropy (Gygi and Schluter, 1990b). This is also confirmed experimentally since, in the limit of large vortex separation at low applied magnetic field, the anisotropy remains undiminished (Hess *et al.*, 1991). By increasing the applied bias voltage to 1.2 mV, corresponding to an energy just above the gap energy Δ, the contrast in the normalized conductance image reverses and the conductance becomes reduced inside the vortex core (Fig. 4.123 and Fig. 4.119). This image no longer corresponds to quasiparticle bound states, but rather to scattering states with energies $E > \Delta$ (Fig. 4.122).

The experimental results obtained for NbSe$_2$ have impressively demonstrated that STM promises to become a powerful new tool for measuring superconductor parameters, such as ξ and Δ, as a function of position on the Fermi surface. In addition, STM allows one to study the influence of material impurities and imperfections on the superconducting state. Thus far, we have discussed STM investigations of pure 2H-NbSe$_2$, a 'clean' superconductor for which the mean free path l is much longer than the coherence length ξ. By substitutionally alloying NbSe$_2$ to become Nb$_{1-x}$Ta$_x$Se$_2$ ($0 < x < 0.2$), the mean free path systematically decreases with increased Ta substitution, leading to a gradual transition from a 'clean' superconductor ($l \gg \xi$) to one approaching the 'dirty' limit ($l \lesssim \xi$). It has been found that the zero-bias conductance peak, associated with the low-lying quasiparticle bound states in a vortex core, is very sensitive to disorder and gradually disappears with increasing impurity concentration x, as shown in Fig. 4.125 (Renner *et al.*, 1991). For $x = 0.2$, the density of states in the vortex center appears to be equal to that in the normal state. This experimental observation can be explained by quasiparticle scattering processes on a scale smaller than the size of the radial potential well for the quasiparticle bound states which is on the order of 2ξ (Fig. 4.122). The scattering events lead to a significant mixing of quasiparticle states, thereby destroying the zero-bias conductance peak. Consequently, by probing the vortex-

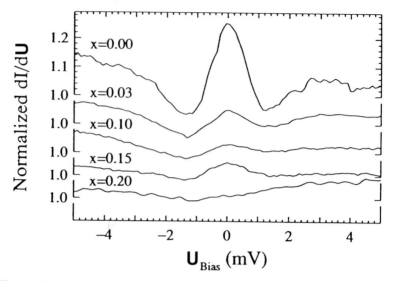

Fig. 4.125. Tunneling spectra taken at the center of a vortex core for various Ta substitutions in the alloy system $Nb_{1-x}Ta_xSe_2$ at 1.3 K and 0.3 T. The spectra are normalized to the differential conductance at high bias (Renner *et al.*, 1991).

core density of states with STM, it is possible to classify the supercon- ductor as being either in the clean or dirty limit, as well as to study quasiparticle scattering caused by sample impurities.

There exist several other competing experimental techniques for investigation of magnetic flux structures in superconductors, such as neutron diffraction (Cribier *et al.*, 1964; Schelten *et al.*, 1972), NMR (Fite and Redfield, 1966), the Bitter method (Essman and Träuble, 1967; Herring, 1974), low-temperature scanning electron microscopy (Bosch *et al.*, 1985), and electron holography (Matsuda *et al.*, 1989, 1991; Hasegawa *et al.*, 1991). STM can be regarded as an outstanding technique because it directly probes the quasiparticle density of states, the fundamental physical quantity, at an extremely high spatial resolu- tion. On the other hand, the ability to study the Abrikosov flux lattice by STM appears to be highly sensitive to the surface quality of the superconducting samples. Therefore, other scanning probe methods, such as magnetic force microscopy (Hug *et al.*, 1991) or scanning Hall probe microscopy (Hess, 1991) are considered as alternative techniques to study the Abrikosov flux lattice of delicate samples, e.g. HTSC, at high spatial resolution. MFM measures the force response of the superconductor to the magnetic stray field of the magnetized sensor tip. Non-destructive imaging of the flux lattice appears to be possible because the magnetostatic interactions between the ferromagnetic tip and the individual vortices induce negligible distortion of the flux lattice (Berthe *et al.*, 1990; Berthe, 1991).

Finally, scanning probe methods can be applied to study the dynamics of vortex motion in type-II superconductors. This has been demonstrated by using a tunnel junction consisting of a niobium sample and a niobium tip, held at a fixed position (Dittrich and Heiden, 1988). Vortex motion is generated by passing an electric transport current through the sample which exceeds a certain critical current value. As the vortices move underneath the tip, the tunnel gap voltage becomes modulated due to the different I–U characteristics of the vortices and the vortex-free regions in between (Fig. 4.126). Therefore, by monitoring the time-dependence of the tunnel gap voltage, the flux flow can directly be characterized on a local scale.

The influence of a transport current on the Abrikosov flux lattice of 2H-NbSe$_2$ has been studied by STM as well (Berthe, 1991). It was found that the microscopic STM observations lead to the same value for the critical current, necessary to induce vortex motion, compared with macroscopic measurements performed by the standard four-probe technique. Switching the transport current I_t between values above and below the critical current I_c did not alter the vortex positions of the flux lattice as observed for $I_t < I_c$. This is indicative of the existence of a pronounced minimum-energy arrangement of the flux lattice.

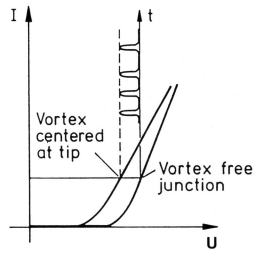

Fig. 4.126. Signal generation due to vortex motion across a quasiparticle tunnel junction (Dittrich and Heiden, 1988).

4.4 Charge density waves (CDW)

Charge density wave formation is associated with a condensed state of matter, having some similarities with the superconducting state (Wilson *et al.*, 1975; Grüner, 1988). It has been shown theoretically that for a one-dimensional metal the system of conduction electrons coupled to the underlying periodic lattice is not stable at low temperatures (Peierls, 1955). The ground state of the coupled electron–phonon system is characterized by an energy gap in the single-particle excitation spectrum and by a collective mode formed by electron–hole pairs involving the wave vector $2k_F$, where k_F is the Fermi wave vector (Fig. 4.127). The spatial modulation of the concentration of conduction electrons $\rho(r)$ associated with this collective mode can be regarded as a standing wave of charge density

$$\rho(r) = \rho_0 + \rho_1 \cos(2k_F r + \phi) \tag{4.2}$$

where ρ_1 and ϕ denote the amplitude and phase of the charge density wave, respectively. The novel condensed state is therefore denoted as the charge density wave (CDW) state. The charge density modulation is accompanied by a periodic lattice distortion (PLD) since ionic displacements $u(r)$ are necessary to screen the electronic modulation:

$$u(r) = u_1 \cos(2k_F r + \phi) \tag{4.3}$$

For an arbitrary band filling, the spatial period of the charge density modulation and the accompanying PLD is incommensurate with the underlying atomic lattice.

In real materials, CDW formation is favored if Fermi surface nesting occurs, because the electron–hole interaction at $2k_F$ can then drive the Fermi surface instability. This is more likely to be the case for quasi-low-dimensional metals, i.e. layered materials or linear chain-like compounds. The amplitude of the charge density modulation depends on the detailed geometry of the Fermi surface, which itself depends on the type of material. The transition from normal to CDW state can occur for a temperature T_p as high as 600 K, which is far above the highest normal–superconducting transition temperature currently known. Within the CDW state, incommensurate (I) to commensurate (C) transitions can often be observed, sometimes well below the CDW transition temperature.

Since CDW formation is directly connected with a modification of the conduction electron wave functions and the local density of states

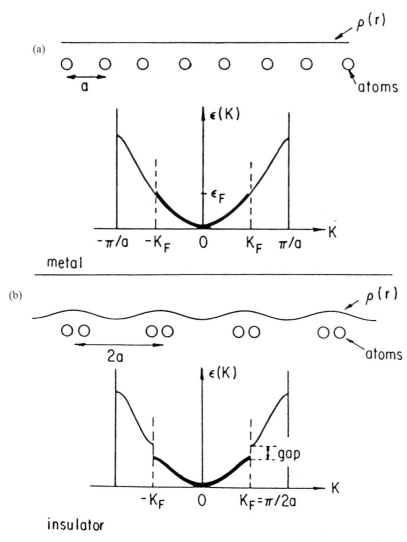

Fig. 4.127. Peierls distortion in a one-dimensional metal with a half-filled band: (a) undistorted metal; (b) Peierls insulator (Grüner, 1988).

at the Fermi level, STM has become a powerful experimental technique for investigation of the local aspects of CDW formation. In contrast, bulk-sensitive methods such as neutron, electron and x-ray diffraction average over a macroscopic sample volume and probe the PLD, rather than the conduction electron density modulation. The applicability of STM to study the CDW state is based on the fact that the charge density waves propagate up to the topmost atom layer of a crystal, which has been proved earlier by means of helium-atom diffraction experiments (Boato *et al.*, 1979; Cantini *et al.*, 1980).

4.4.1 CDW in quasi-two-dimensional materials

STM was first applied to study the superlattices of standing charge density waves on surfaces of quasi-two-dimensional materials, such as transition metal dichalcogenides (TMD), introduced in section 4.1.3.1. The most intensively studied materials have been the 1T phases of TaS_2 and $TaSe_2$ which undergo a CDW transition at a temperature of about 600 K, well above room temperature. While the CDW of 1T-$TaSe_2$ become commensurate at 473 K, the CDW state of 1T-TaS_2 exhibits several intermediate phases before becoming commensurate at 150 K. The first STM observation of CDW on cleaved surfaces of 1T-TaS_2 at 77 K revealed a triangular ($\sqrt{13} \times \sqrt{13}$) superlattice which is generated by three CDW at relative orientations of 120° (Coleman *et al.*, 1985). STM images of 1T-TaS_2 and 1T-$TaSe_2$, which show the CDW superlattice simultaneously with the underlying atomic lattice (Fig. 4.128), allow one to determine the orientational relationship between these two lattices and to check whether they are commensurate or incommensurate.

A large corrugation of 2–4 Å for the CDW superlattice of 1T-TaS_2 and 1T-$TaSe_2$ is typically measured by STM, whereas the corrugation for the underlying atomic lattice usually remains below 1 Å. This anomalous high corrugation of the CDW superlattice may have several origins. Firstly, a large electron condensation into the CDW state can lead to a strong CDW amplitude (Coleman *et al.*, 1988a, 1988b). Secondly, the CDW modulation can additionally be enhanced by several different mechanisms. For instance, anomalous high corrugations can result from electronic enhancement due to a particular structure of the Fermi surface (Tersoff, 1986), or from the elastic response of a soft sample due to the presence of tip–sample interaction forces (Soler *et al.*, 1986).

Besides the anomalous size of the measured STM corrugation for the CDW superlattice on 1T-TaS_2 and 1T-$TaSe_2$, an anomalous distance-dependence has been observed as well. The corrugation of the CDW superlattice was shown to progressively decrease with increasing tunnel current and therefore with decreasing tip-to-sample distance (Coleman *et al.*, 1987). This is in contrast to the distance-dependence of the corrugation for the atomic lattice, where an increase of the corrugation with decreasing tip–surface separation is found (section 1.11.1). A possible explanation for the anomalous distance-dependence might be inferred from a comparative STM and AFM study of 1T-TaS_2 and 1T-

(a)

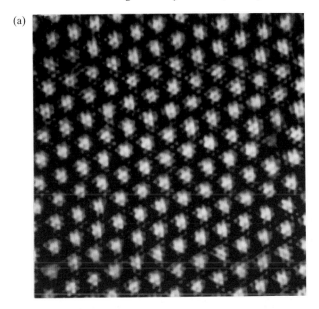

(b)

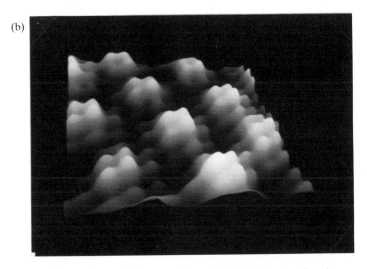

Fig. 4.128. (*a*) STM image (14 nm × 14 nm) of 1T-TaSe$_2$ showing the commensurate ($\sqrt{13} \times \sqrt{13}$) CDW superlattice which is superimposed on the atomic lattice. (*b*) Perspective STM image (33 Å × 39 Å) of the CDW superlattice and the underlying atomic lattice of 1T-TaSe$_2$ (Anselmetti, 1990).

TaSe$_2$ (Meyer *et al.*, 1989a, 1990b; Albrecht, 1989). The increasing tip–sample interaction forces at decreasing tip–surface separation can lead to suppression of the CDW state due to the local applied pressure. An analogous suppression of the CDW state in 1T-TaS$_2$ and 1T-TaSe$_2$ with increasing applied pressure is well known from earlier macroscopic-scale experiments using high-pressure cells. This explanation would then imply that the mean diameter of the spot, over which the pressure is locally applied in a SPM experiment, is at least on the order of the coherence length of the CDW state. Since the coherence length in CDW systems is generally much smaller than in superconductors (McMillan, 1977), this requirement is easily fulfilled.

Large corrugations of up to 4 Å have also been measured in STM studies of the 1T phases of TiS$_2$ and TiSe$_2$ (Slough *et al.*, 1988), where only the latter material exhibits a CDW state below 202 K. The anomalous corrugations have again been explained by electronic structure effects and by additional enhancement due to tip–surface interaction forces because the compressibility perpendicular to the layers is even larger for these Ti-based TMD than for the corresponding Ta-based compounds.

STM has also been applied to study the (3×3) CDW superlattice of the 2H phases of TaS$_2$, TaSe$_2$, and NbSe$_2$ (Fig. 4.129), exhibiting CDW transition temperatures at 75 K, 122 K and 33 K, respectively. For these materials, the measured corrugations of the CDW superlattices were found to be significantly smaller compared with the 1T phases of TaS$_2$ and TaSe$_2$, indicating a smaller charge modulation of the conduction electrons (Slough *et al.*, 1986; Coleman *et al.*, 1987; Giambattista *et al.*, 1988a).

The 4Hb phases of TMD appear to be particularly interesting systems to study by STM because they are composed of alternating sandwiches of octahedral and trigonal prismatic coordination (section 4.1.3.3). Independent CDW formation can be observed in the two different types of sandwiches with CDW wave vectors similar to those of the pure single coordination phases but with slightly reduced transition temperatures. STM studies of the 4Hb phases of TaS$_2$ and TaSe$_2$ indeed revealed two different types of images, depending on which sandwich is located on top of the surface (Slough *et al.*, 1986; Giambattista *et al.*, 1988a, 1988b; Tanaka *et al.*, 1988, 1989a). STM images of the sandwich with octahedral coordination show the strong ($\sqrt{13} \times \sqrt{13}$) CDW superlattice with an anomalous large corrugation, as found for the pure 1T phases. On the other hand, STM images of the sandwich with trigonal prismatic coordination reveal either the absence of a CDW superlattice at temperatures down to 77 K, or an extremely weak (3×3) CDW

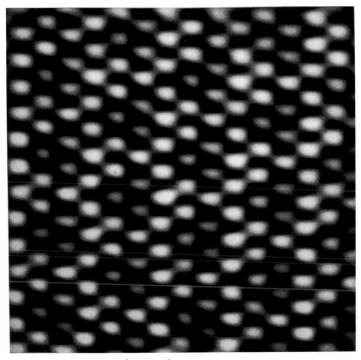

Fig. 4.129. STM image (43 Å × 43 Å) of 2H-NbSe$_2$ obtained at 4.5 K showing the (3×3) CDW superlattice together with the underlying atomic lattice (Pan, Behler and Bernasconi, unpublished work).

superlattice at 4 K which is characteristic for the pure 2H phases. Occasionally, a superposition with the strong ($\sqrt{13} \times \sqrt{13}$) CDW superlattice of the 1T sandwich beneath can be observed, as shown in Fig. 4.130.

4.4.2 CDW defects

STM as a local probe of the conduction electron density is ideally suited to study CDW defects. Point defects in the CDW superlattice appear as single completely quenched charge maxima (Giambattista *et al.*, 1988a, 1990). The perturbation is found to be remarkably localized since the charge-density contour in the vicinity of a CDW point defect shows very little distortion. A STM image revealing an isolated defect as well as a defect complex in the CDW superlattice of 1T-TaSe$_2$, together with the underlying atomic lattice, is presented in Fig. 4.131. It has been proposed that the CDW defects might be caused by atomic vacancies in the subsurface transition metal layer. Unfortunately, even in cases

Fig. 4.130. STM image (70 Å × 70 Å) of the trigonal prismatic layer of 4Hb-TaS$_2$. The atomic lattice is dominating, though a ($\sqrt{13} \times \sqrt{13}$) superlattice is visible originating from the underlying octahedrally coordinated layer. A point defect in the atomic lattice is marked by an arrow (Anselmetti, 1990).

where the underlying atomic lattice can be imaged simultaneously with the CDW superlattice by means of STM, it proves difficult to correlate the defect structures in the CDW superlattice with possible atomic lattice defects, either vacancies or impurities, in the subsurface transition metal layer.

To study the relationship between CDW superlattice defects and atomic lattice impurities in more detail, 1T-TaS$_2$ and 1T-TaSe$_2$ have been substitutionally doped with Ti or Nb, leading to materials of the general form I$_y$Ta$_{1-y}$X$_2$ (I = Ti, Nb; X = S, Se) (Wu et al., 1988a, 1988b; Wu and Lieber, 1989a, 1990a). The local CDW structure was found to become significantly distorted in response to the random lattice potential associated with Ti impurities (Wu et al., 1988a; Wu and Lieber, 1990a). The frequency of localized CDW defects appears to increase linearly with dopant concentration (Fig. 4.132), while several dopant centers seem to be necessary to cause a single CDW defect (Wu and Lieber, 1990a). At low Ti dopant concentration, the localized

Fig. 4.131. STM image (120 Å × 120 Å) of 1T-TaS$_2$ obtained at 77 K. Several defects in the ($\sqrt{13} \times \sqrt{13}$) superlattice are observed (Anselmetti, 1990).

defects consist of a CDW amplitude distortion, while for higher concentrations, the defects appear as a coupled amplitude–phase distortion. CDW twin domains that nucleate at these amplitude–phase defects were directly imaged. Nb doping was found to cause dislocations and small rotations in the CDW superlattice (Dai *et al.*, 1991). The dislocations destroy both translational and orientational order. However, the orientational order appears to decay much more slowly than the translational order, as expected for a hexatic CDW phase.

In the case of Ti doping, the average CDW wavelength was found to increase with increasing dopant concentration (Wu *et al.*, 1988a), which directly proves that the CDW respond on average to the decreased size of the Fermi surface of the Ti-doped samples. (The substitution of Ti for Ta leads to a decrease of the conduction electron density by one electron per Ti atom). In contrast, the CDW wavelength remains the same in the case of Nb doping (Wu and Lieber, 1990a), where substitution of Nb for Ta does not significantly perturb the size of the Fermi surface. Intercalation of lithium into 1T-TaS$_2$, on the other hand, leads

(a)　　　　　　　　　　　　　　(b)

(c)　　　　　　　　　　　　　　(d)

Fig. 4.132. Constant current STM images (220 Å × 220 Å) of (*a*) TaSe$_2$, (*b*) Ti$_{0.02}$Ta$_{0.98}$Se$_2$, (*c*) Ti$_{0.04}$Ta$_{0.96}$Se$_2$ and (*d*) Ti$_{0.07}$Ta$_{0.93}$Se$_2$. The increasing number of CDW amplitude defects from (*a*)–(*d*) is evident (Wu and Lieber, 1990a).

to electron donation to the Ta d band, and therefore to an increase of the Fermi surface. Consequently, a reduction of the CDW wavelength is observed (Wu and Lieber, 1988). As another consequence of increasing dopant concentration, the measured STM corrugation amplitude of the CDW superlattice decreases.

Highly defective CDW superlattices have also been observed in AFM studies of undoped 1T-TaS$_2$ (Barrett *et al.*, 1990). In this case, the perturbation is introduced by the tip–sample interaction which disturbs the CDW state locally. The strength of the perturbation depends on the particular sensor tip as well as on the applied loading force. In contrast to STM, AFM is believed to respond to the PLD, rather than the conduction electron density modulation.

4.4.3 CDW domains

The formation of CDW domains is usually observed in cases where the CDW superlattice is not commensurate with the underlying atomic lattice. For instance, 1T-TaS$_2$ exhibits a nearly commensurate (NC) CDW phase upon cooling from the high-temperature incommensurate (I) phase, which exists above 350 K, to the low-temperature commensurate (C) phase, which exists below 200 K. Upon warming, another nearly commensurate triclinic (T) phase is observed in the temperature range between 220 K and 280 K before entering the NC phase.

Two different models have been proposed for nearly commensurate phases. In one model, the CDW is believed to be uniformly incommensurate. In the other model, a hexagonal domain-like structure of the CDW superlattice is predicted with commensurate CDW within the domains, while at the domain boundaries the CDW phase changes and the CDW amplitude decreases (Nakanishi and Shiba, 1977).

STM studies of the room-temperature NC phase of 1T-TaS$_2$ have indeed revealed the existence of hexagonal CDW domain structures with diffuse domain walls, as shown in Fig. 4.133 (Wu and Lieber, 1989b, 1990b, 1991; Anselmetti, 1990; Burk *et al.*, 1991; Coleman *et al.*, 1992). Within a given domain, the rotation angle of the CDW super-lattice relative to the atomic lattice is close to the commensurate value of 13.9°, independent of temperature. On the other hand, the hexagonal domain structure is rotated by about 6° relative to the CDW superlattice within a single domain. This rotation is opposite to the direction of rotation of the CDW superlattice relative to the atomic lattice. The domain period increases linearly from about 60 Å to about 90 Å between 350 K and 230 K, with a clear discontinuity at the first-order I–NC transition at 350 K. The diffuse domain boundaries are not sharp, but rather comprise more than 50% of the domain dimension, independent of temperature. In these domain boundary regions, the CDW amplitude decreases considerably. The STM observations are generally in excellent agreement with the domain model of Nakanishi and Shiba (1977), as illustrated in Fig. 4.134 (Burk *et al.*, 1991; Coleman *et al.*, 1992).

The effect of doping on the hexagonal domain-like CDW structure in the NC phase of 1T-TaS$_2$ has also been studied by STM. Nb doping was found to lead to a distortion of the hexagonal domains, which can

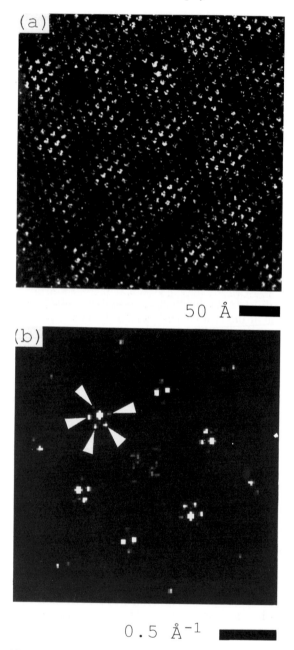

Fig. 4.133. (*a*) STM image of 1T-TaS$_2$ obtained at 295 K in the NC phase. (*b*) Fourier transform of (*a*). Satellite spots near hexagonally arranged CDW peaks are evident; arrows identify five such spots (Burk *et al.*, 1991).

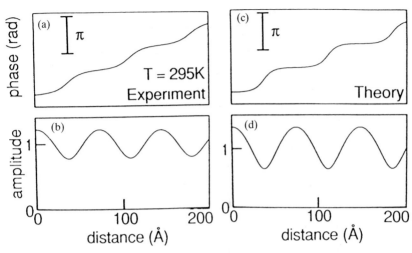

Fig. 4.134. (*a*) Real-space phase variation and (*b*) amplitude modulation of the CDW in 1T-TaS$_2$ in the NC phase reconstructed from the Fourier coefficients and wave vectors extracted from the Fourier transform of Fig. 4.133(*b*). (*c*) Real-space phase variation and (*d*) amplitude modulation of the CDW predicted by the theory of Nakanishi and Shiba (Burk *et al.*, 1991).

be explained by the influence of the random lattice potential introduced by the Nb impurities (Wu and Lieber, 1989a). On the other hand, Ti doping leads to complete destruction of the hexagonal domain structure (Wu *et al.*, 1988a). This might be explained by the fact that the hexagonal domain structure is favored by the electrostatic interaction between the lattice and commensurate CDW domains, and the magnitude of this interaction is proportional to the CDW amplitude. Since Ti doping leads to a uniformly low CDW amplitude, as discussed earlier in section 4.4.2, interactions with the random lattice potential lead to a disordered CDW structure without domains.

STM has also been applied to study CDW domain structures in the nearly commensurate triclinic (T) phase of 1T-TaS$_2$ (Thomson *et al.*, 1988a, 1988b). In contrast to the circular shape of the domains observed in the NC phase (Fig. 4.133), the CDW domains of the T-phase exhibit a long narrow shape. These stripe domains are approximately 60–70 Å across at 230 K and rotated by an angle of about 26° relative to the CDW superlattice.

4.4.4 CDW energy gap

Local tunneling spectroscopy has been applied to study the energy gap of CDW materials locally (Tanaka *et al.*, 1989a; Giambattista *et al.*, 1990; Wang *et al.*, 1990a, 1991). Since the Fermi surfaces of CDW materials are only partially gapped by the CDW state, a substantial density of states remains for applied bias voltages corresponding to energies below the gap value. Measured energy gaps Δ at 4.2 K were found to be on the order of 150 meV for the 1T phases of TaS_2 and $TaSe_2$, yielding values of $2\Delta/k_B T_P \approx 5.8$, whereas the gap values for the 2H phases of TaS_2, $TaSe_2$ and $NbSe_2$ are between 30 and 80 meV, yielding $2\Delta/k_B T_P \approx 15\text{--}25$. These large gap values indicate strong coupling in the above-mentioned CDW systems and require a short-coherence-length model (McMillan, 1977).

A significant variation in the I–U characteristics for different spatial locations has not been detected yet. However, more effort towards spatially resolved tunneling spectroscopy on CDW materials is needed to study a possible spatial variation of the energy gap in CDW systems similar to the case of superconductors (section 4.3.2).

4.4.5 CDW in quasi-one-dimensional materials and CDW dynamics

The CDW state as well as its dynamics can be studied in several different classes of quasi-one-dimensional (1D) materials.

(a) The transition metal trichalcogenides (TMT), such as $NbSe_3$ and TaS_3, are composed of linear chains built up by trigonal prisms stacked along the chain axis. The quasi-one-dimensional electronic band structure results from a strong overlap of the transition metal d orbitals along the chain direction, with no direct d–d overlap perpendicular to the chains. The microscopic linear chain structure of TMT is reflected in the macroscopic shape of the crystals which form thin fibrous ribbons or needles. Cleavage of these crystals is more difficult than for the layered-type compounds. STM studies of $NbSe_3$ and TaS_3 from room temperature down to 4 K clearly reveal the linear chain structure of these materials (Coleman *et al.*, 1988b; Gammie *et al.*, 1988, 1989a, 1989b, 1991; Lyding *et al.*, 1988a; Slough and Coleman, 1989; Slough *et al.*, 1989, 1990). $NbSe_3$ exhibits two CDW transitions at 144 K and 59 K. The CDW superlattices remain incommensurate down to low temperatures. STM images of $NbSe_3$ obtained at 4 K reveal two independent CDW, where the CDW modulation appears to be substantially localized on separate chains for the two different CDW states (Slough *et al.*, 1989, 1990). STM has also been applied to study the CDW state of orthorhombic TaS_3 below the transition temperature at 215 K (Slough and Coleman, 1989; Slough *et al.*, 1990; Gammie *et al.*, 1989b, 1991).

(b) Bronze materials, such as $K_{0.3}MoO_3$ ('blue bronze') and $Rb_{0.3}MoO_3$, are built up from chains of MoO_6 octahedra, separated by chains of alkali metal atoms, leading to a quasi-one-dimensional electronic band structure. STM studies of $K_{0.3}MoO_3$ (Heil *et al.*, 1989; Zettl *et al.*, 1989; Anselmetti *et al.*, 1990c) and $Rb_{0.3}MoO_3$ (Garfunkel *et al.*, 1989) have revealed linear chain-like topographic structures. However, CDW superlattices have not yet been observed at the surface of $K_{0.3}MoO_3$ down to 77 K, well below the CDW transition temperature at 180 K. On the other hand, local tunneling spectroscopy on $K_{0.3}MoO_3$ proved the existence of a CDW energy gap of about 130 K at 77 K (Nomura and Ichimura, 1990).

(c) Quasi-one-dimensional organic conductors, such as TTF-TCNQ, also exhibit CDW transitions. STM studies of TTF-TCNQ have revealed a commensurate CDW modulation at temperatures below 80 K (Pan *et al.*, 1991). In addition, local tunneling spectroscopy has been applied to measure a CDW energy gap of about 100 meV at 10 K.

Besides investigation of the static structure of standing CDW, the dynamic properties of the CDW state are of particular interest. Crystal defects are known to play an important role since, in contrast to the superconducting state, the phase of the CDW condensate can be pinned to the lattice through interaction with impurities, lattice imperfections, grain boundaries and other defect structures. In contrast to the quasi-two-dimensional materials, the pinning energy per electron in the quasi-one-dimensional materials is small enough that the dynamics of the collective CDW mode can be observed in response to finite amplitude d.c. or a.c. excitations. STM seems ideally suited to study the microscopic nature of CDW pinning in the low external field limit and the local aspects of the CDW dynamics for external fields above the threshold value for CDW depinning.

Sliding CDW can be detected with the STM by holding the tip at a fixed position (Fig. 4.135(*a*)), while analyzing the frequency spectrum of the tunnel current (Fig. 4.135(*b*)), as was first demonstrated for $K_{0.3}MoO_3$ (Nomura and Ichimura, 1989, 1990). If a sample bias current exceeding the threshold value for CDW depinning is applied, a charac-

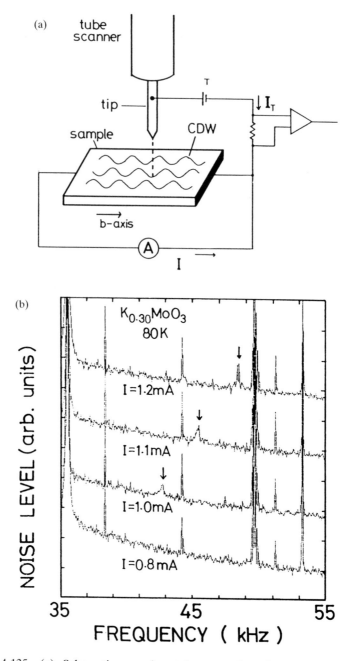

Fig. 4.135. (*a*) Schematic experimental set-up for observation of CDW dynamics. With electric leads attached to both sides of the sample, transport measurements can be performed while monitoring the tunnel current flow between tip and sample. (*b*) Spectra of the tunnel current under different bias currents. The position of the tip was kept fixed while analyzing the tunnel current. For a bias current exceeding a threshold value of 0.95 mA, a new peak (indicated by arrows) appears in the tunnel current spectra (Nomura and Ichimura, 1990).

teristic peak appears in the frequency spectrum of the tunnel current at a frequency f given by

$$f = v_{\mathrm{s}} \cdot q_{\mathrm{CDW}} / 2\pi \tag{4.4}$$

where q_{CDW} and v_{S} denote the CDW wave vector and CDW sliding velocity, respectively. The time-dependence of the CDW phase

$$\phi(t) = -q_{\mathrm{CDW}} \, v_{\mathrm{s}} t + \phi_0 \tag{4.5}$$

leads to a periodic modulation of the tunnel gap due to the sliding motion of the CDW, which explains the modulation in the tunnel current. It was found that the CDW are easily depinned at the surface of $K_{0.3}MoO_3$ and begin to slide with much higher velocity ($v_{\mathrm{S}} \approx 57~\mu\mathrm{m~s}^{-1}$) than in the bulk ($v_{\mathrm{S}} \approx 6.3~\mu\mathrm{m~s}^{-1}$) just above the threshold bias current of 0.95 mA. Combined STM/STS studies of the atomic-scale crystal surface structure and the CDW state will further help to improve our microscopic understanding of CDW transport phenomena in quasi-one-dimensional materials.

5

Chemistry

STM and related scanning probe microscopies (SPM) have become important experimental techniques in chemistry, particularly solid state chemistry, as well as in solid state physics. There exist two major fields where SPM has already made significant contributions.

1. Chemical reactions at the solid–vacuum interface, i.e. surface reactions, which are in the focus of surface science research where the border between surface chemistry and surface physics (section 4.1) has become blurred.
2. Chemical reactions at the solid–liquid interface, particularly those initiated by electrochemical processes.

In the following, we will concentrate on some of the achievements of STM and related SPM techniques towards an atomic-level understanding of chemical reaction processes occurring at the solid–vacuum interface (section 5.1) and the solid–liquid interface (section 5.2).

5.1 Surface reactions

STM, or generally SPM, can contribute in various ways to a detailed investigation of surface chemical reactions.

1. Characterization of the atomic and electronic structure of the as-prepared surface before the initiation of chemical reactions is important to identify the variety of inequivalent surface sites and their structural as well as electronic characteristics.
2. In the initial stage of the reaction process, site-specific modifications of the substrate surface and chemisorbed species have to be characterized. In particular, a correlation between the local reactivity and the atomic and electronic properties of individual surface sites has to be established. As the chemical reaction proceeds, it is also important to study how the reaction at one particular surface site affects the local electronic structure of neighboring sites with possible

468

influences on their local reactivity. This way, detailed information about the mechanism of the surface chemical reaction, from the initial nucleation to the growth of reacted surface regions, can be obtained.

3. Characterization of the final state of the surface after the chemical reaction may additionally allow conclusions to be drawn about the reaction processes which have taken place. In particular, if the reaction processes occur on an extremely short time scale, this is the only way that STM and SPM can provide information about them.

There are several possibilities for the final state of a surface after the chemical reaction.

1. Surface reactions of or between adsorbed species may have taken place without altering the substrate surface. In this case, it is particularly interesting to study the conformation and the local electronic properties of the individual reaction products on top of the surface.
2. Surface reactions can sometimes lead to a structural transformation of the whole surface, for instance, via adsorbate-induced reconstruction. In this case, a detailed characterization of the atomic and electronic structure of the new surface phase is of interest.
3. Surface reactions may eventually cause severe chemical modification of the substrate, e.g. by oxidation. The new chemical state of the surface might have drastically altered physical properties (e.g. electrical conductivity), which has to be taken into account for choice of an appropriate SPM technique to be applied to study the modified surface.

If the surface remains sufficiently conducting as the chemical reaction proceeds, STM combined with local tunneling spectroscopy has been found to be most powerful for atomic-level characterization of surface reactions because this technique directly probes the valence electronic structure of individual surface sites and allows it to be correlated with the local reactivity. On the other hand, STM with its lack of chemical selectivity (section 1.13) generally fails to identify the reaction products. Therefore, a combination of STM with other surface science techniques, which allow surface chemical analysis, is often required to fully understand the reaction process.

To characterize surface chemical reactions at the atomic level, it is preferable to start with well-defined model systems, such as single-crystal surfaces prepared under ultra-high vacuum (UHV) conditions, before attacking the more serious problem of understanding industrial reaction processes involving materials with surfaces of much greater complexity (Langmuir, 1922; Ertl, 1990).

5.1.1 *Chemical reactions at semiconductor surfaces*

For the discussion of chemical reactions at semiconductor surfaces, we
will mainly focus on a very well defined and characterized model substrate, the Si(111) 7×7 surface (section 4.1.1.1). Clean semiconductor
surfaces usually exhibit surface dangling bonds which provide chemically active sites. In the case of Si(111) 7×7, these surface dangling bonds
introduce electronic states in the bulk band gap region, as illustrated in
the upper part of Fig. 5.1 (Avouris and Lyo, 1990). By tunneling in or

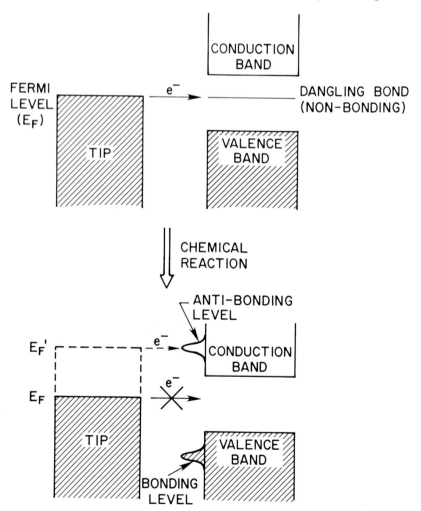

Fig. 5.1. Schematic energy level diagram indicating how STM imaging of surface
dangling-bond sites and of reaction products can be used to study the spatial
distribution of a surface reaction with atomic resolution (Avouris and Lyo,
1990).

out of these surface dangling bond states, the spatial distribution of the chemically active sites can directly be obtained with atomic resolution. As a result of chemical reaction processes at particular surface sites, the corresponding dangling bond states are eliminated (Fig. 5.1, lower part), leading to a drastic reduction of the local density of states (LDOS). Consequently, the reacted surface sites will appear 'dark' in STM images, and the spatial distribution of these 'dark' sites will directly yield the spatial distribution of the reacted surface sites with atomic resolution.

As another consequence of surface chemical reactions, new low-lying occupied or unoccupied electronic states might be introduced, leading to a corresponding increase in the surface LDOS at particular energies. Tunneling in or out of these states will let the reacted sites appear brighter than unreacted sites. In favorable cases, different reaction products introduce states at different energies, so that, by adjusting the sample bias voltage in the STM experiment appropriately, spatial maps of the various reaction products may be obtained. However, a definite identification of particular reaction products can generally be achieved only by combination of STM with additional element-specific surface analytical techniques, and/or with theoretical electronic structure calculations.

As a first example, we will discuss the chemical reaction of the Si(111) 7×7 surface with NH_3 which is used to synthesize silicon nitride, an important electronic material (Wolkow and Avouris, 1988b; Avouris and Wolkow, 1989a; Avouris and Lyo, 1990). NH_3 is known to dissociate on the Si(111) 7×7 surface, producing Si–H and Si–NH_2 groups. STM 'topographs' of a partially reacted Si(111) 7×7–NH_3 surface are presented in Fig. 5.2. About half the number of adatoms appear dark in the 'topographic' image obtained at a sample bias of +0.8 V (Fig. 5.2(a)). These dark adatom sites result from the saturation of their dangling bonds by the chemical reaction process, leading to a decrease of the LDOS near E_F. By increasing the applied sample bias voltage to +3 V, both unreacted and reacted adatom sites become visible, indicating that the (7×7) reconstruction is preserved overall by the chemical reaction (Fig. 5.2(b)). Furthermore, it was found that the products of the surface reaction can directly be imaged at a sample bias voltage of −3 V, probing the occupied electronic states. Two distinct reaction products of different apparent 'size' can be observed, which were tentatively attributed to Si–H and Si–NH_2 groups. The Si–NH_2 group is expected to correspond to the larger reaction product species as seen in 'topographic' STM images because Si–NH_2 contributes more than the Si–H group to the LDOS at −3 eV, and because Si–NH_2 extends further out from the surface plane.

(a)

(b)

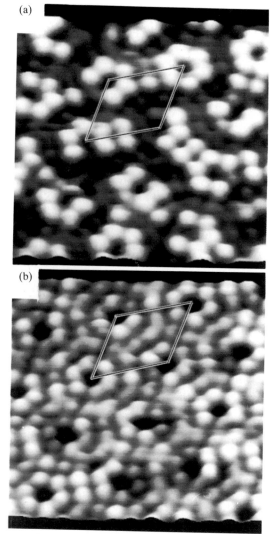

Fig. 5.2. (*a*) STM topograph of the unoccupied states of a Si(111) 7×7 surface partially reacted with NH₃. Sample bias was +0.8 V. Reacted adatoms appear dark. (*b*) STM topograph of the unoccupied states of a partially reacted surface. Sample bias was +3 V. Both reacted and unreacted adatom sites are visible. The (7×7) reconstruction is preserved overall by the reaction (Avouris and Wolkow, 1989a).

A statistical analysis of the STM images obtained at the initial stage of the chemical reaction of the Si(111) 7×7 surface with NH_3 clearly indicates that the majority of unreacted sites are 'corner adatoms' (section 4.1.1.1), while the 'center adatoms' exhibit a significantly higher degree of local reactivity. This can be seen easily from an inspection of Fig. 5.2(*a*), whereas such information would be difficult, if not impossible, to gain from conventional surface analytical techniques, averaging over macroscopic surface areas. Two obvious differences between corner and center adatoms might be responsible for the different local reactivity (section 4.1.1.1).

1. Center adatoms have two rest atom neighbors, while corner adatoms have only one.
2. Corner adatoms are bonded to two silicon dimers while center adatoms are bonded to only one dimer.

Detailed information about the relationship between local surface electronic structure and local reactivity can be obtained by spatially resolved tunneling spectroscopy. By comparing the local $(dI/dU)/(I/U)$–U characteristics above rest atom, corner adatom and center adatom sites before and after chemical reaction of the Si(111) 7×7 surface with NH_3 (Fig. 5.3), the influence of the reaction process on the local electronic structure can directly be studied. It is found that the peak at -0.8 eV, corresponding to a rest atom state (section 4.1.1.1), disappears upon reaction with NH_3. Consequently, the presence or absence of the peak at -0.8 eV can be used as a signature for unreacted and reacted rest atom sites, respectively. This is important because the rest atom sites do not show up as prominent features in 'topographic' STM images. Based on a statistical analysis of combined STM/tunneling spectroscopy data, it was found that the rest atoms are more reactive than adatoms. Interestingly, the reaction at rest atom sites appears to have an influence on the local electronic structure of neighboring unreacted sites. In particular, the tunneling spectra of unreacted corner and center adatoms next to a reacted rest atom site become virtually indistinguishable, as seen in Fig. 5.3. This change of the local electronic structure at unreacted sites, caused by reaction of nearby sites, can have a significant influence on how the surface reaction proceeds. Again, this information is not provided by surface analytical techniques averaging over large surface areas.

As another example, the chemical reaction of the Si(111) 7×7 surface with H_2O has been studied (Avouris and Lyo, 1990). H_2O is known to dissociate on the Si(111) 7×7 surface at 300 K into OH and H, which become bound to silicon surface dangling bonds. STM studies have

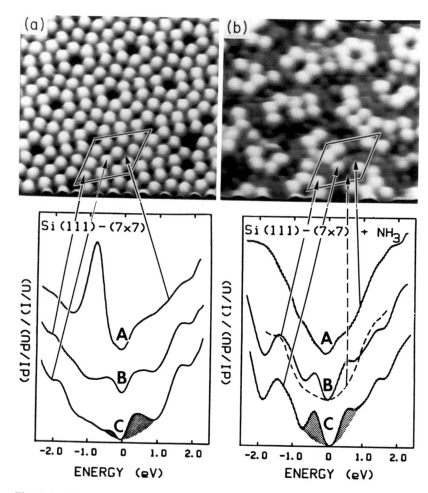

Fig. 5.3. (*a*) STM topograph of the unoccupied states of the clean 7×7 surface (top) and atom-resolved tunneling spectra (below). Curve A gives the spectrum over a rest atom site, curve B gives the spectrum over a corner adatom site, and curve C gives the spectrum over a center adatom site. Negative energies indicate occupied states, while positive energies indicate empty states. (*b*) STM topograph of the unoccupied states (top) and atom-resolved tunneling spectra (below) of an NH₃-exposed surface. Curve A gives the spectrum over a reacted rest atom site, curve B (dashed line) gives the spectrum over a reacted corner adatom, while curves B (solid line) and C give the spectra over unreacted corner and center adatoms, respectively (Avouris and Wolkow, 1989a).

shown that the center adatom sites are more reactive than the corner adatom sites, as for the Si(111) 7×7–NH$_3$ system, with a local reactivity ratio of about 2:1. For the chemical reactions of the Si(111) 7×7 surface with phosphine (PH$_3$) and disilane (Si$_2$H$_6$), it was found that the rest atoms exhibit a higher degree of local reactivity than the adatoms, which again agrees with STM observations made for the Si(111) 7×7–NH$_3$ system (Avouris and Bozso, 1990).

On the other hand, for reactions of the Si(111) 7×7 surface with open shell systems, such as NO or O$_2$ (section 4.1.1.9), adatom sites appear to be more reactive than rest atom sites. This is also true for reaction of the Si(111) 7×7 surface with Cl (Villarrubia and Boland, 1989; Boland and Villarrubia, 1990). In particular, after saturation coverage with Cl, followed by an anneal, the adatoms of the (7×7) structure are stripped away from large areas of the surface and become accumulated in pyramidal silicon structures on top of the surface, which involves extensive mass transport. In the areas where the adatoms have been completely removed, the silicon rest atom layer can directly be studied by STM (Fig. 5.4). A stacking fault, which is characteristic for the

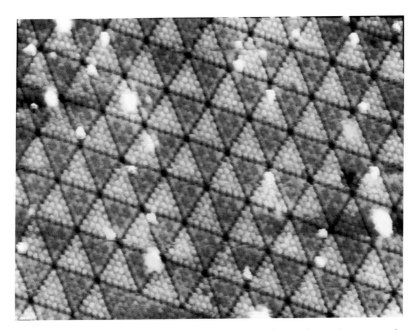

Fig. 5.4. STM topograph of the Si(111) 7×7 surface taken after saturation exposure with Cl and an anneal cycle. The large protrusions near the corner holes are the only remaining adatoms (Villarrubia and Boland, 1989).

Si(111) 7×7 DAS structure (section 4.1.1.1), causes one half of the (7×7) unit cell to appear 0.2 Å higher than the unfaulted half. This height difference, as measured on the rest atom layer, has been attributed to geometric rather than electronic effects.

As an example of a chemical reaction leading to a surface structural transformation, we now focus on the B on Si(111) system. By exposing a Si(111) 7×7 surface to decaborane ($B_{10}H_{14}$) at room temperature, followed by an anneal to temperatures above that of hydrogen desorption (>500 °C), a surface transformation into a ($\sqrt{3} \times \sqrt{3}$)R30° structure can be observed (Lyo *et al.*, 1989; Avouris *et al.*, 1990; Avouris and Lyo, 1990). It was found that, for low annealing temperatures (<<1000 °C), B adsorbs as an adatom on a 'T$_4$-site' (Fig. 5.5(*b*)) of the ($\sqrt{3} \times \sqrt{3}$) surface structure. These B adatoms appear darker in STM images than Si adatoms in T$_4$-sites (Fig. 5.5(*a*)) because, according to electronic structure calculations, B adatoms in T$_4$-sites are about 0.4 Å lower than Si adatoms, and the B wave function extends less far out into the vacuum region. However, in contrast to other group III elements, such as Al, Ga and In (section 4.1.1.11), the T$_4$ adatom site is not the most stable one for B (Bedrossian *et al.*, 1989b, 1990; Lyo *et al.*, 1989). Upon annealing to a temperature of about 1000 °C, B occupies a subsurface substitutional 'S$_5$-site' (Fig. 5.5(*c*)) directly below a Si

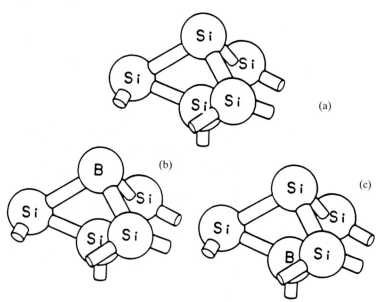

Fig. 5.5. The local structures of (*a*) the Si T$_4$-site, (*b*) the B T$_4$-site and (*c*) the B S$_5$-site (Avouris and Lyo, 1990).

adatom. Charge transfer from this Si adatom to the nearby substitutional B atom, having higher electronegativity, leads to an emptying of the Si adatom dangling-bond state. Consequently, the Si adatoms with B underneath appear darker in STM images of the ($\sqrt{3} \times \sqrt{3}$) surface structure than Si adatoms without a subsurface B atom in an S_5-site (Fig. 5.6). As another consequence of charge transfer between Si and B, the Si top layer of the Si(111)–B($\sqrt{3} \times \sqrt{3}$) surface is insulating, in contrast to the 'metallic' Si(111) 7×7 surface. This in turn leads to completely different chemical properties of the Si(111)–B($\sqrt{3} \times \sqrt{3}$) surface compared with Si(111) 7×7. For instance, NH_3 adsorbs molecularly and reversibly on the ($\sqrt{3} \times \sqrt{3}$) surface, whereas it dissociates on the (7×7) surface, as discussed earlier.

Surface reaction studies have also been performed on Si(100) 2×1 substrates. As for the Si(111) 7×7 surface, NH_3 adsorbs dissociatively on Si(100) at 300 K, while preserving the (2×1) symmetry. STM combined with local tunneling spectroscopy revealed changes in the spatial distribution of electronic states upon reaction, allowing one to distinguish reacted and unreacted Si(100) dimers (Hamers *et al.*, 1987; Avouris *et al.*, 1987). The STM results were found to be consistent with formation of a Si(100)–H(2×1) monohydride surface, leaving the Si

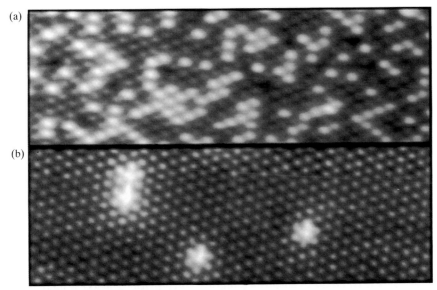

(a)

(b)

Fig. 5.6. (*a*) STM topograph of the surface produced by exposing Si(111) 7×7 to 0.4 L of decaborane and annealing to 1000 °C. (*b*) The surface which results after exposure to 1 L of decaborane and annealing to 1000 °C (Avouris and Lyo, 1990).

dimer bonds intact. The hydrogen saturates the Si adatom dangling bonds, resulting in a very low chemical reactivity of the Si(100)–H(2×1) surface which causes the reaction of Si(100) with NH_3 at 300 K to be self-limiting at the monohydride stage. No evidence for dihydride formation, which would lead to a (1×1) local symmetry was found by STM. The nitrogen atoms reside mostly in subsurface sites, as verified by using additional surface analytical techniques.

Besides prototypical gas-phase reaction studies on clean, UHV-prepared silicon surfaces as discussed so far, STM has also been applied to investigate wet chemical etching processes of oxide-covered silicon surfaces which constitute an integral part of a variety of Si-wafer cleaning procedures (Memmert and Behm, 1991). Several different NH_4F/HF solutions of varying acidity have been tried. It was found that etching Si(111) samples for several minutes in 40% NH_4F solution can be used as an *ex situ* preparation method for obtaining atomically flat Si(111)–H(1×1) surfaces, as shown in Fig. 5.7. The density of steps observed on this surface is determined by the misorientation of the silicon wafer. In contrast, etching in 50% HF solution leads to removal of the surface oxide without affecting the silicon substrate. As a consequence, the resulting surface reflects the structure of the previous Si–SiO_2 interface and exhibits its characteristic roughness.

5.1.2 Chemical reactions at metal surfaces

The adsorption of gases on clean metal substrates often leads to structural transformation of the surface. The question about the nucleation and growth of the new surface phase is of primary interest and can be addressed at a microscopic level by using STM (Behm, 1990). Two fundamentally different structural transformations can be distinguished.

1. Adsorbate-induced surface reconstruction as observed, for instance, for O on Cu(110) (section 4.1.2.6).
2. Adsorbate-induced removal of a surface reconstruction. In the following, we will concentrate on the second type of structural transformation, which is found, for instance, for the CO on Pt(110) system (Gritsch *et al.*, 1989a, 1989b).

A clean Pt(110) surface exhibits a (1×2) 'missing row'-type reconstruction (section 4.1.2.2). Adsorption of CO leads to transition into a non-reconstructed (1×1) surface phase, which involves a 100% increase in surface atomic density. Since this transformation is known to occur at temperatures as low as 250 K, where surface self-diffusion is almost completely inhibited, the mechanism of transport of surface metal

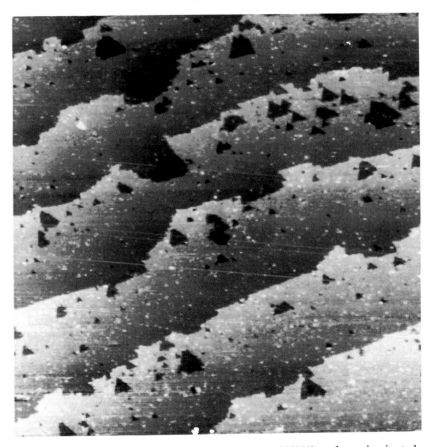

Fig. 5.7. STM topograph (500 nm × 500 nm) of a Si(111) wafer, misoriented by 0.2° in the [11$\bar{2}$] direction, after etching in 40% NH₄F (Memmert and Behm, 1991).

atoms during the (1×2) → (1×1) surface transformation has long been an unsolved issue. STM studies of the initial stage of the CO-induced structural transformation of the Pt(110) surface at 300 K have revealed formation of small square surface features (Fig. 5.8(a), left-hand side), which can be identified as nuclei of the new (1×1) phase. A structural model for the observed squares is presented in Fig. 5.8(a) (right-hand side). Formation of these squares involves migration of Pt atoms over very short distances only, being triggered by destabilization of the close-packed rows in the (1×2) phase by adsorption of CO which remains 'invisible' in the STM images. Further CO exposure leads to an increase of the density of the square features, rather than to growth of the already existing ones. This is explained by the limited mobility of Pt

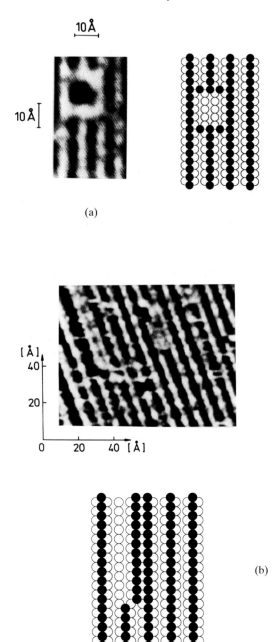

Fig. 5.8. (*a*) High-magnification STM image of a 'hole' formed by CO-induced structural transformation of the Pt(110) (1×2) surface and corresponding ball model. (*b*) STM image recorded at 350 K showing partial 1×2 → 1×1 transformation by parallel displacement of pieces of [1$\bar{1}$0] rows and corresponding ball model (Gritsch *et al.*, 1989a).

atoms at room temperature, being restricted to a few lattice sites. In contrast, at elevated temperatures ($\gtrsim 350$ K), correlated jumps of Pt atoms can cause a lateral displacement of relatively long strings of the (1×2) phase, leading to formation of larger, strongly anisotropic (1×1) nuclei (Fig. 5.8(*b*)).

Microscopic STM investigations of the $(1\times2) \to (1\times1)$ transformation of the Pt(110) surface induced by the adsorption of CO have clearly shown that this transition is not of the order–disorder type, as might be judged on the basis of integral surface analysis studies. The formation of the square (1×1) nuclei at 300 K rather preserves atomic-scale local order (Fig. 5.8(*a*)), while the random spatial distribution of these nuclei leads to the appearance of a disordered surface on a larger scale.

STM has also been applied to study the CO-induced lifting of the hexagonal reconstruction of the Pt(100) surface, where the nucleation of small (1×1) domains has directly been observed as well (Hösler *et al.*, 1986; Ritter *et al.*, 1987). However, in contrast to the CO on Pt(110) system, the growth rate of the (1×1) domains was found to be much faster than the nucleation rate.

Despite remarkable progress in atomic-level characterization of surface chemical reactions at surfaces of well-defined model systems, which has been achieved by application of local probe methods, it is fair to say that we still remain a long way from understanding chemical reaction processes occurring at surfaces of 'real-world' samples, including industrial catalysts (Schlögl *et al.*, 1987; Walz *et al.*, 1988; Komiyama *et al.*, 1988, 1990; Yeung and Wolf, 1991).

5.2 Electrochemistry

So far, we have mainly concentrated on the applications of scanning probe microscopies to investigation of the solid–vacuum or solid–gas interface (sections 4.1 and 5.1). However, many important processes occur at the solid–liquid interface, such as electrochemical reactions, corrosion or lubrication. In this section, we will concentrate on a key issue in modern electrochemistry, namely, the microscopic structure of the solid–electrolyte interface and its relation to other properties of the electrochemical system, which have traditionally been studied by voltammetric techniques and electrochemical impedance analysis.

Structural investigations of electrochemical electrodes can be performed in two different ways.

1. By using *ex situ* methods which are well established in surface science, such as electron diffraction or photoemission spectroscopy, the surface structure of the electrode after controlled emersion from the

electrochemical cell and subsequent transfer into the UHV surface analysis chamber can be studied in detail (Kolb, 1986). A significant limitation of this experimental procedure, however, lies in the uncontrollable possible change of the electrode upon emersion from the electrochemical cell.

2. By using *in situ* methods, the structure of the solid–electrolyte interface can directly be probed. However, most *in situ* spectroscopical and diffraction techniques applicable to the solid–liquid interface, such as classical reflectance spectroscopy in the visible and near ultraviolet (UV) regime, infrared (IR) and x-ray spectroscopy, Raman spectroscopy, non linear optical spectroscopy (e.g. second-harmonic generation), and x-ray diffraction, all provide information averaged over macroscopic regions of the solid–electrolyte interface. In contrast, scanning probe methods allow *in situ* studies of the solid–electrolyte interface with nanometer to atomic resolution. *In situ* characterization of the solid–electrolyte interface at high spatial resolution is of interest because structural and chemical inhomogeneities of this interface can have a significant influence on the local mechanism of electrochemical reactions, for instance, through establishment of non-homogeneous transport regions of electrochemically active species in the electrolyte (Siegenthaler and Christoph, 1990; Cataldi *et al.*, 1990; Siegenthaler, 1992).

Three different types of scanning probe microscopes have already found applications for investigation of electrochemical processes at a microscopic level: the scanning electrochemical microscope (SECM), the scanning tunneling microscope (STM) and the atomic force microscope (AFM).

5.2.0.1 Scanning electrochemical microscopy (SECM)

As already discussed in section 3.5, the scanning electrochemical microscope (SECM) probes the Faradaic current that flows through a small electrode probe in close proximity (1 nm or more) to an electrically conducting or semiconducting substrate, both being immersed in an electrolytic solution. Since the Faradaic current is only weakly dependent on the probe-to-substrate spacing, the SECM achieves a spatial resolution not better than about 10 nm. However, this already constitutes an improvement by at least two orders of magnitude compared with other conventional *in situ* experimental methods.

5.2.0.2 Scanning tunneling microscopy (STM)

The application of STM in electrochemistry should be based on measurement of a tunnel current flowing between a sharp probe tip and a plane electrode at a distance of typically 1 nm or less, both being immersed in an electrolytic solution. Since the tunnel current is exponentially dependent on the tip-to-substrate spacing, spatial resolution down to the atomic scale can be expected. However, the total current measured with the STM in an electrochemical environment is given by the sum of the tunnel current and the Faradaic current, the latter results from electrochemical processes both at the substrate–electrolyte and the exposed tip–electrolyte interface. To decrease the unwanted contribution from the Faradaic current to a small percentage, the exposed tip–electrolyte interfacial area has to be reduced. This can be achieved by coating the tip with an electrically insulating material (Fig. 5.9), leaving only 0.01–10 μm of the tip end exposed. Several different coating materials have already been tried: glass, epoxy resin, Apiezon wax, polymers, or even nail polish (Heben *et al.*, 1988, 1989; Christoph *et al.*, 1989; Nagahara *et al.*, 1989). Upon appropriate insulation of the tip shaft, the residual Faradaic current can usually be kept below 100 pA, which is less than 10% of the typically demanded current of 1 nA for an STM experiment. The use of glass-coated tips allowed the first atomic-resolution STM imaging under aqueous solution in 1986, which already indicated the potential of the STM technique for *in situ* studies of electrochemical processes down to the atomic scale (Sonnenfeld and Hansma, 1986; Sonnenfeld and Schardt, 1986; Schneir *et al.*, 1986, 1988; Sonnenfeld *et al.*, 1987, 1990; Drake *et al.*, 1987). It was also realized

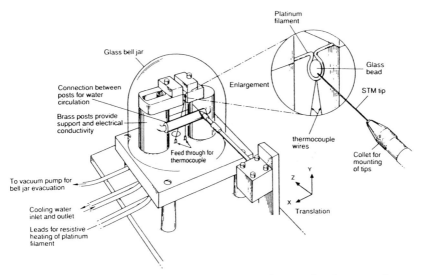

Fig. 5.9. Schematic diagram of an apparatus used to apply glass or polymer layers to etched Pt–Ir wires (Heben *et al.*, 1988).

at this time that keeping a sample under a clean liquid provides an alternative way to perform atomic-resolution surface studies by STM, besides investigations under UHV conditions (Giambattista *et al.*, 1987).

To fully explore the potential of STM for *in situ* studies of electro-chemical processes, it was soon recognized that it is necessary to control the electrochemical reactions at the substrate and the exposed part of the tip by introducing a potentiostatic STM concept (Morita *et al.*, 1987, 1989; Lev *et al.*, 1988; Lustenberger *et al.*, 1988; Wiechers *et al.*, 1988; Itaya and Tomita, 1988; Itaya *et al.*, 1988; Otsuka and Iwasaki, 1988; Green *et al.*, 1988, 1989; Uosaki & Kita, 1989). This concept involves provision for independent adjustment of the potentials E_S and E_T of substrate and tip relative to a reference electrode (RE), thereby defin-ing a tunnel bias voltage of $U = E_S - E_T$. Within a conventional potent-iostatic three-electrode circuitry, the reference electrode carries the total electrolytic current, this being the sum of the substrate and tip currents, whereas a four-electrode assembly offers a complete potent-iostatic circuitry with potentiostat (P), currentless reference electrode (RE), and current-carrying counter electrode (CE), as illustrated in Fig. 5.10 (Siegenthaler, 1992). In this way, the STM becomes part of an electrochemical cell.

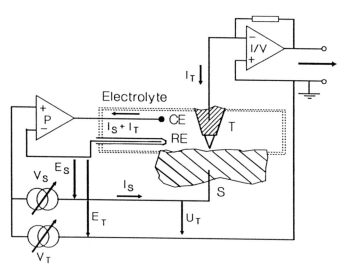

Fig. 5.10. Schematic presentation of the bipotentiostatic principle for independ-ent control of E_S and E_T, the potential difference of substrate and tip, respect-ively, versus a reference electrode RE, by means of a potentiostat P and voltage sources V_S and V_T. The tunneling voltage U is defined by the difference of E_S and E_T (Siegenthaler, 1992).

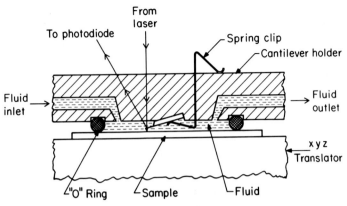

Fig. 5.11. Cross-section of an AFM fluid cell used for electrochemistry (Manne *et al.*, 1991a).

5.2.0.3 Atomic force microscopy (AFM)

More recently, AFM has also been applied for *in situ* investigations of electrochemical processes (Manne *et al.*, 1991a, 1991b; Gewirth, 1992; Chen and Gewirth, 1992). For this purpose, the AFM is incorporated in a fluid cell, as shown in Fig. 5.11. In contrast to the STM, no additional electrode is introduced by an AFM, and therefore voltammetry in a fluid cell equipped with an AFM can be performed in a manner identical to that in dedicated electrochemical cells. On the other hand, AFM offers atomic resolution on metal electrodes under aqueous solution, as does STM (Manne *et al.*, 1990).

In the following, we will discuss three representative examples of microscopical studies of electrochemical processes by means of *in situ* scanning probe microscopy.

5.2.1 Electrochemical oxidation and reduction of surfaces

In situ characterization of the electrochemical formation and reduction of surface oxide layers at high spatial resolution is one of the important applications of scanning probe methods in the field of electrochemistry. The onset of surface oxidation or reduction in an electrochemical cell can be monitored by cyclic voltammetry, which involves measurement of the electrochemical current as a function of the electrochemical potential of the substrate. This is illustrated in Fig. 5.12(*a*) for the case of a Au(111) sample immersed in a 0.1 M $HClO_4$ solution. The two peaks at $+1.25$ V and $+1.45$ V have been attributed to formation of a monolayer of gold hydroxide and oxide, respectively, whereas the nega-

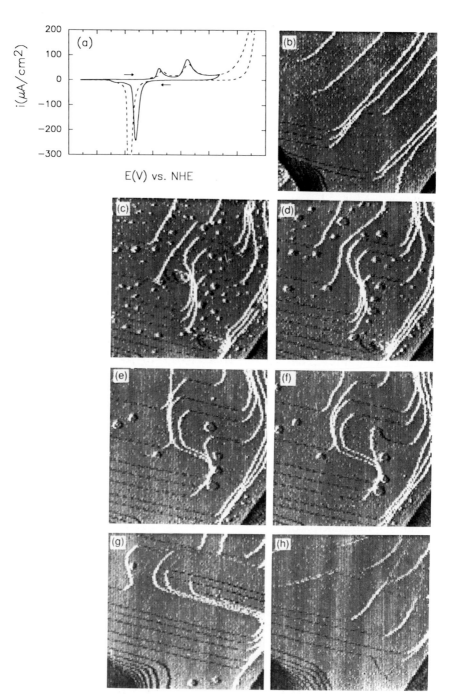

Fig. 5.12. (*a*) Current versus potential plots for scans from +0.7 V to +1.7 V (solid) or +2.0 V (dashed) and back to +0.7 V for a Au(111) film on mica in 0.1 M HClO$_4$. (*b*) STM image (200 nm × 200 nm) of the Au(111) film on mica at an electrochemical potential of +0.7 V in 0.1 M HClO$_4$ before cycling the electrochemical potential; (*c*) after cycling to +1.9 V and back to +0.7 V. (*d*) 5 min later; (*e*) 13 min later; (*f*) 17 min later; (*g*) 38 min after cycling to +2.0 V; (*h*) 15 min after adding dilute HCl to make the electrolyte 5×10^{-5} M HCl, 0.1 M HClO$_4$ (Trevor *et al.*, 1989).

tive peak near +1.05 V corresponds to re-reduction of the surface to elemental gold.

Upon cycling of the gold sample potential to +1.9 V and back to +0.7 V, which corresponds to a complete electrochemical surface oxidation–reduction sequence, a significant change in the surface morphology has been revealed by *in situ* STM observations (Trevor *et al.*, 1989). Numerous pits one monolayer deep as well as new terraces have been created (Fig. 5.12(*c*)), which were not present on the Au(111) surface before potential cycling (Fig. 5.12(*b*)). The surface roughening is believed to be the consequence of place-exchange reactions of gold and oxygen, which must occur whenever more than a monolayer of oxide is formed or reduced. However, the Au(111) surface anneals within minutes at moderate potentials due to surface diffusion, as can be seen in Fig. 5.12(*c*)–(*f*). The annealing process involves pit fusion, step edge motion, and island shrinking which are driven by diffusion of step adatoms along the step edges, rather than diffusion of adatoms across terraces (Trevor and Chidsey, 1991). Interestingly, the mobility of gold surface atoms can greatly be enhanced by adding a chloride-containing solution, such as HCl, thereby accelerating the surface annealing process (Wiechers *et al.*, 1988; Trevor *et al.*, 1989).

STM has also been applied to study electrochemical oxide formation and reduction on various other metal surfaces, including Au(100) (Nichols *et al.*, 1990), Ag (Sakamaki *et al.*, 1990b; Chen *et al.*, 1991), and Pt (Uosaki and Kita, 1990; Itaya *et al.*, 1990; Sashikata *et al.*, 1991).

5.2.2 Potential-dependent reconstruction at electrochemical surfaces

Another application of *in situ* STM in electrochemistry is given by potential-dependent studies of surface reconstructions. By using a Au(100) single crystal immersed in a 0.1 M $HClO_4$ solution, it has directly been shown that a reversible transition between a non-reconstructed (1×1) and a reconstructed '(1×5)' surface can be induced by appropriately adjusting the electrochemical sample potential (Gao *et al.*, 1991a). This transition occurs over a time period of about 10 min for the Au(100) surface. Since the reconstructed surface exhibits a 20% increase in atomic density compared with the unreconstructed surface, the question of the source of the additional gold atoms arises. The STM observations indicate that the reconstruction preferentially starts near step edges, suggesting that the additional gold atoms are supplied from the step sites. It is believed that the surface reconstruction is driven by potential-induced changes in the surface electrical state, specifically the accumulation of excess electronic charge density at the metal surface.

Similar STM observations have been made for the Au(110) surface immersed in 0.1 M HClO₄ solution, where a potential-induced reversible transition between a non-reconstructed (1×1) and a reconstructed surface, exhibiting (1×2) (Fig. 5.13(a)) and (1×3) (Fig. 5.13(b)) symmetries occurs (Gao *et al.*, 1991b). However, transformations were found to be much more rapid on the Au(110) surface than on the Au(100) surface. This might suggest that only short-range atomic motion is required for the structural transformation on the Au(110) surface, in contrast to the long-range diffusional transport of atoms necessary for a transformation on the Au(100) surface. The STM results obtained for the Au(110) 2×1 surface structure are largely consistent with the missing-row model (section 4.1.2.2), though indications for some surface relaxation towards a slightly asymmetric structure have been found.

5.2.3 Electrochemical deposition under potentiostatic control

The initial stage of growth of thin metal films by electrochemical deposition is of particular interest for a profound understanding of electrochemical phase formation. However, in contrast to gas-phase-deposited adlayers (section 4.1.2.8), appropriate *in situ* characterization methods have been lacking for a long time. Scanning probe microscopes now provide the ideal *in situ* techniques to study structure and growth of metal adlayers, as well as their dependence on experimental conditions, such as the electrochemical substrate potential and the chosen electrolytic solution. The initial stage of adlayer formation can best be studied in the so-called 'underpotential deposition' (UPD) regime: by adjusting the electrochemical substrate potential appropriately, the substrate–adatom bond becomes stronger than subsequent adatom–adatom bonds which form as deposition is continued. Consequently, formation of a uniform monolayer of adatoms on a smooth substrate surface can be expected prior to any island growth.

The first atomic-resolution study of electrochemical deposition under potentiostatic control by means of *in situ* STM has been performed using gold substrates immersed in a 0.05 M H_2SO_4 + 5 mM $CuSO_4$ solution from which Cu adlayers have been deposited (Magnussen *et al.*, 1990, 1991). For a Au(111) substrate, a hexagonal $(\sqrt{3} \times \sqrt{3})R30°$ Cu adlayer structure was observed at the onset of the first pronounced voltammetric adsorption peak (Fig. 5.14). In the presence of trace amounts (10^{-5} M) of Cl⁻, this $(\sqrt{3} \times \sqrt{3})$ structure is converted into a more densely packed (5×5) structure. The larger Cu–Cu distances in these adlayer structures, as compared with the pseudomorphic (1×1)

(a)

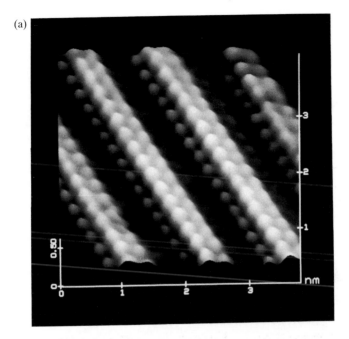

(b)

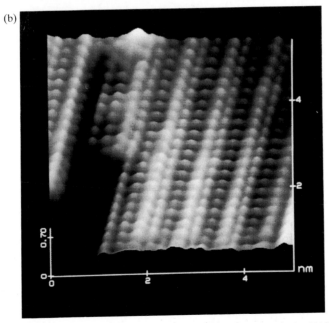

Fig. 5.13. (*a*) STM image of a (1×2)-reconstructed domain on a Au(110) surface in 0.1 M HClO₄ at an electrochemical potential of −0.3 V. (*b*) STM image of a (1×3)-reconstructed domain (Gao *et al.*, 1991b).

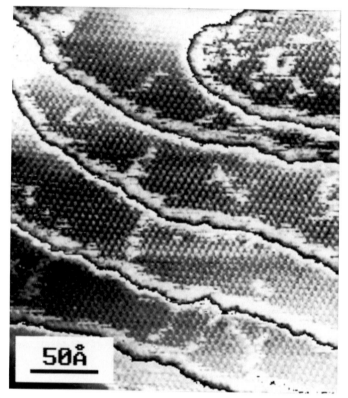

Fig. 5.14. STM image of a Cu-covered Au(111) surface recorded in 0.05 M H_2SO_4 + 10^{-3} M $CuSO_4$. After extended cleaning cycles of the electrochemical cell the ($\sqrt{3} \times \sqrt{3}$)R30 ° phase dominates the surface (Magnussen *et al.*, 1991).

Cu islands formed in vacuum deposits even at lower coverages, have clearly demonstrated the structure-determining role of co-adsorbed anions originating from the electrolytic solution which, in the present case of sulfate anions, introduce a net repulsive adatom–adatom interaction within the Cu adlayer. Since the type of anions changes with the electrolytic solution, the structure of the adlayer is expected to depend on the chosen electrolyte. This has indeed been verified by *in situ* AFM studies, where the structure of the electrochemically deposited Cu adlayers on a Au(111) substrate immersed either in a sulfate (0.1 M H_2SO_4) or perchloric acid (0.1 M $HClO_4$) electrolyte have been compared (Manne *et al.*, 1991a; Gewirth, 1992). For the sulfate electrolyte, the ($\sqrt{3} \times \sqrt{3}$)R30° structure, as already observed by STM, was confirmed, whereas an incommensurate close-packed Cu adlayer structure was found in the presence of the perchloric acid electrolyte.

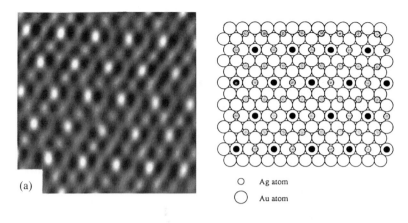

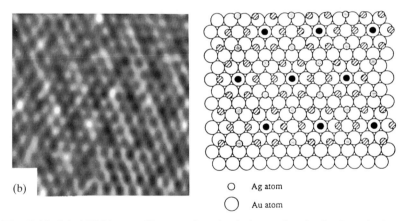

Fig. 5.15. (a) AFM image (5 nm × 5 nm) of electrochemically deposited Ag on Au(111) in 0.1 M H₂SO₄ and schematic diagram of the monolayer showing a (3×3) structure. (b) AFM image (5 nm × 5 nm) of electrochemically deposited Ag on Au(111) in 0.1 M HNO₃ and schematic diagram of the monolayer showing a (4×4) structure (Gewirth, 1992).

A dependence of adlayer structure on the chosen electrolyte has also been found by *in situ* AFM studies of electrochemical deposition of Ag onto Au(111) substrates (Gewirth, 1992; Chen *et al.*, 1992). The Ag adlayer exhibits a (3×3) structure in a sulfuric acid electrolyte (Fig. 5.15(a)), whereas a (4×4) structure is observed in carbonate and nitrate containing electrolytes (Fig. 5.15(*b*)). Different adlayer structures depending on the chosen electrolyte have also been observed for electrochemically deposited Hg on Au(111) substrates by means of *in situ* AFM (Chen and Gewirth, 1992). Besides determination of the structure of deposited adlayers, the scanning probe methods allow detailed investigation of the dynamics of electrochemical deposition and dissolution processes by *in situ* real-time imaging (Robinson, 1989, 1990).

In the previous sections, we have focused on applications of SPM to electrochemical studies at metal electrodes. However, *in situ* scanning probe microscopy can also successfully be performed on semiconductor substrates, including determination of their atomic surface structure under electrolytic solutions (Houbertz *et al.*, 1991; Itaya *et al.*, 1992), as well as investigation of *in situ* photoelectrochemical processes (Sakamaki *et al.*, 1990a).

Despite the success of SPM in electrochemistry, there remain important issues to be addressed in more detail, for instance, the mechanism of electron tunneling through electrolytic solutions, including a theoretical understanding of the measured tunnel barrier heights (Sass *et al.*, 1991; Siegenthaler, 1992). Furthermore, a detailed description of the diffusional transport for the complex tip–substrate geometry is required for a profound understanding of the local aspects of electrochemical deposition or dissolution processes being simultaneously probed by either *in situ* STM or AFM. In the case of STM studies, knowledge of the electric potential distribution between tip and substrate might be important as well. However, lack of information about the detailed tip shape often prevents an appropriate theoretical description. On the other hand, consistent experimental results obtained from different SPM techniques, such as *in situ* STM and AFM, indicate that the influence of the electric field between tip and substrate in electrolytic STM studies does not play a major role.

6

Organic material

The application of scanning probe microscopy (SPM) to organic material, from small molecules to supramolecular assemblies, is a challenging task, which requires to address the following key issues.

1. Suitable substrates have to be found exhibiting a surface roughness considerably less than the size of the molecular species to be deposited in order to allow clear distinction between substrate and molecular features in SPM images. For SPM studies under ambient conditions, substrate surfaces must be chemically inert. The applicability of STM additionally requires electrically conducting substrates.
2. Deposited molecular species have to be immobilized in some way to allow stable SPM imaging.
3. Interpretation of STM images of molecular species requires profound understanding of their electronic structure and transport properties. In addition, the elastic properties must be known for interpretation of STM as well as AFM results.

A variety of different substrates has already been tried for SPM studies of molecules and biological specimens. At first sight, layered materials (section 4.1.3) seem to be particularly attractive because they usually provide large, atomically flat terraces after sample cleavage. In addition, the surfaces of graphite or transition metal dichalcogenides (TMD), for instance, are relatively inert, which is favorable for SPM studies in ambient air. On the other hand, the binding strength of molecules to these surfaces tends to be low in the absence of surface defects. Consequently, the molecular species exhibit either an intrinsic surface mobility or can be dragged by the tip during scanning as a result of tip–molecule interactions. Furthermore, it has been shown that as-cleaved surfaces of graphite and other layered materials often exhibit structural features which mimic the presence of biomolecules (Clemmer and

Beebe, 1991). Therefore, care has to be taken in order to distinguish 'real' molecules from features of the native substrate.

Owing to these difficulties with substrates of layered structure, alternative substrate materials have been tried as well. For instance, Au(111) films epitaxially grown on mica or Au(111) single-crystal substrates also provide large, atomically flat terraces, and can be used under ambient as well as under ultra-high vacuum (UHV) conditions. Additionally, biological specimens tend to be bound more strongly to gold than to graphite surfaces. However, artifactual periodic structures may be found on bare Au(111) as well as on bare graphite surfaces (Wilkins *et al.*, 1992). To exclude possible imaging artifacts inherent in the use of crystalline substrates, several different amorphous substrates have been considered, including amorphous carbon or platinum–carbon films, amorphous silicon oxide or ion-etched surfaces of glassy metals. Surfaces of 'amorphous' substrates usually exhibit cluster-like features, as revealed by real-space STM images. The typical size of these clusters depends on the preparation conditions and determines the surface roughness of the amorphous substrate. While amorphous substrates offer several advantages for investigation of larger biomolecules, they are not useful for SPM studies of small molecules because of this inherent surface roughness, in contrast to the atomically flat terraces provided by single-crystal substrates.

For stable high-resolution imaging of molecular species deposited on reasonably flat substrates, it is important to eliminate the surface mobility of the molecules (Chiang, 1992). Their immobilization can be achieved in several different ways, for instance, by

strong chemisorption on appropriate substrates offering a high binding strength,

co-adsorption with other molecules, thereby stabilizing particular surface structures,

physisorption at low temperatures,

self-organization of molecular structures, or

deposition of solid molecular films using the Langmuir–Blodgett (LB) technique (Blodgett, 1935).

In addition, the tip–molecule interaction has to be kept as small as possible, particular in view of the typically strong elastic response of organic materials in contrast to inorganic ones. This means that STM experiments should preferably be performed with low tunneling currents (1 pA range), while for AFM experiments the loading force should be reduced as much as possible (10^{-10} N and below). On the other hand, it might sometimes be of interest to study the tip–molecule interaction as a function of an external control parameter, e.g. tip–molecule spacing, applied electric field or locally applied pressure.

The most serious limitation on useful application of SPM to organic materials is often encountered with the problem of image interpretation. To understand the contrast in STM images, for instance, an improved knowledge of the electronic structure and transport properties of the molecular species would be required. Furthermore, the three-dimensional structure of larger biomolecules appears to be another intrinsic problem because SPM studies of three-dimensional structures are likely to be susceptible to imaging artifacts resulting from the presence of multiple tips. These and other reasons might explain the fact that, in contrast to applications in solid state physics and chemistry, SPM has not yet made any really significant contribution to structural biology (Baumeister, 1988; Baumeister and Zeitler, 1992; Guckenberger *et al.*, 1992).

6.1 Thin molecular layers

Thin molecular films of monolayer height deposited onto electrically conducting single crystal substrates with atomically flat terraces can be studied by STM down to the atomic scale. For these systems, STM can provide valuable information about

binding sites of individual molecules,
the orientation of individual molecules with respect to the substrate lattice,
the conformational state of individual molecules,
the periodicity of ordered molecular surface structures, as well as defects and domains that are present in ordered molecular surface structures.

For appropriate interpretation of STM images, it is necessary to consider various possible contrast mechanisms, for instance:

tunneling of electrons via electronic states of the molecule, or
direct electron tunneling between tip and substrate not involving molecular electronic states.

In the second case, the molecules might still be 'seen' by STM due to their influence on the electronic structure of the substrate, e.g. via hybridization of molecular electronic states with metallic states of the substrate, or by affecting the local tunneling barrier height. In any case, detailed knowledge about the molecular states from theoretical electronic structure calculations is of great importance for a profound understanding of observed STM images of molecular species.

6.1.1 Chemisorbed molecules on metal substrates

As a first example, we will discuss STM studies of chemisorbed molecules on metal substrates performed under well-defined UHV conditions.

A system which has been extensively studied by STM is benzene (C_6H_6) co-adsorbed with CO on clean Rh(111) substrates (Ohtani *et al.*, 1988; Chiang *et al.*, 1988b, 1990). Co-adsorption of the two different molecules from the vapor phase leads to an extremely stable surface structure with strongly chemisorbed molecules, allowing atomic-resolution STM imaging at room temperature. Depending on the surface concentration of CO, either a (3×3) or a c($2\sqrt{3}$×4) rect superlattice structure is obtained with two CO molecules or one CO molecule per unit cell, respectively. For the (3×3) surface structure (Fig. 6.1), a well-ordered triangular array of ring-like features was revealed by STM, which has been attributed to the flat-lying benzene molecules, while the upright CO molecules appear to be invisible (Ohtani *et al.*, 1988). Since the lowest unoccupied molecular orbitals (LUMO) of benzene (and CO) are located >2.5 eV above the Fermi energy E_F, no contributions to the tunneling current through these states can be expected under the low-bias conditions employed for the STM studies (<0.5 V). However, inverse photoemission studies indicate an enhanced density of states near E_F in the presence of adsorbed benzene, which may be associated with hybridized metal–benzene states, giving rise to the observed 'topographic' maxima at the benzene sites in STM images (Netzer and Frank, 1989). In contrast to the expected sixfold symmetry, the benzene molecules tend to appear threefold-symmetric in STM images (Fig. 6.1(*b*)). The observed maxima of the benzene features are located above the bridge sites between two rhodium substrate atoms. It is believed that interaction of electronic states of the molecule and the substrate is responsible for the observed threefold symmetry. On the Rh(111) c($2\sqrt{3}$×4) rect (C_6H_6 + CO) surface (Fig. 6.2), both the benzene and the CO molecules have been resolved by STM (Chiang *et al.*, 1988b). In addition, rotational and translational domain boundaries have been observed, as shown in Fig. 6.2(*c*) (Chiang *et al.*, 1990). It was found that step edges on the metal substrate often coincide with domain boundaries. Furthermore, the domain boundaries tend to intersect isolated defects or steps.

Another STM study has focused on naphthalene ($C_{10}H_8$) chemisorbed on a Pt(111) substrate (Hallmark *et al.*, 1991a, 1991b). The naphthalene

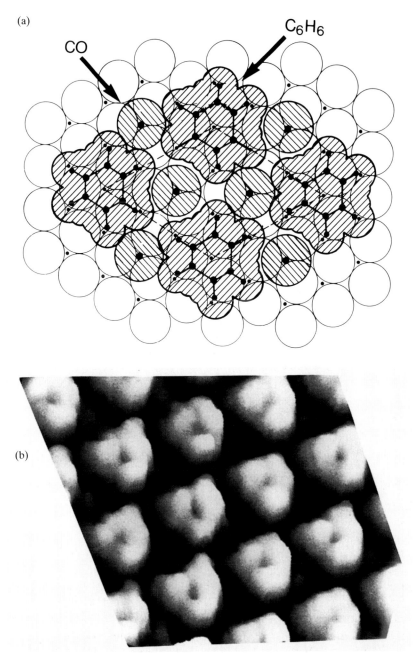

Fig. 6.1. (*a*) The structure for Rh(111)–(3×3)(C₆H₆ + 2CO) determined by LEED crystallography. Large circles and small dots represent the first-and second-layer metal atoms, respectively. (*b*) STM image showing the benzene molecules on the Rh(111) substrate (Ohtani *et al.*, 1988).

(a)

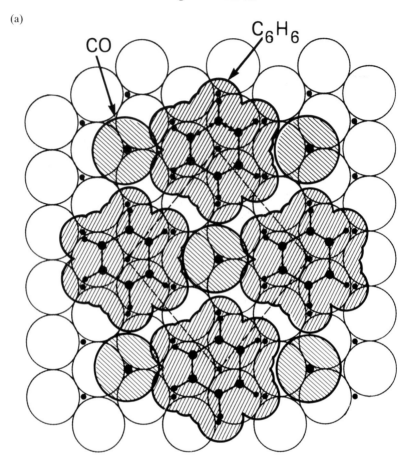

Fig. 6.2. (*a*) Structural model of the c(2√3×4)rect ordered array of co-adsorbed benzene and CO on Rh(111), as determined by dynamical LEED analysis. Large circles and small dots represent first- and second-layer Rh atoms, respectively. The dashed lines show the primitive unit cell, with benzene molecules at the corners and one CO molecule in the center. (*b*) STM image of the c(2√3×4)rect ordered array of benzene and CO molecules on Rh(111). The mesh, with large (small) diamonds indicating top (second) layer Rh atoms, has been overlaid on the data according to the dynamical LEED model (Chiang *et al.*, 1988b). (*c*) STM images of the Rh(111)–c(2√3×4)rect(C₆H₆ + CO) surface showing the three possible rotational domains in the benzene overlayer, with rotational domain boundaries (R) near an atomic step (S) of the Rh(111) substrate. The rotational domain boundary on the left is also indicated by a line (Chiang *et al.*, 1990).

(b)

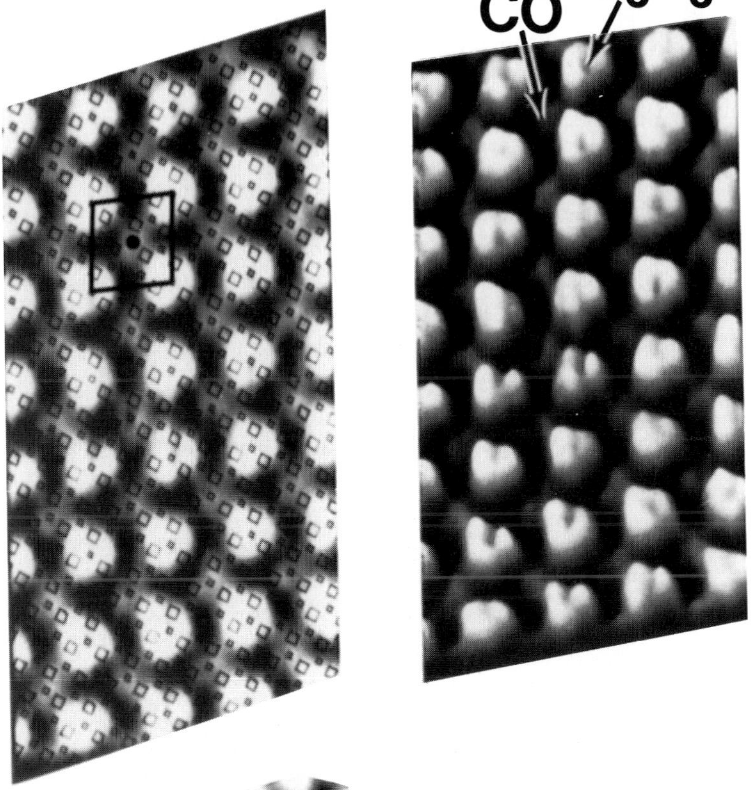

CO   C₆H₆

(c)

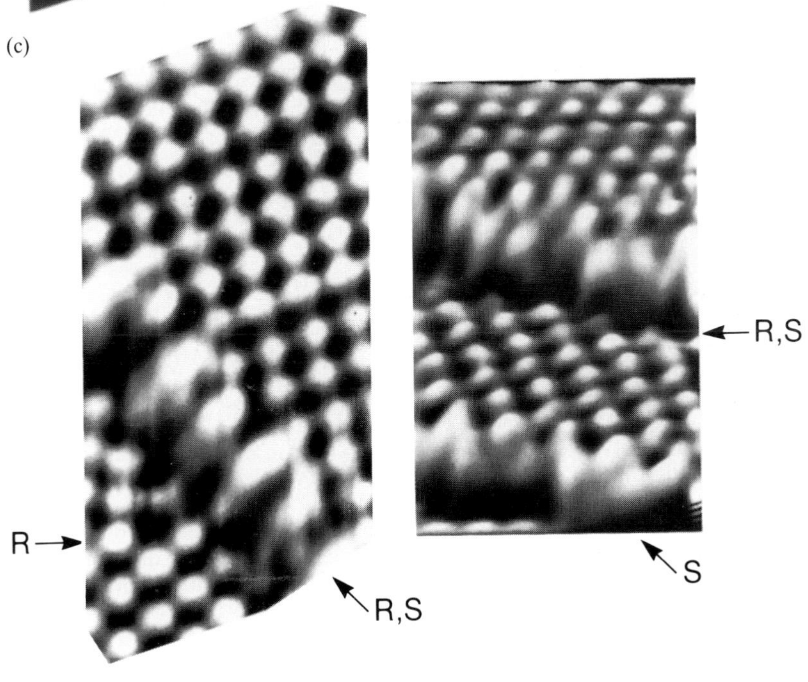

R,S

R

R,S

S

molecules appear as bi-lobed elongated features in STM images with three equivalent rotational orientations with respect to the Pt(111) substrate lattice and a preference for on-top binding sites (Fig. 6.3). Small domains (typically <50 Å diameter) with pseudo-(6×3) molecular ordering have been observed, rather than a perfect long-range periodic superlattice, in contrast to the Rh(111)–(C_6H_6 + CO) system. For high coverage (approaching one ML), the naphthalene molecules showed limited motion, occurring only at domain boundaries. In contrast, for low-coverage adsorption (about 0.2 ML) the naphthalene molecules were found to undergo discrete rotations among the three equivalent allowed orientations, as well as translational motion by single Pt substrate lattice spacings.

STM has also been applied to study isolated Cu-phthalocyanine (Cu-phth) molecules on various substrates, including Au, Ag, Cu, Si

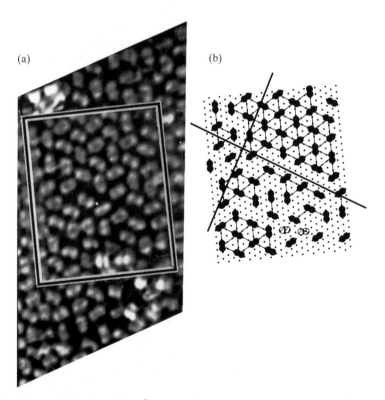

(a)   (b)

Fig: 6.3. (*a*) STM image (90 Å × 150 Å) of ordered naphthalene on Pt(111). The image has been corrected for thermal drift. (*b*) Schematic diagram of the overlay of a Pt(111) lattice with molecular positions. Glide symmetries are indicated by dashed lines (Hallmark *et al.*, 1991a).

and GaAs (Gimzewski *et al.*, 1987b; Lippel *et al.*, 1989; Möller *et al.*, 1990). The best results were obtained by using Cu(100) substrates, for which strong chemisorption bonds with the Cu-phth molecules are formed, thereby reducing molecular motion, in contrast to the behavior on Au and Ag substrates (Lippel *et al.*, 1989). The STM images revealed two different rotational orientations of the flat-lying Cu-phth molecules with respect to the Cu(100) substrate lattice, as well as the internal structure of isolated molecules (Fig. 6.4). Four protrusions on each of the four lobes of the Cu-phth molecule were clearly resolved (Fig. 6.4(*e*)). By comparison with Hückel molecular orbital calculations, these protrusions have been assigned to highly populated carbon atom p states. A negligible dependence of low-bias voltage images on the bias-voltage polarity was found during the STM studies which is explained by the similarity of the HOMO and LUMO charge density distributions (Fig. 6.4(*b-d*)).

In all STM studies of chemisorbed molecules on metal substrates discussed above, simultaneous observation of the molecules and the atomic lattice of the substrate was not achieved. To clearly resolve the atomic lattice on a close-packed metal substrate usually requires small tip–surface separations which are accompanied by strong tip–substrate interaction forces. As a consequence, the molecules are likely to be moved by the scanned tip under these conditions. On the other hand, by increasing the tip-to-sample spacing to allow non-destructive imaging of the molecules, the corrugation of the metal substrate lattice becomes too small to be detectable in most cases.

6.1.2 Liquid crystal molecules

Liquid crystals are of interest both technologically (liquid crystal displays) and scientifically because they appear in many different phases which exhibit a higher degree of order than a liquid, but are less ordered than a crystal. Liquid crystals are built up by elongated molecules. If these molecules form a one-dimensionally ordered structure with alignment along a common axis, a nematic liquid crystal results. In contrast, smectic liquid crystals exhibit a layer-type structure with two-dimensional ordering of the molecules within each layer, but no registration between molecules of adjacent layers.

To study liquid crystal molecules by STM under ambient conditions, three different preparation procedures have been tried.

1. A drop of the liquid crystal can be placed directly onto a suitable substrate.

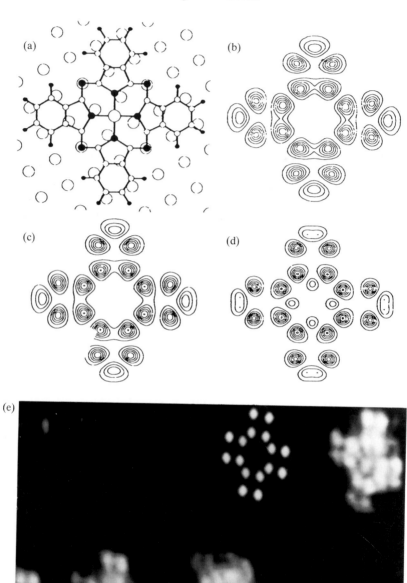

(f)

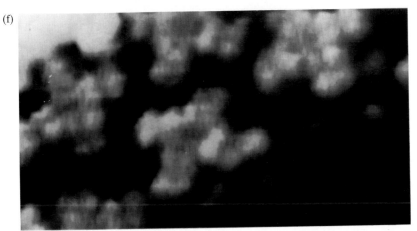

Fig. 6.4. (*a*) Model of the Cu-phth molecule above a Cu(100) surface. Small (large) open circles are C (Cu) atoms and small (large) filled circles are H (N) atoms. The Cu(100) lattice is shown rotated by 26.5°. In (*b*) and (*c*) are shown contour plots of the charge densities of the highest occupied molecular orbital (HOMO) and lowest unoccupied molecular orbital (LUMO) 2 Å above the molecular plane. (*d*) Charge density of the HOMO, 1 Å above the molecular plane. (*e*) STM image of Cu-phth molecules on Cu(100) at submonolayer coverage. A gray scale representation of the HOMO, evaluated 2 Å above the molecular plane, has been embedded in the image. (*f*) STM image obtained near 1 ML coverage (Lippel *et al.*, 1989).

2. The liquid crystal can be heated until a sublimation of molecules sets in, which can finally condense on a nearby substrate.
3. The liquid crystal molecules can be dissolved in a polar solvent. A drop of the solution can then be placed onto the substrate, from which the solvent finally evaporates.

The best conditions for STM studies of molecular films have been obtained by using freshly cleaved layered materials, such as graphite or transition metal dichalcogenides (TMD), as substrates. The ambient temperature during the experiments should be 10–20 K below the bulk transition temperature between the liquid crystal and the isotropic phase. For higher temperatures, the films become disordered, whereas for lower temperatures, the films no longer wet the graphite substrate (Smith *et al.*, 1990).

STM images of the molecular films on graphite substrates generally reveal well-ordered arrays of flat-lying molecules over distances of thousand of Ångström units (Foster and Frommer, 1988; Spong *et al.*, 1989a, 1989b; Smith *et al.*, 1989a, 1990; Smith, 1991). Remarkably, the STM images were found to be independent of the film thickness. It is believed that the STM tip penetrates through the electrically insulating molecular

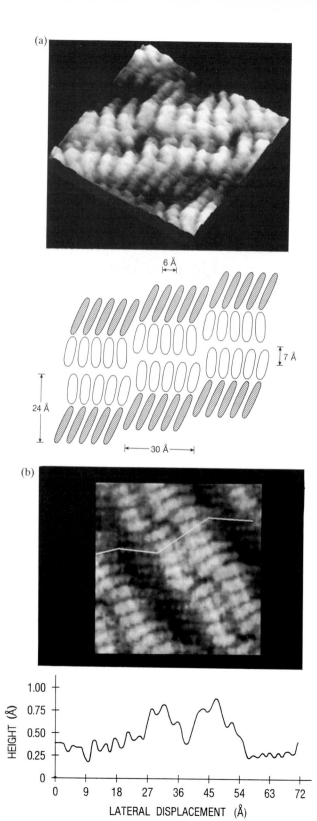

(a)

6 Å

7 Å

24 Å

30 Å

(b)

HEIGHT (Å)

1.00

0.75

0.50

0.25

0

0   9   18   27   36   45   54   63   72

LATERAL DISPLACEMENT (Å)

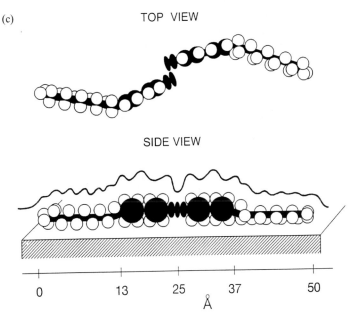

(c)   TOP VIEW

SIDE VIEW

Fig. 6.5. (*a*) STM image of 10CB on graphite and a model showing the packing of the 10CB molecules. The shaded and unshaded segments represent the alkyl tails and the cyanobiphenyl head groups, respectively. (*b*) Contour line taken along a pair of 10CB molecules. The contour chosen is shown by the white line overlaying the image in (*a*). (*c*) Model showing the orientation of two molecules of 10CB. A trace is shown representing a possible STM contour (Smith *et al.*, 1989a).

film and only images the molecular layer being in direct contact with the substrate. As a result of the molecule–substrate interaction, the observed structure represents a well-ordered two-dimensional molecular crystal, rather than a liquid crystal. Self-organization of the molecules at the substrate–film interface into a two-dimensional molecular crystal leads to their immobilization, which is required for high-resolution STM studies of individual molecules. On the other hand, the arrangement of the molecules in this two-dimensional molecular crystal may be quite different from that found in bulk liquid crystals.

The most widely studied system has been *n*-alkylcyanobiphenyl (*n*CB, where *n* = 6,8,10 or 12) deposited on graphite substrates. The *n*CB molecules form smectic liquid crystal bulk phases. Upon deposition onto a graphite substrate, a two-dimensional crystalline state is formed which is more highly ordered than the bulk smectic phase. Apparently, the graphite substrate directs the orientation of the adsorbed molecules. A STM image of 10CB on graphite together with a structural model is presented in Fig. 6.5 (Smith *et al.*, 1989a). The unit cell of the observed ordered molecular array consists of ten molecules arranged into five molecular pairs. The cyanobiphenyl head groups and the alkyl tails are slightly interdigitated, forming a 'double-row' structure, presumably due to a repulsive interaction between the head groups facing each other.

The different appearance of the head groups and the tails, as typically found in STM images of liquid crystal molecules, might have several explanations.

1. The observed contrast might be primarily of topographic origin because the hydrogen orbitals of the phenyl head groups are higher above the substrate and have a larger component normal to the substrate than those of the alkyl tails (Smith *et al.*, 1989a).

2. Alternatively, it has been proposed that the observed contrast in STM images of liquid crystal molecules on graphite substrates arises from a modulation of the local work function of the substrate by the polarizable molecular adsorbates, leading to a modulation of the local tunneling barrier height ϕ:

$$\phi = \phi_0 - e\mu / \varepsilon_0 \qquad (6.1)$$

where ϕ_0 denotes the barrier height for the case of a bare substrate surface, μ is the dipole moment density of the adsorbate, e is the electronic charge and ε_0 is the permittivity of free space (Spong *et al.*, 1989a). Since the alkyl tails of the liquid crystal molecules are less polarizable than the head groups, they are expected to modulate the local work function and therefore the local tunneling barrier height more weakly.

3. Another explanation for the different appearance of head groups and tails of liquid crystal molecules in STM images might be based on an electronic density-of-states effect (Fig. 6.6). The head groups, containing aromatic carbons, typically exhibit higher molecular orbital densities for the LUMO and HOMO than the alkyl tails containing tetrahedrally bonded carbon atoms (Smith *et al.*, 1990). However, it has then to be explained why these molecular orbitals are accessible for the tunneling electrons at small sample bias voltages (typically below 1 V), despite the fact that, for an isolated nCB molecule, the energy gap between the HOMO and the LUMO can be larger than 10 eV. It has been argued that a molecular energy level shift caused by the molecule–substrate interaction could make the molecular states accessible to the tunneling electrons (Nejoh, 1990). In this case, resonant tunneling processes (section 1.2.8)

a

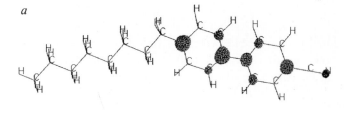

b

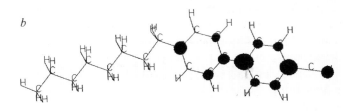

c

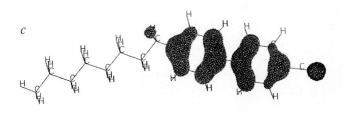

d

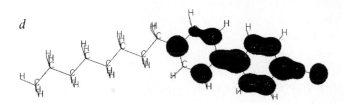

Fig. 6.6. Molecular orbital calculations of a free 8CB molecule: (*a*) and (*c*) HOMO calculation showing the surfaces of two different molecular orbital densities; (*b*) and (*d*) corresponding LUMO calculation. The fact that the LUMO and HOMO contours are very similar is consistent with the experimental observation that reversing the polarity of the tunnel bias does not significantly alter the STM images (Smith *et al.*, 1990).

might play an important role for understanding the STM image contrast, including its bias-voltage dependence. In particular, it has been found that reversible switching between molecular imaging at a sample bias of about 1 V and imaging of the underlying graphite substrate lattice at a bias of about 0.1 V can be achieved without touching or disturbing the molecular ordering by the STM tip (Mizutani *et al.*, 1990). By changing the tunnel current in these STM experiments, it has been verified that the observed bias-dependence of the STM images does not result from an accompanying variation of tip-to-substrate spacing. The experimental observations have been explained by assuming that, at an elevated sample bias voltage, the majority of electrons tunnel resonantly via the molecular energy levels, while at lower bias voltages these states are no longer accessible and the electrons tunnel directly between tip and substrate as if there were no molecules in between.

Consideration of molecule–substrate interactions is certainly important for elucidating the electronic structure of the film–substrate interface and therefore for understanding the contrast mechanism in STM studies of such systems. In addition, molecule–substrate interactions can have a strong influence on the structure of the molecular layer, which is experimentally verified by STM studies of liquid crystal molecules deposited on different substrates. For instance on graphite substrates, 8CB molecules form a 'double-row' structure as also observed for 10CB molecules (Fig. 6.5), whereas a 'single-row' structure is revealed in STM images of 8CB molecules on MoS_2 substrates (Hara *et al.*, 1990). On the other hand, 10CB molecules deposited on MoS_2 substrates exhibit the same 'double-row' structure as on graphite substrates. To explain these experimental observations, it has been proposed that the resulting molecular lattice is predominantly determined by registry with the substrate lattice, maximizing the number of substrate sites filled by the molecules (Smith and Heckl, 1990). Experimentally, the registry between the molecular and substrate lattices can be studied by switching the sample bias voltage between high and low values, allowing preferential imaging of either the molecular arrangement or the substrate lattice, respectively, as described above.

Owing to the different symmetries of the molecular and substrate lattices, rotational domain boundaries can occasionally be observed by STM as shown in Fig. 6.7(*a*) (Smith, 1991). The allowed angles between the rotational domains are determined by the structure of the two-dimensional molecular crystal. Steps on the substrate often coincide with rotational domain boundaries, suggesting that the growth of the two-dimensional molecular crystal starts independently on each terrace.

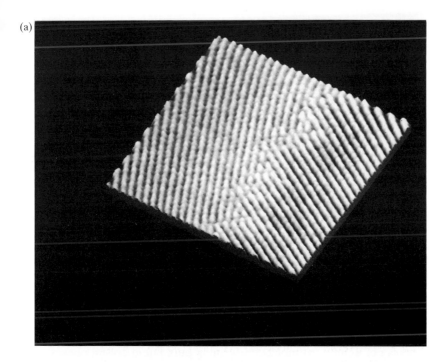

(a)

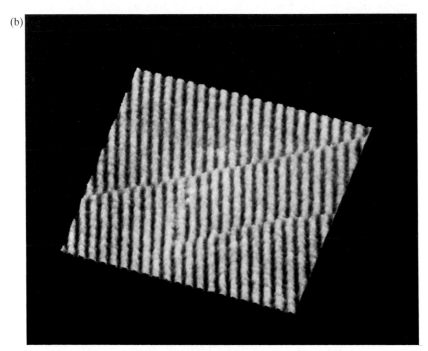

(b)

Fig. 6.7. (*a*) STM image (56 nm × 56 nm) of a boundary between two 10CB grains on graphite. The angle between the two domains is measured to be 36°. (*b*) STM image (112 nm × 112 nm) of 12CB on graphite showing two screw dislocations caused by shear stress in the molecular film (Smith, 1991).

Defects in the two-dimensional molecular crystal, such as dislocations, have been observed by STM as well, as shown in Fig. 6.7(*b*). On the other hand, the molecular rows of the two-dimensional crystal often appear uninterrupted across a variety of substrate defects, demonstrating its stability. Temperature-dependent STM studies of the two-dimensional molecular crystal also indicate a high degree of stability, since the two-dimensional crystalline state is observed to remain intact even upon heating to temperatures 10–15 K above the bulk transition temperature from the liquid crystal into the isotropic phase (Spong *et al.*, 1989a).

More recently, non-contact force microscopy (section 2.7) has proven to be a useful technique to study thick liquid crystalline samples (Terris *et al.*, 1992). Remarkably, their surfaces do not appear as featureless as one would expect with regard to the liquid character of these materials.

6.1.3 *Alkanes and alkane-derived molecules*

The two-dimensional molecular ordering at the substrate–solution interface is not unique for molecules forming bulk liquid crystal phases, but can be observed for other molecules, e.g. alkanes and alkane derivatives, as well. For instance, an organic solution of didodecylbenzene $H_{25}C_{12}(C_6H_4)C_{12}H_{25}$ (DDB) was found to form a two-dimensional polycrystal of flat-lying DDB molecules with crystallite sizes in the range of a few nanometers at a graphite substrate–solution interface (Rabe and Buchholz, 1991; Buchholz and Rabe, 1991). The contrast in STM

(a)

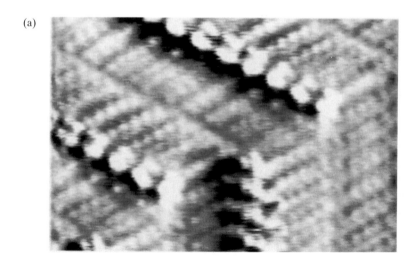

(b)

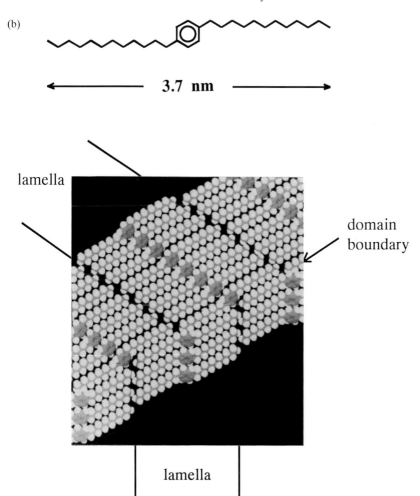

Fig. 6.8. (*a*) STM image (7.2 nm × 4.7 nm) of a domain boundary in a mono-layer of didodecylbenzene adsorbed at the interface between an organic solution and the basal plane of graphite. The strongest contrast results from the phenyl stacks. Evidently the side chains are tilted relative to the lamella boundary and parallel to each other in the adjacent domains. (*b*) Molecular model for the domain boundary of (*a*) (Rabe and Buchholz, 1991).

images appeared to be higher at the benzene sites than for the alkyl side chains (Fig. 6.8), following the trend already found in STM studies of liquid crystal molecules. Bias-dependent STM studies again allowed either imaging preferentially the molecular layer at elevated sample bias, or the underlying substrate lattice at low sample bias, thereby facilitating determination of the registration of the DDB molecular layer with respect to the graphite substrate lattice. The alkyl side chains were found to orient predominantly parallel to the graphite lattice axis. Domain boundaries, where the orientation of the DDB molecules changes by 180°, have frequently been observed by STM. Cooperative motion of several molecules can lead to changes in the domain boundary configuration. Fast STM imaging allowed resolution of the molecular dynamics involved in such processes on a time scale of 100 ms.

STM has also been applied to study molecular layers of n-alkanes at the graphite–solution interface (McGonigal *et al.*, 1990, 1991). The n-alkanes are saturated linear hydrocarbons with their carbon atoms arranged in a zig-zag chain. For a chain length of up to 17 carbons, alkanes are liquid at room temperature, whereas for longer chain lengths solids are formed. Adsorption of n-$C_{32}H_{66}$ (melting temperature 70 °C) from a liquid solvent on a graphite substrate leads to a highly ordered two-dimensional layer of flat-lying molecules, as directly revealed by STM (Fig. 6.9). However, in contrast to the liquid crystal and DDB molecular layers discussed above, the atomic structure seen in STM images of n-alkane layers is dominated by structural features associated with the substrate, rather than with the adsorbed molecules, indicating that the tunneling process does not involve molecular states. On the other hand, the n-alkane molecular layer appears to influence the electronic structure of the interface in such a way that the tunnel current becomes enhanced above particular graphite substrate sites. By determination of the periodicity of enhanced sites, the commensurability of the adsorbed layer with respect to the substrate lattice can directly be checked, though the molecules themselves are not directly visible to the STM.

While electron tunneling through a monolayer of flat-lying molecules can easily be understood, STM images of alkane crystals 100 nm thick, which are known to be excellent bulk insulators, appeared as a great surprise (Michel *et al.*, 1989). Direct observation of structural changes near the known melting temperature of the alkane crystals proved the chemical nature of the imaged objects. To explain these experimental STM results, several different contrast mechanisms have been considered, such as electron transfer processes through defect-related states or the presence of electrically conducting surface layers associated with the ambient experimental environment.

6.1.4 Langmuir–Blodgett (LB) films

Langmuir–Blodgett (LB) films are built up by two-dimensional layers of chain-like molecules with a hydrophilic head group and a hydro-

(a)

1 nm ⊢⊣

(b)

Fig. 6.9. (*a*) STM image (10 nm × 10 nm) of n-$C_{32}H_{66}$ molecules adsorbed on graphite. (*b*) Schematic depiction of the graphite lattice, proposed location of n-$C_{32}H_{66}$ molecules, and expected enhancement of tunneling current due to these molecules. Small dots correspond to graphite sites predominantly imaged by the STM with no adsorbed layer present. Upper part: superimposed n-C_{32} H_{66} molecules with the proposed commensurate registration. Lower part: schematic representation of the expected STM image. Large dots mark graphite sites enhanced by the presence of n-$C_{32}H_{66}$ molecules. Hence, the dots correspond to the bright sites in the experimental image shown in (*a*) (McGonigal *et al.*, 1990).

phobic tail, which can be transferred from an air–water interface to a variety of substrates according to the procedure illustrated in Fig. 6.10. In contrast to the molecular layers discussed so far, LB films usually exhibit orientation of the chain-like molecules perpendicular to the substrate surface, although, in some cases, molecules can be found to lie parallel to the substrate as well. For hydrophobic substrates, such as graphite (Fig. 6.10), one molecular monolayer is deposited while the

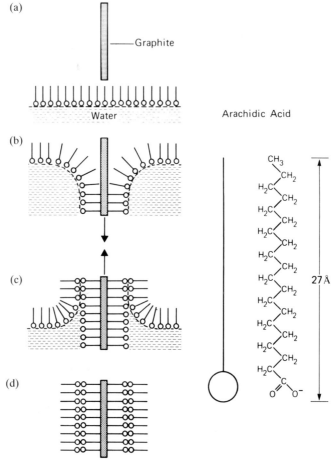

Fig. 6.10. Schematic depiction of the Langmuir–Blodgett (LB) technique. Arachidic acid dissolved in chloroform is spread on a water surface (*a*). The graphite substrate is lowered into the water, and because graphite is hydrophobic, it gains a monolayer of molecules on the way down (*b*). A second monolayer is transferred on the way up (*c*), giving a molecular bilayer on graphite in air (Smith *et al.*, 1987a).

substrate is lowered into the water and another monolayer is added as the substrate is withdrawn. Subsequent local probe studies of such LB films, as schematically illustrated in Fig. 6.11, can provide important information about the local structure, orientation, and packing of the molecular layers, which have potential applications in diverse fields, such as molecular electronics, microelectronics, integrated optics, microlithography, tribology and as biological sensors (Roberts *et al.*, 1981; Sugi, 1985; Peterson, 1987; Agarwal, 1988).

The first STM studies of LB films have concentrated on cadmium arachidate bilayers on graphite substrates (Smith *et al.*, 1987a). The packing of the molecular layer was found to be partially ordered with a triclinic unit cell. The mean molecular density appeared to be about one molecule per 20 Å^2. Time-dependent STM studies revealed a significant drift of the imaged molecular features, indicating that the LB film might be actually sliding across the graphite substrate which is possibly related to the weak bonding of the molecules to the substrate. The ability to study monolayer or bilayer LB films with molecular resolution by using STM has subsequently been demonstrated with other molecular systems and substrates as well (Hörber *et al.*, 1988; Fuchs, 1988). WSe$_2$ substrates have been found to be particularly suitable for high-resolution STM investigations of stable monolayer LB films (Fuchs *et al.*, 1990a, 1990b).

For the bilayer system, e.g. cadmium arachidate with a total thickness of about 54 Å, the question of the STM contrast mechanism becomes

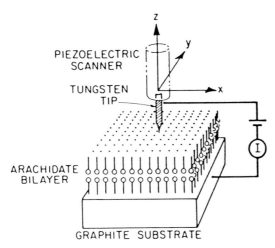

Fig. 6.11. Schematic diagram of an STM experiment on a LB bilayer film (Smith *et al.*, 1987a).

important again, since tunneling is not expected to occur over such large tip-to-substrate distances. Earlier tunneling experiments performed with planar metal–LB-film–metal junctions have indicated a high resistivity of the LB films, suggesting that they are good electrical insulators (Mann *et al.*, 1971; Mann and Kuhn, 1971; Polymeropoulos, 1977). However, at that time the planar tunnel junctions were fabricated with at least one Al electrode, which is known to become easily oxidized, thereby contributing to the high measured resistivity of the junctions. More recent tunneling experiments performed with Ag or Au electrodes have indicated that multilayer LB films are likely to exhibit good electrical conductivity (Couch *et al.*, 1986), which would better agree with the experimental STM results. On the other hand, one cannot exclude that the STM tip is at least partially penetrating into the LB film until the demanded current flow is reached at a sufficiently small tip-to-substrate spacing.

The structure of LB films has also been studied by AFM for bilayer and four-layer films either submerged in a buffer solution (Weisenhorn *et al.*, 1991), or exposed to air (Meyer *et al.*, 1991c). For AFM studies of soft molecular films, the applied loading force has to be reduced as much as possible ($<1 \times 10^{-8}$ N) to prevent deep penetration of the probe tip into these films. On the other hand, by intentionally increasing the applied loading force, the substrate lattice can be imaged, rather than the adsorbed molecular layer (Weisenhorn *et al.*, 1991).

AFM investigations of four-layer films of cadmium arachidate deposited on amorphous silicate substrates revealed periodic structures of the molecular films over large distances (several hundred Ångström units), as shown in Fig. 6.12 (Meyer *et al.*, 1991c). The molecular order in the absence of a periodic substrate lattice indicates that the adsorbed molecules near the interface are driven to self-organize primarily, if not solely, by intermolecular forces rather than by molecule–substrate interactions. AFM has also been used to study defects present in LB films, such as pairs of dislocations (Bourdieu *et al.*, 1991), as well as domain boundaries between regions of different crystallographic orientation in LB multilayer films (Garnaes *et al.*, 1992).

Frictional force microscopy (section 2.5) is becoming increasingly important for studying the interesting tribological properties of LB films. The frictional forces become reduced above the LB film compared with the bare substrate, as can directly be revealed by simultaneous topographic and frictional force studies of the same surface area (Meyer *et al.*, 1992). The frictional forces were additionally shown to depend on particular molecular groups of the LB film molecules (Fig. 6.13).

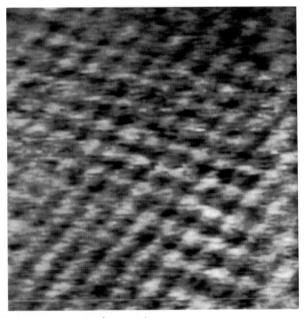

Fig. 6.12. AFM image (53 Å × 51 Å) of a four-layer cadmium arachidate film deposited on an amorphous silicate substrate showing periodic arrangement of the molecules (Meyer *et al.*, 1991c).

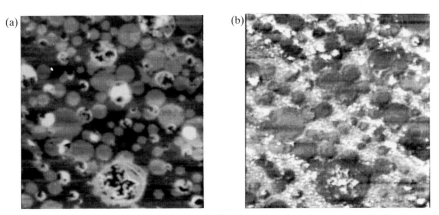

Fig. 6.13. (*a*) Topographic AFM image (5 μm × 5 μm) of the surface of a bilayer LB film prepared from a 1:1 molar mixture of fluorocarbon and hydrocarbon carboxylates. The circular domains have been assigned to the hydrocarbon component and the surrounding flat film to the partially fluorinated component. (*b*) Friction force image obtained simultaneously with the topography. The measured lateral (friction) force appears to be higher (brighter contrast) over the fluorinated regions (Overney *et al.*, 1992).

6.1.5 Polymers

STM and AFM can also be applied to study polymers, such as polyoctadecylacrylate (PODA) and polymethylmethacrylate (PMMA), deposited with the LB technique on graphite substrates (Albrecht *et al.*, 1988a; Dovek *et al.*, 1988). Depending on the local surface coverage, PODA films exhibited isolated narrow fibril structures or parallel groups of fibrils, which have been interpreted as individual polymer chains or small bundles of parallel chains. In addition, ordered structures attributed to the alkyl side-chain crystallization known from the bulk have been revealed by topographic SPM studies.

Besides imaging individual polymer chains on flat substrates or polymer crystals by SPM (Reneker *et al.*, 1990; Piner *et al.*, 1990; Patil *et al.*, 1990), AFM in particular has the potential to study a variety of properties of polymeric liquid films, such as their thickness at a particular location with a lateral resolution on the order of the AFM tip radius, or the strength of the meniscus force acting on the AFM tip as a function of depth into the liquid film (Mate *et al.*, 1989b).

6.1.6 Fullerene films

The discovery of an effective preparation procedure for obtaining macroscopic quantities of C_{60} molecules, a third form of carbon besides graphite and diamond, has triggered a significant research effort to understand the physical and chemical properties of C_{60} and C_{60}-related 'fullerene' molecules (Krätschmer *et al.*, 1990a, 1990b). Thin films of C_{60} as well as a solid form, called fullerite, have become available. This has allowed bulk-structure sensitive techniques to be applied, leading to direct confirmation of the 'soccer-ball' structure of the C_{60} molecule with its icosahedral symmetry (Kroto *et al.*, 1985), as well as determination of the bulk crystal structure of fullerite.

The two-dimensional ordering of the C_{60} molecules in thin films deposited from the gas phase onto a variety of different substrates, including Au(111) and GaAs(110), has been studied directly by STM (Wilson *et al.*, 1990; Wragg *et al.*, 1990; Li *et al.*, 1991b, 1991c). On UHV-prepared Au(111) substrates, a nearly hexagonal array of clusters with an intercluster spacing of about 11 Å has been revealed, in reasonable agreement with theoretical expectations for a two-dimensional close-packed lattice of C_{60} molecules. The molecular species exhibited a relatively high degree of surface mobility on the Au(111) substrate (Wilson *et al.*, 1990). In contrast, C_{60} monolayer films deposited onto UHV-cleaved GaAs(110) substrates were found to be much more stable though the bonding of C_{60} molecules to GaAs(110) is van der Waals in character and therefore again relatively weak (Li *et al.*, 1991b, 1991c). The C_{60} molecules on GaAs(110) appear to be nearly close-packed. However, they do not show a hexagonal arrangement. The nearest-neighbor distance of C_{60} molecules along the [$\bar{1}1\bar{1}$] or [$1\bar{1}\bar{1}$] directions of the substrate lattice was found to be 9.8 Å, whereas the separation along the [001] direction was found to be 11.3 Å. The STM results on thin C_{60} films have demonstrated that the film structure is determined by a balance between molecule–substrate and intermolecular interactions. The different degree of surface mobility of C_{60} molecules on Au(111) and GaAs(110) substrates has been attributed to a different strength of molecule–substrate interactions.

STM has also been applied to study the growth behavior of C_{60} multi-layer films on GaAs(110) substrates held at temperatures between 300 K and 470 K. Point defects, dislocations, domain boundaries, and surface faceting have directly been imaged in real space (Li *et al.*, 1991b). A preferred bonding and nucleation of C_{60} molecules at step edges of the substrate has been found. The observation of two-dimensional grain boundaries (Fig. 6.14) indicates that C_{60} islands nucleate separately and finally grow together. Commensurate as well as incommensurate growth structures have been revealed by STM on C_{60} monolayer films.

The internal structure of individual C_{60} molecules is usually not resolved in STM studies performed at room temperature. This has been explained by a smearing of intramolecular features due to rapid rotation of C_{60} molecules at room temperature (Wilson *et al.*, 1990). These rotations can be frozen in either by reducing the temperature or by using a more reactive substrate. Strong tip–molecule interactions can addition-ally have an influence on the rotational state of C_{60} molecules since applied pressure can induce orientational ordering of the molecules (Samara *et al.*, 1991).

The fact that STM can image the relatively 'thick' C_{60} molecules at all, even at low sample bias voltages (0.1 V), might seem surprising at first sight by virtue of the 1.5–2.0 eV energy gap between the LUMO and HOMO of the isolated C_{60} molecule. However, the LUMO of the C_{60} molecule is expected to lie not far from the Fermi level E_F of the substrate. In addition, molecule–substrate as well as intermolecular interactions will shift, split and broaden the orbitals of the isolated molecule, thereby providing some density of states at E_F, as is required for STM investigations (Wilson *et al.*, 1990). On the other hand, thick films of pure C_{60} or even fullerite crystals have to be studied by AFM (Snyder *et al.*, 1991; Dietz *et al.*, 1992).

(a)

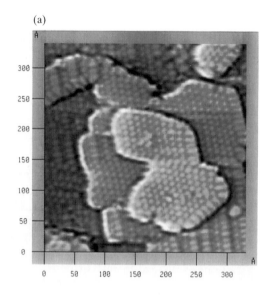

(b)

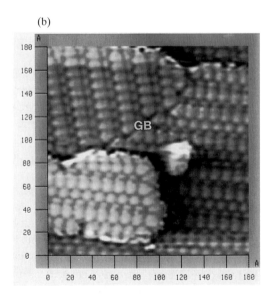

(c)

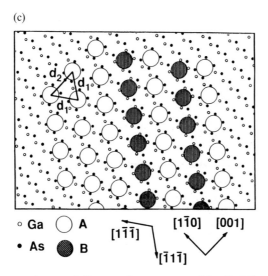

° Ga ◯ A   [1̄10] [001]
• As ● B   [11̄1̄]
            [1̄11̄]

Fig. 6.14. (*a*) STM image of C$_{60}$ growth structures of the first three layers on a GaAs(110) substrate. (*b*) A grain boundary (GB) in the second layer appears in the center of the image (Li *et al.*, 1991b). (*c*) Model of the C$_{60}$ overlayer growth structure on GaAs(110) (Li *et al.*, 1991c).

6.2 Molecular crystals

Before the invention of the scanning probe microscopes, molecular crystals were studied extensively by bulk-sensitive experimental methods. Little was known about the surface structure. This information can now be provided by STM and AFM down to the molecular level.

As an example of an electrically conducting organic molecular crystal, TTF-TCNQ (tetrathiafulvalene-tetracyanoquinodimethane) has been studied in detail by STM (Sleator and Tycko, 1988). Its crystal structure is schematically illustrated in Fig. 6.15. Individual molecules at the surface of TTF-TCNQ were resolved by STM (Fig. 6.16(*a*)), and the surface structure was found to agree well with the bulk crystal structure without indications of a surface reconstruction. The experimental results were compared with image simulations based on calculated contours of constant electron density from the molecular orbitals that are involved in conduction (Fig. 6.16(*b*)). The electrical conductivity of TTF-TCNQ results from the overlap of molecular orbitals producing bands of delocalized states, accompanied by charge transfer from the TTF 'valence band' to the TCNQ 'conduction band'. This gives rise to partially filled metallic-like bands contributing to the tunneling current. As a direct consequence of the metallic character of TTF-TCNQ, a

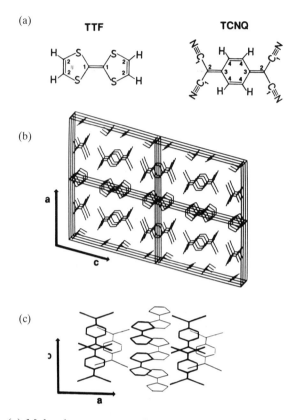

Fig. 6.15. (*a*) Molecular structures of TTF and TCNQ. (*b*) Crystal structure, showing 12 unit cells as viewed slightly off the *b* axis. (*c*) Two layers of molecules viewed perpendicular to the *a–b* plane. Molecules in heavy lines are above those in light lines (Sleator and Tycko, 1988).

significant bias-polarity dependence of the STM images was not observed.

A particularly interesting class of organic molecular crystals, the BEDT-TTF (bis(ethylenedithio)-tetrathiafulvalene) family, exhibiting transition temperatures to superconducting phases as high as about 10 K, has recently received considerable attention. STM has been applied to study the atomic and electronic structure of such materials, motivated by the desire for an improved understanding of the relationship between the molecular structure and superconductivity (Yoshimura *et al.*, 1990, 1991a, 1991b, 1991c; Kawazu *et al.*, 1991). The BEDT-TTF molecular crystals exhibit layer-type structures with an alternating stacking of

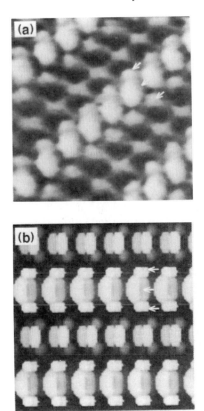

Fig. 6.16. (*a*) STM image (25 Å × 25 Å) of TTF-TCNQ. Arrows indicate a row of triplets of balls assigned to individual TCNQ molecules. (*b*) A simulated image (25 Å × 25 Å). Arrows indicate a row of TCNQ molecules along the *b* direction (Sleator and Tycko, 1988).

cation layers (BEDT-TTF) and anion layers (e.g. $Cu(NCS)_2^-$, $KHg(SCN)_4^-$ or I_3^-). It is known that the transition temperature to the superconducting state of these materials depends strongly on the structure of the anion layers which can be found on top of the surface and are therefore directly accessible for STM investigations. For $(BEDT-TTF)_2$ $[Cu(NCS)_2]$, $(BEDT-TTF)_2[(NH_4)Hg(SCN)_4]$ and $(BEDT-TTF)_2[KHg(SCN)_4]$, the surface structure as directly revealed by STM images was found to be in good agreement with the bulk structure as determined by x-ray diffraction. The insulating anion layers on top of the surfaces obviously influenced the STM image contrast. The surfaces appeared to be stable and defect-free over large surface areas.

In contrast, a high defect-density as well as a surface reconstruction were found in STM studies of (BEDT-TTF)$_2$I$_3$, which has been attributed to the unstable structure of the anion layers (I$_3$) of this molecular crystal (Yoshimura *et al.*, 1991c).

In contrast to the examples discussed so far, most molecular crystals are electrically insulating and can therefore be studied only by AFM. For instance, the surface of tetracene as the four-ring analog in the naphthalene–anthracene series of linear, fused benzene rings (Fig. 6.17) has directly been imaged with molecular resolution by means of contact force microscopy using an applied force of typically 15 nN (Overney *et al.*, 1991). The intermolecular spacings at the surface as determined from the AFM data were found to correspond closely to those in the bulk. In addition, AFM apparently distinguishes between the two translationally inequivalent molecules within the unit cell of tetracene (Fig. 6.18). The recorded AFM images of the molecular crystals appeared to be widely independent of the applied loading force.

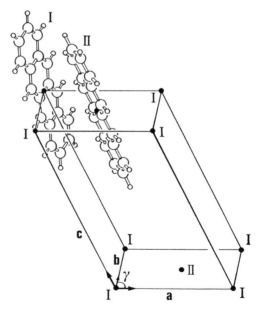

Fig. 6.17. The unit cell of crystalline tetracene in a perspective view. Representative molecules of types I and II are drawn in full to illustrate their orientation in the bulk (Overney *et al.*, 1991).

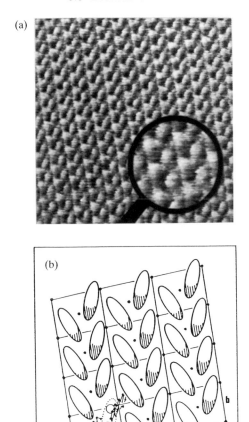

Fig. 6.18. (*a*) AFM image (10 nm × 10 nm) of the *a–b* plane of tetracene. The inequivalent molecules, I and II, create a superstructure of rows spaced 8.0 Å apart. The magnified circle is of an area about 20 Å in diameter. (*b*) Schematic view of the bulk structure as viewed perpendicular to the *a–b* plane. Each ellipse represents the two physically topmost electron orbitals of one molecule with the lower of the two being cross-hatched. A herringbone pattern is visibly built up by the ellipses. The dots represent the symmetry center of each molecule. One molecule of each type is outlined at the bottom left of the sketch (Overney *et al.*, 1991).

6.3 Biomacromolecules

Based on the successful application of scanning probe microscopies (SPM) to molecular thin films as well as molecular crystals, one might next ask about the potential of SPM techniques for investigation of large biomacromolecules. SPM studies of biological systems appear to be promising for several reasons.

1. SPM can be applied to hydrated specimens under ambient conditions. Even the operation of SPM under liquids has become routine and causes no major difficulties. This principally allows one to study biological specimens in their native environment, rather than in a frozen hydrated state under high-vacuum conditions as is typical for high-resolution transmission electron microscopy (TEM) studies.
2. Time-dependent SPM studies using fast imaging techniques allow direct observation of the *in vivo* dynamics of biological systems. SPM investigations of the structure–function relationship of biomacromolecules appear to be particularly promising.
3. SPM techniques, applied under appropriate operating conditions, are less destructive to biological specimens than is high-resolution electron microscopy, which involves bombardment of specimens with high-energy electrons.
4. SPM studies provide three-dimensional 'topographs' of biomacromolecules, whereas TEM images represent two-dimensional projections. Though the three-dimensional structure of the object can be reconstructed by tomographic techniques based on a series of two-dimensional projections with the specimen tilted at different angles during the TEM studies, the vertical resolution is usually considerably less than the lateral resolution. In contrast, a particularly high vertical resolution is offered by SPM.

However, there are major challenges involved in successful application of SPM techniques to biological systems, which are mainly related to issues of appropriate specimen preparation procedures and theoretical understanding of SPM image contrast for thick biomacromolecules (Engel, 1991; Guckenberger *et al.*, 1992; Nawaz *et al.*, 1992). In particular, the following problems have to be solved.

1. The biomacromolecules have to be fixed to a suitable substrate. The surface roughness of the chosen substrate has to be much smaller than the size of the biomacromolecules for clear identification of the observed features. Furthermore, the biomacromolecules must be strongly attached to the substrate to allow high-resolution imaging by SPM. However, strong adsorption forces can have a significant influence on the appearance of the biomacromolecules which has to be taken into account for image interpretation. To be able to image biomacromolecules that are only weakly bound to the substrate, a new 'hopping mode' of SPM operation has been introduced (Jericho

et al., 1989). In this particular scanning mode, the tip is retracted from the sample whenever the tip moves laterally to a new position, where a new approach is made. This way, lateral forces due to tip–sample interactions, which might cause lateral displacement of the biomacromolecule to be imaged, can be minimized.

2. To facilitate the search for biomacromolecules deposited onto a suitable substrate, combination of an SPM instrument with a light microscope is highly favorable. In addition, the SPM instrument has to allow for coarse sample positioning within the x–y plane (Michel and Travaglini, 1988; Stemmer *et al.*, 1988; Guckenberger *et al.*, 1988, 1992; Putman *et al.*, 1992).

3. Most biomacromolecules are electrically insulating. Therefore, the applicability of STM appears to be limited. STM studies of freeze–fracture replicas or biological specimens coated by electrically conducting Pt/C or Pt/Ir/C films have been proposed as possible solutions. However, this would prevent direct investigation of biological systems in their native environment as well as observation of the dynamics of biological processes, and one ends up with exactly the same limitations as are already encountered in conventional high-resolution electron microscopy studies. Fortunately, there exist numerous examples of STM studies where thick uncoated organic material has successfully been imaged, though in some cases it has been necessary to operate the STM with extremely low tunnel currents (on the order of 1 pA) and with elevated sample bias voltages (5–10 V) (Guckenberger *et al.*, 1989, 1991). To understand the STM results obtained for large biomacromolecules, an improved understanding of their electronic structure as well as their microscopic transport properties is required (DeVault, 1981; Joachim and Sautet, 1990; García and García, 1990; Lindsay *et al.*, 1990; Lindsay and Sankey, 1992). In addition, the influence of electrically conducting surface films, the high inhomogeneous electric fields present between STM tip and sample, and the locally applied pressure due to tip–sample interaction forces have to be considered for appropriate interpretation of the observed STM image contrast.

4. The application of force microscopy does not require electrically conducting specimens. However, another problem arises because large biomacromolecules tend to be soft and can easily be deformed by significant tip–sample interaction forces. Non-contact force microscopy performed in the attractive force regime (section 2.7) with applied forces on the order of 10^{-11} N or even less is expected to be the most appropriate SFM technique to study biological specimens (Persson, 1987). On the other hand, contact force microscopy per-

formed in the repulsive force regime is likely to induce significant deformations of soft specimens leading to limitations in the spatial resolution which can be achieved (Weihs *et al.*, 1991). It has been proposed to freeze the biological specimens, in order to increase their stiffness, and to perform low-temperature AFM studies (Prater *et al.*, 1991b). However, this experimental procedure again prevents investigation of the structure and dynamics of biological systems in their native environment.

Other SPM techniques, such as scanning near-field optical microscopy (Betzig *et al.*, 1986), scanning ion conductance microscopy (Hansma *et al.*, 1989), or scanning electrochemical microscopy (Wang *et al.*, 1989; Bard *et al.*, 1990) appear to be promising for application to biological systems as well. However, the spatial resolution offered by these techniques on biological specimens needs to be further improved before significant contributions to biology can be expected.

6. Perhaps the most serious limitation for the application of SPM techniques to biological systems is given by the three-dimensional shape of biomacromolecules which makes them highly susceptible to tip imaging artifacts (Guckenberger *et al.*, 1992).

Despite the problems encountered in SPM studies of biological systems, it might currently be the most rapidly growing field of application of SPM techniques.

6.3.1 Nucleic acids

Most STM studies of biological macromolecules have concentrated on DNA (deoxyribonucleic acid), which has been imaged in air (Beebe *et al.*, 1989; Lee *et al.*, 1989; Arscott *et al.*, 1989; Dunlap and Bustamante, 1989; Li *et al.*, 1991a; De Stasio *et al.*, 1991), in aqueous solutions (Lindsay *et al.*, 1989a, 1989b, 1989c), as well as under ultra-high vacuum (UHV) conditions (Driscoll *et al.*, 1990; Youngquist *et al.*, 1991, 1992). The primary focus is on DNA for several reasons.

1. DNA is a particularly important biomacromolecule. It contains the information which widely determines the properties of living organisms.

2. The structure of DNA has already been well characterized, e.g. by x-ray diffraction. Therefore, the structural parameters derived from STM data can directly be checked.

3. The thickness of a DNA double helix is only on the order of 2 nm, which might allow direct tunneling through the molecule.

4. The information contained in the sequence of nucleotides is linearly encoded and therefore, at least in principle, accessible to surface-sensitive SPM techniques.
5. The potential of SPM to sequence DNA has stimulated increasing efforts towards atomic-scale imaging of DNA.

STM studies of DNA have almost exclusively been performed using graphite substrates. This might not be the optimum choice because as-cleaved graphite surfaces sometimes exhibit structural features which can mimic the presence of DNA (Clemmer and Beebe, 1991). Therefore, STM images of 'DNA' on graphite substrates generally have to be interpreted with great care.

The highest resolution STM images of DNA were obtained under UHV conditions (Driscoll *et al.*, 1990). The double-helical structure, the base pairs, as well as atomic-scale features have directly been revealed in real-space, as shown in Fig. 6.19. The experimental STM profiles along particular directions were found to be in excellent agreement with atomic contours of the van der Waals surface of DNA as derived from x-ray crystallography. STM data were obtained at a constant tunneling current of 1 nA and a sample bias voltage of $+100$ mV. Since the DNA adsorbate wave functions lie far from the Fermi level, it is unlikely that they support electron conduction for these tunneling parameters. However, the interaction of the adsorbate wave functions with the tails of the substrate wave functions may locally enhance the probability for electron tunneling between tip and substrate at the adsorption site, which would explain the observed STM image contrast (Youngquist *et al.*, 1991, 1992).

With further improvements in image quality, STM-based DNA sequencing may indeed become possible. Two key issues would then have to be addressed, i.e. the limit of sequencing speed and the error rate associated with a particular sequencing speed.

6.3.2 Proteins

The structure of proteins, for which three- or two-dimensional crystals are available, can be determined with high accuracy by well-established techniques such as x-ray or electron crystallography. Randomly distributed biomacromolecules such as proteins can be studied by electron tomography, although with limited vertical resolution. SPM appears to be promising for directly revealing the three-dimensional shape of proteins at high vertical as well as lateral resolution. However, several difficulties arise.

(a)

(b)

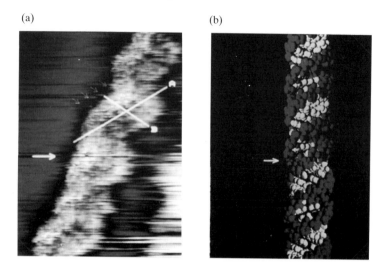

(c)

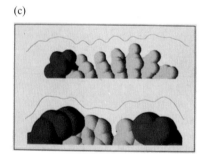

Fig. 6.19. (*a*) STM image (80 Å × 120 Å) of DNA. (*b*) Model of the van der Waals surface of A-DNA derived from x-ray crystallographic data, scaled to (*a*). (*c*) Experimental STM tip trajectories following lines marked A and B in (*a*), compared with the corresponding atomic contours of an A-DNA van der Waals surface model. In each case, the experimental cross-section is placed above the corresponding region of the model (Driscoll *et al.*, 1990).

1. Accurate determination of the three-dimensional shape of globular proteins by tip-based microscopies is often prevented by the finite size of the probe tip as well as tip-induced imaging artifacts.
2. STM studies of uncoated thick proteins prove to be difficult because of their low electrical conductivity.
3. Thick proteins tend to be extremely soft and therefore require exceptionally small loading forces in AFM experiments.

To be able to perform STM studies of proteins, two different experimental procedures are commonly used.

1. Coating of freeze-dried proteins with electrically conducting thin films (e.g. Pt–Ir–C) allows stable electron tunneling (Amrein *et al.*, 1988, 1991; Stemmer *et al.*, 1989). However, the finite grain size of the coating film sets a limit for the spatial resolution practically achieved for the specimen.

2. STM images of uncoated proteins may be obtained at sufficiently high humidity (Amrein *et al.*, 1989; Guckenberger *et al.*, 1989; Mou *et al.*, 1991). In addition, extremely low tunneling currents (< 1 pA) as well as high sample bias voltages (up to 10 V) may sometimes help to obtain stable experimental conditions, resulting in STM images of improved quality (Guckenberger *et al.*, 1989).

Detailed STM studies have been performed on coated as well as uncoated recA–DNA complexes (Amrein *et al.*, 1988, 1989; Travaglini *et al.*, 1988). In Fig. 6.20, STM images of such complexes with their helical structure and a diameter of about 10 nm are shown. The thin filaments in the overview image were interpreted as free DNA. To achieve stable imaging of the uncoated specimens, it was necessary to keep the samples humid.

A humidity in the range between 30% and 45% was required to obtain stable STM images of an uncoated HPI (hexagonally packed intermediate) layer (Fig. 6.21), a protein which in nature forms a two-dimensional ordered array in the cell wall of the bacterium *Deinococcus radiodurans* (Guckenberger *et al.*, 1989). In these experiments, tunneling currents as low as 0.1 pA and sample bias voltages as high as 10 V were used. For metal-coated specimens (Fig. 6.22), a tunneling current of 10 pA and a bias voltage of 1 V proved to be suitable (Amrein *et al.*, 1991).

Even under optimized experimental conditions, the spatial resolution in STM images of globular proteins is usually limited to a few nanometers and therefore does not exceed that obtained by competing experimental techniques. Only STM studies of filamentous proteins having diameters on the order of 1–2 nm appear to be more promising (McMaster *et al.*, 1990; Miles *et al.*, 1991). Another application of SPM, which may become particularly important, is given by direct visualization of the dynamics of biological processes. A first model study was performed using AFM to observe the thrombin-catalyzed polymerization of the protein fibrin (Drake *et al.*, 1989). Polymerization of this protein is known to be responsible for blood clotting involved in the healing of wounds as well as the initiation of heart attacks. Fibrinogen

(a)

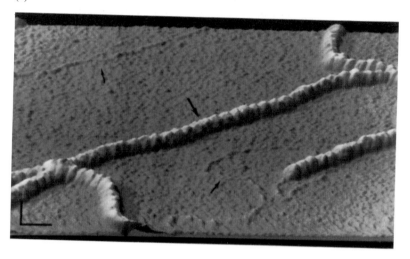

(b)

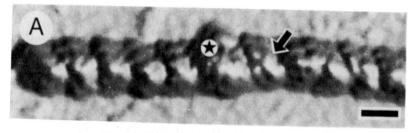

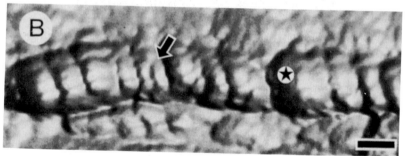

Fig. 6.20. (*a*) STM image (236 nm × 192 nm) of freeze-dried recA-DNA complexes coated with an electrically conducting Pt–Ir–C film. Large arrows indicate recA-DNA complex; small arrows indicate free DNA (Amrein *et al.*, 1988). (*b*) Comparison of STM images of (A) a freeze-dried and metal-coated recA-DNA complex and (B) an uncoated complex. The characteristic features appear in good agreement for the two methods of sample preparation, giving confidence in the reproducibility of STM imaging of such highly corrugated specimens. However, the uncoated sample appears to be better resolved. The scale bar corresponds to 10 nm (Amrein *et al.*, 1989).

(a)

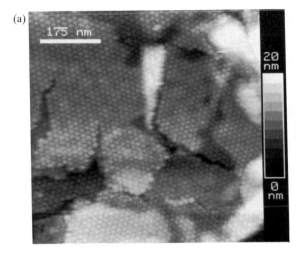

(b)

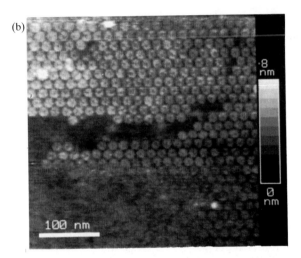

Fig. 6.21. (*a*) STM image of an uncoated HPI layer. Several sheets are locally piled up on top of each other. (*b*) Another STM image of the uncoated HPI layer at higher magnification. The pores approximately 3 nm wide in the hexameric complexes can be seen clearly (Guckenberger *et al.*, 1989).

molecules were first dissolved in an appropriate solution and a few drops of the solution were placed on a mica substrate. While imaging the mobile fibrinogen molecules on the mica substrate, a few drops of a second solution containing the enzyme thrombin were added, inducing polymerization of the fibrinogen molecules. As a consequence of the

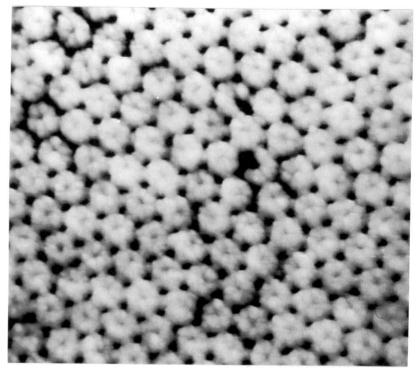

Fig. 6.22. STM image of the outer surface of an HPI layer, freeze-dried and coated with a Pt–Ir–C film. Most of the protein mass of the hexameric complexes is concentrated in a core region that contains a central cavity. The cores are interconnected via extensions, the spokes, across the twofold axis, leaving open large holes around the threefold axis (Amrein *et al.*, 1991).

polymerization process, large immobile aggregates were formed, which further grew to chain-like structures and finally to a fibrin net, as directly revealed by a temporal sequence of AFM images (Fig. 6.23). Though the spatial resolution was moderate in this AFM study, it nevertheless has shown the potential of real-time and real-space SPM observations of biological processes.

6.3.3 Membranes

Membranes play an important role in separation of different parts of a biological system, though allowing for transfer of specific species. In principle, the quasi-two-dimensional structures of membranes make them suitable for high-resolution SPM studies. Two different types of membranes have predominantly been studied by SPM.

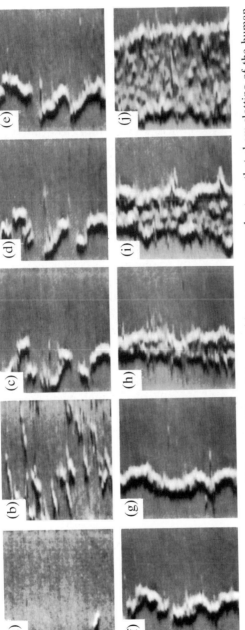

Fig. 6.23. Ten AFM images (450 nm × 450 nm) from a video cassette recorder tape that show clotting of the human blood protein fibrinogen in real time. The images were selected from before introduction of the clotting enzyme thrombin (a), and at various times after its introduction: 9 min, 10 min 20 s, 10 min 30 s, 11 min 20 s, 12 min 10s, 12 min 40 s, 14 min 50 s, 17 min 10 s, 33 min for (b) through (j) (Drake et al., 1989).

1. Model membranes built up from synthetic layers of fatty acids or lipids, usually prepared using the LB technique (section 6.1.4).
2. Membrane crystals which occur naturally or have been synthesized.

A particularly well-studied natural membrane crystal is the purple membrane from *Halobacterium halobium*. It consists of a two-dimensional crystalline array of a single protein called bacteriorhodopsin, forming a hexagonal lattice with a lattice constant of 6.3 nm. The structure of bacteriorhodopsin is known from electron crystallography to a resolution of 0.35 nm parallel and 0.7 nm perpendicular to the membrane. The thickness of purple membrane is about 4.8 nm according to x-ray diffraction studies of membrane stacks. To be able to perform STM studies of such thick uncoated membrane crystals, it has been found favorable to operate the STM at extremely low tunnel currents (< 1 pA) and exceptionally high sample bias voltages (5–15 V) (Guckenberger et al., 1991). However, by using these tunneling parameters, only moderate lateral resolution has been achieved, preventing observation of the hexagonal lattice of period 6.3 nm. On the other hand, AFM studies of purple membranes adsorbed on a mica surface in buffer solution directly revealed the hexagonal arrangement of the bacteriorhodopsin molecules with a lateral resolution of about 1.1 nm (Butt, et al., 1991).

More promising applications of SPM to membranes may be found in determination of ionic channel structures (Kolomytkin et al., 1991), as well as direct probing of ion currents through membrane channels by the SICM (Hansma et al., 1989).

7

Metrology and standards

7.1 Nanometrology

Nanometrology is defined as the science of measuring the dimensions of objects or object features to uncertainties of 1nm or less. The demand for nanometrology comes together with advances in integrated circuit technology where uncertainty requirements, e.g. in mask alignment, will soon approach the length scale 1nm (Teague, 1992). The achievement of atomic-resolution real-space imaging of single-crystal surfaces by SPM has opened up novel opportunities in the field of nanometrology.

1. The highly-ordered atomic lattice of a single-crystal surface can serve as a reference against which the position and motion of an object can be measured and controlled. This idea has triggered the development of a dual tunnel-unit STM (Fig. 7.1), where one tunnel unit is used to provide a crystal reference for the second unit (Kawakatsu and Higuchi, 1990; Kawakatsu et al., 1991). To obtain a reference lattice over technologically relevant areas, there is a strong need for single-crystal surfaces being atomically flat over extended surface regions without the presence of steps or dislocations.
2. Piezoelectric crystals have proved to allow highly accurate and repeatable motion down to a subatomic length scale. This is clearly demonstrated by SPM images showing regular two-dimensional crystal lattices with measured corrugation amplitudes below 0.1 Å.

However, several problems have to be solved before successful application of SPM in nanometrology can be achieved.

1. Tip instabilities associated with switching of either the vertical or lateral position of the tip with respect to the substrate must be eliminated by preparation of highly stable tips with reliable long-term performance.

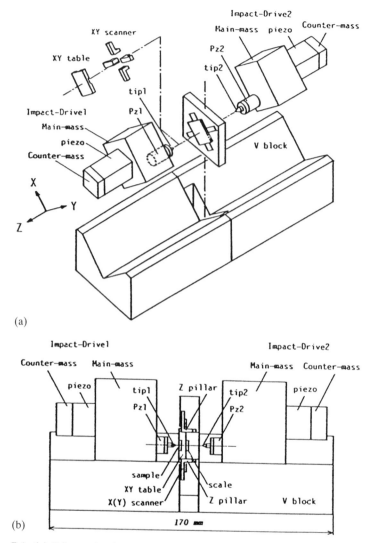

Fig. 7.1. (*a*) Schematic view of the dual-tunneling-unit STM. (*b*) Cross-section of the dual-tunneling-unit STM (Kawakatsu and Higuchi, 1990).

2. Possible tip and sample deformations due to tip–sample interaction forces have to be controlled and/or reduced below an acceptable level.

3. Hysteresis, creep and other nonlinearities of the piezoelectric positioning devices have to be controlled and/or reduced below an acceptable level.

4. Thermal drifts have to be kept as small as possible by a drift-compensated SPM design as well as a temperature-controlled environment.
5. A high degree of mechanical stability without disturbances caused by external vibration sources over extended time periods is needed.
6. The SPM has to be incorporated into an appropriate metrology reference frame.

Solutions to some of the above-mentioned problems have been proposed within the 'molecular measuring machine' project started in 1987 (Teague, 1989). An ultra-high-accuracy measuring machine has been designed with the goal of obtaining a point-to-point spatial resolution of 0.1 nm for the distance between any two points within a 50 nm $\times$ 50 nm $\times$ 100 μm volume, with a net uncertainty for point-to-point measurements of 1.0 nm. A spherical core structure accommodating a SPM instrument was chosen for its high mechanical stiffness and ease of temperature control. This core structure is located at the center of mass and in the plane of the vibrationally supported components, thereby minimizing any cross-coupling between translational and rotational excitations (Fig. 7.2(a)). An ultra-high-resolution heterodyne interferometer combined with the SPM unit (Fig. 7.2(b)) allows measurements of the probe tip position relative to the specimen which is mounted on a metrology box. This way, any remaining drift in the tip position and nonlinearities in the piezo-driven tip motion can directly be measured and controlled. It remains to be explored whether the 'molecular measuring machine' will meet all requirements necessary to become a valuable tool for nanometrology and therefore for the rapidly growing field of nanotechnology (chapter 8) as well.

7.2 Quantum standards

SPM techniques could possibly have an impact on the refined definition or exploitation of several different standards (Teague, 1992).

1. An improved quantum standard of mass could be based on a more accurate knowledge of Avogadro's number by determining the number of atoms in a cubic centimeter of pure silicon. Scanning probe techniques have the potential to measure the dimensions of that cube with unprecedented accuracy.
2. An improved quantum standard of current (Guinea and García, 1990) could be based on single-electron tunneling effects, as discussed in section 1.3.3, which can be observed by STM, even up to room temperature.

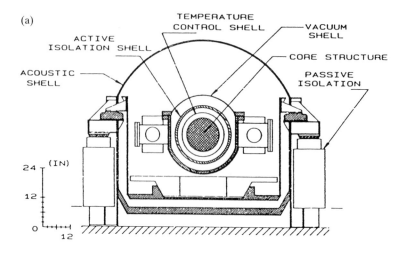

(a)

TEMPERATURE
CONTROL SHELL

VACUUM
SHELL

ACTIVE
ISOLATION SHELL

CORE STRUCTURE

ACOUSTIC
SHELL

PASSIVE
ISOLATION

24 (IN)

12

0

12

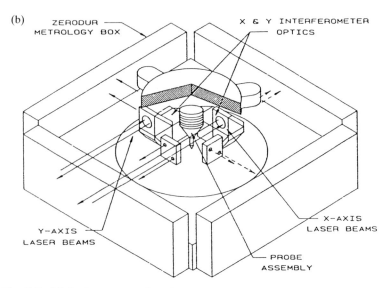

(b)

ZERODUR
METROLOGY BOX

X & Y INTERFEROMETER
OPTICS

Y-AXIS
LASER BEAMS

X-AXIS
LASER BEAMS

PROBE
ASSEMBLY

Fig. 7.2. (*a*) Scale cross-section drawing of the environmental isolation system for the molecular measuring machine. (*b*) Schematic drawing of the metrology reference system and the *x* and *y* interferometer arrangement. The interferometer beams for *x* and *y* displacement straddle the probe tip to minimize Abbé's offset errors (Teague, 1989).

3. The time standard is based on a ^{133}Cs transition associated with a frequency of approximately 9 GHz. In this regime, accurate direct frequency measurements may be performed. On the other hand, accurate wavelength measurements are restricted to the visible portion of the electromagnetic spectrum, corresponding to frequencies of several hundred terahertz. In principle, complex chains of harmonic generation and mixing of maser and laser signals can be formed to link the visible laser frequency to the ^{133}Cs frequency. An improved understanding of the phenomena involved in the mixing of two frequency beams by an adjustable metal – vacuum – metal diode, as provided by an STM (section 1.17.2), could allow improvement of this link between the two different frequency regimes (Sullivan and Cutler, 1992).

8
Nanotechnology

With the increasing degree of miniaturization in microelectronics, we unquestionably become confronted with structures of matter on a scale below 100 nm. The transition from microtechnology (lateral dimensions of 0.1–100 µm) to nanotechnology (lateral dimensions of 0.1–100 nm) requires the ability to fabricate smaller structures as well as the exploration and application of new physical phenomena occurring on the refined scale which becomes comparable to the characteristic lengths associated with the elementary processes in physics (e.g. the electron mean free path). In particular, entering the nanotechnology age involves the following tasks:

nanopositioning and nanocontrol of processes,
nanoprecision machining,
finding natural supersmooth surfaces and/or creating them,
fabrication of nanometer-scale structures,
analysis of nanometer-scale structures,
understanding the physical properties of matter on a nanometer scale,
development of nanometer-scale devices,
finding appropriate architectures for nanometer-scale structures, and
linking the 'nanoscopic' to the macroscopic world.

With the invention of STM and related scanning probe methods, we have been equipped in good time with the appropriate tools to attack most of these tasks, and there can be no doubt that STM-based technology and nanotechnology will interact closely in the coming decades.

STM-based technology has enabled us to control the position and motion of arbitrarily small objects down to a sub-Ångström unit scale (section 1.10).

STM and AFM can nowadays routinely be used to control the roughness of surfaces from a millimeter down to the atomic scale.

SPM instruments allow fabrication of nanometer-scale structures down to the atomic level by using a variety of different methods (section 8.1). It has even proven feasible to build up well-defined nanometer structures by appropriately arranging individual atoms.

Local probes offer the appropriate experimental tools to analyze nanometer-scale structures individually.

SPM contributes significantly to our improved understanding of nanometer-scale properties of matter, as is well documented by numerous examples presented in chapters 4 and 5.

Initial trials to realize nanometer-scale devices based on STM technology have already been reported (section 8.2).

The close interaction between scanning probe methods and silicon-based microfabrication appears to be very fruitful (section 8.3) and will help to link the 'nanoscopic' to the macroscopic world.

The importance of the emerging field of nanotechnology is certain (Franks, 1987).

Nanotechnology can be regarded as an 'enabling' technology because it provides the basis for many other technological developments.

Nanotechnology is also a 'cross-sectional' technology because one particular technique may, with slight variations, be applicable in widely differing fields.

8.1 Fabrication of nanometer-scale structures

The applications of SPM discussed in chapters 4–7 have concentrated on characterization of given surface structures with the intention of being non-destructive to the specimens under investigation. Non-destructive SPM analysis can be achieved if the tip–sample interaction is kept sufficiently small by appropriately adjusting the experimental parameters. The sets of parameters leading to a non-destructive SPM operation generally depend on the choice of sample and tip as well as on the type of interaction considered. By increasing the tip–sample interaction above a critical value, local modification of the sample surface results. These local modifications, which are undesirable during normal SPM operation, can advantageously be used for fabrication of small structures on a nanometer length scale (Shedd and Russell, 1990; Quate, 1990; Wiesendanger, 1992).

To be able to distinguish the 'written' features from structural features of the substrate, its surface has to be sufficiently 'flat' with a local surface roughness considerably less than the size of the written features. Substrates which have most widely been used include:

cleaved layered materials (e.g. graphite, transition metal dichalcogenides, mica, layered copper oxides etc.),

single-crystal semiconductors prepared under ultra-high vacuum (UHV) conditions (e.g. Si, Ge etc.),

single-crystal metals prepared under UHV conditions (e.g. Au, Pt, Ni etc.),

metal thin films (e.g. Au epitaxially grown on mica),

gold balls prepared by melting a gold wire in the flame of an oxyacetylene torch,

amorphous materials, either semiconductors or metals, after surface oxide removal by ion etching.

The technological potential of a given writing procedure can be judged by considering the following key issues:

size of the written structures,

error rate in the writing process,

writing speed,

environmental conditions for the writing process,

temporal stability of the written structures for a given environment.

The efforts towards SPM-based fabrication of nanometer-scale structures are primarily motivated by the smallness of the features which can directly be written onto various solid substrates. Conventional high-resolution electron beam lithography is based on using electron resist films (e.g. PMMA) which are exposed by well-focused high-energy electrons. These high-energy electrons can penetrate several micrometers into the solid substrates of these films, thereby generating showers of secondary electrons which spread out into a volume having a size much larger than the diameter of the primary electron beam. Consequently, the resist also becomes exposed by these secondary electrons which are spatially much less confined and therefore set a lower limit to the size of the written structures (Broers, 1988).

In contrast, STM-based writing, for instance, involves the use of low-energy electrons, and electron resist films are not necessarily required for successful fabrication of small structures. Moreover, STM allows one to address individual atomic sites, thereby offering the opportunity to manipulate matter down to the atomic level! On the other hand, writing with SPM instruments as well as with scanned beam methods has an inherent drawback: it involves a slow serial process rather than a fast parallel process as, for instance, in photolithography. Introducing parallelism by fabricating arrays of simultaneously operating miniaturized SPM instruments (section 8.3) may partially solve this problem.

In any case, atomic-scale writing by SPM offers a unique opportunity to fabricate atomically well-defined test structures for investigation of physical phenomena in small dimensions. Understanding these phenomena is of crucial importance for development of nanodevices, whether they are to be fabricated by SPM instruments or by any other means (Chang *et al.*, 1988). In the following, we will describe various methods of fabrication for nanometer-scale structures using SPM instruments. These methods can be grouped according to the microscopic mechanism involved in the writing process.

8.1.1 Mechanical surface modifications

Local surface modifications can be achieved, for instance, by using the force interaction between tip and sample, either in an STM-type or an AFM-type configuration. Depending on the amount of applied force and mode of SPM operation, different types of 'mechanical' surface modifications from sub-micrometer down to the atomic scale have been demonstrated.

8.1.1.1 Scratching

By scanning the tip of an STM or AFM while tip and substrate are in contact, sub-micrometer-wide lines can be drawn simply by scratching the surface. This was first demonstrated by using the STM as a micromechanical tool with a tungsten tip penetrating into a thin insulating film of polycrystalline calcium fluoride (thickness 20 nm) deposited on a silicon substrate (McCord and Pease, 1987a). By operating the STM at a constant tunneling current of 2 nA and a sample bias voltage of 1 V, the tip was found to be in contact with the insulating film and therefore produced scratches by locally removing the film during the scanning process. These scratches appeared as parallel lines 0.36 μm wide with a spacing of about 1.4 μm. Neither the tungsten tip nor the silicon substrate appeared to be damaged as a result of the surface modification procedure.

The STM scratching technique has also been applied to conducting oxide substrates such as $Rb_{0.3}MoO_3$ (Garfunkel *et al.*, 1989; Rudd *et al.*, 1991). By lowering the sample bias voltage to about 10 mV, the tip was forced to penetrate the first atomic layer to maintain the required current (0.25 nA). Consequently, lines of less than 1 nm depth and 10 nm width could be drawn by scanning the tip at constant current. Square-shaped pits have been produced as well by scanning a corresponding surface area. Similar local modifications by means of surface abrasion with an STM tip have been demonstrated for a variety of layered-type substrates, such as transition metal dichalcogenides (Parkinson, 1990) or high-temperature superconducting copper oxides (Albrecht, 1989; Harmer *et al.*, 1991).

Alternatively, the AFM can also be used as a micromechanical tool to produce scratches and pits on surfaces. In contrast to the STM, for which the strength of the tip–sample interaction is indirectly controlled via the current and the applied sample bias voltage, the AFM allows direct measurement and control of the tip–sample force interaction. The first demonstration of controlled surface modification with the AFM was given by the local removal of a monomolecular film of octadecyl phosphate ($C_{18}P$) which had been deposited on a mica substrate (Albrecht, 1989). This was accomplished by increasing the applied force above 10^{-6} N while scanning the surface. The lateral resolution of the writing process appeared to be on the order of 10 nm. Lines of width 70 nm and depth 10 nm were produced on a polycarbonate substrate also by using the AFM scratching method (Jung *et al.*, 1992). In these experiments, the applied force on the Si_3N_4 cantilever had to be set to a value above 5×10^{-8} N for successful surface modification, thereby directly indicating the limit for local plastic deformation of the sample which depends on the tip geometry as well as on the elastic and plastic properties of the polycarbonate substrate.

8.1.1.2 Indentations

By increasing the tip–sample interaction at a particular surface location, indentations can be produced with an STM or AFM tip. This was demonstrated, for instance, in STM experiments performed with a tungsten tip and clean silicon surfaces prepared under UHV conditions where holes of 2–10 nm diameter have been created (van Loenen *et al.*, 1989, 1990b). Instead of operating the STM at a constant tunnel current, the feedback circuit was interrupted while the tip was moved vertically towards the sample over an apparent distance of typically 2 nm. Upon retracting the tip to its original level, the feedback circuit was activated again, allowing the tip to be moved controllably laterally to a new position. By choosing distances between indentations which are much smaller than their lateral dimensions, continuous lines were obtained having a width of 2–5 nm (Fig. 8.1). The single indentations as well as the lines were found to be stable over extended time periods.

Indentations can also be performed by locally increasing the applied force in an AFM experiment. By modulating the force acting on the cantilever during scanning, continuous lines can be drawn as a result of

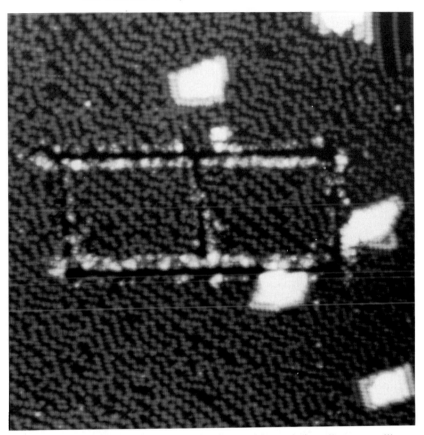

Fig. 8.1. STM image of a character '8' written with continuous lines on a silicon substrate. The dimensions of the character '8' are 70 nm × 35 nm. The width of the lines is 2–5 nm (van Loenen *et al.*, 1989).

a sequence of local indentations (Jung *et al.*, 1992). This was demonstrated for a polycarbonate surface of a compact disc (CD), where 'HEUREKA' has been written in between two information pits of the CD separated by 1.6 μm (Fig. 8.2). The type of nanometer-scale structure resulting from a local indentation process depends critically on the chosen substrate. For instance, atomic-resolution STM studies have shown that indentation of a tungsten tip into a transition metal dichalcogenide (WSe_2) surface can lead to a nanometer-scale plastic surface deformation without altering the in-plane atomic order of the deformed surface areas (Fuchs *et al.*, 1990b). To perform the indentation, a voltage pulse a few milliseconds in length was applied to the z piezodrive, leading to tip excursions of up to 100 Å. However, the depth of the

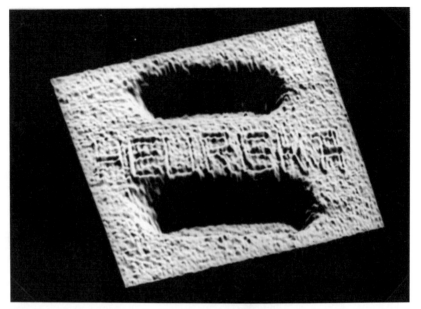

Fig. 8.2. 'HEUREKA' written with an AFM between the information pits of a compact disk. The letter height is 700 nm and the indentation depth is 10 nm (Jung *et al.*, 1992).

resulting indentation was found to be less than 1 nm, whereas the width was typically on the order of 3 nm. The indentations appeared to have a hexagonal shape, reflecting the symmetry of the substrate lattice.

Indentation experiments have also been performed using metal substrates such as thick polycrystalline silver films prepared under UHV conditions (Gimzewski *et al.*, 1987a). If either the STM tip or the sample surface were covered by a surface oxide layer, the resulting nanometer-scale structure was found to be a local surface depression. On the other hand, if both the metal tip and the metal substrate are clean, the metal–metal adhesion forces can lead to formation of a thin continuous neck on slightly retracting the tip from the substrate. Upon further tip retraction, the neck finally breaks up and a small nanometer-scale protrusion is left on the surface, rather than a local depression.

8.1.1.3 Pulling and pushing single atoms

By appropriately adjusting the tip–sample force interaction, the ultimate in 'mechanical' surface modification can be achieved: pulling and pushing of an individual atom on a substrate surface. This was first demonstrated using a low-temperature UHV STM (Eigler and

Schweizer, 1990). The system studied was Xe adsorbed on Ni(110) which could be imaged non-destructively with a tunnel current of 1 nA and a bias voltage of 10 mV. To move an adsorbed Xe atom along the substrate surface, the STM tip was first positioned above the selected atom (Fig. 8.3(*a*)). The force interaction of the tip with this Xe atom, being primarily due to van der Waals interaction, was subsequently increased by making the tip approach toward the atom (Fig. 8.3(*b*)). This was achieved by increasing the demanded tunnel current to about 16 nA. The appropriate adjustment of the tip–adsorbate force interaction for the sliding process is critical. On the one hand, this interaction has to be strong enough to allow the adsorbed atom to overcome the energy barrier to the neighboring substrate lattice position. On the other hand, the tip–adsorbate interaction has to be kept considerably smaller than the adsorbate–substrate interaction in order to prevent transfer of the adsorbate from the substrate to the tip. The tip was then moved laterally under constant current conditions while dragging the Xe atom with it (Fig. 8.3(*c*)). Upon reaching the desired destination (Fig. 8.3(*d*)), the tip was finally withdrawn by decreasing the set-value of the tunnel current, leaving the Xe atom at the new position (Fig. 8.3(*e*)).

By applying the sliding procedure to other adsorbed atoms as well,

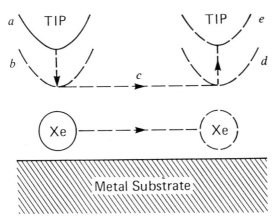

Fig. 8.3. Schematic illustration of the process for sliding an atom across a surface by means of an STM tip. The atom is located and the tip is placed directly over it (*a*). The tip is lowered to position (*b*), where the atom–tip attractive force is sufficient to keep the atom located beneath the tip when the tip is subsequently moved across the surface (*c*) to the desired destination (*d*). Finally, the tip is withdrawn to a position (*e*) where the atom–tip interaction is negligible, leaving the atom bound to the surface at a new location (Eigler and Schweizer, 1990).

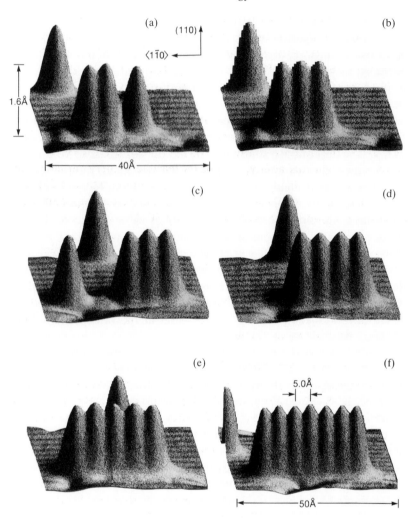

Fig. 8.4. Various stages in the construction of a linear chain of xenon atoms on a Ni(110) surface. The individual xenon atoms appear as protrusions 1.6 Å high in these images. The rows of nickel atoms are visible as alternating light and dark stripes. (*a*) The assembled xenon dimer. To the right of the dimer, a xenon atom has been moved into position to form a xenon trimer. (*b*) Formation of the xenon linear trimer. (*c*)–(*f*) Various stages in construction of the linear heptamer, a process that can be completed in an hour. The xenon atoms are 5 Å apart, occupying every other unit cell of the nickel surface (Eigler and Schweizer, 1990).

overlayer structures of ones own design can be fabricated, atom-by-atom. This is illustrated in Fig. 8.4, showing a linear chain of seven Xe atoms along the [1$\bar{1}$0] direction of the Ni(110) substrate which has successively been built up. The spacing between the Xe atoms along the linear chain is determined by the corresponding lattice constant of the underlying Ni(110) substrate.

This type of atomic-scale modification experiment has indicated that the 'great time', envisioned by Feynman (1961) has already come: 'we can arrange the atoms the way we want; the very atoms, all the way down!' Based on this ability, it appears feasible to perform chemical synthesis by appropriately arranging individual atoms with an STM tip. Well-defined nanometer-scale test structures can be built up, and their properties can directly be probed locally by SPM.

8.1.2 Chemical surface modifications

The finely focused 'beam' of low-energy electrons offered by an STM is ideally suited to induce local chemical reaction processes, which provides another method for fabrication of artificial surface structures from sub-micrometer down to atomic scale.

8.1.2.1 Exposure of electron beam resists

The first investigation of any controlled fabrication of nanometer-scale structures by STM was performed using amorphous metal substrates covered by about two monolayers of hydrocarbons and carbon–oxygen species (Ringger *et al.*, 1985, 1986). By operating the STM with a constant tunneling current of about 10 nA and a bias voltage of 100 mV, it was possible to draw conducting carbon lines, presumably as the result of a polymerization of the hydrocarbon film acting as an electron beam resist. Fig. 8.5(*a*) shows a scanning electron microscope (SEM) image of a regular line pattern which was written by STM on a glassy $Pd_{81}Si_{19}$ substrate. The average spacing between the lines is about 16 nm, whereas the line width is less than 2 nm. As an initial attempt to fabricate a nanojunction, crossed lines were generated as well (Fig. 8.5(*b*)).

A variety of electron beam resists, such as hydrocarbons, metal halides, PMMA, and LB films, have been exposed by STM operated in the field emission mode with bias voltages between 5 V and 200 V and currents between 10 pA and several micro-amperès (McCord and Pease, 1986a, 1986b, 1987b, 1988; McCord, 1987; Zhang *et al.*, 1989; Marrian and Colton, 1990; Marrian *et al.*, 1990). Line widths below 100 nm were routinely achieved and features down to 20 nm size were generated with an electron beam energy just above the exposure thresh-

(a)

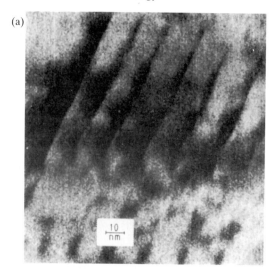

(b)

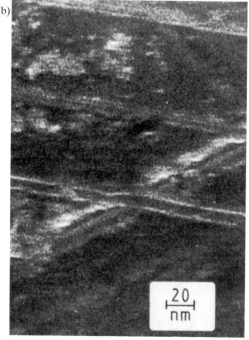

Fig. 8.5. (*a*) SEM image of lines written by the STM on a glassy $Pd_{81}Si_{19}$ substrate. (*b*) SEM image of a junction made with the STM on glassy $Pd_{81}Si_{19}$
(Ringger *et al.*, 1985).

old of the resist. A significant advantage of STM-based lithography over conventional electron beam lithography is gained if the STM is operated with 'reversed' polarity so that electrons are field-emitted from the sample up through the resist (McCord and Pease, 1988). This avoids problems associated with resist exposure by secondary or backscattered electrons which limit resolution in conventional electron beam lithography.

8.1.2.2 Local oxidation

By using an STM operated in air, local oxidation of passivated semi-conductor surfaces, such as hydrogen-passivated silicon, sulfur-passivated GaAs or arsenic-capped III–V semiconductors, can be induced leading to permanent nanometer-scale oxide patterns on these substrates (Dagata *et al.*, 1990a, 1990b, 1991a, 1991c). This modification procedure clearly requires an oxygen ambient. The lateral resolution and thickness of the oxide patterns were found to depend on the strength of the electric field present between tip and substrate surface, which is ultimately determined by the bias voltage, the tip–substrate spacing (and therefore the set-point current), and the tip geometry. The localized chemical change can be confined to a depth of a few nanometers below the surface, and typical line widths are on the order of 30–50 nm. Pattern transfer can be accomplished, for instance, by selective chemical etching of the STM-modified areas, or by selective-area growth of GaAs on patterned silicon substrates. Alternatively, the STM can be operated under a HF solution which directly dissolves the oxide that is locally grown on silicon or gallium arsenide substrate surfaces underneath the STM tip. Etch features of 20–100 nm in line width have been fabricated under these experimental conditions (Nagahara *et al.*, 1990).

8.1.2.3 Electron-enhanced etching

Another STM modification process successfully applied in air has led to the formation of holes on a graphite surface (Albrecht, 1989; Albrecht *et al.*, 1989). By applying voltage pulses (3–8 V, 10–100 μs) across the tunnel gap, one or more layers of graphite in a small region of nanometer size were removed. The writing process was found to depend on the presence of water on the graphite surface. It has been proposed that an electron-enhanced reaction of water molecules with the graphite surface leads to formation of CH and H_2 molecules, as well as removal of carbon atoms from the graphite surface (Quate, 1990). Arrays of hundreds of holes have been written with a success rate as high as 99.6% (Fig. 8.6). By continuously moving the tip at an elevated

3150 Å

Fig. 8.6. Nearly 500 individual holes were written by using voltage pulses applied to an STM tip to record this message on graphite. 99.6% of the intended dots were successfully written. The letters are about 300 Å tall (Albrecht *et al.*, 1989).

bias voltage, it is also possible to draw continuous lines on the graphite surface, having a line width below 10 nm (Roberts *et al.*, 1991). Finally, etching experiments were also performed with the graphite surface in contact with pure water (Penner *et al.*, 1991). A well-defined voltage threshold of (4.0 ± 0.2) V for a successful surface modification was found under these conditions. Near this threshold voltage, protrusions of 7 Å diameter and 2 Å height were generated, whereas pulses of larger voltage led to formation of pits, in agreement with the experimental results obtained in air.

8.1.2.4 Electron-stimulated desorption

Bombardment of a surface with low-energy field-emitted electrons from an STM tip can lead to local desorption of adsorbed species. This was demonstrated for the hydrogen terminated Si(111) surface (Becker *et al.*, 1990), where atomic hydrogen was selectively removed at room temperature by field-emitted electrons with kinetic energies of 2–10 eV. (The electron kinetic energy is the difference between the applied bias voltage and the sample work function). It was proposed that the desorption process is mediated by promoting electrons from the σ-bonding to the σ*-antibonding band. The resultant increase in surface free energy then drives a spontaneous local phase transition from the H-stabilized (1×1) to a (2×1) π-bonded chain structure which is generally found on the clean as-cleaved Si(111) surface (section 4.1.1.1). Individual (2×1) regions as small as ten atomic sites were converted from the (1×1) H-terminated surface by the local electron bombardment procedure.

8.1.2.5 Dissociation of molecules

An STM can also be used to dissociate individual adsorbed molecules by electron bombardment, as was demonstrated for decaborane ($B_{10}H_{14}$) adsorbed on a Si(111) 7×7 substrate surface (Dujardin *et al.*, 1992). To induce dissociation of an isolated $B_{10}H_{14}$ molecule adsorbed next to a substrate defect site (Fig. 8.7), the selected surface region was scanned at an elevated bias voltage of 8 V, corresponding to the field emission regime. It was estimated that the adsorbed molecule received a dose of about 10^5 electrons during a single scan, and it has been proposed that this high flux of low-energy electrons incident on the adsorbed molecule may be responsible for the observed dissociation process.

8.1.2.6 Decomposition of organometallic gases

By dissociating molecules of an organometallic gas in the spatial gap between tip and substrate, local metal deposition can be achieved (Silver *et al.*, 1987; Ehrichs *et al.*, 1988a, 1988b). This requires that the STM be operated in the field emission mode. It is assumed that a microscopic plasma between tip and substrate is formed due to the high electric field, leading to dissociation of molecules and deposition of metal from the organometallic gas onto the substrate. Since the field-emitted electrons must have sufficient energy to induce the dissociation, a well-defined voltage threshold for the onset of metal deposition is typically observed, the value of which depends on the particular organo-metallic gas. Arrays of metallic dots as well as lines of 10 nm width have been fabricated by the decomposition of $W(CO)_6$ and $Fe(CO)_5$ on various substrates (McCord *et al.*, 1988; McCord and Awschalom, 1990). By directly monitoring the height rather than the deposition time during the growth of dots, a greater uniformity of their size is achieved. As an interesting application, an array of nanometer-scale magnets,

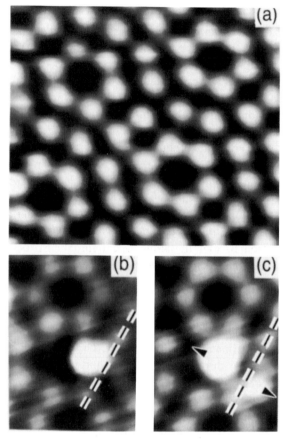

Fig. 8.7. (*a*) An STM image of the clean Si(111) 7×7 surface with the sample biased to +2 V. (*b*) An isolated $B_{10}H_{14}$ molecule (large white spot) adsorbed next to a dark defect. (*c*) The result of electron bombardment at a bias voltage of +8 V. A large molecular species is now located over the original defect, and an additional fragment (small white spot) is seen to the lower right (Dujardin *et al.*, 1992).

obtained by decomposition of $Fe(CO)_5$, was directly written within the superconducting planar input coil of an integrated d.c. SQUID micro-susceptometer, which allowed study of low-temperature spin dynamics in small magnetic systems (Awschalom *et al.*, 1990).

Deposits from organometallic gases as small as 1 nm have been achieved by using laser pulses while operating the STM in the tunneling mode (Yau *et al.*, 1990). The laser radiation serves to generate ions, which are then guided to the surface by the electric field between the STM tip and the substrate.

8.1.2.7 Electrochemical etching and deposition

By using a scanning electrochemical microscope (SECM), Faradaic processes can be exploited to produce local modifications of electrode surfaces, either by electrochemical etching or deposition (Lin *et al.*, 1987; Craston *et al.*, 1988; Hüsser *et al.*, 1988, 1989; Mandler and Bard, 1989, 1990a, 1990b). Since the tip–substrate spacing in SECM (typically 0.1–1 μm) is considerably larger than in STM (typically 1 nm), the etched or deposited lines have a width on the order of 0.1–2 μm. As an alternative to conventional SECM experiments with the tip immersed in an electrolyte solution, it has been proposed to use a thin ionically conductive polymer film spin-coated onto a metal substrate surface and to bring the tip into slight contact with this polymer film (contact area of a few nanometers). If a current is then passed through the ionic conductor, metal deposition in the film underneath the tip as well as localized etching of the metal substrate are achieved (Fig. 8.8). By scanning the tip at a sufficiently high speed, sub-micrometer width lines can be drawn.

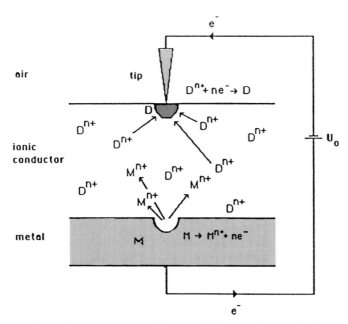

Fig. 8.8. Schematic representation of a method to simultaneously deposit species D in an ionic conductor and etch the substrate M. Both deposition and etching processes result as a consequence of passing a current through an ionic conductor. Metal deposition occurs in the film at the tip electrode and localized etching is observed on the metal substrate (Hüsser *et al.*, 1988).

Local electrochemical modifications have also been demonstrated by means of an STM operated in air (Utsugi, 1990). Localized etching of a surface of a solid electrolyte was achieved by an STM-induced reduction reaction which occurred at elevated bias voltages. Line widths as small as a few nanometers were obtained using this method.

8.1.3 Electric-field-induced surface modifications

By raising the bias voltage in an STM experiment up to several volts, an electric field strength on the order of 1 V $Å^{-1}$ can easily be reached. Electric fields of this magnitude are strong enough to cause local modifications on the sample surface and/or at the tip apex.

8.1.3.1 Electric-field-induced surface diffusion

We first consider manipulation of adsorbed atoms and molecules under the influence of the inhomogeneous electric field present between tip and substrate. It was demonstrated that, by applying an appropriate voltage pulse (1–3 V) between tip and sample, adsorbed atoms (e.g. Cs on GaAs(110) or Cs on InSb(110)) can be induced to diffuse into the region beneath the tip (Whitman *et al.*, 1991c). Field-induced diffusion occurs preferentially toward the tip during the voltage pulse because of the local potential energy gradient dE/dr arising from interaction of the adsorbate dipole moment with the electric field gradient dF/dr at the surface:

$$\frac{\mathrm{d}E}{\mathrm{d}r} \simeq -(\mu + \alpha F)\frac{\mathrm{d}F}{\mathrm{d}r} \tag{8.1}$$

where μ and α are the static dipole moment and the polarizability of the adsorbate, respectively. Depending on the particular system and the pulse parameters, structures from 1 nm down to atomic scale (Avouris and Lyo, 1991) can be created as a result of directional surface diffusion of the adsorbates.

8.1.3.2 Electric-field-induced desorption

At a sufficiently large bias voltage, electric-field-induced local desorption of adsorbates beneath the STM tip can be initiated. This was demonstrated, for instance, for a H_2O-exposed Si(111) 7×7 surface for which scanning at a sample bias of +3 V has proven to lead to a removal of individual adsorbates (Lyo and Avouris, 1990; Avouris and Lyo, 1991, 1992). It was proposed that the combined effects of chemical tip–adsorbate interactions and strong electric fields on the order of 1 V $Å^{-1}$

are responsible for the rupture of adsorbate bonds. A desorption process due to electronic excitation by the tunneling electrons was eliminated because in this experiment the electron energy was below the energy required for such excitations.

8.1.3.3 Electric-field-induced evaporation of substrate atoms

For bare substrates without adsorbates, enhanced electric fields can cause a local field evaporation of surface atoms. For instance, by using a Si (111) 7×7 substrate, reproducible transfer of silicon atoms and silicon clusters up to tens of atoms from the surface to the tip has been demonstrated (Lyo and Avouris, 1991). The electric field strength was varied either by applying voltage pulses of variable height at a fixed tip–sample separation or by applying the same voltage pulse at variable tip-to-sample spacing. The indication for a field-induced evaporation process was provided by the existence of a threshold field of about 1 V Å^{-1} for a successful surface modification. This threshold field is, however, significantly lower than that observed in FIM studies, which has been explained by the additional influence of strong chemical and mechanical tip–sample interactions in STM at small tip–sample distances ('chemically-assisted field-evaporation').

Local field evaporation of individual surface atoms at room temperature has also been demonstrated for a variety of transition metal dichalcogenide substrates, such as MoS_2 (Watt, 1991) and WSe_2 (García-García, 1992). In the case of a MoS_2 substrate, it was even possible to write a message from letters having a line width of a single atomic row (Fig. 8.9).

For metal substrates, such as gold or silver, local field evaporation processes have been observed as well, as a result of bias voltage pulses (McBride and Wetsel, 1990; Rabe and Buchholz, 1991). The craters left on the surface typically had a size of about 10 nm diameter.

8.1.3.4 Electric-field-induced evaporation of tip atoms

Large electric field strengths can also cause field evaporation of tip atoms, resulting in a deposition of tip material onto the substrate surface. Atomic-scale deposition upon application of a voltage pulse (4 V) was first reported for a tungsten tip and a Ge(111) surface (Becker *et al.*, 1987). It was assumed that germanium atoms previously transferred to the tungsten tip by contact with the sample had been redeposited under the electrical stimulation.

A redeposition of substrate material upon increasing the electric field strength has also been observed in STM experiments performed with a tungsten tip and a Si(111) 7×7 surface (Lyo and Avouris, 1991). By

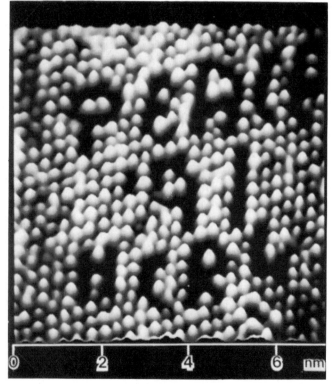

Fig. 8.9. Atoms at the surface of a MoS_2 crystal, from which selected sulfur atoms have been 'field evaporated' by means of an STM, to write the message 'PEACE '91 HCRL' (Hosoki *et al.*, 1992).

appropriately adjusting the tip-to-substrate spacing and the height of the applied voltage pulse, it was possible to remove and redeposit even a single silicon atom, as illustrated in Fig. 8.10.

The material to be deposited may not necessarily originate from the substrate. For instance, atomic emission from a gold STM tip, caused by application of voltage pulses, can be used to write several thousand gold mounds onto a substrate with no apparent degradation of the tip's ability to emit atoms (Mamin *et al.*, 1990, 1991). The threshold voltage for a successful deposition of tip material was found to be between 3.5 and 4.0 V for a gold tip and a gold substrate. The linear dependence of this threshold voltage on tip–surface separation indicated the existence of a critical electric field strength, which was derived to be on the

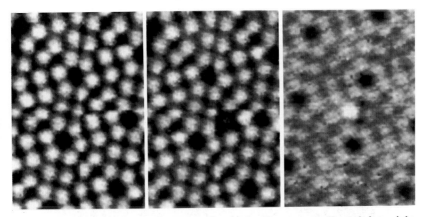

Fig. 8.10. Removal and redeposition of a single silicon atom. From left to right the images are taken before and after a voltage pulse (+1 V and tip displacement of 5 Å) was applied and, finally, after redeposition of the atom back to the surface (Lyo and Avouris, 1991).

order of 0.4 V Å^{-1}. This value is significantly lower than that inferred from FIM studies, which has been explained by the influence of the close proximity of tip and substrate in the STM experiments. The emission process is highly reproducible, allowing one to write complex patterns (Fig. 8.11). It is also fast since pulses with a width as low as 10 ns can be used. The gold mounds deposited on a gold substrate had a typical base diameter of about 10 nm and were found to be stable over periods of weeks. They even appeared relatively unaffected by heat treatment (Fig. 8.12). Similar modification experiments have also been performed with other tip and substrate materials (McBride and Wetsel, 1991).

8.1.3.5 Liquid metal ion source (LMIS)

As an alternative to field evaporation of STM tip atoms, a liquid metal ion source (LMIS), brought in close proximity (about 100 nm) to a substrate, can be used for local metal deposition (Ben Assayag et al., 1987; Bell et al., 1988). When a positive bias voltage is applied to a liquid-metal-covered tungsten needle, exceeding a critical value corresponding to the balance between electric field stress and surface tension, ion emission from the apex of a 'Taylor cone', formed by the liquid metal, sets in. Proximity focusing can then lead to deposition of sub-micrometer width metallic lines. A further reduction of written feature

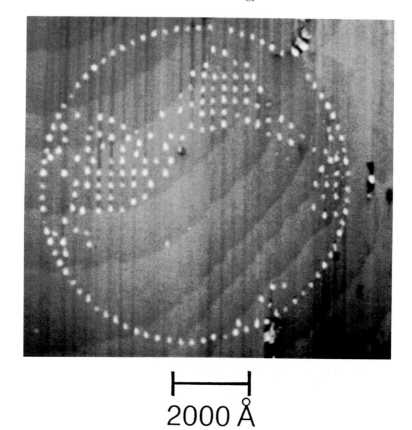

2000 Å

Fig. 8.11. Simplified map of the world fabricated by an STM on a Au(111) substrate. The diameter is about 1 μm, resulting in a scale of $1:10^{13}$ (Mamin *et al.*, 1991).

size would require a corresponding decrease of the tip–substrate separation which, however, may not be practicable due to attractive tip–substrate forces that act on the liquid metal.

8.1.4 Thermally induced surface modifications

Effects caused by thermal heating are unimportant in STM experiments on crystalline substrates if a tunnel current I of about 1 nA and a bias voltage U on the order of 1 V are used (Persson and Demuth, 1986;

Elevated Temperature Studies

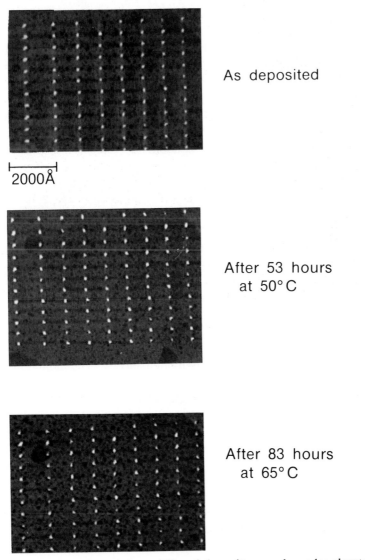

As deposited

2000Å

After 53 hours
at 50°C

After 83 hours
at 65°C

Fig. 8.12. Demonstration of the stability of the written marks under elevated temperatures. Testing was done over the course of 17 days. The marks appear relatively unaffected by heating. (Mamin *et al.*, 1991).

Flores *et al.*, 1986; Marella and Pease, 1989). The local temperature
rise ΔT can roughly be estimated by

$$\Delta T \simeq \frac{I\,U}{4\,\pi\,\kappa\,\lambda} \tag{8.2}$$

where κ and λ denote the thermal conductivity and electron mean free
path of the substrate material. For STM experiments on a typical metal
($\kappa \approx 300$ W K^{-1} m^{-1}; $\lambda \approx 30$ nm), ΔT is less than 1 mK and therefore
entirely negligible. However, if the tunnel current and sample bias volt-
age are significantly raised and, in addition, an amorphous metal is used
as substrate, for which both thermal conductivity and electron mean
free path are lower by a factor of about 100 compared with crystalline
materials, a temperature rise up to several hundred kelvin can be
obtained, assuming the validity of Eq. (8.2). Depending on the current
and the electric field strength, enhanced diffusion, local crystallization
of the glassy substrate or even local melting may occur.

Evidence for an STM-induced local melting process was indeed found
in surface modification experiments with amorphous metal substrates
(Wiesendanger, 1987; Staufer *et al.*, 1987, 1988, 1989, 1991; Staufer,
1990). At a fixed tip position above the substrate surface, the sample
bias voltage was first increased from about 0.1 V, used for non-
destructive STM imaging, to > 1 V. Subsequently, the current was
increased from 1 nA to several hundred nano-ampères or even micro-
ampères, until oscillations in the current appeared, resulting from
instabilities related to the locally molten surface. The molten material
is drawn towards the tip under the influence of the high electrostatic
forces acting between tip and sample surface, thereby forming a bridge
between the two electrodes. Upon tip retraction, this liquid metal
bridge breaks up and a cone of rapidly solidified material is left on the
surface which can subsequently be imaged with the STM under normal
operation conditions (Fig. 8.13(*a*)). A depression is usually observed
around the cone, which can be explained by the material transport
during cone formation.

The bias voltage applied during the surface modification process
affects the power density as well as the electric field strength, both of
which have an influence on the size of the written cones. Therefore, by
using different values of applied bias voltage, the size of the cones can
directly be controlled (Fig. 8.13(*b*)). The dimensions of the cones, their
diameter as well as their height, were found to scale linearly with
applied bias voltage during the modification process (Staufer *et al.*,
1991). The smallest cones fabricated by this thermally induced modification process have a base diameter of about 3 nm (Staufer *et al.*,
1989).

(a)

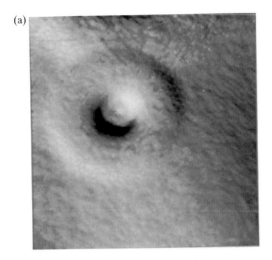

(b)

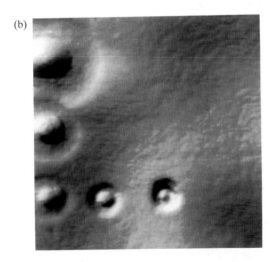

Fig. 8.13. (*a*) STM image (150 nm × 150 nm) of a cone written thermally on the surface of a glassy $Co_{35}Tb_{65}$ substrate. The FWHM value of the cone is 29 nm and the topographic height is 23 nm. (*b*) STM image (300 nm × 300 nm) of a sequence of cones written thermally at different sample bias voltages on the surface of glassy $Co_{35}Tb_{65}$. The FWHM values of the written cones from the upper left to the lower right part of the image are 44, 30, 23, 16 and 10 nm, whereas the topographic heights are 31, 22, 11, 8 and 7 nm (Staufer *et al.*, 1989).

Lines, letters and even messages (Fig. 8.14) can be written by scanning the tip during local melting of the surface at elevated current and bias voltage. The line width varies between 10 and 50 nm, depending on the sharpness of the tip. At the end of each line, a cone is left, resulting from the break-up of the liquid metal bridge between tip and substrate upon tip retraction.

By using seven different amorphous metal substrates, it has experimentally been verified that the power needed to induce local melting of the surface scales correctly with the product of the thermal conductivity and the melting temperature of these materials (Staufer *et al.*, 1991).

Thermally induced surface modifications may also be performed on amorphous semiconductor substrates. For instance, localized amorphous to crystalline transformation of a phosphorus-doped hydrogenated amorphous silicon (a-Si : H(P)) thin film was reported (Jahanmir *et al.*, 1989). Spectroscopic STM measurements before and after the application of voltage pulses (10 V, 35 μs) indicated an increase in local conductivity which was attributed to local crystallization of the a-Si : H material. Lines of width 140 nm were achieved by applying multiple voltage pulses.

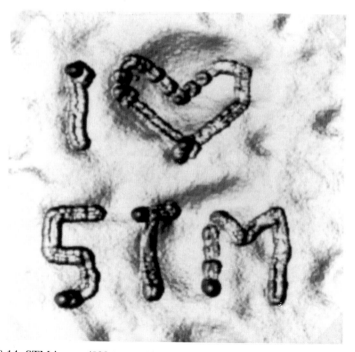

Fig. 8.14. STM image (800 nm × 800 nm) of a message written on a surface of a glassy $Rh_{25}Zr_{75}$ substrate (Staufer *et al.*, 1989).

Another thermally induced modification was reported for an a-Si: H(P) layer deposited on a heavily doped p-type Si(111) substrate, forming a p–n junction (Hartmann *et al.*, 1991). By increasing the current from 1 to 300 nA and the applied bias voltage from 4 to 10 V, the structure of the a-Si:H(P) layer was locally changed as a result of a local temperature rise which was estimated to be on the order of 100 K. As a consequence of the structural change of the a-Si:H(P) layer, the barrier height of the p–n junction became reduced, leading to features 50 nm high in constant current STM images. The lateral size of these features, which were verified to be of electronic rather than topographic origin, is on the order of 50–500 nm. They appear to decay as a function of time on a time scale of hours.

Further applications of the STM used as a local heating source appear to be promising because the possibility of inducing structural and phase transformations on a local scale can quite generally be exploited to modify the physical properties of small sample volume elements.

8.1.5 Electrostatic surface modifications

By using a scanning capacitance microscope (SCAM, section 3.4), nanometer-scale writing with charges on a nitride–oxide–silicon (NOS) system was demonstrated (Barrett and Quate, 1991b). Upon application of a voltage pulse to the NOS sample, charge tunnels from the silicon substrate through the oxide layer and gets trapped in the top nitride film (Fig. 8.15). This trapped charge induces a depletion region in the

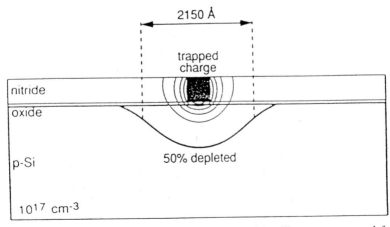

Fig. 8.15. Electrostatic simulation of a nitride–oxide–silicon system used for charge storage. Equipotential contours are shown at 0.3 V intervals. The dark contour represents the region where the semiconductor is 50% depleted (Barrett and Quate, 1991b).

silicon substrate, which can be detected by the resulting depletion capacitance between the SCAM tip and the NOS sample. Fig. 8.16(*a*) shows a SCAM image of nine 'bits' of charge which were written by using −40 V and 100 μs voltage pulses. The FWHM size of the charge bits, as seen by the SCAM, is about 170 nm. This value is predominantly determined by the size of that part of the SCAM tip which is in actual

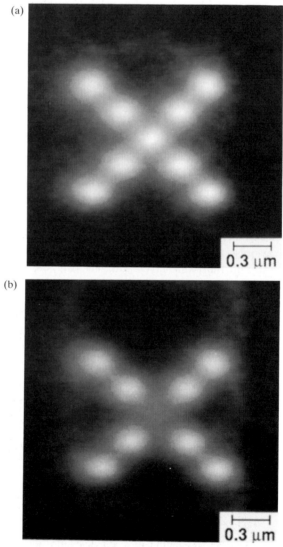

Fig. 8.16. An example of charge writing and erasing. An 'X' pattern of nine dots was written with −40 V pulses of 20 μs width. The center dot was subsequently erased with a +40 V pulse (Barrett and Quate, 1991b).

contact with the nitride layer whereas the aspect ratio of the tip appears to be unimportant. The smallest bits that were written had a FWHM of 75 nm. By simultaneously recording capacitance and force data, it was experimentally verified that the stored information is not the result of any topographic change to the surface. On the other hand, the stored charge was observed to be stable over extended time periods. The charge bits can, however, be erased by applying a voltage pulse of opposite sign (e.g. +40 V, 1 ms), as demonstrated for the central bit in Fig. 8.16(*b*). The charge writing process had been found to be fast and highly reliable, allowing one to write complex patterns (Fig. 8.17) as well as a complete text page consisting of several thousand separate bits.

8.1.6 Magnetic surface modifications

There exist several possibilities for magnetic surface modifications by using scanning probe microscopes, covering the sub-micrometer down to the atomic length scale.

8.1.6.1 Magnetic surface modifications by thermal heating

Magnetic bits can be written by using either phase-change or magneto-optical systems. In phase-change systems, localized heating causes a transition between crystalline and amorphous states which exhibit different magnetic properties. This is usually accomplished by means of a

Fig. 8.17. An example of a charge pattern written with 202 voltage pulses (Barrett and Quate, 1991b).

focused laser beam with a spot size of typically 1 μm, which sets the size limit for the written bits. By using an STM as a local heating source (section 8.1.4) rather than a laser beam, a significantly smaller bit size on the nanometer scale may be achieved. However, effective heating by STM requires an amorphous substrate, thereby limiting the application to local amorphous-to-crystalline or amorphous-to-amorphous transformations.

A promising application for STM used as a local heating source would be thermomagnetic recording on magneto-optical amorphous substrates, such as heavy rare earth transition metal alloy films. By locally increasing the temperature of the film to the point where an externally applied magnetic field can reverse the magnetization in the region being heated, a magnetic bit is obtained. In the conventional thermomagnetic recording process, a focused laser beam is used for local heating, which again limits the size of the written bits to the micrometer length scale. In contrast, STM allows one to confine the locally heated region of an amorphous metal substrate to nanometer-scale dimensions (section 8.1.4). Thermal recording by STM on amorphous CoTb substrates has already been demonstrated (Staufer *et al.*, 1989). However, the magnetic properties of the written bits could not be characterized in these STM experiments.

8.1.6.2 Magnetic surface modifications by the stray field of a magnetic tip

The magnetic stray field of a magnetized tip can cause local changes of the substrate surface magnetization at sufficiently small tip–substrate separations as a consequence of magnetic dipole interactions. By using this modification method, magnetic bits were written on the surface of a hard disk by means of a tunneling-stabilized magnetic force microscope (TSMFM, section 1.22) with a magnetized iron thin film tip, which also served to read the magnetically recorded information (Moreland and Rice, 1990). The size of the written bits was on the order of several hundred nanometers.

8.1.6.3 Atomic-scale modifications of surface spin structures

By using a spin-polarized scanning tunneling microscope (SPSTM, section 1.14), modifications of the spin structure of a magnetite (001) surface were observed as localized to a near-atomic scale (Wiesendanger *et al.*, 1992c). It is assumed that such localized changes of the magnetic surface structure are caused by the exchange interaction between the magnetic tip and the magnetic substrate at short distances (<3 Å), rather than by long-range dipolar interaction.

8.2 Nanometer-scale electronic devices

The various methods for the fabrication of nanometer-scale structures based on SPM instruments, as discussed in section 8.1, could probably lead to novel nanometer-scale electronic devices and ultimately to a novel 'nanoelectronics'.

The first working device, made by using an STM to expose PMMA and subsequently lifting off a 13.5 nm film of gold–palladium, was a thin-film resistor 2 μm long and 120 nm in width (Fig. 8.18). The room-temperature resistance of the device was 2.5 kΩ, yielding a resistivity of 200 μΩ cm (McCord and Pease, 1988).

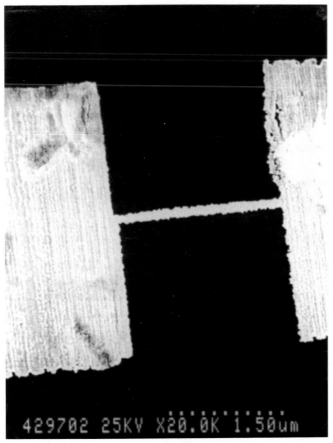

Fig. 8.18. A thin-film resistor made by using the STM to expose PMMA and subsequently lifting off a 13.5 nm film of gold–palladium. The device showed a room-temperature resistance of 2.5 kΩ (McCord and Pease, 1988).

More recently, prototypes of atomic-scale electronic devices, based on STM instruments, have been constructed. For instance, an atomic switch deriving its function from the controlled reversible motion of a single atom between an STM tip and a substrate surface (Fig. 8.19(*a*)) has been demonstrated (Eigler *et al.*, 1991b). The experiment was performed at low temperatures (4 K) under UHV conditions. A tungsten tip was first positioned above an adsorbed atom (Xe) on a single-crystal Ni substrate. The height of the tip was adjusted to yield a tunnel junction resistance of 1.5 MΩ (tunnel current of 13 nA, tip bias of −0.02 V). By applying a voltage pulse (+0.8 V, 64 ms) to the tip, the Xe atom was transferred to the tip, resulting in a 0.22 MΩ resistance state (Fig. 8.19(*b*)). To switch back to the initial high-resistance state with the Xe atom adsorbed on the substrate surface, another voltage pulse (−0.8 V, 64 ms) had to be applied. The motion of the Xe atom was found to be always towards the positively biased electrode and could be reversed by changing the polarity of the applied voltage. It was proposed that heating-assisted electromigration may be the physical mechanism responsible for the observed behavior of the switch. The function of this 'atom switch' still relies on the physical properties of the macroscopic leads, i.e. tip and substrate, in particular their electronic and thermal transport properties. To realize an 'atomic-scale switch' would require miniaturization of the leads as well, and probably a new physical mechanism may then have to be exploited to operate such a kind of switch.

Another prototype of an STM-based electronic device has been demonstrated, with properties resembling those of a tunnel diode (section 1.5). A tunnel diode exhibits a negative differential resistance (NDR), i.e. a negative slope in the *I–U* curve, resulting from the existence of a bias range for which tunneling is forbidden or suppressed following a

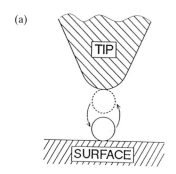

(a)

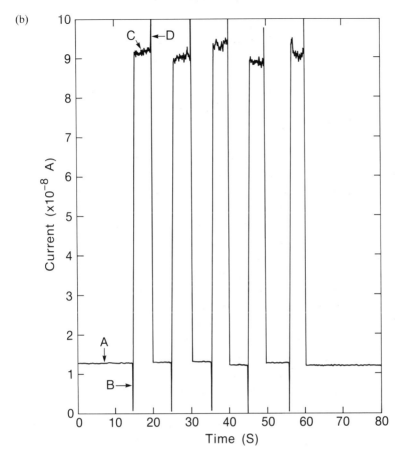

Fig. 8.19. (*a*) A two-terminal atom switch. The atom is reversibly transferred between the surface and the tip by application of voltage pulses. The conductance of the tunnel junction switches state according to whether the atom is located on the surface or the tip. (*b*) The time-dependence of the current through the switch during operation. The low-conductance state (A) is established with the tip biased at −0.02 V and the Xe atom bound to the surface. The transient current spike (B) is due to application of a +0.8 V pulse to the tip for a duration of 64 ms, after which the −0.02 V bias to the tip was re-established. This results in transfer of the Xe atom to the tip and establishment of the high-conductance state (C). Applying a reverse polarity −0.8 V pulse for 64 ms gives rise to the transient current spike (D) followed by re-establishment of the low-conductance state (Eigler *et al.*, 1991b).

bias for which tunneling is strongly favored. This NDR behavior has been observed in STM studies of a boron-doped Si(111) surface with the tip positioned above particular surface sites as shown in Fig. 8.20 (Bedrossian *et al.*, 1989a; Lyo and Avouris, 1989; Avouris *et al.*, 1990).

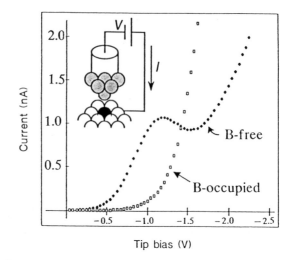

Tip bias (V)

Fig. 8.20. *I–U* curves for the case of negative tip bias, acquired with the tip over boron-occupied and boron-free sites of a boron-rich Si(111) surface, in regions of substantial mixing of the two types of sites. Inset: geometry of the tunneling device (Bedrossian *et al.*, 1989a).

The surface regions exhibiting NDR were found to be localized to a diameter of only 1 nm and were often related to individual surface defect sites (Fig. 8.21). However, depending on the state of the tip, NDR can also be observed over sites of a clean surface, e.g. Si(111) 7×7. In contrast to the Esaki tunnel diode (section 1.5), NDR in STM experiments arises as a result of tunneling between localized quasi-atomic states. The existence of such localized states gives rise to allowed and suppressed energies for tunneling.

Generally, it may be possible in the future not to use just circuits, but some systems involving the quantized energy levels, or the interactions of quantized spins, leading to a new generation of atomic-scale electronic devices, as envisioned by Feynman as early as 1961.

8.3 Scanning probe methods combined with microfabrication

There exist various fields where the combination of scanning probe methods with microfabrication techniques based on silicon technology has proven to be extremely fruitful or appears to be highly promising for future applications.

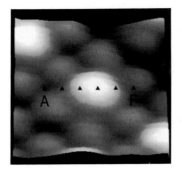

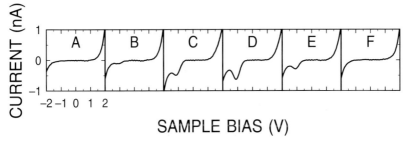

Fig. 8.21. Top: STM topograph showing a boron defect site whose *I–U* curve shows negative differential resistance. Bottom: *I–U* curves obtained at different points A to F across the above defect site. The points are spaced 3 Å apart (Avouris *et al.*, 1990).

8.3.1 Microfabrication of SPM sensors

For many SPM techniques, such as AFM, FFM, MFM, SNOM, SCAM and SICM, considerable progress has come along with the use of micro-fabricated sensors having a small mass and reproducible shape.

8.3.2 Microfabrication of multiple sensor tips

The writing (and reading) speed has been found to be one of the major limitations for the application of SPM to nanometer-scale recording (section 8.1). Therefore, introduction of parallel operating systems has been suggested. This could be accomplished, for instance, by using multiple tip arrangements. As an initial step, a twin-probe STM with two tips at a distance of 0.1 μm was constructed (Tsukamoto *et al.*, 1991). Since the lengths of the two tips always differed, the twin-probe STM not only allowed control of the spacing between the twin sensor and the sample surface, but also of the angular orientation of the sample, which was required to establish electron tunneling for both junctions simultaneously.

The next step would be to use multiple tip arrays (Busta *et al.*, 1989). For arrays with more than three tips, simultaneous probing of the sample surface by these tips can no longer be established by appropriately tilting the sample. Therefore, an independent positioning system for each tip is required. This can be achieved, for instance, by using piezoelectric thin-film technology combined with microfabrication techniques.

8.3.3 Microfabrication of SPM instruments

By using an integrated circuit (IC)-compatible process, an array of STM instruments has been microfabricated on a silicon wafer substrate (Akamine *et al.*, 1989; Albrecht, 1989; Albrecht *et al.*, 1990b). The STM design was based on a thin-film piezoelectric actuator forming a cantilever bimorph (Fig. 8.22(*a*)). This bimorph was constructed from alternating layers of metal electrodes, dielectric films and piezoelectric zinc oxide films (Fig. 8.22(*b*)) and allowed four independent types of motion as illustrated in Fig. 8.23. The dimensions of the device were as small as 1000 μm × 200 μm × 8μm, providing excellent resistance against thermal and vibrational disturbances.

Since the fabrication of the device is compatible with conventional integrated circuit technology, the STM control electronics can easily be added to the device, leading to an array of complete miniature STM units. Arrays of other SPM-type instruments can similarly be fabricated. For instance, by scanning the cantilever bimorph of the integrated STM over the surface of a sample with a light repulsive force, the bimorph can be used for detection of cantilever deflections, thereby serving as an integrated AFM sensor. The fabrication of arrays of closely spaced miniature SPM units will help to solve the problem of limited writing and reading speed in SPM-based recording by the parallel operation of many SPM instruments.

8.3.4 Field emission microprobe system

An STM aligned field emission (SAFE) microprobe has been developed, incorporating an exceptionally high-brightness electron microsource and a low-aberration electron optical system (McCord *et al.*, 1989; Chang *et al.*, 1989, 1990, 1991, 1992). The SAFE microprobe, as

MICROFABRICATED STM

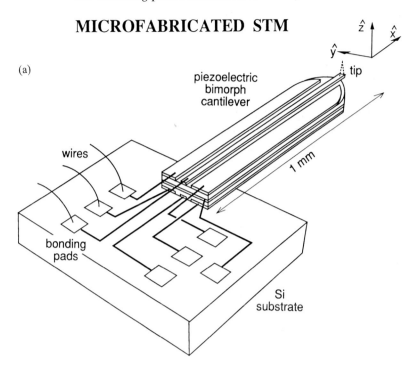

(a)

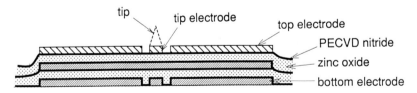

(b) ## CANTILEVER CROSS SECTION

Fig. 8.22. Design of a microfabricated STM. (*a*) The tunneling tip is located on the free end of a piezoelectric cantilever bimorph. The cantilever extends from the edge of a silicon block which contains bonding pads for external signal connections. (*b*) The cantilever cross-section contains two piezoelectric ZnO layers sandwiched between three layers of electrodes. The top and bottom electrodes are divided into independent left and right sections, and the narrow center conductor is provided for the STM tip. Dielectric layers of PECVD nitride reduce leakage and increase breakdown voltage (Albrecht *et al.*, 1990b).

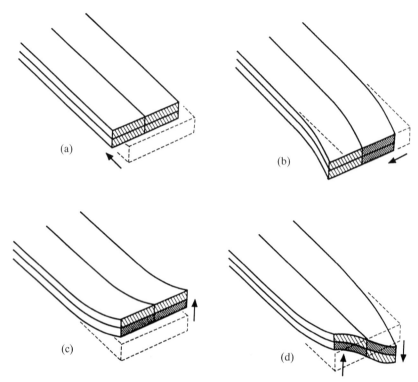

Fig. 8.23. Four independent motions are available with the piezoelectric bimorph actuator. (*a*) By applying the same electric field in all four piezoelectric sections, lengthwise expansion or contraction occurs. (*b*) Opposite fields on the left and right sides cause lateral bending. (*c*) Opposite fields on the top and bottom sections cause vertical bending. (*d*) Opposite fields on diagonal pairs result in twisting (Akamine *et al.*, 1989).

schematically illustrated in Fig. 8.24, consists of an STM-aligned field emission tip and a microlens system forming a microsource which, combined with additional microlenses, can provide a highly focused low-energy (about 100 eV) electron beam with 1–10 nm diameter. The microlenses are fabricated from silicon substrates by using standard integrated circuit technology. A small size of the lenses is desirable because lens aberrations generally scale with lens dimensions. The use of a single-atom tip as an electron point source (Fink, 1986, 1988) can additionally improve the performance of the SAFE source which has great potential for a variety of applications, including electron microscopy, electron holography, electron beam testing and electron lithography. Since the total length of the SAFE column is only a few milli-

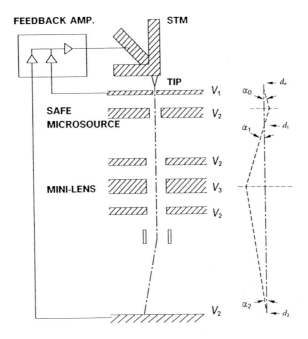

SAFE MICROPROBE SYSTEM

Fig. 8.24. Basic configuration of an STM aligned field emission (SAFE) microprobe system (Chang *et al.*, 1991).

meters, an array of SAFE microprobe systems can be envisioned, thereby improving throughput and offering further applications.

8.3.5 Microfabrication of SPM-based sensors

The scanning probe methods appear to be ideally suited for a wide variety of sensor applications. The high sensitivity of SPM and the small effective area being probed allow sensing of extremely small environment-induced changes in the physical properties of extremely small objects approaching the atomic level. Therefore, the ultimate in sensor miniaturization can be attained by microfabrication of integrated SPM units. In particular, the tunneling probe as a low-noise transducer of displacement to electrical current has a great potential (Bocko *et al.*, 1988; Bocko and Stephenson, 1991; Kenny *et al.*, 1991), for instance, as a gravitational-wave sensor (Niksch and Binnig, 1988; Bocko *et al.*,

1988), as an acceleration sensor (Baski *et al.*, 1988; Waltman and Kaiser, 1989), as a magnetic-field sensor (Wandass *et al.*, 1989; Witek and Onn, 1991), and as a temperature or pressure sensor. The possibilities for incorporating a tunneling transducer into microfabricated integrated sensors seem to be limitless because any physical effect causing a change of the dimension or position of a test object as a result of a change of the environment can be exploited to sense these environmental changes with high sensitivity.

8.4 Conclusion

Looking back to chapters 4–8 of this book, the various applications of scanning probe microscopes used as imaging, measuring, modifying and sensing instruments have undoubtedly proven the strong impact of SPM on modern science and technology, and the future of SPM certainly appears to be bright. However, predictions about future developments tend to fail. Therefore, it seems more appropriate to close by citing a final remark of one of the inventors of STM (Rohrer, 1990): 'The prospects of science and technology reside with the scientists themselves, their ideas, visions and, of course, their work . . . '

References

Abbé, E. (1873). *Archiv. Microskopische Anal.* **9**, 413.
Abraham, D. L., Veider, A., Schönenberger, Ch., Meier, H. P., Arent, D.
    J. and Alvarado, S. F. (1990). *Appl. Phys. Lett.* **56**, 1564.
Abraham, D. W. and McDonald, F. A. (1990). *Appl. Phys. Lett.* **56**, 1181.
Abraham, D. W., Sattler, K., Ganz, E., Mamin, H. J., Thomson, R. E. and
    Clarke, J. (1986). *Appl. Phys. Lett.* **49**, 853.
Abraham, D. W., Williams, C., Slinkman, J. and Wickramasinghe, H. K.
    (1991). *J. Vac. Sci. Technol.* B **9**, 703.
Abraham, D. W., Williams, C. C. and Wickramasinghe, H. K. (1988a).
    Appl. Phys. Lett. **53**, 1503.
Abraham, D. W., Williams, C. C. and Wickramasinghe, H. K. (1988b). *J.*
    Microsc. **152**, 599.
Abraham, D. W., Williams, C. C. and Wickramasinghe, H. K. (1988c).
    Appl. Phys. Lett. **53**, 1446.
Abraham, F. F. and Batra, I. P. (1989). *Surf. Sci.* **209**, L125.
Abraham, F. F., Batra, I. P. and Ciraci, S. (1988d). *Phys. Rev. Lett.* **60**,
    1314.
Adams, A., Wyss, J. C. and Hansma, P. K. (1979). *Phys. Rev. Lett.* **42**, 912.
Adkins, C. J. and Phillips, W. A. (1985). *J. Phys.* C **18**, 1313.
Agarwal, V. K. (1988). *Phys. Today*, June, p. 40.
Akamine, S., Albrecht, T. R., Zdeblick, M. J. and Quate, C. F. (1989).
    IEEE Electron Device Lett. **10**, 490.
Akari, S., Lux-Steiner, M. Ch., Vögt, M., Stachel, M. and Dransfeld, K.
    (1991). *J. Vac. Sci. Technol.* B **9**, 561.
Akari, S., Stachel, M., Birk, H., Schreck, E., Lux, M. and Dransfeld, K.
    (1988). *J. Microsc.* **152**, 521.
Albrecht, T. R. (1989) Ph.D. Thesis, Stanford University, Stanford,
    California.
Albrecht, T. R., Akamine, S., Carver, T. E. and Quate, C. F. (1990a). *J.*
    Vac. Sci. Technol. A **8**, 3386.
Albrecht, T. R., Akamine, S., Zdeblick, M. J. and Quate, C. F. (1990b). *J.*
    Vac. Sci. Technol. A **8**, 317.
Albrecht, T. R., Dovek, M. M., Kirk, M. D., Lang, C. A., Quate, C. F.
    and Smith, D. P. E. (1989). *Appl. Phys. Lett.* **55**, 1727.
Albrecht, T. R., Dovek, M. M., Lang, C. A., Grütter, P., Quate, C. F.,
    Kuan, S. W. J., Frank, C. W. and Pease, R. F. W. (1988a). *J. Appl.*
    Phys. **64**, 1178.

Albrecht, T. R., Mizes, H. A., Nogami, J., Park, S.-I. and Quate, C. F. (1988b). *Appl. Phys. Lett.* **52**, 362.

Albrecht, T. R. and Quate, C. F. (1987). *J. Appl. Phys.* **62**, 2599.

Albrecht, T. R. and Quate, C. F. (1988). *J. Vac. Sci. Technol.* A **6**, 271.

Albrektsen, O., Arent, D. J., Meier, H. P. and Salemink, H. W. M. (1990). *Appl. Phys. Lett.* **57**, 31.

Alexander, S., Hellemans, L., Marti, O., Schneir, J., Elings, V., Hansma, P. K., Longmire, M. and Gurley, J. (1989). *J. Appl. Phys.* **65**, 164.

Allenspach, R. and Bischof, A. (1989). *Appl. Phys. Lett.* **54**, 587.

Allenspach, R., Salemink, H., Bischof, A. and Weibel, E. (1987). *Z. Phys.* B **67**, 125.

Allenspach, R., Stampanoni, M. and Bischof, A. (1990). *Phys. Rev. Lett.* **65**, 3344.

Alvarado, S. F. and Renaud, P. (1992). *Phys. Rev. Lett.* **68**, 1387.

Alvarado, S. F., Renaud, Ph., Abraham, D. L., Schönenberger, Ch., Arent, D. J. and Meier, H. P. (1991). *J. Vac. Sci. Technol.* B **9**, 409.

Amer, N. M., Skumanich, A. and Ripple, D. (1986). *Appl. Phys. Lett.* **49**, 137.

Amrein, M., Dürr, R., Stasiak, A., Gross, H. and Travaglini, G. (1989). *Science* **243**, 1708.

Amrein, M., Stasiak, A., Gross, H., Stoll, E. and Travaglini, G. (1988). *Science* **240**, 514.

Amrein, M., Wang, Z. and Guckenberger, R. (1991). *J. Vac. Sci. Technol.* B **9**, 1276.

Anders, M., Mück, M. and Heiden, C. (1988). *Ultramicroscopy* **25**, 123.

Anders, M., Thaer, M. and Heiden, C. (1987). *Surf. Sci.* **181**, 176.

Anderson, P. W. and Dayem, A. H. (1964). *Phys. Rev. Lett.* **13**, 95.

Anderson, P. W. and Rowell, J. M. (1963). *Phys. Rev. Lett.* **10**, 230.

Ando, Y. and Itoh, T. (1987). *J. Appl. Phys.* **61**, 1497.

Anselmetti, D. (1990). Ph.D. Thesis, Basel University, Basel, Switzerland.

Anselmetti, D., Geiser, V., Brodbeck, D., Overney, G., Wiesendanger, R. and Güntherodt, H.-J. (1990a). *Synth. Met.* **38**, 157.

Anselmetti, D., Geiser, V., Overney, G., Wiesendanger, R. and Güntherodt, H.-J. (1990b). *Phys. Rev.* B **42**, 1848.

Anselmetti, D., Wiesendanger, R., Geiser, V., Hidber, H. R. and Güntherodt, H.-J. (1988). *J. Microsc.* **152**, 509.

Anselmetti, D., Wiesendanger, R. and Güntherodt, H.-J. (1989). *Phys. Rev.* B **39**, 11 135.

Anselmetti, D., Wiesendanger, R., Güntherodt, H.-J. and Grüner, G. (1990c). *Europhys. Lett.* **12**, 241.

Appelbaum, J. (1966). *Phys. Rev. Lett.* **17**, 91.

Arnold, L., Krieger, W. and Walther, H. (1987). *Appl. Phys. Lett.* **51**, 786.

Arnold, L., Krieger, W. and Walther, H. (1988). *J. Vac. Sci. Technol.* A **6**, 466.

Arscott, P. G., Lee, G., Bloomfield, V. A. and Evans, D. F. (1989). *Nature* **339**, 484.

Ash, E. A. and Nicholls, G. (1972). *Nature* **237**, 510.

Averin, D. V. and Likharev, K. K. (1986). *J. Low Temp. Phys.* **62**, 345.

Avouris, Ph. and Bozso, F. (1990). *J. Phys. Chem.* **94**, 2243.

Avouris, Ph., Bozso, F. and Hamers, R. J. (1987). *J. Vac. Sci. Technol.* B **5**, 1387.

Avouris, P. and Lyo, I.-W. (1990). In: *Chemistry and Physics of Solid*

Surfaces VIII (ed. Vanselow, R. and Howe, R.), Springer Series in Surface Sciences 22, p. 371 (Berlin, Heidelberg, New York: Springer).

Avouris, Ph. and Lyo, I.-W. (1991). *Surf. Sci.* **242**, 1.

Avouris, Ph. and Lyo, I.-W. (1992). In: *Scanned Probe Microscopy* (ed. Wickramasinghe, H. K.), AIP Conf. Proc. 241, p. 283 (New York: AIP).

Avouris, Ph., Lyo, I.-W. and Bozso, F. (1991). *J. Vac. Sci. Technol.* B **9**, 424.

Avouris, Ph., Lyo, I.-W., Bozso, F. and Kaxiras, E. (1990). *J. Vac. Sci. Technol.* A **8**, 3405.

Avouris, Ph. and Wolkow, R. (1989a). *Phys. Rev.* B **39**, 5091.

Avouris, Ph. and Wolkow, R. (1989b). *Appl. Phys. Lett.* **55**, 1074.

Awana, V. P. S., Samanta, S. B., Dutta, P. K., Gmelin, E. and Narlikar, A. V. (1991). *J. Phys.: Condens. Matter* **3**, 8893.

Awschalom, D. D., McCord, M. A. and Grinstein, G. (1990). *Phys. Rev. Lett.* **65**, 783.

Bando, H., Kashiwaya, S., Tokumoto, H., Anzai, H., Kinoshita, N. and Kajimura, K. (1990). *J. Vac. Sci. Technol.* A **8**, 479.

Bando, H., Morita, N., Tokumoto, H., Mizutani, W., Watanabe, K., Homma, A., Wakiyama, S., Shigeno, M., Endo, K. and Kajimura, K. (1988). *J. Vac. Sci. Technol.* A **6**, 344.

Bando, H., Tokumoto, H., Mizutani, W., Watanabe, K., Okano, M., Ono, M., Murakami, H., Okayama, S., Ono, Y., Wakiyama, S., Sakai, F., Endo, K. and Kajimura, K. (1987). *Jpn. J. Appl. Phys.* **26**, L41.

Baratoff, A. (1984). *Physica* B **127**, 143.

Baratoff, A., Binnig, G., Fuchs, H., Salvan, F. and Stoll, E. (1986). *Surf. Sci.* **168**, 734.

Baratoff, A. and Persson, B. N. J. (1988). *J. Vac. Sci. Technol.* A **6**, 331.

Bard, A. J., Denuault, G., Lee, C., Mandler, D. and Wipf, D. O. (1990). *Acc. Chem. Res.* **23**, 357.

Bard, A. J., Fan, F. F., Kwak, J. and Lev, O. (1989). *Anal. Chem.* **61**, 132.

Bard, A. J., Unwin, P. R., Wipf, D. O. and Zhou, F. (1992). In: *Scanned Probe Microscopy* (ed. Wickramasinghe, H. K.), AIP Conf. Proc. 241, p. 235 (New York: AIP).

Bardeen, J. (1961). *Phys. Rev. Lett.* **6**, 57.

Bardeen, J., Cooper, L. N. and Schrieffer, J. R. (1957). *Phys. Rev.* **108**, 1175.

Baró, A. M., Bartolome, A., Vazquez, L., García, N., Reifenberger, R., Choi, E. and Andres, R. P. (1987). *Appl. Phys. Lett.* **51**, 1594.

Barrett, R. C., Nogami, J. and Quate, C. F. (1990). *Appl. Phys. Lett.* **57**, 992.

Barrett, R. C. and Quate, C. F. (1991a). *Rev. Sci. Instrum.* **62**, 1393.

Barrett, R. C. and Quate, C. F. (1991b). *J. Appl. Phys.* **70**, 2725.

Barth, J. V., Brune, H., Ertl, G. and Behm, R. J. (1990). *Phys. Rev.* B **42**, 9307.

Baski, A. A., Albrecht, T. R. and Quate, C. F. (1988). *J. Microsc.* **152**, 73.

Batra, I. P. and Ciraci, S. (1988). *J. Vac. Sci. Technol.* A **6**, 313.

Batra, I. P., García, N., Rohrer, H., Salemink, H., Stoll, E. and Ciraci, S. (1987). *Surf. Sci.* **181**, 126.

Bauer, E., Mundschau, M., Swiech, W. and Telieps, W. (1989). *Ultramicroscopy* **31**, 49.

Baum, G., Kisker, E., Mahan, A. H., Raith, W. and Reihl, B. (1977). *Appl. Phys.* **14**, 149.

Baumeister, W. (1988). *Ultramicroscopy* **25**, 103.

Baumeister, W. and Zeitler, E. (1992). In: *Scanned Probe Microscopy* (ed. Wickramasinghe, H. K.), AIP Conf. Proc. 241, p. 544 (New York: AIP).

Bäumer, M., Cappus, D., Kuhlenbeck, H., Freund, H.-J., Wilhelmi, G., Brodde, A. and Neddermeyer, H. (1991). *Surf. Sci.* **253**, 116.

Baz', A. I. (1967a). *Sov. J. Nucl. Phys.* **4**, 182.

Baz', A. I. (1967b). *Sov. J. Nucl. Phys.* **5**, 161.

Becker, R. S., Golovchenko, J. A., Hamann, D. R. and Swartzentruber, B. S. (1985a). *Phys. Rev. Lett.* **55**, 2032.

Becker, R. S., Golovchenko, J. A., Higashi, G. S. and Swartzentruber, B. S. (1986). *Phys. Rev. Lett.* **57**, 1020.

Becker, R. S., Golovchenko, J. A., McRae, E. G. and Swartzentruber, B. S. (1985b). *Phys. Rev. Lett.* **55**, 2028.

Becker, R. S., Golovchenko, J. A. and Swartzentruber, B. S. (1985c). *Phys. Rev. Lett.* **54**, 2678.

Becker, R. S., Golovchenko, J. A. and Swartzentruber, B. S. (1985d). *Phys. Rev. Lett.* **55**, 987.

Becker, R. S., Golovchenko, J. A. and Swartzentruber, B. S. (1987). *Nature* **325**, 419.

Becker, R. S., Higashi, G. S., Chabal, Y. J. and Becker, A. J. (1990). *Phys. Rev. Lett.* **65**, 1917.

Becker, R. S., Klitsner, T. and Vickers, J. S. (1988). *Phys. Rev.* B **38**, 3537.

Becker, R. S., Kortan, A. R., Thiel, F. A. and Chen, H. S. (1991). *J. Vac. Sci. Technol.* B **9**, 867.

Becker, R. S., Swartzentruber, B. S., Vickers, J. S. and Klitsner, T. (1989). *Phys. Rev.* B **39**, 1633.

Bednorz, J. G. and Müller, K. A. (1986). *Z. Phys.* B **64**, 189.

Bednorz, J. G. and Müller, K. A. (1988). *Rev. Mod. Phys.* **60**, 585.

Bedrossian, P., Chen, D. M., Mortensen, K. and Golovchenko, J. A. (1989a). *Nature* **342**, 258.

Bedrossian, P. and Klitsner, T. (1991). *Phys. Rev.* B **44**, 13783.

Bedrossian, P. and Klitsner, T. (1992). *Phys. Rev. Lett.* **68**, 646.

Bedrossian, P., Meade, R. D., Mortensen, K., Chen, D. M., Golovchenko, J. A. and Vanderbilt, D. (1989b). *Phys. Rev. Lett.* **63**, 1257.

Bedrossian, P., Mortensen, K., Chen, D. M. and Golovchenko, J. A. (1990). *Phys. Rev.* B **41**, 7545.

Beebe, Jr., T. P., Wilson, T. E., Ogletree, D. F., Katz, J. E., Balhorn, R., Salmeron, M. B. and Siekhaus, W. J. (1989). *Science* **243**, 370.

Behm, R. J. (1990). In: *Scanning Tunneling Microscopy and Related Methods* (ed. Behm, R. J., Garcia, N. and Rohrer, H.), NATO ASI Series E: Appl. Sci. Vol. 184, p. 173 (Dordrecht: Kluwer).

Bell, A. E., Rao, K. and Swanson, L. W. (1988). *J. Vac. Sci. Technol.* B **6**, 306.

Bell, L. D., Hecht, M. H., Kaiser, W. J. and Davis, L. C. (1990). *Phys. Rev. Lett.* **64**, 2679.

Bell, L. D. and Kaiser, W. J. (1988). *Phys. Rev. Lett.* **61**, 2368.

Ben Assayag, G., Sudraud, P. and Swanson, L. W. (1987). *Surf. Sci.* **181**, 362.

Bendersky, L. (1985). *Phys. Rev. Lett.* **55**, 1461.

Ben-Jacob, E. & Gefen, Y. (1985). *Phys. Lett.* A **108**, 289.

Berghaus, Th., Brodde, A., Neddermeyer, H. and Tosch, St. (1987). *Surf. Sci.* **184**, 273.

Berghaus, Th., Brodde, A., Neddermeyer, H. and Tosch, St. (1988a). *J. Vac. Sci. Technol.* A **6**, 478.

Berghaus, Th., Brodde, A., Neddermeyer, H. and Tosch, St. (1988b). *J. Vac. Sci. Technol.* A **6**, 483.

Berghaus, Th., Brodde, A., Neddermeyer, H. and Tosch, St. (1988c). *Surf. Sci.* **193**, 235.

Berndt, R., Gimzewski, J. K. and Johansson, P. (1991a). *Phys. Rev. Lett.* **67**, 3796.

Berndt, R., Schlittler, R. R. and Gimzewski, J. K. (1991b). *J. Vac. Sci. Technol.* B **9**, 573.

Berthe, R. (1991). Ph.D. Thesis, Research Center Jülich, Jülich, Germany.

Berthe, R., Hartmann, U. and Heiden, C. (1990). *Appl. Phys. Lett.* **57**, 2351.

Besenbacher, F., Jensen, F., Laegsgaard, E., Mortensen, K. and Stensgaard, I. (1991). *J. Vac. Sci. Technol.* B **9**, 874.

Beskrovnyi, A. I., Dlouhá, M., Jirák, Z., Vratislav, S. and Pollert, E. (1990). *Physica* C **166**, 79.

Betzig, E., Finn, P. L. and Weiner, J. S. (1992). *Appl. Phys. Lett.* **60**, 2484.

Betzig, E., Isaacson, M. and Lewis, A. (1987). *Appl. Phys. Lett.* **51**, 2088.

Betzig, E., Lewis, A., Harootunian, A., Isaacson, M. and Kratschmer, E. (1986). *Biophys. J.* **49**, 269.

Biegelsen, D. K., Bringans, R. D., Northrup, J. E. and Swartz, L.-E. (1990a). *Phys. Rev.* B **41**, 5701.

Biegelsen, D. K., Bringans, R. D., Northrup, J. E. and Swartz, L.-E. (1990b). *Phys. Rev. Lett.* **65**, 452.

Biegelsen, D. K., Ponce, F. A. and Tramontana, J. C. (1989). *Appl. Phys. Lett.* **54**, 1223.

Biegelsen, D. K., Ponce, F. A., Tramontana, J. C. and Koch, S. M. (1987). *Appl. Phys. Lett.* **50**, 696.

Binh, V. T. and Marien, J. (1988). *Surf. Sci.* **202**, L539.

Binnig, G. (1992). *Ultramicroscopy* **42–44**, 7.

Binnig, G., Frank, K. H., Fuchs, H., Garcia, N., Reihl, B., Rohrer, H., Salvan, F. and Williams, A. R. (1985a). *Phys. Rev. Lett.* **55**, 991.

Binnig, G., Fuchs, H., Gerber, Ch., Rohrer, H., Stoll, E. and Tosatti, E. (1986a). *Europhys. Lett.* **1**, 31.

Binnig, G., Fuchs, H. and Stoll, E. (1986b). *Surf. Sci.* **169**, L295.

Binnig, G., Garcia, N. and Rohrer, H. (1985b). *Phys. Rev.* B **32**, 1336.

Binnig, G., Garcia, N., Rohrer, H., Soler, J. M. and Flores, F. (1984a). *Phys. Rev.* B **30**, 4816.

Binnig, G., Gerber, Ch., Stoll, E., Albrecht, T. R. and Quate, C. F. (1987a). *Europhys. Lett.* **3**, 1281.

Binnig, G., Gerber, Ch., Stoll, E., Albrecht, T. R. and Quate, C. F. (1987b). *Surf. Sci.* **189/190**, 1.

Binnig, G., Quate, C. F. and Gerber, Ch. (1986c). *Phys. Rev. Lett.* **56**, 930.

Binnig, G. and Rohrer, H. (1982). *Helv. Phys. Acta* **55**, 726.

Binnig, G. and Rohrer, H. (1983). *Surf. Sci.* **126**, 236.

Binnig, G. and Rohrer, H. (1987). *Rev. Mod. Phys.* **59**, 615.

Binnig, G. K., Rohrer, H., Gerber, Ch. and Stoll, E. (1984b). *Surf. Sci.* **144**, 321.

Binnig, G., Rohrer, H., Gerber, Ch. and Weibel, E. (1982a). *Appl. Phys. Lett.* **40**, 178.

Binnig, G., Rohrer, H., Gerber, Ch. and Weibel, E. (1982b). *Phys. Rev. Lett.* **49**, 57.

Binnig, G., Rohrer, H., Gerber, Ch. and Weibel, E. (1982c). *Physica B* **109** & **110**, 2075.

Binnig, G., Rohrer, H., Gerber, Ch. and Weibel, E. (1983a). *Phys. Rev. Lett.* **50**, 120.

Binnig, G., Rohrer, H., Gerber, Ch. and Weibel, E. (1983b). *Surf. Sci.* **131**, L379.

Binnig, G. and Smith, D. P. E. (1986). *Rev. Sci. Instrum.* **57**, 1688.

Bitter, F. (1931) *Phys. Rev.* **38**, 1903.

Blackman, G. S., Mate, C. M. and Philpott, M. R. (1990). *Phys. Rev. Lett.* **65**, 2270.

Bleaney, B. (1984). *Contemp. Phys.* **25**, 320.

Blodgett, K. B. (1935). *J. Am. Chem. Soc.* **57**, 1007.

Blonder, G. E. and Tinkham, M. (1983). *Phys. Rev.* B **27**, 112.

Blonder, G. E., Tinkham, M. and Klapwijk, T. M. (1982). *Phys. Rev.* B **25**, 4515.

Blügel, S., Pescia, D. and Dederichs, P. H. (1989). *Phys. Rev.* B **39**, 1392.

Boato, G., Cantini, P. and Colella, R. (1979). *Phys. Rev. Lett.* **42**, 1635.

Bocko, M. F. and Stephenson, K. A. (1991). *J. Vac. Sci. Technol.* B **9**, 1363.

Bocko, M. F., Stephenson, K. A. and Koch, R. H. (1988). *Phys. Rev. Lett.* **61**, 726.

Boland, J. J. (1990). *Phys. Rev. Lett.* **65**, 3325.

Boland, J. J. (1991a). *Surf. Sci.* **244**, 1.

Boland, J. J. (1991b). *J. Vac. Sci. Technol.* B **9**, 764.

Boland, J. J. (1992). *Surf. Sci.* **261**, 17.

Boland, J. J. and Villarrubia, J. S. (1990). *Phys. Rev.* B **41**, 9865.

Bono, J. and Good, Jr, R. H. (1987). *Surf. Sci.* **188**, 153.

Bosch, J., Gross, R., Koyanagi, M. and Huebener, R. P. (1985). *Phys. Rev. Lett.* **54**, 1448.

Bourdieu, L., Silberzan, P. and Chatenay, D. (1991). *Phys. Rev. Lett.* **67**, 2029.

Bowden, F. P. and Tabor, D. (1950). *The Friction and Lubrication of Solids*, Vol. 1 (Oxford: Clarendon).

Briceno, G. and Zettl, A. (1989). *Solid State Commun.* **70**, 1055.

Bringans, R. D., Biegelsen, D. K. and Swartz, L.-E. (1991). *Phys. Rev.* B **44**, 3054.

Broers, A. N. (1988). *IBM J. Res. Develop.* **32**, 502.

Brune, H., Wintterlin, J., Behm, R. J. and Ertl, G. (1992). *Phys. Rev. Lett.* **68**, 624.

Bryant, A., Smith, D. P. E., Binnig, G., Harrison, W. A. and Quate, C. F. (1986b). *Appl. Phys. Lett.* **49**, 936.

Bryant, A., Smith, D. P. E. and Quate, C. F. (1986a). *Appl. Phys. Lett.* **48**, 832.

Bryant, P. J., Kim, H. S., Zheng, Y. C. and Yang, R. (1987). *Rev. Sci. Instrum.* **58**, 1115.

Buchholz, S. and Rabe, J. P. (1991). *J. Vac. Sci. Technol.* B **9**, 1126.

Buckel, W. (1984). *Supraleitung* (Weinheim: Physik-Verlag).

Buckley, J. E., Wragg, J. L., White, H. W., Bruckdorfer, A. and Worcester, D. L. (1991). *J. Vac. Sci. Technol.* B **9**, 1079.

Burk, B., Thomson, R. E., Zettl, A. and Clarke, J. (1991). *Phys. Rev. Lett.* **66**, 3040.

Burnham, N. A. and Colton, R. J. (1989). *J. Vac. Sci. Technol.* A **7**, 2906.

Burnham, N. A. and Colton, R. J. and Pollock, H. M. (1991). *J. Vac. Sci. Technol.* A **9**, 2548.

Burnham, N. A., Dominguez, D. D., Mowery, R. L. and Colton, R. J. (1990). *Phys. Rev. Lett.* **64**, 1931.

Burstein, E. and Lundquist, S. (1969). *Tunneling Phenomena in Solids* (New York: Plenum).

Bussian, B., Frankowski, I. and Wohlleben, D. (1982). *Phys. Rev. Lett.* **49**, 1026.

Busta, H. H., Shadduck, R. R. and Orvis, W. J. (1989). *IEEE Trans. Electron Devices* **36**, 2679.

Butt, H.-J., Prater, C. B. and Hansma, P. K. (1991). *J. Vac. Sci. Technol.* B **9**, 1193.

Büttiker, M. (1983). *Phys. Rev.* **27**, 6178.

Büttiker, M. and Landauer, R. (1982). *Phys. Rev. Lett.* **49**, 1739.

Cahill, D. G. and Avouris, Ph. (1992). *Appl. Phys. Lett.* **60**, 326.

Cahill, D. G. and Hamers, R. J. (1991a). *J. Vac. Sci. Technol.* B **9**, 564.

Cahill, D. G. and Hamers, R. J. (1991b). *Phys. Rev.* B **44**, 1387.

Cantini, P., Boato, G. and Colella, R. (1980). *Physica* B **99**, 59.

Cataldi, T. R. I., Blackham, I. G., Briggs, G. A. D., Pethica, J. B. and Hill, H. A. O. (1990). *J. Electroanal. Chem.* **290**, 1.

Celotta, R. J. and Pierce, D. T. (1986). *Science* **234**, 333.

Chadi, D. J. (1987). *Phys. Rev. Lett.* **59**, 1691.

Chambliss, D. D. and Chiang, S. (1992). *Surf. Sci. Lett.* **264**, L187.

Chambliss, D. D. and Wilson, R. J. (1991). *J. Vac. Sci. Technol.* B **9**, 928.

Chambliss, D. D., Wilson, R. J. and Chiang, S. (1991a). *Phys. Rev. Lett.* **66**, 1721.

Chambliss, D. D., Wilson, R. J. and Chiang, S. (1991b). *J. Vac. Sci. Technol.* B **9**, 933.

Chang, L. L., Esaki, L. and Tsu, R. (1974). *Appl. Phys. Lett.* **24**, 593.

Chang, T. H. P., Kern, D. P., Kratschmer, E., Lee, K. Y., Luhn, H. E., McCord, M. A., Rishton, S. A. and Vladimirsky, Y. (1988). *IBM J. Res. Develop.* **32**, 462.

Chang, T. H. P., Kern, D. P. and McCord, M. A. (1989). *J. Vac. Sci. Technol.* B **7**, 1855.

Chang, T. H. P., Kern, D. P., McCord, M. A. and Muray, L. P. (1991). *J. Vac. Sci. Technol.* B **9**, 438.

Chang, T. H. P., Kern, D. P. and Muray, L. P. (1990). *J. Vac. Sci. Technol.* B **8**, 1698.

Chang, T. H. P., Kern, D. P., Muray, L. P. and Staufer, U. (1992). In: *Scanned Probe Microscopy* (ed. Wickramasinghe, H. K.), AIP Conf. Proc. 241, p. 420 (New York: AIP).

Chen, C. and Gewirth, A. A. (1992). *Phys. Rev. Lett.* **68**, 1571.

Chen, C., Vesecky, S. M. and Gewirth, A. A. (1992). *J. Am. Chem. Soc.* **114**, 451.

Chen, C. J. (1988). *J. Vac. Sci. Technol.* A **6**, 319.

Chen, C. J. (1990a). *Phys. Rev.* B **42**, 8841.

Chen, C. J. (1990b). *Phys. Rev. Lett.* **65**, 448.

Chen, C. J. (1991a). *J. Vac. Sci. Technol.* A **9**, 44.

Chen, C. J. (1991b). *J. Phys.: Condens. Matter* **3**, 1227.

Chen, C. J. (1992a). *Appl. Phys. Lett.* **60**, 132.

Chen, C. J. (1992b). *Ultramicroscopy* **42–44**, 1653.

Chen, C. J. (1992c). *Ultramicroscopy* **42–44**, 147.

Chen, C. J. and Hamers, R. J. (1991). *J. Vac. Sci. Technol.* B **9**, 503.

Chen, C. J. and Tsuei, C. C. (1989). *Solid State Commun.* **71**, 33.

Chen, J.-S., Devine, T. M., Ogletree, D. F. and Salmeron, M. (1991). *Surf. Sci.* **258**, 346.

Chen, Q. and Ng, K.-W. (1992). *Phys. Rev.* B **45**, 2569.

Chiang, S. (1992). In: *Scanning Tunneling Microscopy I* (ed. Güntherodt, H.-J. & Wiesendanger, R.), Springer Series in Surface Sciences 21, p. 181 (Berlin, Heidelberg, New York: Springer).

Chiang, S., Wilson, R. J., Gerber, Ch. and Hallmark, V. M. (1988a). *J. Vac. Sci. Technol.* A **6**, 386.

Chiang, S., Wilson, R. J., Mate, C. M. and Ohtani, H. (1988b). *J. Microsc.* **152**, 567.

Chiang, S., Wilson, R. J., Mate, C. M. and Ohtani, H. (1990). *Vacuum* **41**, 118.

Christoph, R., Siegenthaler, H., Rohrer, H. and Wiese, H. (1989). *Electrochim. Acta* **34**, 1011.

Chynoweth, A. G. and McKay, K. G. (1957). *Phys. Rev.* **106**, 418.

Ciccacci, F., Molinari, E. and Christensen, N. E. (1987). *Solid State Commun.* **62**, 1.

Ciraci, S. (1990). In: *Scanning Tunneling Microscopy and Related Methods* (ed. Behm, R. J., Garcia, N. and Rohrer, H.), NATO ASI Series E: Appl. Sci. Vol. 184, p. 113 (Dordrecht: Kluwer).

Ciraci, S., Baratoff, A. and Batra, I. P. (1990a). *Phys. Rev.* B **41**, 2763.

Ciraci, S., Baratoff, A. and Batra, I. P. (1990b). *Phys. Rev.* B **42**, 7618.

Ciraci, S. and Batra, I. P. (1987). *Phys. Rev.* B **36**, 6194.

Ciraci, S. and Tekman, E. (1989). *Phys. Rev.* B **40**, 11969.

Cites, J., Sanghadasa, M. F. M., Sung, C. C., Reddick, R. C., Warmack, R. J. and Ferrell, T. L. (1992). *J. Appl. Phys.* **71**, 7.

Clemmer, C. R. and Beebe, Jr, T. P. (1991). *Science* **251**, 640.

Cline, J. A., Barshatzky, H. and Isaacson, M. (1991). *Ultramicroscopy* **38**, 299.

Cohen, M. H., Falicov, L. M. and Phillips, J. C. (1962). *Phys. Rev. Lett.* **8**, 316.

Cohen, S. R., Neubauer, G. and McClelland, G. M. (1990). *J. Vac. Sci. Technol.* A **8**, 3449.

Coleman, R. V., Drake, B., Giambattista, B., Johnson, A., Hansma, P. K., McNairy, W. W. and Slough, G. (1988a). *Phys. Scripta* **38**, 235.

Coleman, R. V., Drake, B., Hansma, P. K. and Slough, G. (1985). *Phys. Rev. Lett.* **55**, 394.

Coleman, R. V., Giambattista, B., Hansma, P. K., Johnson, A., McNairy, W. W. and Slough, C. G. (1988b). *Adv. Phys.* **37**, 559.

Coleman, R. V., Giambattista, B., Johnson, A., McNairy, W. W., Slough, G., Hansma, P. K. and Drake, B. (1988c). *J. Vac. Sci. Technol.* A **6**, 338.

Coleman, R. V., McNairy, W. W. and Slough, C. G. (1992). *Phys. Rev.* B **45**, 1428.

Coleman, R. V., McNairy, W. W., Slough, C. G., Hansma, P. K. and Drake, B. (1987). *Surf. Sci.* **181**, 112.

Coombs, J. H. and Gimzewski, J. K. (1988). *J. Microsc.* **152**, 841.

Coombs, J. H. and Gimzewski, J. K., Reihl, B., Sass, J. K. and Schlittler, R. R. (1988b). *J. Microsc.* **152**, 325.

Coombs, J. H. and Pethica, J. B. (1986). *IBM J. Res. Develop.* **30**, 443.

Coombs, J. H., Welland, M. E. and Pethica, J. B. (1988a). *Surf. Sci.* **198**, L353.

Coratger, R., Claverie, A., Ajustron, F. and Beauvillain, J. (1990). *Surf. Sci.* **227**, 7.

Corb, B. W., Ringger, M. and Güntherodt, H.-J. (1985). *J. Appl. Phys.* **58**, 3947.

Couch, N. R., Montgomery, C. M. and Jones, R. (1986). *Thin Solid Films* **135**, 173.

Coulman, D., Wintterlin, J., Barth, J. V., Ertl, G. and Behm, R. J. (1990a). *Surf. Sci.* **240**, 151.

Coulman, D., Wintterlin, J., Behm, R. J. and Ertl, G. (1990b). *Phys. Rev. Lett.* **64**, 1761.

Courjon, D. (1990). In: *Scanning Tunneling Microscopy and Related Methods* (ed. Behm, R. J., Garcia, N. and Rohrer, H.) NATO ASI Series E: Appl. Sci. Vol. 184, p. 497 (Dordrecht: Kluwer).

Courjon, D., Sarayeddine, K. and Spajer, M. (1989). *Opt. Commun.* **71**, 23.

Courjon, D., Vigoureux, J.-M., Spajer, M., Sarayeddine, K. and Leblanc, S. (1990). *Appl. Opt.* **29**, 3734.

Cox, G., Graf, K. H., Szynka, D., Poppe, U. and Urban, K. (1990a). *Vacuum* **41**, 591.

Cox, G., Szynka, D., Poppe, U., Graf, K. H., Urban, K., Kisielowski-Kemmerich, C., Krüger, J. and Alexander, H. (1990b). *Phys. Rev. Lett.* **64**, 2402.

Cox, G., Szynka, D., Poppe, U., Graf, K. H., Urban, K., Kisielowski-Kemmerich, C., Krüger, J. and Alexander, H. (1991). *J. Vac. Sci. Technol.* B **9**, 726.

Craston, D. H., Lin, C. W. and Bard, A. J. (1988). *J. Electrochem. Soc.* **135**, 785.

Cribier, D., Jacrot, B., Rao, L. M. and Farnoux, B. (1964). *Phys. Lett.* **9**, 106.

Crommie, M. F., Bourne, L. C., Zettl, A., Cohen, M. L. and Stacy, A. (1987). *Phys. Rev.* B **35**, 8853.

Cutler, P. H., Feuchtwang, T. E., Tsong, T. T., Nguyen, H. and Lucas, A. A. (1987). *Phys. Rev.* B **35**, 7774.

Dagata, J. A., Schneir, J., Harary, H. H., Bennett, J. and Tseng, W. (1991a). *J. Vac. Sci. Technol.* B **9**, 1384.

Dagata, J., Schneir, J., Harary, H. H., Evans, C. J., Pastek, M. T. and Bennett, J. (1990a). *Appl. Phys. Lett.* **56**, 2001.

Dagata, J. A., Tseng, W., Bennett, J., Evans, C. J., Schneir, J. and Harary, H. H. (1990b). *Appl. Phys. Lett.* **57**, 2437.

Dagata, J. A., Tseng, W., Bennett, J., Schneir, J. and Harary, H. H. (1991b). *Appl. Phys. Lett.* **59**, 3288.

Dagata, J. A., Tseng, W., Bennett, J., Schneir, J. and Harary, H. H. (1991c). *J. Appl. Phys.* **70**, 3661.

Dahn, D. C., Watanabe, M. O., Blackford, B. L. and Jericho, M. H. (1988). *J. Appl. Phys.* **63**, 315.

Dai, H., Chen, H. and Lieber, C. M. (1991). *Phys. Rev. Lett.* **66**, 3183.

d'Ambrumenil, N. and White, R. M. (1984). *Solid State Commun.* **50**, 1043.

Das, B. and Mahanty, J. (1987). *Phys. Rev.* B **36**, 898.

Daumas, N. and Hérold, A. (1969). *C. R. Acad. Sci. Paris* **268**, C 373.
Davis, L. C., Everson, M. P., Jaklevic, R. C. and Shen, W. (1991). *Phys. Rev. B* **43**, 3821.
Deb, B. M. (1973). *Rev. Mod. Phys.* **45**, 22.
de Broglie, L. (1923). *Nature* **112**, 540.
de Groot, R. A. (1991). *Physica B* **172**, 45.
de Groot, R. A., Mueller, F. M., van Engen, P. G. and Buschow, K. H. J. (1983). *Phys. Rev. Lett.* **50**, 2024.
de Lozanne, A. L., Elrod, S. A. and Quate, C. F. (1985). *Phys. Rev. Lett.* **54**, 2433.
Delsing, P., Claeson, T., Likharev, K. K. and Kuzmin, L. S. (1990). *Phys. Rev. B* **42**, 7439.
Demuth, J. E., Hamers, R. J., Tromp, R. M. and Welland, M. E. (1986a). *J. Vac. Sci. Technol. A* **4**, 1320.
Demuth, J. E., Hamers, R. J., Tromp, R. M. and Welland, M. E. (1986b). *IBM J. Res. Develop.* **30**, 396.
Demuth, J. E., Koehler, U. and Hamers, R. J. (1988). *J. Microsc.* **152**, 299.
den Boef, A. J. (1989). *Appl. Phys. Lett.* **55**, 439.
den Boef, A. J. (1990). *Appl. Phys. Lett.* **56**, 2045.
den Boef, A. J. (1991). *Rev. Sci. Instrum.* **62**, 88.
De Stasio, G., Rioux, D., Margaritondo, G., Mercanti, D., Trasatti, L. and Moore, C. (1991). *J. Vac. Sci. Technol. A* **9**, 2319.
DeVault, D. (1981). *Quantum-mechanical Tunnelling in Biological Systems* (Cambridge: Cambridge University Press).
Dietz, P., Fostiropoulos, K., Krätschmer, W. and Hansma, P. K. (1992). *Appl. Phys. Lett.* **60**, 62.
DiNardo, N. J., Wong, T. M. and Plummer, E. W. (1990). *Phys. Rev. Lett.* **65**, 2177.
Dittrich, R. and Heiden, C. (1988). *J. Vac. Sci. Technol. A* **6**, 263.
Dovek, M. M., Albrecht, T. R., Kuan, S. W. J., Lang, C. A., Emch, R., Grütter, P., Frank, C. W., Pease, R. F. W. and Quate, C. F. (1988). *J. Microsc.* **152**, 229.
Dovek, M. M., Lang, C. A., Nogami, J. and Quate, C. F. (1989). *Phys. Rev. B* **40**, 11973.
Doyen, G., Drakova, D., Kopatzki, E. and Behm, R. J. (1988). *J. Vac. Sci. Technol. A* **6**, 327.
Dragoset, R. A., First, P. N., Stroscio, J. A., Pierce, D. T. and Celotta, R. J. (1989). *Mat. Res. Soc. Symp. Proc.* **151**, 193.
Drake, B., Prater, C. B., Weissenhorn, A. L., Gould, S. A. C., Albrecht, T. R., Quate, C. F., Cannell, D. S., Hansma, H. G. and Hansma, P. K. (1989). *Science* **243**, 1586.
Drake, B., Sonnenfeld, R., Schneir, J. and Hansma, P. K. (1987). *Surf. Sci.* **181**, 92.
Dransfeld, K. and Xu, J. (1988). *J. Microsc.* **152**, 35.
Dresselhaus, M. S. and Dresselhaus, G. (1981). *Adv. Phys.* **30**, 139.
Driscoll, R. J., Youngquist, M. G. and Baldeschwieler, J. D. (1990). *Nature* **346**, 294.
Ducker, W. A., Cook, R. F. and Clarke, D. R. (1990). *J. Appl. Phys.* **67**, 4045.
Dujardin, G., Walkup, R. E. and Avouris, Ph. (1992). *Science* **255**, 1232.
Duke, C. B. (1969). *Tunneling in Solids* (New York, London: Academic).

Dunlap, D. D. and Bustamante, C. (1989). *Nature* **342**, 204.

Dürig, U., Gimzewski, J. K. and Pohl, D. W. (1986a). *Phys. Rev. Lett.* **57**, 2403.

Dürig, U., Pohl, D. W. and Rohner, F. (1986b). *J. Appl. Phys.* **59**, 3318.

Dürig, U., Pohl, D. W. and Rohner, F. (1986c). *IBM J. Res. Develop.* **30**, 478.

Dürig, U. and Züger, O. (1990). *Vacuum* **41**, 382.

Dürig, U., Züger, O. and Pohl, D. W. (1988). *J. Microsc.* **152**, 259.

Dürig, U., Züger, O. and Pohl, D. W. (1990). *Phys. Rev. Lett.* **65**, 349.

Dynes, R. C., Narayanamurti, V. and Garno, J. P. (1978). *Phys. Rev. Lett.* **41**, 1509.

Ehrichs, E. E., Silver, R. M. and de Lozanne, A. L. (1988a). *J. Vac. Sci. Technol.* A **6**, 540.

Ehrichs, E. E., Yoon, S. and de Lozanne, A. L. (1988b). *Appl. Phys. Lett.* **53**, 2287.

Eigler, D. M., Lutz, C. P. and Rudge, W. E. (1991b). *Nature* **352**, 600.

Eigler, D. M. and Schweizer, E. K. (1990). *Nature* **344**, 524.

Eigler, D. M., Weiss, P. S., Schweizer, E. K. and Lang, N. D. (1991a). *Phys. Rev. Lett.* **66**, 1189.

Eliashberg, G. M. (1960). *Sov. Phys. – JETP* **11**, 696.

Elrod, S. A., de Lozanne, A. L. and Quate, C. F. (1984). *Appl. Phys. Lett.* **45**, 1240.

Engel, A. (1991). *Ann. Rev. Biophys. Biophys. Chem.* **20**, 79.

Erlandsson, R., Hadziioannou, G., Mate, C. M., McClelland, G. M. and Chiang, S. (1988a). *J. Chem. Phys.* **89**, 5190.

Erlandsson, R., McClelland, G. M., Mate, C. M. and Chiang, S. (1988b). *J. Vac. Sci. Technol.* A **6**, 266.

Ertl, G. (1990). In: *Chemistry and Physics of Solid Surfaces VIII* (ed. Vanselow, R. and Howe, R.), Springer Series in Surface Sciences 22, p. 1 (Berlin, Heidelberg, New York: Springer).

Esaki, L. (1958). *Phys. Rev.* **109**, 603.

Esaki, L. (1969). In: *Tunneling Phenomena in Solids* (ed. Burstein, E. and Lundquist, S.), p. 47 (New York: Plenum).

Esaki, L. (1974). *Rev. Mod. Phys.* **46**, 237.

Esaki, L. and Stiles, P. J. (1966). *Phys. Rev. Lett.* **16**, 1108.

Escudero, R., Guarner, E. and Morales, F. (1990). *Physica* C **166**, 15.

Essmann, U. and Träuble, H. (1967). *Phys. Lett.* A **24**, 526.

Everson, M. P., Jaklevic, R. C. and Shen, W. (1990). *J. Vac. Sci. Technol.* A **8**, 3662.

Everson, M. P., Davis, L. C., Jaklevic, R. C. and Shen, W. (1991). *J. Vac. Sci. Technol.* B **9**, 891.

Feenstra, R. M. (1989a). *J. Vac. Sci. Technol.* B **7**, 925.

Feenstra, R. M. (1989b). *Phys. Rev. Lett.* **63**, 1412.

Feenstra, R. M. (1990). In: *Scanning Tunneling Microscopy and Related Methods* (ed. Behm, R. J., Garcia, N. and Rohrer, H.), NATO ASI Series E: Appl. Sci. Vol. 184, p. 211 (Dordrecht: Kluwer).

Feenstra, R. M. (1991). *Phys. Rev.* B **44**, 13791.

Feenstra, R. M. and Lutz, M. A. (1990). *Phys. Rev.* B **42**, 5391.

Feenstra, R. M. and Lutz, M. A. (1991a). *Surf. Sci.* **243**, 151.

Feenstra, R. M. and Lutz, M. A. (1991b). *J. Vac. Sci. Technol.* B **9**, 716.

Feenstra, R. M. and Mårtensson, P. (1988). *Phys. Rev. Lett.* **61**, 447.

Feenstra, R. M. and Slavin, A. J. (1991). *Surf. Sci.* **251/252**, 401.

Feenstra, R. M. and Slavin, A. J., Held, G. A. and Lutz, M. A. (1991). *Phys. Rev. Lett.* **66**, 3257.

Feenstra, R. M. and Stroscio, J. A. (1987a). *Phys. Scripta* T **19**, 55.

Feenstra, R. M. and Stroscio, J. A. (1987b). *Phys. Rev. Lett.* **59**, 2173.

Feenstra, R. M. and Stroscio, J. A. (1987c). *J. Vac. Sci. Technol.* B **5**, 923.

Feenstra, R. M., Stroscio, J. A. and Fein, A. P. (1987a). *Surf. Sci.* **181**, 295.

Feenstra, R. M., Stroscio, J. A., Tersoff, J. and Fein, A. P. (1987b). *Phys. Rev. Lett.* **58**, 1192.

Feenstra, R. M., Thompson, W. A. and Fein, A. P. (1986). *Phys. Rev. Lett.* **56**, 608.

Feidenhans'l, R., Grey, F., Nielsen, M., Besenbacher, F., Jensen, F., Laegsgaard, E., Stensgaard, I., Jacobsen, K. W., Nørskov, J. K. and Johnson, R. L. (1990). *Phys. Rev. Lett.* **65**, 2027.

Fein, A. P., Kirtley, J. R. and Feenstra, R. M. (1987). *Rev. Sci. Instrum.* **58**, 1806.

Fein, A. P., Kirtley, J. R. and Shafer, M. W. (1988). *Phys. Rev.* B **37**, 9738.

Ferrell, T. L., Goundonnet, J. P., Reddick, R. C., Sharp, S. L. and Warmack, R. J. (1991). *J. Vac. Sci. Technol.* B **9**, 525.

Ferrer, J., Flores, F. and Martín-Rodero, A. (1989). *Phys. Rev.* B **39**, 11320.

Ferrer, J., Martín-Rodero, A. and Flores, F. (1988). *Phys. Rev.* B **38**, 10113.

Fertig, H. A. (1990). *Phys. Rev. Lett.* **65**, 2321.

Feuchtwang, T. E., Leipold, W. C. and Martino, R. C. (1977). *Surf. Sci.* **64**, 109.

Feynman, R. P. (1939). *Phys. Rev.* **56**, 340.

Feynman, R. P. (1961). In: *Miniaturization* (ed. Gilbert, H. D.), p. 282 (New York: Reinhold).

Fink, H.-W. (1986). *IBM J. Res. Develop.* **30**, 460.

Fink, H.-W. (1988). *Phys. Scripta* **38**, 260.

First, P. N., Dragoset, R. A., Stroscio, J. A., Celotta, R. J. and Feenstra, R. M. (1989a). *J. Vac. Sci. Technol.* A **7**, 2868.

First, P. N., Stroscio, J. A., Dragoset, R. A., Pierce, D. T. and Celotta, R. J. (1989b). *Phys. Rev. Lett.* **63**, 1416.

First, P. N., Stroscio, J. A., Pierce, D. T., Dragoset, R. A. and Celotta, R. J. (1991). *J. Vac. Sci. Technol.* B **9**, 531.

Fischer, U. Ch. (1985). *J. Vac. Sci. Technol.* B **3**, 386.

Fischer, U. Ch., Dürig, U. T. and Pohl, D. W. (1988). *Appl. Phys. Lett.* **52**, 249.

Fischer, U. Ch. and Pohl, D. W. (1989). *Phys. Rev. Lett.* **62**, 458.

Fischer, J. C. and Giaever, I. (1961). *J. Appl. Phys.* **32**, 172.

Fiske, M. D. (1964). *Rev. Mod. Phys.* **36**, 221.

Fite, II, W. and Redfield, A. G. (1966). *Phys. Rev. Lett.* **17**, 381.

Flores, F., Echenique, P. M. and Ritchie, R. H. (1986). *Phys. Rev.* B **34**, 2899.

Forbes, R. G. (1990). In: *Scanning Tunneling Microscopy and Related Methods* (ed. Behm, R. J., Garcia, N. and Rohrer, H.), NATO ASI Series E: Appl. Sci. Vol. 184, p. 163 (Dordrecht: Kluwer).

Foster, J. S. and Frommer, J. E. (1988). *Nature* **333**, 542.

Fowler, R. H. and Nordheim, L. (1928). *Proc. Roy. Soc. (London)* A **119**, 173.

Franks, A. (1987). *J. Phys.* E **20**, 1442.

Frenkel, J. (1930). *Phys. Rev.* **36**, 1604.

Frenken, J. W. M., Hamers, R. J. and Demuth, J. E. (1990). *J. Vac. Sci. Technol.* A **8**, 293.

Frohn, J., Giesen, M., Poensgen, M., Wolf, J. F. and Ibach, H. (1991). *Phys. Rev. Lett.* **67**, 3543.

Frohn, J., Wolf, J. F., Besocke, K. and Teske, M. (1989). *Rev. Sci. Instrum.* **60**, 1200.

Fuchs, H. (1988). *Phys. Scripta* **38**, 264.

Fuchs, H., Akari, S. and Dransfeld, K. (1990a). *Z. Phys.* B **80**, 389.

Fuchs, H. and Laschinski, R. (1990). *Scanning* **12**, 126.

Fuchs, H., Laschinski, R. and Schimmel, Th. (1990b). *Europhys. Lett.* **13**, 307.

Fuchs, H. and Tosatti, E. (1987). *Europhys. Lett.* **3**, 745.

Fulton, T. A. and Dolan, G. J. (1987). *Phys. Rev. Lett.* **59**, 109.

Fulton, T. A., Gammel, P. L., Bishop, D. J., Dunkleberger, L. N. and Dolan, G. J. (1989). *Phys. Rev. Lett.* **63**, 1307.

Gadzuk, J. W. and Plummer, E. W. (1971). *Phys. Rev. Lett.* **26**, 92.

Gallagher, M. C. and Adler, J. G. (1990). *J. Vac. Sci. Technol.* A **8**, 464.

Gallagher, M. C., Adler, J. G., Jung, J. and Franck, J. P. (1988). *Phys. Rev.* B **37**, 7846.

Gammie, G., Hubacek, J. S., Skala, S. L., Brockenbrough, R. T., Tucker, J. R. and Lyding, J. W. (1989a). *Phys. Rev.* B **40**, 9529.

Gammie, G., Hubacek, J. S., Skala, S. L., Brockenbrough, R. T., Tucker, J. R. and Lyding, J. W. (1989b). *Phys. Rev.* B **40**, 11965.

Gammie, G., Hubacek, J. S., Skala, S. L., Tucker, J. R. and Lyding, J. W. (1991). *J. Vac. Sci. Technol.* B **9**, 1027.

Gammie, G., Skala, S., Hubacek, J. S., Brockenbrough, R., Lyons, W. G., Tucker, J. R. and Lyding, J. W. (1988). *J. Microsc.* **152**, 497.

Gamov, G. (1928). *Z. Phys.* **51**, 204.

Ganz, E., Hwang, I.-S., Xiong, F., Theiss, S. K. and Golovchenko, J. (1991a). *Surf. Sci.* **257**, 259.

Ganz, E., Sattler, K. and Clarke, J. (1988a). *Phys. Rev. Lett.* **60**, 1856.

Ganz, E., Sattler, K. and Clarke, J. (1988b). *J. Vac. Sci. Technol.* A **6**, 419.

Ganz, E., Sattler, K. and Clarke, J. (1989). *Surf. Sci.* **219**, 33.

Ganz, E., Theiss, S. K., Hwang, I.-S. and Golovchenko, J. (1992). *Phys. Rev. Lett.* **68**, 1567.

Ganz, E., Xiong, F., Hwang, I.-S. and Golovchenko, J. (1991b). *Phys. Rev.* B **43**, 7316.

Gao, X., Hamelin, A. and Weaver, M. J. (1991a). *Phys. Rev. Lett.* **67**, 618.

Gao, X., Hamelin, A. and Weaver, M. J. (1991b). *Phys. Rev.* B **44**, 10983.

García, N. (1986). *IBM J. Res. Develop.* **30**, 533.

García, N. and Escapa, L. (1989). *Appl. Phys. Lett.* **54**, 1418.

García, N., Flores, F. and Guinea, F. (1988). *J. Vac. Sci. Technol.* A **6**, 323.

García, N., Ocal, C. and Flores, F. (1983). *Phys. Rev. Lett.* **50**, 2002.

García, R., Sáenz, J. J., Soler, J. M. and García, N. (1986). *J. Phys.* C **19**, L131.

García, R. and García, N. (1990). In: *Scanning Tunneling Microscopy and Related Methods* (ed. Behm, R. J., Garcia, N. and Rohrer, H.), NATO ASI Series E: Appl. Sci. Vol. 184, p. 391 and (Dordrecht: Kluwer).

García-García, R. (1992). *Appl. Phys. Lett.* **60**, 1960.

García-García, R. and García, N. (1991). *Surf. Sci.* **251/252**, 408.

Garfunkel, E., Rudd, G., Novak, D., Wang, S., Ebert, G., Greenblatt, M., Gustafsson, T. and Garofalini, S. H. (1989). *Science* **246**, 99.

Garnaes, J., Kragh, F., Mørch, K. A. and Thölén, A. R. (1990). *J. Vac. Sci. Technol.* A **8**, 441.

Garnaes, J., Schwartz, D. K., Viswanathan, R. and Zasadzinski, J. A. N. (1992). *Nature* **357**, 54.

Gauthier, S., Rousset, S., Klein, J., Sacks, W. and Belin, M. (1988). *J. Vac. Sci. Technol.* A **6**, 360.

Geerligs, L. J., Anderegg, V. F., Romijn, J. and Mooij, J. E. (1990). *Phys. Rev. Lett.* **65**, 377.

Gerber, C., Anselmetti, D., Bednorz, J. G., Mannhart, J. and Schlom, D. G. (1991). *Nature* **350**, 279.

Gerber, Ch., Binnig, G., Fuchs, H., Marti, O. and Rohrer, H. (1986). *Rev. Sci. Instrum.* **57**, 221.

Germano, C. P. (1959). *IRE Trans. Audio*, July–August, p. 96.

Gewirth, A. A. (1992). In: *Scanned Probe Microscopy* (ed. Wickramasinghe, H. K.), AIP Conf. Proc. 241, p. 253 (New York: AIP).

Giaever, I. (1960a). *Phys. Rev. Lett.* **5**, 147.

Giaever, I. (1960b). *Phys. Rev. Lett.* **5**, 464.

Giaever, I. (1974). *Rev. Mod. Phys.* **46**, 245.

Giaever, I. and Megerle, K. (1961). *Phys. Rev.* **122**, 1101.

Giamarchi, T., Béal-Monod, M. T. and Valls, O. T. (1990). *Phys. Rev.* B **41**, 11 033.

Giambattista, B., Johnson, A., Coleman, R. V., Drake, B. and Hansma, P. K. (1988a). *Phys. Rev.* B **37**, 2741.

Giambattista, B., Johnson, A., McNairy, W. W., Slough, C. G. and Coleman, R. V. (1988b). *Phys. Rev.* B **38**, 3545.

Giambattista, B., McNairy, W. W., Slough, C. G., Johnson, A., Bell, L. D., Coleman, R. V., Schneir, J., Sonnenfeld, R., Drake, B. and Hansma, P. K. (1987). *Proc. Natl. Acad. Sci. USA* **84**, 4671.

Giambattista, B., Slough, C. G., McNairy, W. W. and Coleman, R. V. (1990). *Phys. Rev.* B **41**, 10082.

Gimzewski, J. K., Berndt, R. and Schlittler, R. R. (1991a). *Surf. Sci.* **247**, 327.

Gimzewski, J. K., Berndt, R. and Schlittler, R. R. (1991b). *J. Vac. Sci. Technol.* B **9**, 897.

Gimzewski, J. K., Berndt, R. and Schlittler, R. R. (1992). *Phys. Rev.* B **45**, 6844.

Gimzewski, J. K., Humbert, A., Bednorz, J. G. and Reihl, B. (1985). *Phys. Rev. Lett.* **55**, 951.

Gimzewski, J. K. and Möller, R. (1987). *Phys. Rev.* B **36**, 1284.

Gimzewski, J. K., Möller, R., Pohl, D. W. and Schlittler, R. R. (1987a). *Surf. Sci.* **189/190**, 15.

Gimzewski, J. K., Reihl, B., Coombs, J. H. and Schlittler, R. R. (1988). *Z. Phys.* B **72**, 497.

Gimzewski, J. K., Sass, J. K., Schlittler, R. R. and Schott, J. (1989). *Europhys. Lett.* **8**, 435.

Gimzewski, J. K., Stoll, E. and Schlittler, R. R. (1987b). *Surf. Sci.* **181**, 267.

Göddenhenrich, T., Hartmann, U., Anders, M. and Heiden, C. (1988). *J. Microsc.* **152**, 527.

Göddenhenrich, T., Lemke, H., Hartmann, U. and Heiden, C. (1990a). *J. Vac. Sci. Technol.* A **8**, 383.

Göddenhenrich, T., Lemke, H., Hartmann, U. and Heiden, C. (1990b). *Appl. Phys. Lett.* **56**, 2578.

Göddenhenrich, T., Lemke, H., Mück, M., Hartmann, U. and Heiden, C. (1990c). *Appl. Phys. Lett.* **57**, 2612.

Goldberg, J. L., Wang, X.-S., Wei, J., Bartelt, N. C. and Williams, E. D. (1991). *J. Vac. Sci. Technol.* A **9**, 1868.

Golubok, A. O. and Tarasov, N. A. (1990). *Sov. Tech. Phys. Lett.* **16**, 418.

Gomez, R. D., Burke, E. R., Adly, A. A. and Mayergoyz, I. D. (1992). *Appl. Phys. Lett.* **60**, 906.

Gómez-Rodriguez, J. M., Gómez-Herrero, J. and Baró, A. M. (1989). *Surf. Sci.* **220**, 152.

Goodman, F. O. and García, N. (1991). *Phys. Rev.* B **43**, 4728.

Gould, S. A. C., Burke, K. and Hansma, P. K. (1989). *Phys. Rev.* B **40**, 5363.

Grafström, S., Kowalski, J., Neumann, R., Probst, O. and Wörtge, M. (1991). *J. Vac. Sci. Technol.* B **9**, 568.

Green, M. P., Hanson, K. J., Scherson, D. A., Xing, X., Richter, M., Ross, P. N., Carr, R. and Lindau, I. (1989). *J. Phys. Chem.* **93**, 2181.

Green, M. P., Richter, M., Xing, X., Scherson, D., Hanson, K. J., Ross, Jr., P. N., Carr, R. and Lindau, I. (1988). *J. Microsc.* **152**, 823.

Greenwood, J. A. (1967). *Trans. ASME* **89**, 81.

Gregory, S. (1990). *Phys. Rev. Lett.* **64**, 689.

Griffith, J. E. and Kochanski, G. P. (1990a). *Ann. Rev. Mat. Sci.* **20**, 219.

Griffith, J. E. and Kochanski, G. P. (1990b). *Crit. Rev. Solid State Mat. Sci.* **16**, 255.

Griffith, J. E., Kochanski, G. P., Kubby, J. A. and Wierenga, P. E. (1989). *J. Vac. Sci. Technol.* A **7**, 1914.

Griffith, J. E., Miller, G. L., Green, C. A., Grigg, D. A. and Russell, P. E. (1990). *J. Vac. Sci. Technol.* B **8**, 2023.

Gritsch, T., Coulman, D., Behm, R. J. and Ertl, G. (1989a). *Phys. Rev. Lett.* **63**, 1086.

Gritsch, T., Coulman, D., Behm, R. J. and Ertl, G. (1989b). *Appl. Phys.* A **49**, 403.

Gritsch, T., Coulman, D., Behm, R. J. and Ertl, G. (1991). *Surf. Sci.* **257**, 297.

Grüner, G. (1988). *Rev. Mod. Phys.* **60**, 1129.

Grütter, P. (1989). Ph.D. Thesis, Basel University, Basel, Switzerland.

Grütter, P., Jung, Th., Heinzelmann, H., Wadas, A., Meyer, E., Hidber, H. R. and Güntherodt, H. J. (1990a). *J. Appl. Phys.* **67**, 1437.

Grütter, P., Rugar, D., Mamin, H. J., Castillo, G., Lambert, S. E., Lin, C.-J., Valletta, R. M., Wolter, O., Bayer, T. and Greschner, J. (1990b). *Appl. Phys. Lett.* **57**, 1820.

Grütter, P., Wadas, A., Meyer, E., Hidber, H.-R. and Güntherodt, H.-J. (1989). *J. Appl. Phys.* **66**, 6001.

Guckenberger, R., Hacker, B., Hartmann, T., Scheybani, T., Wang, Z., Wiegräbe, W. and Baumeister, W. (1991). *J. Vac. Sci. Technol.* B **9**, 1227.

Guckenberger, R., Hartmann, T., Wiegräbe, W. and Baumeister, W. (1992). In: *Scanning Tunneling Microscopy II* (ed. Wiesendanger, R. and Güntherodt, H.-J.), Springer Series in Surface Sciences 28, p. 51 (Berlin, Heidelberg, New York: Springer).

Guckenberger, R., Kösslinger, C., Gatz, R., Breu, H., Levai, N. and Baumeister, W. (1988). *Ultramicroscopy* **25**, 111.

Guckenberger, R., Wiegräbe, W., Hillebrand, A., Hartmann, T., Wang, Z. and Baumeister, W. (1989). *Ultramicroscopy* **31**, 327.

Guéret, P. (1991). *Solid State Commun.* **79**, 403.

Guerra, J. M. (1990). *Appl. Opt.* **29**, 3741.

Guethner, P., Fischer, U. Ch. and Dransfeld, K. (1989). *Appl. Phys.* B **48**, 89.

Guethner, P., Schreck, E., Dransfeld, K. and Fischer, U. Ch. (1990). In: *Scanning Tunneling Microscopy* (ed. Behm, R. J., Garcia, N. and Rohrer, H.), NATO ASI Series E: Appl. Sci. Vol. 184, p. 507 (Dordrecht: Kluwer).

Guinea, F. and García, N. (1990). *Phys. Rev. Lett.* **65**, 281.

Günther, C. and Behm, R. J. (1992). In press.

Gundlach, K. H. (1966). *Solid State Electronics* **9**, 949.

Güntherodt, G., Thompson, W. A., Holtzberg, F. and Fisk, Z. (1982). *Phys. Rev. Lett.* **49**, 1030.

Gurney, R. W. and Condon, E. U. (1928). *Nature* **122**, 439.

Gygi, F. and Schluter, M. (1990a). *Phys. Rev.* B **41**, 822.

Gygi, F. and Schluter, M. (1990b). *Phys. Rev. Lett.* **65**, 1820.

Gygi, F. and Schluter, M. (1991). *Phys. Rev.* B **43**, 7609.

Haas, G. A. and Thomas, R. E. (1963). *J. Appl. Phys.* **34**, 3457.

Haas, G. A. and Thomas, R. E. (1966). *Surf. Sci.* **4**, 64.

Hadley, M. J. and Tear, S. P. (1991). *Surf. Sci. Lett.* **247**, L221.

Haefke, H., Meyer, E., Güntherodt, H.-J., Gerth, G. and Krohn, M. (1991). *J. Imaging Sci.* **35**, 290.

Halbritter, J. (1982). *Surf. Sci.* **122**, 80.

Halbritter, J. (1985). *J. Appl. Phys.* **58**, 1320.

Hale, M. E., Fuller, H. W. and Rubinstein, H. (1959). *J. Appl. Phys.* **30**, 789.

Hallmark, V. M., Chiang, S., Brown, J. K. and Wöll, Ch. (1991a). *Phys. Rev. Lett.* **66**, 48.

Hallmark, V. M., Chiang, S., Rabolt, J. F., Swalen, J. D. and Wilson, R. J. (1987). *Phys. Rev. Lett.* **59**, 2879.

Hallmark, V. M., Chiang, S. and Wöll, Ch. (1991b). *J. Vac. Sci. Technol.* B **9**, 1111.

Hamers, R. J. (1988). *J. Vac. Sci. Technol.* B **6**, 1462.

Hamers, R. J. (1989a). *Ann. Rev. Phys. Chem.* **40**, 531.

Hamers, R. J. (1989b). *Phys. Rev.* B **40**, 1657.

Hamers, R. J., Avouris, Ph. and Bozso, F. (1987). *Phys. Rev. Lett.* **59**, 2071.

Hamers, R. J. and Cahill, D. G. (1990). *Appl. Phys. Lett.* **57**, 2031.

Hamers, R. J. and Cahill, D. G. (1991). *J. Vac. Sci. Technol.* B **9**, 514.

Hamers, R. J. and Demuth, J. E. (1988). *Phys. Rev. Lett.* **60**, 2527.

Hamers, R. J. and Köhler, U. K. (1989). *J. Vac. Sci. Technol.* A **7**, 2854.

Hamers, R. J., Köhler, U. K. and Demuth, J. E. (1989). *Ultramicroscopy* **31**, 10.

Hamers, R. J., Köhler, U. K. and Demuth, J. E. (1990). *J. Vac. Sci. Technol.* A **8**, 195.

Hamers, R. J. and Markert, K. (1990a). *Phys. Rev. Lett.* **64**, 1051.

Hamers, R. J. and Markert, K. (1990b). *J. Vac. Sci. Technol.* A **8**, 3524.

Hamers, R. J., Tromp, R. M. and Demuth, J. E. (1986a). *Phys. Rev. Lett.* **56**, 1972.

Hamers, R. J., Tromp, R. M. and Demuth, J. E. (1986b). *Phys. Rev.* B **34**, 5343.

Hansma, P. K. (1977). *Phys. Rep.* **30**, 145.

Hansma, P. K. (ed.), (1982). *Tunneling Spectroscopy* (New York, London: Plenum).

Hansma, P. K., Drake, B., Marti, O., Gould, S. A. C. and Prater, C. B. (1989). *Science* **243**, 641.

Hansma, P. K. and Tersoff, J. (1987). *J. Appl. Phys.* **61**, R1.

Hao, X., Moodera, J. S. and Meservey, R. (1990). *Phys. Rev.* B **42**, 8235.

Hara, M., Iwakabe, Y., Tochigi, K., Sasabe, H., Garito, A. F. and Yamada, A. (1990). *Nature* **344**, 228.

Harmer, M. A., Fincher, C. R. and Parkinson, B. A. (1991). *J. Appl. Phys.* **70**, 2760.

Harootunian, A., Betzig, E., Isaacson, M. and Lewis, A. (1986). *Appl. Phys. Lett.* **49**, 674.

Harten, U., Lahee, A. M., Toennies, J. P. and Wöll, Ch. (1985). *Phys. Rev. Lett.* **54**, 2619.

Hartmann, E., Behm, R. J., Krötz, G., Müller, G. and Koch, F. (1991). *Appl. Phys. Lett.* **59**, 2136.

Hartmann, U. (1988). *J. Appl. Phys.* **64**, 1561.

Hartmann, U. (1989a). *Phys. Lett.* A **137**, 475.

Hartmann, U. (1989b). *Phys. Rev.* B **40**, 7421.

Hartmann, U. (1990). *Phys. Rev.* B **42**, 1541.

Hartmann, U. (1991). *Phys. Rev.* B **43**, 2404.

Hartmann, U., Göddenhenrich, T., Lemke, H. and Heiden, C. (1990). *IEEE Trans. Magn.* **26**, 1512.

Hartmann, U. and Heiden, C. (1988). *J. Microsc.* **152**, 281.

Hasegawa, S., Matsuda, T., Endo, J., Osakabe, N., Igarashi, M., Kobayashi, T., Naito, M., Tonomura, A. and Aoki, R. (1991). *Phys. Rev.* B **43**, 7631.

Hasegawa, T., Nantoh, M., Suzuki, H., Motohira, N., Kishio, K. and Kitazawa, K. (1990). *Physica* B **165** & **166**, 1563.

Hasegawa, T., Suzuki, H., Yaegashi, S., Takagi, H., Kishio, K., Uchida, S., Kitazawa, K. and Fueki, K. (1989). *Jpn. J. Appl. Phys.* **28**, L179.

Hashizume, T., Hasegawa, Y., Sumita, I. and Sakurai, T. (1991a). *Surf. Sci.* **246**, 189.

Hashizume, T., Motai, K., Hasegawa, Y., Sumita, I., Tanaka, H., Amano, S., Hyodo, S. and Sakurai, T. (1991b). *J. Vac. Sci. Technol.* B **9**, 745.

Hauge, E. H. and Støvneng, J. A. (1989). *Rev. Mod. Phys.* **61**, 917.

Hawley, M. E., Gray, K. E., Capone, II, D. W. and Hinks, D. G. (1987). *Phys. Rev.* B **35**, 7224.

Hawley, M., Raistrick, I. D., Beery, J. G. and Houlton, R. J. (1991). *Science* **251**, 1587.

Hebard, A. F., Rosseinsky, M. J., Haddon, R. C., Murphy, D. W., Glarum, S. H., Palstra, T. T. M., Ramirez, A. P. and Kortan, A. R. (1991). *Nature* **350**, 600.

Heben, M. J., Dovek, M. M., Lewis, N. S., Penner, R. M. and Quate, C. F. (1988). *J. Microsc.* **152**, 651.

Heben, M. J., Penner, R. M., Lewis, N. S., Dovek, M. M. and Quate, C. F. (1989). *Appl. Phys. Lett.* **54**, 1421.

Hecht, M. H., Bell, L. D., Kaiser, W. J. and Davis, L. C. (1990). *Phys. Rev.* B **42**, 7663.

Heil, J., Wesner, J., Lommel, B., Assmus, W. and Grill, W. (1989). *J. Appl. Phys.* **65**, 5220.

Heinzelmann, H. (1989). Ph.D. Thesis, Basel University, Basel, Switzerland.

Heinzelmann, H., Meyer, E., Güntherodt, H.-J. and Steiger, R. (1989). *Surf. Sci.* **221**, 1.

Hellmann, H. (1937). *Einführung in die Quantentheorie* (Leipzig, Wien: Franz Deuticke).

Hembree, G. G., Unguris, J., Celotta, R. J. and Pierce, D. T. (1987). *Scanning Microsc. Suppl.* **1**, 229.

Henson, T. D., Sarid, D. and Bell, L. S. (1988). *J. Microsc.* **152**, 467.

Herring, C. P. (1974). *Phys. Lett.* A **47**, 105.

Hess, H. F. (1991). *Physica* C **185–189**, 259.

Hess, H. F., Robinson, R. B., Dynes, R. C., Valles, Jr, J. M. and Waszczak, J. V. (1989). *Phys. Rev. Lett.* **62**, 214.

Hess, H. F., Robinson, R. B., Dynes, R. C., Valles, Jr, J. M. and Waszczak, J. V. (1990a). *J. Vac. Sci. Technol.* A **8**, 450.

Hess, H. F., Robinson, R. B. and Waszczak, J. V. (1990b). *Phys. Rev. Lett.* **64**, 2711.

Hess, H. F., Robinson, R. B. and Waszczak, J. V. (1991). *Physica* B **169**, 422.

Hirschorn, E. S., Lin, D. S., Leibsle, F. M., Samsavar, A. and Chiang, T.-C. (1991). *Phys. Rev.* B **44**, 1403.

Hobbs, P. C. D., Abraham, D. W. and Wickramasinghe, H. K. (1989). *Appl. Phys. Lett.* **55**, 2357.

Hoeven, A. J., Dijkkamp, D., Lenssinck, J. M. and van Loenen, E. J. (1990). *J. Vac. Sci. Technol.* A **8**, 3657.

Hoeven, A. J., Dijkkamp, D., van Loenen, E. J. and van Hooft, P. J. G. M. (1989a). *Surf. Sci.* **211/212**, 165.

Hoeven, A. J., Lenssinck, J. M., Dijkkamp, D., van Loenen, E. J. and Dieleman, J. (1989b). *Phys. Rev. Lett.* **63**, 1830.

Holm, R. and Meißner, W. (1932). *Z. Phys.* **74**, 715.

Holm, R. and Meißner, W. (1933). *Z. Phys.* **86**, 787.

Holonyak, Jr., N., Lesk, I. A., Hall, R. N., Tiemann, J. J. and Ehrenreich, H. (1959). *Phys. Rev. Lett.* **3**, 167.

Hörber, J. K. H., Lang, C. A., Hänsch, T. W., Heckl, W. M. and Möhwald, H. (1988). *Chem. Phys. Lett.* **145**, 151.

Hosaka, S., Hosoki, S., Takata, K., Horiuchi, K. and Natsuaki, N. (1988). *Appl. Phys. Lett.* **53**, 487.

Hösler, W., Ritter, E. and Behm, R. J. (1986). *Ber. Bunsenges. Phys. Chem.* **90**, 205.

Hosoki, S., Hosaka, S. and Hasegawa, T. (1992). *Appl. Surf. Sci.* **60/61**, 643.

Houbertz, R., Memmert, U. and Behm, R. J. (1991). *Appl. Phys. Lett.* **58**, 1027.

Huang, Z. H., Feuchtwang, T. E., Cutler, P. H. and Kazes, E. (1990). *J. Vac. Sci. Technol.* A **8**, 177.

Hug, H. J., Jung, Th., Güntherodt, H.-J. and Thomas, H. (1991). *Physica* C **175**, 375.

Humbert, A., Pierrisnard, R., Sangay, S., Chapon, C., Henry, C. R. and Claeys, C. (1989). *Europhys. Lett.* **10**, 533.

Hüsser, O. E., Craston, D. H. and Bard, A. J. (1988). *J. Vac. Sci. Technol.* B **6**, 1873.

Hüsser, O. E., Craston, D. H. and Bard, A. J. (1989). *J. Electrochem. Soc.* **136**, 3222.

Hwang, D. M., Parker, N. W., Utlaut, M. and Crewe, A. V. (1983). *Phys. Rev.* B **27**, 1458.

Hwang, R. Q., Schröder, J., Günther, C. and Behm, R. J. (1991a). *Phys. Rev. Lett.* **67**, 3279.

Hwang, R. Q., Zeglinski, D. M., Lopez Vazquez-de-Parga, A., Ogletree, D. F., Somorjai, G. A., Salmeron, M. and Denley, D. R. (1991b). *Phys. Rev.* B **44**, 1914.

Ibe, J. P., Bey, Jr, P. P., Brandow, S. L., Brizzolara, R. A., Burnham, N. A., DiLella, D. P., Lee, K. P., Marrian, C. R. K. and Colton, R. J. (1990). *J. Vac. Sci. Technol.* A **8**, 3570.

Ichihashi, T. and Matsui, S. (1988). *J. Vac. Sci. Technol.* B **6**, 1869.

Ichinokawa, T., Ichinose, T., Tohyama, M. and Itoh, H. (1990). *J. Vac. Sci. Technol.* A **8**, 500.

Ichinokawa, T., Miyazaki, Y. and Koga, Y. (1987). *Ultramicroscopy* **23**, 115.

Ihm, J. (1988). *Phys. Scripta* **38**, 269.

Ikeda, K., Tomeno, I., Takamuku, K., Yamaguchi, K., Itti, R. and Koshizuka, N. (1992). *Physica* C **191**, 505.

Imamura, T. and Hasuo, S. (1991). *Appl. Phys. Lett.* **58**, 645.

Israelachvili, J. N. (1985). *Intermolecular and Surface Forces* (London: Academic).

Israelachvili, J. N. (1987). *Proc. Natl. Acad. Sci. USA* **84**, 4722.

Israelachvili, J. N. and Adams, G. E. (1976). *Nature* **262**, 774.

Israelachvili, J. N. and McGuiggan, P. M. (1988). *Science* **241**, 795.

Israelachvili, J. N. and McGuiggan, P. M. (1990). *J. Mat. Res.* **5**, 2223.

Israelachvili, J. N. and Tabor, D. (1972). *Proc. Roy. Soc. (London)* A **331**, 19.

Isshiki, N., Kobayashi, K. and Tsukada, M. (1990). *Surf. Sci. Lett.* **238**, L439.

Itaya, K., Sugawara, R., Morita, Y. and Tokumoto, H. (1992). *Appl. Phys. Lett.* **60**, 2534.

Itaya, K., Sugawara, S. and Higaki, K. (1988). *J. Phys. Chem.* **92**, 6714.

Itaya, K., Sugawara, S., Sashikata, K. and Furuya, N. (1990). *J. Vac. Sci. Technol.* A **8**, 515.

Itaya, K. and Tomita, E. (1988). *Surf. Sci.* **201**, L507.

Iwatsuki, M., Murooka, K., Kitamura, S., Takayanagi, K. and Harada, Y. (1991). *J. Electron Microsc.* **40**, 48.

Iwawaki, F., Tomitori, M. and Nishikawa, O. (1991). *Surf. Sci. Lett.* **253**, L411.

Jahanmir, J., West, P. E., Hsieh, S. and Rhodin, T. N. (1989). *J. Appl. Phys.* **65**, 2064.

Jaklevic, R. C. and Elie, L. (1988). *Phys. Rev. Lett.* **60**, 120.

Jaklevic, R. C. and Lambe, J. (1966). *Phys. Rev. Lett.* **17**, 1139.

Jaklevic, R. C. and Lambe, J. (1973). *Surf. Sci.* **37**, 922.

Jansen, A. G. M., van Gelder, A. P. and Wyder, P. (1980). *J. Phys.* C **13**, 6073.

Jansen, A. G. M., Mueller, F. M. and Wyder, P. (1977). *Phys. Rev.* B **16**, 1325.

Jansen, A. G. M., Wyder, P. and van Kempen, H. (1987). *Europhys. News* **18**, 21.

Jarrell, J. A., King, J. G. and Mills, J. W. (1981). *Science* **211**, 277.

Jennings, P. J. and Jones, R. O. (1988). *Adv. Phys.* **37**, 341.

Jensen, F., Besenbacher, F., Laegsgaard, E. and Stensgaard, I. (1990a). *Phys. Rev.* B **41**, 10233.

Jensen, F., Besenbacher, F., Laegsgaard, E. and Stensgaard, I. (1990b). *Phys. Rev.* B **42**, 9206.

Jensen, F., Besenbacher, F., Laegsgaard, E. and Stensgaard, I. (1991). *Surf. Sci. Lett.* **259**, L774.

Jericho, M. H., Blackford, B. L. and Dahn, D. C. (1989). *J. Appl. Phys.* **65**, 5237.

Jiang, M. H., Ventrice Jr, C. A., Scoles, K. J., Tyagi, S., DiNardo, N. J. and Rothwarf, A. (1991). *Physica* C **183**, 39.

Joachim, C. and Sautet, P. (1990). In: *Scanning Tunneling Microscopy and Related Methods* (ed. Behm, R. J., Garcia, N. and Rohrer, H.), NATO ASI Series E: Appl. Sci. Vol. 184, p. 377 (Dordrecht: Kluwer).

Johansson, P. and Monreal, R. (1991). *Z. Phys.* B **84**, 269.

Johansson, P., Monreal, R. and Apell, P. (1990). *Phys. Rev.* B **42**, 9210.

Johnson, M. and Clarke, J. (1990). *J. Appl. Phys.* **67**, 6141.

Josephson, B. D. (1962). *Phys. Lett.* **1**, 251.

Josephson, B. D. (1965). *Adv. Phys.* **14**, 419.

Josephson, B. D. (1974). *Ref. Mod. Phys.* **46**, 251.

Julliere, M. (1975). *Phys. Lett.* A **54**, 225.

Jung, T. A., Moser, A., Hug, H. J., Brodbeck, D., Hofer, R., Hidber, H. R. and Schwarz, U. D. (1992). *Ultramicroscopy* **42–44**, 1446.

Kaiser, W. J. and Bell, L. D. (1988). *Phys. Rev. Lett.* **60**, 1406.

Kaiser, W. J. and Jaklevic, R. C. (1986). *IBM J. Res. Develop.* **30**, 411.

Kaiser, W. J. and Jaklevic, R. C. (1988). *Rev. Sci. Instrum.* **59**, 537.

Kaizuka, H. (1989). *Rev. Sci. Instrum.* **60**, 3119.

Kaizuka, H. and Siu, B. (1988). *Jpn. J. Appl. Phys.* **27**, L773.

Kajimura, K., Bando, H., Endo, K., Mizutani, W., Murakami, H., Okano, M., Okayama, S., Ono, M., Ono, Y., Tokumoto, H., Sakai, F., Watanabe, K. and Wakiyama, S. (1987). *Surf. Sci.* **181**, 165.

Kämper, K. P., Schmitt, W., Güntherodt, G., Gambino, R. J. and Ruf, R. (1987). *Phys. Rev. Lett.* **59**, 2788.

Kane, E. O. (1961). *J. Appl. Phys.* **32**, 83.

Kaneko, R. and Oguchi, S. (1990). *Jpn. J. Appl. Phys.* **29**, 1854.

Kariotis, R. and Lagally, M. G. (1991). *Surf. Sci.* **248**, 295.

Kawakatsu, H. and Higuchi, T. (1990). *J. Vac. Sci. Technol.* A **8**, 319.

Kawakatsu, H., Hoshi, Y., Higuchi, T. and Kitano, H. (1991). *J. Vac. Sci. Technol.* B **9**, 651.

Kawazu, A., Yoshimura, M., Shigekawa, H., Mori, H. and Saito, G. (1991). *J. Vac. Sci. Technol.* B **9**, 1006.

Keldysh, L. V. (1958). *Sov. Phys. – JETP* **6**, 763.

Keldysh, L. V. (1965). *Sov. Phys. – JETP* **20**, 1018.

Keller, D. (1991). *Surf. Sci.* **253**, 353.

Kelty, S. P. and Lieber, C. M. (1989a). *J. Phys. Chem.* **93**, 5983.

Kelty, S. P. and Lieber, C. M. (1989b). *Phys. Rev.* B **40**, 5856.

Kelty, S. P. and Lieber, C. M. (1991). *J. Vac. Sci. Technol.* B **9**, 1068.

Kelty, S. P., Lu, Z. and Lieber, C. M. (1991). *Phys. Rev.* B **44**, 4064.

Kenny, T. W., Waltman, S. B., Reynolds, J. K. and Kaiser, W. J. (1991). *Appl. Phys. Lett.* **58**, 100.

Kent, A. D., Maggio-Aprile, I., Niedermann, Ph. and Fischer, Ø. (1989). *Phys. Rev.* B **39**, 12363.

Kern, K., Niehus, H., Schatz, A., Zeppenfeld, P., Goerge, J. and Comsa, G. (1991). *Phys. Rev. Lett.* **67**, 855.

Kim, Y., Huang, J.-L. and Lieber, C. M. (1991). *Appl. Phys. Lett.* **59**, 3404.

Kirk, M. D., Smith, D. P. E., Mitzi, D. B., Sun, J. Z., Webb, D. J., Char, K., Hahn, M. R., Naito, M., Oh, B., Beasley, M. R., Geballe, T. H., Hammond, R. H., Kapitulnik, A. and Quate, C. F. (1987). *Phys. Rev.* B **35**, 8850.

Kirk, M. D., Nogami, J., Baski, A. A., Mitzi, D. B., Kapitulnik, A., Geballe, T. H. and Quate, C. F. (1988). *Science* **242**, 1673.

Kirtley, J. (1978). In: *Inelastic Electron Tunneling Spectroscopy* (ed. Wolfram, T.), p. 80 (Berlin, Heidelberg, New York: Springer).

Kirtley, J. R., Collins, R. T., Schlesinger, Z., Gallagher, W. J., Sandstrom, R. L., Dinger, T. R. and Chance, D. A. (1987a). *Phys. Rev.* B **35**, 8846.

Kirtley, J. R., Feenstra, R. M., Fein, A. P., Raider, S. I., Gallagher, W. J., Sandstrom, R., Dinger, T., Shafer, M. W., Koch, R., Laibowitz, R. and Bumble, B. (1988c). *J. Vac. Sci. Technol.* A **6**, 259.

Kirtley, J. R., Raider, S. I., Feenstra, R. M. and Fein, A. P. (1987b). *Appl. Phys. Lett.* **50**, 1607.

Kirtley, J. and Soven, P. (1979). *Phys. Rev.* B **19**, 1812.

Kirtley, J. R., Tsuei, C. C., Park, S. I., Chi, C. C., Rozen, J. and Shafer, M. W. (1987c). *Phys. Rev.* B **35**, 7216.

Kirtley, J. R., Washburn, S. and Brady, M. J. (1988a). *Phys. Rev. Lett.* **60**, 1546.

Kirtley, J. R., Washburn, S. and Brady, M. J. (1988b). *IBM J. Res. Develop.* **32**, 414.

Kisker, E., Baum, G., Mahan, A. H., Raith, W. and Schröder, K. (1976). *Phys. Rev. Lett.* **36**, 982.

Kisker, E. Baum, G., Mahan, A. H., Raith, W. and Reihl, B. (1978), *Phys. Rev.* B **18**, 2256.

Kitamura, S., Sato, T. and Iwatsuki, M. (1991). *Nature* **351**, 215.

Klein, U. (1989). *Phys. Rev.* B **40**, 6601.

Klein, U. (1990). *Phys. Rev.* B **41**, 4819.

Klitsner, T., Becker, R. S. and Vickers, J. S. (1990). *Phys. Rev.* B **41**, 3837.

Klitsner, T., Becker, R. S. and Vickers, J. S. (1991). *Phys. Rev.* B **44**, 1817.

Knall, J. and Pethica, J. B. (1992). *Surf. Sci.* **265**, 156.

Knall, J., Pethica, J. B., Todd, J. D. and Wilson, J. H. (1991). *Phys. Rev. Lett.* **66**, 1733.

Knauer, H., Richter, J. and Seidel, P. (1977). *Phys. Stat. Sol.* (a) **44**, 303.

Kobayashi, K. and Tsukada, M. (1990). *J. Vac. Sci. Technol.* A **8**, 170.

Koch, R. H. and Hartstein, A. (1985). *Phys. Rev. Lett.* **54**, 1848.

Kochanski, G. P. (1989). *Phys. Rev. Lett.* **62**, 2285.

Kochanski, G. P. and Bell, R. F. (1992). *Surf. Sci. Lett.* **273**, L435.

Kochanski, G. P. and Griffith, J. E. (1991). *Surf. Sci. Lett.* **249**, L293.

Köhler, U. K., Demuth, J. E. and Hamers, R. J. (1988). *Phys. Rev. Lett.* **60**, 2499.

Köhler, U. K., Demuth, J. E. and Hamers, R. J. (1989). *J. Vac. Sci. Technol.* A **7**, 2860.

Köhler, U., Jusko, O., Pietsch, G., Müller, B. and Henzler, M. (1991). *Surf. Sci.* **248**, 321.

Koike, K., Matsuyama, H., Todokoro, H. and Hayakawa, K. (1987). *Scanning Microsc. Suppl.* **1**, 241.

Kolb, D. M. (1986). *J. Vac. Sci. Technol.* A **4**, 1294.

Kolomytkin, O. V., Golubok, A. O., Davydov, D. N., Timofeev, V. A., Vinogradova, S. A. and Tipisev, S. Y. (1991). *Biophys. J.* **59**, 889.

Koltun, R., Hoffmann, M., Splittgerber-Hünnekes, P. C., Jarchow, Ch., Güntherodt, G., Moshchalkov, V. V. and Leonyuk, L. I. (1991). *Z. Phys.* B **82**, 53.

Komiyama, M., Morita, S. and Mikoshiba, N. (1988). *J. Microsc.* **152**, 197.

Komiyama, M., Kobayashi, J. and Morita, S. (1990). *J. Vac. Sci. Technol.* A **8**, 608.

Kopatzki, E. and Behm, R. J. (1991). *Surf. Sci.* **245**, 255.

Kopatzki, E., Doyen, G., Drakova, D. and Behm, R. J. (1988). *J. Microsc.* **152**, 687.

Kordic, S., van Loenen, E. J. and Walker, A. J. (1991). *Appl. Phys. Lett.* **59**, 3154.

Kordic, S., van Loenen, E. J., Dijkkamp, D. Hoeven, A. J. and Moraal, H. K. (1990). *J. Vac. Sci. Technol.* A **8**, 549.

Kortan, A. R., Becker, R. S., Thiel, F. A. and Chen, H. S. (1990). *Phys. Rev. Lett.* **64**, 200.

Kranz, J. and Hubert, A. Z. (1963). *Z. Angew. Phys.* **15**, 220.

Krätschmer, W., Fostiropoulos, K. and Huffman, D. R. (1990a). *Chem. Phys. Lett.* **170**, 167.

Krätschmer, W., Lamb, L. D., Fostiropoulos, K. and Huffman, D. R. (1990b). *Nature* **347**, 354.

Krieger, W., Koppermann, H., Suzuki, T. and Walther, H. (1989). *IEEE Trans. Instrum. Meas.* **38**, 1019.

Krieger, W., Suzuki, T., Völcker, M. and Walther, H. (1990). *Phys. Rev.* B **41**, 10229.

Kroó, N., Szentirmay, Zs. and Félszerfalvi, J. (1980). *Phys. Stat. Sol.* B **102**, 227.

Kroó, N., Thost, J.-P., Völcker, M., Krieger, W. and Walther, H. (1991). *Europhys. Lett.* **15**, 289.

Kroto, H. W., Heath, J. R., O'Brien, S. C., Curl, R. F. and Smalley, R. E. (1985). *Nature* **318**, 162.

Kubby, J. A., Griffith, J. E., Becker, R. S. and Vickers, J. S. (1987). *Phys. Rev.* B **36**, 6079.

Kubby, J. A., Wang, Y. R. and Greene, W. J. (1990). *Phys. Rev. Lett.* **65**, 2165.

Kubby, J. A., Wang, Y. R. and Greene, W. J. (1991). *Phys. Rev.* B **43**, 9346.

Kuk, Y., Becker, R. S., Silverman, P. J. and Kochanski, G. P. (1990a). *Phys. Rev. Lett.* **65**, 456.

Kuk, Y., Becker, R. S., Silverman, P. J. and Kochanski, G. P. (1991). *J. Vac. Sci. Technol.* B **9**, 545.

Kuk, Y., Chua, F. M., Silverman, P. J. and Meyer, J. A. (1990b). *Phys. Rev.* B **41**, 12393.

Kuk, Y., Jarrold, M. F., Silverman, P. J., Bower, J. E. and Brown, W. L. (1989). *Phys. Rev.* B **39**, 11168.

Kuk, Y. and Silverman, P. J. (1986). *Appl. Phys. Lett.* **48**, 1597.

Kuk, Y. and Silverman, P. J. (1989). *Rev. Sci. Instrum.* **60**, 165.

Kuk, Y. and Silverman, P. J. (1990). *J. Vac. Sci. Technol.* A **8**, 289.

Kuk, Y., Silverman, P. J. and Chua, F. M. (1988b). *J. Microsc.* **152**, 449.
Kuk, Y., Silverman, P. J. and Nguyen, H. Q. (1988a). *J. Vac. Sci. Technol.* A **6**, 524.
Küppers, J., Wandelt, K. and Ertl, G. (1979). *Phys. Rev. Lett.* **43**, 928.
Kurtin, S., McGill, T. C. and Mead, C. A. (1970). *Phys. Rev. Lett.* **25**, 756.
Kuwabara, M., Clarke, D. R. and Smith, D. A. (1990). *Appl. Phys. Lett.* **56**, 2396.
Kuwabara, M., Lo, W. and Spence, J. C. H. (1989). *J. Vac. Sci. Technol.* A **7**, 2745.
Kwak, J. and Bard, A. J. (1989). *Anal. Chem.* **61**, 1794.
Lagally, M. G., Kariotis, R., Swartzentruber, B. S. and Mo, Y.-W. (1989). *Ultramicroscopy* **31**, 87.
Laguës, M., Fischer, J. E., Marchand, D. and Fretigny, C. (1988). *Solid State Commun.* **67**, 1011.
Lambe, J. and Jaklevic, R. C. (1968). *Phys. Rev.* **165**, 821.
Lambe, J. and McCarthy, S. L. (1976). *Phys. Rev. Lett.* **37**, 923.
Land, T. A., Michely, T., Behm, R. J., Hemminger, J. C. and Comsa, G. (1992). *Surf. Sci.* **264**, 261.
Landauer, R. (1987). *Z. Phys.* B **68**, 217.
Landman, U., Luedtke, W. D. and Nitzan, A. (1989a). *Surf. Sci.* **210**, L177.
Landman, U., Luedtke, W. D. and Ribarsky, M. W. (1989b). *J. Vac. Sci. Technol.* A **7**, 2829.
Lang, C. A., Dovek, M. M., Nogami, J. and Quate, C. F. (1989). *Surf. Sci.* **224**, L947.
Lang, C. A., Quate, C. F. and Nogami, J. (1991). *Appl. Phys. Lett.* **59**, 1696.
Lang, H. P., Frey, T. and Güntherodt, H.-J. (1991). *Europhys. Lett.* **15**, 667.
Lang, H. P., Haefke, H., Leemann, G. and Güntherodt, H.-J. (1992a). *Physica* C **194**, 81.
Lang, H. P., Wiesendanger, R., Thommen-Geiser, V. and Güntherodt, H.-J. (1992b). *Phys. Rev.* B **45**, 1829.
Lang, N. D. (1985). *Phys. Rev. Lett.* **55**, 230.
Lang, N. D. (1986a). *Phys. Rev. Lett.* **56**, 1164.
Lang, N. D. (1986b). *Phys. Rev.* B **34**, 5947.
Lang, N. D. (1986c). *IBM J. Res. Develop.* **30**, 374.
Lang, N. D. (1987a). *Phys. Rev. Lett.* **58**, 45.
Lang, N. D. (1987b). *Phys. Rev.* B **36**, 8173.
Lang, N. D. (1988). *Phys. Rev.* B **37**, 10395.
Langmuir, I. (1992). *Trans. Faraday Soc.* **17**, 607.
Lea, C. and Gomer, R. (1970). *Phys. Rev. Lett.* **25**, 804.
Leavens, C. R. and Aers, G. C. (1986). *Solid State Commun.* **59**, 285.
LeDuc, H. G., Kaiser, W. J., Hunt, B. D., Bell, L. D., Jaklevic, R. C. and Youngquist, M. G. (1989). *Appl. Phys. Lett.* **54**, 946.
LeDuc, H. G., Kaiser, W. J. and Stern, J. A. (1987). *Appl. Phys. Lett.* **50**, 1921.
Lee, G., Arscott, P. G., Bloomfield, V. A. and Evans, D. F. (1989). *Science* **244**, 475.
Leibsle, F. M., Samsavar, A. and Chiang, T.-C. (1988). *Phys. Rev.* B **38**, 5780.
Lemke, H., Göddenhenrich, T., Bochem, H. P., Hartmann, U. and Heiden, C. (1990). *Rev. Sci. Instrum.* **61**, 2538.

Le Page, Y., McKinnon, W. R., Tarascon, J.-M. and Barboux, P. (1989). *Phys. Rev.* B **40**, 6810.

Lev, O., Fan, F.-R. and Bard, A. J. (1988). *J. Electrochem. Soc.* **135**, 783.

Levine, D. and Steinhardt, P. J. (1984). *Phys. Rev. Lett.* **53**, 2477.

Levine, J. D. (1973). *Surf. Sci.* **34**, 90.

Levinstein, H. J., Chirba, V. G. and Kunzler, J. E. (1967). *Phys. Lett.* A **24**, 362.

Levinstein, H. J., and Kunzler, J. E. (1966). *Phys. Lett.* **20**, 581.

Lewis, A. and Lieberman, K. (1991). *Nature* **354**, 214.

Li, M.-Q., Zhu, J., Zhu, J.-Q., Hu, J., Gu, M.-M., Xu, Y.-L., Zhang, L.-P., Huang, Z.-Q., Xu, L.-Z. and Yao, X.-W. (1991a). *J. Vac. Sci. Technol.* B **9**, 1298.

Li, Y. Z., Chander, M., Patrin, J. C., Weaver, J. H., Chibante, L. P. F. and Smalley, R. E. (1991b). *Science* **253**, 429.

Li, Y. Z., Patrin, J. C., Chander, M., Weaver, J. H., Chibante, L. P. F. and Smalley, R. E. (1991c). *Science* **252**, 547.

Lieberman, K., Harush, S., Lewis, A. and Kopelman, R. (1990). *Science* **247**, 59.

Lilienfeld, J. E. (1922). *Z. Phys.* **23**, 506.

Lin, C. W., Fan, F. F. and Bard, A. J. (1987). *J. Electrochem. Soc.* **134**, 1038.

Lin, T.-S. and Chung, Y.-W. (1989). *Surf. Sci.* **207**, 539.

Lindsay, S. M., Nagahara, L. A., Thundat, T., Knipping, U., Rill, R. L., Drake, B., Prater, C. B., Weisenhorn, A. L., Gould, S. A. C. and Hansma, P. K. (1989a). *J. Biomol. Struct. Dyn.* **7**, 279.

Lindsay, S. M., Nagahara, L. A., Thundat, T. and Oden, P. (1989b). *J. Biomol. Struct. Dyn.* **7**, 289.

Lindsay, S. M. and Sankey, O. F. (1992). In: *Scanned Probe Microscopy* (ed. Wickramasinghe, H. K.), AIP Conf. Proc. 241, p. 235 (New York: AIP).

Lindsay, S. M., Sankey, O. F., Li, Y., Herbst, C. and Rupprecht, A. (1990). *J. Phys. Chem.* **94**, 4655.

Lindsay, S. M., Thundat, T., Nagahara, L., Knipping, U. and Rill, R. L. (1989c). *Science* **244**, 1063.

Lippel, P. H., Wilson, R. J., Miller, M. D., Wöll, Ch. and Chiang, S. (1989). *Phys. Rev. Lett.* **62**, 171.

Liu, H.-Y., Fan, F. F., Lin, C. W. and Bard, A. J. (1986). *J. Am. Chem. Soc.* **108**, 3838.

Liu, J.-X., Wan, J.-C., Goldman, A. M., Chang, Y. C. and Jiang, P. Z. (1991). *Phys. Rev. Lett.* **67**, 2195.

Logan, R. A. (1969). In: *Tunneling Phenomena in Solids* (ed. Burstein, E. and Lundquist, S.), p. 149 (New York: Plenum).

Lucas, A. A., Morawitz, H., Henry, G.R., Vigneron, J.-P., Lambin, Ph., Cutler, P. H. and Feuchtwang, T. E. (1988). *Phys. Rev.* B **37**, 10708.

Lustenberger, P., Rohrer, H., Christoph, R. and Siegenthaler, H. (1988). *J. Electroanal. Chem.* **243**, 225.

Lyding, J. W., Hubacek, J.S., Gammie, G., Skala, S., Brockenbrough, R., Shapley, J.R. and Keyes, M. P. (1988a). *J. Vac. Sci. Technol.* A **6**, 363.

Lyding, J. W., Skala, S., Hubacek, J.S., Brockenbrough, R. and Gammie, G. (1988b). *Rev. Sci. Instrum.* **59**, 1897.

Lyo, I.-W. and Avouris, Ph. (1989). *Science* **245**, 1369.

Lyo, I.-W. and Avouris, Ph. (1990). *J. Chem. Phys.* **93**, 4479.

Lyo, I.-W. and Avouris, Ph. (1991). *Science* **253**, 173.

Lyo, I.-W., Avouris, Ph., Schubert, B. and Hoffmann, R. (1990). *J. Phys. Chem.* **94**, 4400.

Lyo, I.-W., Kaxiras, E. and Avouris, Ph. (1989). *Phys. Rev. Lett.* **63**, 1261.

Maekawa, S. and Gäfvert, U. (1982). *IEEE Trans. Magn.* **18**, 707.

Magnussen, O. M., Hotlos, J., Beitel, G., Kolb, D. M. and Behm, R. J. (1991). *J. Vac. Sci. Technol.* B **9**, 969.

Magnussen, O. M., Hotlos, J., Nichols, R. J., Kolb, D. M. and Behm, R. J. (1990). *Phys. Rev. Lett.* **64**, 2929.

Maivald, P., Butt, H. J., Gould, S. A. C., Prater, C. B., Drake, B., Gurley, J. A., Elings, V. B. and Hansma, P. K. (1991). *Nanotechnology* **2**, 103.

Mamin, H. J., Abraham, D. W., Ganz, E. and Clarke, J. (1985). *Rev. Sci. Instrum.* **56**, 2168.

Mamin, H. J., Chiang, S., Birk, H., Guethner, P. H. and Rugar, D. (1991). *J. Vac. Sci. Technol.* B **9**, 1398.

Mamin, H. J., Ganz, E., Abraham, D. W., Thomson, R. E. and Clarke, J. (1986). *Phys. Rev.* B **34**, 9015.

Mamin, H. J., Guethner, P. H. and Rugar, D. (1990). *Phys. Rev. Lett.* **65**, 2418.

Mamin, H. J., Rugar, D., Stern, J. E., Fontana, Jr, R. E. and Kasiraj, P. (1989). *Appl. Phys. Lett.* **55**, 318.

Mamin, H. J., Rugar, D., Stern, J. E., Terris, B. D. and Lambert, S. E. (1988). *Appl. Phys. Lett.* **53**, 1563.

Manassen, Y., Hamers, R. J., Demuth, J. E. and Castellano, Jr, A. J. (1989). *Phys. Rev. Lett.* **62**, 2531.

Mandler, D. and Bard, A. J. (1989). *J. Electrochem. Soc.* **136**, 3143.

Mandler, D. and Bard, A. J. (1990a). *J. Electrochem. Soc.* **137**, 1079.

Mandler, D. and Bard, A. J. (1990b). *J. Electrochem. Soc.* **137**, 2468.

Mann, B. and Kuhn, H. (1971). *J. Appl. Phys.* **42**, 4398.

Mann, B., Kuhn, H. and v. Szentpály, L. (1971). *Chem. Phys. Lett.* **8**, 82.

Manne, S., Butt, H. J., Gould, S. A. C. and Hansma, P. K. (1990). *Appl. Phys. Lett.* **56**, 1758.

Manne, S., Hansma, P. K., Massie, J., Elings, V. B. and Gewirth, A. A. (1991a). *Science* **251**, 183.

Manne, S., Massie, J., Elings, V. B., Hansma, P. K. and Gewirth, A. A. (1991b). *J. Vac. Sci. Technol.* B **9**, 950.

Mannhart, J., Anselmetti, D., Bednorz, J.G., Catana, A., Gerber, Ch., Müller, K. A. and Schlom, D.G. (1992). *Z. Phys.* B **86**, 177.

Marchon, B., Bernhardt, P., Bussell, M. E., Somorjai, G. A., Salmeron, M. and Siekhaus, W. (1988a). *Phys. Rev. Lett.* **60**, 1166.

Marchon, B., Ogletree, D. F., Bussell, M. E., Somorjai, G. A., Salmeron, M. and Siekhaus, W. (1988b). *J. Microsc.* **152**, 427.

Marella, P. F. and Pease, R. F. (1989). *Appl. Phys. Lett.* **55**, 2366.

Marrian, C.R. K. and Colton, R. J. (1990). *Appl. Phys. Lett.* **56**, 755.

Marrian, C.R. K., Dobisz, E. A. and Colton, R. J. (1990). *J. Vac. Sci. Technol.* A **8**, 3563.

Mårtensson, P. and Feenstra, R. M. (1989). *Phys. Rev.* B **39**, 7744.

Marti, O., Gould, S. and Hansma, P. K. (1988). *Rev. Sci. Instrum.* **59**, 836.

Marti, O., Colchero, J. and Mlynek, J. (1990). *Nanotechnology*, **1**, 141.

Martin, Y., Abraham, D. W. and Wickramasinghe, H. K. (1988a). *Appl. Phys. Lett.* **52**, 1103.

Martin, Y., Rugar, D. and Wickramasinghe, H. K. (1988b). *Appl. Phys. Lett.* **52**, 244.

Martin, Y. and Wickramasinghe, H. K. (1987). *Appl. Phys. Lett.* **50**, 1455.
Martin, Y., Williams, C. C. and Wickramasinghe, H. K. (1987). *J. Appl. Phys.* **61**, 4723.
Martin, Y., Williams, C. C. and Wickramasinghe, H. K. (1988c). *Scanning Microsc.* **2**, 3.
Maserjian, J. (1974). *J. Vac. Sci. Technol.* **11**, 996.
Maserjian, J. and Zamani, N. (1982). *J. Appl. Phys.* **53**, 559.
Mate, C. M., Erlandsson, R., McClelland, G. M. and Chiang, S. (1989a). *Surf. Sci.* **208**, 473.
Mate, C. M., Lorenz, M.R. and Novotny, V. J. (1989b). *J. Chem. Phys.* **90**, 7550.
Mate, C. M., McClelland, G. M., Erlandsson, R. and Chiang, S. (1987). *Phys. Rev. Lett.* **59**, 1942.
Mate, C. M. and Novotny, V. J. (1991). *J. Chem. Phys.* **94**, 8420.
Matey, J.R. and Blanc, J. (1985). *J. Appl. Phys.* **57**, 1437.
Matsuda, T., Fukuhara, A., Yoshida, T., Hasegawa, S., Tonomura, A. and Ru, Q. (1991). *Phys. Rev. Lett.* **66**, 457.
Matsuda, T., Hasegawa, S., Igarashi, M., Kobayashi, T., Naito, M., Kajiyama, H., Endo, J., Osakabe, N., Tonomura, A. and Aoki, R. (1989). *Phys. Rev. Lett.* **62**, 2519.
Maurice, V. and Marcus, P. (1992). *Surf. Sci. Lett.* **262**, L59.
McBride, S. E. and Wetsel, Jr, G. C. (1991). *Appl. Phys. Lett.* **59**, 3056.
McClelland, G. M. (1989). In: *Adhesion and Friction* (ed. Grunze, M. and Kreuzer, H. J.), Springer Series in Surface Sciences 17, p. 1 (Berlin, Heidelberg, New York: Springer).
McClelland, G. M. and Cohen, S.R. (1990). In: *Chemistry and Physics of Solid Surfaces VIII* (ed. Vanselow, R. and Howe, R.), Springer Series in Surface Sciences 22, p. 419 (Berlin, Heidelberg, New York: Springer).
McClelland, G. M., Erlandsson, R. and Chiang, S. (1987). In: *Review of Progress in Quantitative Non-Destructive Evaluation* (ed. Thompson, D.O. and Chimenti, D. E.) Vol. 6, p. 307 (New York: Plenum).
McCord, M. A. (1987). Ph.D. Thesis, Stanford University, Stanford, California.
McCord, M. A. and Awschalom, D. D. (1990). *Appl. Phys. Lett.* **57**, 2153.
McCord, M. A., Chang, T. H. P., Kern, D. P. and Speidell, J. L. (1989). *J. Vac. Sci. Technol.* B **7**, 1851.
McCord, M. A., Kern, D. P. and Chang, T. H. P. (1988). *J. Vac. Sci. Technol.* B **6**, 1877.
McCord, M. A. and Pease, R. F. W. (1985). *J. Vac. Sci. Technol.* B **3**, 198.
McCord, M. A. and Pease, R. F. W. (1986a). *J. Vac. Sci. Technol.* B **4**, 86.
McCord, M. A. and Pease, R. F. W. (1986b). *J. Physique* C **2**, 485.
McCord, M. A. and Pease, R. F. W. (1987a). *Appl. Phys. Lett.* **50**, 569.
McCord, M. A. and Pease, R. F. W. (1987b). *J. Vac. Sci. Technol.* B **5**, 430.
McCord, M. A. and Pease, R. F. W. (1988). *J. Vac. Sci. Technol.* B **6**, 293.
McElfresh, M., Miller, T.G., Schaefer, D. M., Reifenberger, R., Muenchausen, R. E., Hawley, M., Foltyn, S.R. and Wu, X. D. (1992). *J. Appl. Phys.* **71**, 5099.
McGonigal, G. C., Bernhardt, R. H. and Thomson, D. J. (1990). *Appl. Phys. Lett.* **57**, 28.
McGonigal, G. C., Bernhardt, R. H., Yeo, Y. H. and Thomson, D. J. (1991). *J. Vac. Sci. Technol.* B **9**, 1107.

McMaster, T. J., Carr, H., Miles, M. J., Cairns, P. and Morris, V. J. (1990). *J. Vac. Sci. Technol.* A **8**, 648.

McMillan, W. L. (1977). *Phys. Rev.* B **16**, 643.

McMillan, W. L. and Rowell, J. M. (1965). *Phys. Rev. Lett.* **14**, 108.

Mead, C. A. (1969). In: *Tunneling Phenomena in Solids* (ed. Burstein, E. and Lundquist, S.), p. 127 (New York: Plenum).

Melmed, A. J. (1991). *J. Vac. Sci. Technol.* B **9**, 601.

Melmed, A. J., Kaufman, M. J. and Fowler, H. A. (1986). *J. Physique* C **7**, 35.

Memmert, U. and Behm, R. J. (1991). In: *Advances in Solid State Physics 31* (ed. Rössler, U.), p. 189 (Braunschweig: Vieweg).

Mercereau, J. E. (1969). In: *Superconductivity* (ed. Parks), vol. 1, p. 393 (New York: Dekker).

Meservey, R., Tedrow, P. M. and Fulde, P. (1970). *Phys. Rev. Lett.* **25**, 1270.

Meyer, E. (1990). Ph.D. Thesis, Basel University, Basel, Switzerland.

Meyer, E., Anselmetti, D., Wiesendanger, R., Güntherodt, H.-J., Lévy, F. and Berger, H. (1989a). *Europhys. Lett.* **9**, 695.

Meyer, E., Güntherodt, H.-J., Haefke, H., Gerth, G. and Krohn, M. (1991a). *Europhys. Lett.* **15**, 319.

Meyer, E., Heinzelmann, H., Brodbeck, D., Overney, G., Overney, R., Howald, L., Hug, H., Jung, T., Hidber, H.-R. and Güntherodt, H.-J. (1991b). *J. Vac. Sci. Technol.* B **9**, 1329.

Meyer, E., Heinzelmann, H., Grütter, P., Hidber, H.-R., Güntherodt, H.-J. and Steiger, R. (1989b). *J. Appl. Phys.* **66**, 4243.

Meyer, E., Heinzelmann, H., Grütter, P., Jung, Th., Weisskopf, Th., Hidber, H.-R., Lapka, R., Rudin, H. and Güntherodt, H.-J. (1988). *J. Microsc.* **152**, 269.

Meyer, E., Heinzelmann, H., Rudin, H. and Güntherodt, H.-J. (1990a). *Z. Phys.* B **79**, 3.

Meyer, E., Howald, L., Overney, R. M., Heinzelmann, H., Frommer, J., Güntherodt, H.-J., Wagner, T., Schier, H. and Roth, S. (1991c). *Nature* **349**, 398.

Meyer, E., Overney, R., Brodbeck, D., Howald, L., Lüthi, R., Frommer, J. and Güntherodt, H.-J. (1992). *Phys. Rev. Lett.* **69**, 1777.

Meyer, E., Wiesendanger, R., Anselmetti, D., Hidber, H.R., Güntherodt, H.-J., Lévy, F. and Berger, H. (1990b). *J. Vac. Sci. Technol.* A **8**, 495.

Meyer, G. and Amer, N. M. (1988). *Appl. Phys. Lett.* **53**, 1045.

Meyer, G. and Amer, N. M. (1990a). *Appl. Phys. Lett.* **56**, 2100.

Meyer, G. and Amer, N. M. (1990b). *Appl. Phys. Lett.* **57**, 2089.

Michel, B. and Travaglini, G. (1988). *J. Microsc.* **152**, 681.

Michel, B., Travaglini, G., Rohrer, H., Joachim, C. and Amrein, M. (1989). *Z. Phys.* B **76**, 99.

Michely, Th., Besocke, K. H. and Comsa, G. (1990). *Surf. Sci. Lett.* **230**, L135.

Michely, Th., Besocke, K. H. and Teske, M. (1988). *J. Microsc.* **152**, 77.

Michely, Th. and Comsa, G. (1991a). *Surf. Sci.* **256**, 217.

Michely, Th. and Comsa, G. (1991b). *Phys. Rev.* B **44**, 8411.

Michely, Th. and Comsa, G. (1991c). *J. Vac. Sci. Technol.* B **9**, 862.

Miles, M. J., Carr, H. J., McMaster, T. C., I'Anson, K. J., Belton, P.S., Morris, V. J., Field, J. M., Shewry, P.R. and Tatham, A.S. (1991). *Proc. Natl Acad. Sci. USA* **88**, 68.

Miyamoto, I., Ezawa, T. and Itabashi, K. (1991). *Nanotechnology* **2**, 52.

Mizes, H. A. and Foster, J.S. (1989). *Science* **244**, 559.

Mizes, H. A. and Harrison, W. A. (1988). *J. Vac. Sci. Technol.* A **6**, 300.

Mizes, H. A., Loh, K.-G., Miller, R. J. D., Ahuja, S. K. and Grabowski, E. F. (1991). *Appl. Phys. Lett.* **59**, 2901.

Mizes, H. A., Park, S.-I. and Harrison, W. A. (1987). *Phys. Rev.* B **36**, 4491.

Mizutani, W., Shigeno, M., Ohmi, M., Suginoya, M., Kajimura, K. and Ono, M. (1991). *J. Vac. Sci. Technol.* B **9**, 1102.

Mizutani, W., Shigeno, M., Ono, M. and Kajimura, K. (1990). *Appl. Phys. Lett.* **56**, 1974.

Mo, Y.-W., Kleiner, J., Webb, M. B. and Lagally, M.G. (1991). *Phys. Rev. Lett.* **66**, 1998.

Mo, Y.-W., Kleiner, J., Webb, M. B. and Lagally, M.G. (1992). *Surf. Sci.* **268**, 275.

Mo, Y.-W. and Lagally, M.G. (1991). *Surf. Sci.* **248**, 313.

Mo, Y.-W., Savage, D. E., Swartzentruber, B.S. and Lagally, M.G. (1990). *Phys. Rev. Lett.* **65**, 1020.

Mo, Y.-W., Swartzentruber, B.S., Kariotis, R., Webb, M. B. and Lagally, M.G. (1989). *Phys. Rev. Lett.* **63**, 2393.

Moiseev, Yu.N., Mostepanenko, V. M., Panov, V.I. and Sokolov, I.Yu. (1988). *Phys. Lett.* A **132**, 354.

Möller, R., Albrecht, U., Boneberg, J., Koslowski, B., Leiderer, P. and Dransfeld, K. (1991a). *J. Vac. Sci. Technol.* B **9**, 506.

Möller, R., Baur, C., Esslinger, A. and Kürz, P. (1991b). *J. Vac. Sci. Technol.* B **9**, 609.

Möller, R., Coenen, R., Esslinger, A. and Koslowski, B. (1990). *J. Vac. Sci. Technol.* A **8**, 659.

Möller, R., Coenen, R., Koslowski, B. and Rauscher, M. (1989b). *Surf. Sci.* **217**, 289.

Möller, R., Esslinger, A. and Koslowski, B. (1989a). *Appl. Phys. Lett.* **55**, 2360.

Molotkov, S.N. (1992). *Surf. Sci.* **264**, 235.

Moodera, J.S., Hao, X., Gibson, G. A. and Meservey, R. (1988). *Phys. Rev. Lett.* **61**, 637.

Moreland, J., Alexander, S., Cox, M., Sonnenfeld, R. and Hansma, P. K. (1983). *Appl. Phys. Lett.* **43**, 387.

Moreland, J. and Rice, P. (1990). *Appl. Phys. Lett.* **57**, 310.

Moreland, J. and Rice, P. (1991). *J. Appl. Phys.* **70**, 520.

Moreland, J., Rice, P., Russek, S. E., Jeanneret, B., Roshko, A., Ono, R. H. and Rudman, D. A. (1991). *Appl. Phys. Lett.* **59**, 3039.

Morishita, S. and Okuyama, F. (1991). *J. Vac. Sci. Technol.* A **9**, 167.

Morita, S., Okada, T. and Mikoshiba, N. (1989). *Jpn. J. Appl. Phys.* **28**, 535.

Morita, S., Otsuka, I., Okada, T., Yokoyama, H., Iwasaki, T. and Mikoshiba, N. (1987). *Jpn. J. Appl. Phys.* **26**, L1853.

Mortensen, K., Chen, D. M., Bedrossian, P. J., Golovchenko, J. A. and Besenbacher, F. (1991). *Phys. Rev.* B **43**, 1816.

Moser, M., Wachter, P., Hulliger, F. and Etourneau, J.R. (1985). *Solid State Commun.* **54**, 241.

Mou, J., Sun, W., Yan, J., Yang, W.S., Liu, C., Zhai, Z., Xu, Q. and Xie, Y. (1991). *J. Vac. Sci. Technol.* B **9**, 1566.

Moyer, P. J., Jahncke, C. L., Paesler, M. A., Reddick, R. C. and Warmack, R. J. (1990). *Phys. Lett.* A **145**, 343.

Mulhern, P. J., Hubbard, T., Arnold, C.S., Blackford, B. L. and Jericho, M. H. (1991). *Rev. Sci. Instrum.* **62**, 1280.

Müller, E. W. (1937). *Z. Phys.* **106**, 541.

Müller, E. W. and Tsong, T.T. (1969). *Field Ion Microscopy* (New York: American Elsevier).

Müller, N., Eckstein, W., Heiland, W. and Zinn, W. (1972). *Phys. Rev. Lett.* **29**, 1651.

Muralt, P. (1986). *Appl. Phys. Lett.* **49**, 1441.

Muralt, P., Meier, H., Pohl, D. W. and Salemink, H. W. M. (1987). *Appl. Phys. Lett.* **50**, 1352.

Muralt, P. and Pohl, D. W. (1986). *Appl. Phys. Lett.* **48**, 514.

Muralt, P., Pohl, D. W. and Denk, W. (1986). *IBM J. Res. Develop.* **30**, 443.

Musselman, I. H. and Russell, P. E. (1990). *J. Vac. Sci. Technol.* A **8**, 3558.

Nagahara, L. A., Thundat, T. and Lindsay, S. M. (1989). *Rev. Sci. Instrum.* **60**, 3128.

Nagahara, L. A., Thundat, T. and Lindsay, S. M. (1990). *Appl. Phys. Lett.* **57**, 270.

Naito, M., Smith, D. P. E., Kirk, M. D., Oh, B., Hahn, M.R., Char, K., Mitzi, D. B., Sun, J. Z., Webb, D. J., Beasley, M.R., Fischer, O., Geballe, T. H., Hammond, R. H., Kapitulnik, A. and Quate, C. F. (1987). *Phys. Rev.* B **35**, 7228.

Nakanishi, K. and Shiba, H. (1977). *J. Phys. Soc. Jpn* **43**, 1839.

Nawaz, Z., Cataldi, T.R.I., Knall, J., Somekh, R. and Pethica, J. B. (1992). *Surf. Sci.* **265**, 139.

Nejoh, H. (1990). *Appl. Phys. Lett.* **57**, 2907.

Netzer, F. P. and Frank, K.-H. (1989). *Phys. Rev.* B **40**, 5223.

Neubauer, G., Cohen, S.R., McClelland, G. M., Horne, D. and Mate, C. M. (1990). *Rev. Sci. Instrum.* **61**, 2296.

Nguyen, H.Q., Feuchtwang, T. E. and Cutler, P. H. (1986). *J. Physique* C **2**, 37.

Nichols, R. J., Magnussen, O. M., Hotlos, J., Twomey, T., Behm, R. J. and Kolb, D. M. (1990). *J. Electroanal. Chem.* **290**, 21.

Nicol, J., Shapiro, S. and Smith, P. H. (1960), *Phys. Rev. Lett.* **5**, 461.

Niedermann, Ph., Emch, R. and Descouts, P. (1988). *Rev. Sci. Instrum.* **59**, 368.

Niedermann, Ph., Renner, Ch., Kent, A. D. and Fischer, Ø. (1990). *J. Vac. Sci. Technol.* A **8**, 594.

Niehus, H., Raunau, W., Besocke, K., Spitzl, R. and Comsa, G. (1990). *Surf. Sci. Lett.* **225**, L8.

Niksch, M. and Binnig, G. (1988). *J. Vac. Sci. Technol.* A **6**, 470.

Nishitani, R., Kasuya, A., Kubota, S. and Nishina, Y. (1991). *J. Vac. Sci. Technol.* B **9**, 806.

Nogami, J., Park, S.-I. and Quate, C. F. (1989). *J. Vac. Sci. Technol.* A **7**, 1919.

Noguera, C. (1988). *J. Microsc.* **152**, 3.

Noguera, C. (1989). *J. Physique* **50**, 2587.

Noguera, C. (1990). *Phys. Rev.* B **42**, 1629.

Nomura, K. and Ichimura, K. (1989). *Solid State Commun.* **71**, 149.

Nomura, K. and Ichimura, K. (1990). *J. Vac. Sci. Technol.* A **8**, 504.

Norton, D. P., Lowndes, D. H., Zheng, X.-Y., Zhu, S. and Warmack, R. J. (1991). *Phys. Rev.* B **44**, 9760.

Oden, P.I., Thundat, T., Nagahara, L. A., Lindsay, S. M., Adams, G. B. and Sankey, O. F. (1991). *Surf. Sci. Lett.* **254**, L454.

Ogletree, D. F., Ocal, C., Marchon, B., Somorjai, G. A., Salmeron, M., Beebe, T. and Siekhaus, W. (1990). *J. Vac. Sci. Technol.* A **8**, 297.

Ohnishi, S. and Tsukada, M. (1989). *Solid State Commun.* **71**, 391.

Ohnishi, S. and Tsukada, M. (1990). *J. Vac. Sci. Technol.* A **8**, 174.

Ohtani, H., Wilson, R. J., Chiang, S. and Mate, C. M. (1988). *Phys. Rev. Lett.* **60**, 2398.

Okayama, S., Bando, H., Tokumoto, H. and Kajimura, K. (1985). *Jpn. J. Appl. Phys.* **24**, 152.

Olk, C. H., Heremans, J., Dresselhaus, M.S., Speck, J.S. and Nicholls, J.T. (1990). *Phys. Rev.* B **42**, 7524.

Olk, C. H., Heremans, J., Dresselhaus, M.S., Speck, J.S. and Nicholls, J.T. (1991). *J. Vac. Sci. Technol.* B **9**, 1055.

Omori, T., Kurihara, Y., Nakanishi, T., Aoyagi, H., Baba, T., Furuya, T., Itoga, K., Mizuta, M., Nakamura, S., Takeuchi, Y., Tsubata, M. and Yoshioka, M. (1991). *Phys. Rev. Lett.* **67**, 3294.

Onnes, H. K. (1911). *Comm. Leiden* **120**b.

Oppenheimer, J.R. (1928). *Phys. Rev.* **13**, 66.

Otsuka, I. and Iwasaki, T. (1988). *J. Microsc.* **152**, 289.

Overney, R.M. (1992). Ph.D. Thesis, Basel University, Basel, Switzerland.

Overney, R. M., Howald, L., Frommer, J., Meyer, E. and Güntherodt, H.-J. (1991). *J. Chem. Phys.* **94**, 8441.

Overney, R. M., Meyer, E., Frommer, J., Brodbeck, D., Lüthi, R., Howald, L., Güntherodt, H.-J., Fujihira, M., Takano, H. and Gotoh, Y. (1992). *Nature* **359**, 133.

Packard, W. E., Dai, N., Dow, J. D., Jaklevic, R. C., Kaiser, W. J. and Tang, S. L. (1990). *J. Vac. Sci. Technol.* A **8**, 3512.

Paesler, M. A., Moyer, P. J., Jahncke, C. J. and Johnson, C. E. (1990). *Phys. Rev.* B **42**, 6750.

Paik, S. M., Kim, S. and Schuller, I. K. (1991). *Phys. Rev.* B **44**, 3272.

Pan, S., de Lozanne, A. L. and Fainchtein, R. (1991). *J. Vac. Sci. Technol.* B **9**, 1017.

Pan, S., Ng, K. W., de Lozanne, A. L., Tarascon, J. M. and Greene, L. H. (1987). *Phys. Rev.* B **35**, 7220.

Pandey, K. C. (1981). *Phys. Rev. Lett.* **47**, 1913.

Pappas, R. A., Hunt, E.R. and Ulloa, S. E. (1992). *Ultramicroscopy* **42–44**, 679.

Park, S.-I., Nogami, J. and Quate, C. F. (1987). *Phys. Rev.* B **36**, 2863.

Park, S.-I., Nogami, J., Mizes, H. A. and Quate, C. F. (1988). *Phys. Rev.* B **38**, 4269.

Park, S.-I., and Quate, C. F. (1987). *Rev. Sci. Instrum.* **58**, 2004.

Parkinson, B. (1990). *J. Am. Chem. Soc.* **112**, 7498.

Pashley, M. D., Haberern, K. W. and Friday, W. (1988a). *J. Vac. Sci. Technol.* A **6**, 488.

Pashley, M. D., Haberern, K. W., Friday, W., Woodall, J. M. and Kirchner, P. D. (1988b). *Phys. Rev. Lett.* **60**, 2176.

Pashley, M. D., Haberern, K. W. and Gaines, J. M. (1991a). *Appl. Phys. Lett.* **58**, 406.

Pashley, M. D., Haberern, K. W. and Gaines, J. M. (1991b). *J. Vac. Sci. Technol.* B **9**, 938.

Pashley, M. D., Haberern, K. W. and Gaines, J. M. (1992). *Surf. Sci.* **267**, 153.

Pashley, M. D., Haberern, K. W. and Woodall, J. M. (1988c). *J. Vac. Sci. Technol.* B **6**, 1468.

Pashley, M. D. and Pethica, J. B. (1985). *J. Vac. Sci. Technol.* A **3**, 757.

Patil, R., Kim, S.-J., Smith, E., Reneker, D. H. and Weisenhorn, A. L. (1990). *Polymer Commun.* **31**, 455.

Patrin, J. C., Li, Y. Z. and Weaver, J. H. (1992). *Phys. Rev.* B **45**, 1756.

Pauling, L. (1987). *Phys. Rev. Lett.* **58**, 294.

Payne, M. C. (1986). *J. Phys.* C **19**, 1145.

Payne, M. C. and Inkson, J. C. (1985). *Surf. Sci.* **159**, 485.

Peierls, R. E. (1955). *Quantum Theory of Solids* (Oxford: Oxford University Press).

Pelz, J. P. (1991). *Phys. Rev.* B **43**, 6746.

Pelz, J. P. and Koch, R. H. (1989). *Rev. Sci. Instrum.* **60**, 301.

Pelz, J. P. and Koch, R. H. (1990a). *Phys. Rev.* B **41**, 1212.

Pelz, J. P. and Koch, R. H. (1990b). *Phys. Rev.* B **42**, 3761.

Pelz, J. P. and Koch, R. H. (1991). *J. Vac. Sci. Technol.* B **9**, 775.

Penley, J. C. (1962). *Phys. Rev.* **128**, 596.

Penner, R. M., Heben, M. J., Lewis, N.S. and Quate, C. F. (1991). *Appl. Phys. Lett.* **58**, 1389.

Persson, B.N. J. (1987). *Chem. Phys. Lett.* **141**, 366.

Persson, B.N. J. (1988). *Phys. Scripta* **38**, 282.

Persson, B.N. J. and Baratoff, A. (1987). *Phys. Rev. Lett.* **59**, 339.

Persson, B.N. J. and Demuth, J. E. (1986). *Solid State Commun.* **57**, 769.

Peterson, I.R. (1987). *J. Molecul. Electron.* **3**, 103.

Pethica, J. B. (1986). *Phys. Rev. Lett.* **57**, 3235.

Pethica, J. B., Hutchings, R. and Oliver, W. C. (1983). *Phil. Mag.* A **48**, 593.

Pethica, J. B. and Sutton, A. P. (1988). *J. Vac. Sci. Technol.* A **6**, 2494.

Pfluger, P. and Güntherodt, H.-J. (1981). In: *Advances in Solid State Physics 21* (ed. Treusch, J.), p. 271 (Braunschweig: Vieweg).

Phillips, W. A. (1984). *Physica* B **127**, 112.

Piner, R., Reifenberger, R., Martin, D. C., Thomas, E. L. and Apkarian, R. P. (1990). *J. Polymer Sci.* C: *Polymer Lett.* **28**, 399.

Pitarke, J. M., Echenique, P. M. and Flores, F. (1989). *Surf. Sci.* **217**, 267.

Pitarke, J. M., Flores, F. and Echenique, P. M. (1990). *Surf. Sci.* **234**, 1.

Pohl, D. W. (1986). *IBM J. Res. Develop.* **30**, 417.

Pohl, D. W. (1987). *Rev. Sci. Instrum.* **58**, 54.

Pohl, D. W. (1991). *Adv. Opt. Electron Microsc.* **12**, 243.

Pohl, D. W., Denk, W. and Lanz, M. (1984). *Appl. Phys. Lett.* **44**, 651.

Pohl, D. W. and Möller, R. (1988). *Rev. Sci. Instrum.* **59**, 840.

Polymeropoulos, E. E. (1977). *J. Appl. Phys.* **48**, 2404.

Pötschke, G. and Behm, R. J. (1992). In press.

Pötschke, G., Schröder, J., Günther, C., Hwang, R.Q. and Behm, R. J. (1991). *Surf. Sci.* **251/252**, 592.

Prater, C. B., Hansma, P. K., Tortonese, M. and Quate, C. F. (1991a). *Rev. Sci. Instrum.* **62**, 2634.

Prater, C. B., Wilson, M.R., Garnaes, J., Massie, J., Elings, V. B. and Hansma, P. K. (1991b). *J. Vac. Sci. Technol.* B **9**, 989.

Price, P. J. and Radcliffe, J. M. (1959). *IBM J. Res. Develop.* **3**, 364.

Prietsch, M. and Ludeke, R. (1991a). *Phys. Rev. Lett.* **66**, 2511.

Prietsch, M. and Ludeke, R. (1991b). *Surf. Sci.* **251/252**, 413.
Putman, C. A. J., van der Werf, K.O., de Grooth, B.G., van Hulst, N. F., Segerink, F. B. and Greve, J. (1992). *Rev. Sci. Instrum.* **63**, 1914.
Qian, L.Q. and Wessels, B. W. (1991a). *Appl. Phys. Lett.* **58**, 1295.
Qian, L.Q. and Wessels, B. W. (1991b). *Appl. Phys. Lett.* **58**, 2538.
Qin, X. and Kirczenow, G. (1989). *Phys. Rev.* B **39**, 6245.
Qin, X. and Kirczenow, G. (1990). *Phys. Rev.* B **41**, 4976.
Quate, C. F. (1990). In: *Scanning Tunneling Microscopy and Related Methods* (ed. Behm, R. J., Garcia, N. and Rohrer, H.), NATO ASI Series E: Appl. Sci. Vol. 184, p. 281 (Dordrecht: Kluwer).
Rabe, J. P. and Buchholz, S. (1991). *Phys. Rev. Lett.* **66**, 2096.
Rabe, J. P., Sano, M., Batchelder, D. and Kalatchev, A. A. (1988). *J. Microsc.* **152**, 573.
Radhakrishnan, V. (1970). *Wear* **16**, 325.
Ramos, M. A., Vieira, S., Buendia, A. and Baro, A. M. (1988). *J. Microsc.* **152**, 137.
Reddick, R. C., Warmack, R. J. and Ferrell, T. L. (1989). *Phys. Rev.* B **39**, 767.
Reihl, B., Coombs, J. H. and Gimzewski, J. K. (1989). *Surf. Sci.* **211/212**, 156.
Reihl, B. and Gimzewski, J. K. (1987). *Surf. Sci.* **189/190**, 36.
Reihl, B., Gimzewski, J. K., Nicholls, J. M. and Tosatti, E. (1986). *Phys. Rev.* B **33**, 5770.
Reiss, G., Schneider, F., Vancea, J. and Hoffmann, H. (1990b). *Appl. Phys. Lett.* **57**, 867.
Reiss, G., Vancea, J., Wittmann, H., Zweck, J. and Hoffmann, H. (1990a). *J. Appl. Phys.* **67**, 1156.
Rendell, R. W., Scalapino, D. J. and Mühlschlegel, B. (1978). *Phys. Rev. Lett.* **41**, 1746.
Reneker, D. H., Schneir, J., Howell, B. and Harary, H. (1990). *Polymer Commun.* **31**, 167.
Renner, Ch., Kent, A. D., Niedermann, Ph., Fischer, Ø. and Lévy, F. (1991). *Phys. Rev. Lett.* **67**, 1650.
Rice. O. K. (1929). *Phys. Rev.* **34**, 1451.
Rice, P. and Moreland, J. (1991). *Rev. Sci. Instrum.* **62**, 844.
Ringger, M., Corb, B. W., Hidber, H.-R., Schlögl, R., Wiesendanger, R., Stemmer, A., Rosenthaler, L., Brunner, A. J., Oelhafen, P. C. and Güntherodt, H.-J. (1986). *IBM J. Res. Develop.* **30**, 500.
Ringger, M., Hidber, H.R., Schlögl, R., Oelhafen, P. and Güntherodt, H.-J. (1985). *Appl. Phys. Lett.* **46**, 832.
Ritter, E., Behm, R. J., Pötschke, G. and Wintterlin, J. (1987). *Surf. Sci.* **181**, 403.
Roberts, C. J., Davies, M. C., Jackson, D. E., Tendler, S. J. B. and Williams, P. M. (1991). *J. Phys.: Condens. Matter* **3**, 7213.
Roberts, G.G., Vincett, P.S. and Barlow, W. A. (1981). *Phys. Technol.* **12**, 69.
Robinson, R.S. (1989). *J. Electrochem. Soc.* **136**, 3145.
Robinson, R.S. (1990). *J. Vac. Sci. Technol.* A **8**, 511.
Rohrer, H. (1990). In: *Scanning Tunneling Microscopy and Related Methods* (ed. Behm, R. J., Garcia, N. and Rohrer, H.), NATO ASI Series E: Appl. Sci. Vol. 184, p. 1 (Dordrecht: Kluwer).

Rotermund, H. H., Ertl, G. and Sesselmann, W. (1989). *Surf. Sci.* **217**, L383.

Rowell, J. M. (1963). *Phys. Rev. Lett.* **11**, 200.

Roy, D. K. (1986). *Quantum Mechanical Tunnelling and its Applications* (Singapore: World Scientific).

Rudd, G., Novak, D., Saulys, D., Bartynski, R. A., Garofalini, S., Ramanujachary, K.V., Greenblatt, M. and Garfunkel, E. (1991). *J. Vac. Sci. Technol.* B **9**, 909.

Rugar, D. and Hansma, P. (1990). *Phys. Today* October, p. 23.

Rugar, D., Mamin, H. J., Erlandsson, R., Stern, J. E. and Terris, B. D. (1988). *Rev. Sci. Instrum.* **59**, 2337.

Rugar, D., Mamin, H. J. and Guethner, P. (1989). *Appl. Phys. Lett.* **55**, 2588.

Rugar, D., Mamin, H. J., Guethner, P., Lambert, S. E., Stern, J. E., McFadyen, I. and Yogi, T. (1990). *J. Appl. Phys.* **68**, 1169.

Rybachenko, V. F. (1967). *Sov. J. Nucl. Phys.* **5**, 635.

Sáenz, J. J., García, N., Grütter, P., Meyer, E., Heinzelmann, H., Wiesendanger, R., Rosenthaler, L., Hidber, H.R. and Güntherodt, H.-J. (1987). *J. Appl. Phys.* **62**, 4293.

Sáenz, J. J., García, N. and Slonczewski, J. C. (1988). *Appl. Phys. Lett.* **53**, 1449.

Sakamaki, K., Hinokuma, K., Hashimoto, K. and Fujishima, A. (1990a). *Surf. Sci. Lett.* **237**, L383.

Sakamaki, K., Itoh, K., Fujishima, A. and Gohshi, Y. (1990b). *J. Vac. Sci. Technol.* A **8**, 525.

Sakurai, T., Hashizume, T., Kamiya, I., Hasegawa, Y., Ide, T., Miyao, M., Sumita, I., Sakai, A. and Hyodo, S. (1989). *J. Vac. Sci. Technol.* A **7**, 1684.

Salemink, H. and Albrektsen, O. (1991). *J. Vac. Sci. Technol.* B **9**, 779.

Salemink, H. W. M., Albrektsen, O. and Koenraad, P. (1992). *Phys. Rev.* B **45**, 6946.

Salemink, H. W. M., Meier, H. P., Ellialtioglu, R., Gerritsen, J. W. and Muralt, P.R. M. (1989). *Appl. Phys. Lett.* **54**, 1112.

Samara, G. A., Schirber, J. E., Morosin, B., Hansen, L.V., Loy, D. and Sylwester, A. P. (1991). *Phys. Rev. Lett.* **67**, 3136.

Samsavar, A., Hirschorn, E.S., Miller, T., Leibsle, F. M., Eades, J. A. and Chiang, T.-C. (1990). *Phys. Rev. Lett.* **65**, 1607.

Sarid, D., Henson, T. D., Armstrong, N.R. and Bell, L.S. (1988a). *Appl. Phys. Lett.* **52**, 2252.

Sarid, D., Iams, D. A., Weissenberger, V. and Bell, L.S. (1988b). *Opt. Lett.* **13**, 1057.

Sashikata, K., Furuya, N. and Itaya, K. (1991). *J. Vac. Sci. Technol.* B **9**, 457.

Sass, J. K., Gimzewski, J. K., Haiss, W., Besocke, K. H. and Lackey, D. (1991). *J. Phys.: Condens. Matter* **3**, S121.

Saurenbach, F. and Terris, B. D. (1990). *Appl. Phys. Lett.* **56**, 1703.

Scalapino, D. J. and Marcus, S. M. (1967). *Phys. Rev. Lett.* **18**, 459.

Scheinfein, M.R., Unguris, J., Pierce, D.T. and Celotta, R. J. (1990). *J. Appl. Phys.* **67**, 5932.

Schelten, J., Ullmaier, H. and Lippmann, G. (1972). *Z. Phys.* **253**, 219.

Schlögl, R., Wiesendanger, R. and Baiker, A. (1987). *J. Catalysis* **108**, 452.

Schlom, D.G., Anselmetti, D., Bednorz, J.G., Broom, R. F., Catana, A., Frey, T., Gerber, Ch., Güntherodt, H.-J., Lang, H. P. and Mannhart, J. (1992). *Z. Phys.* B **86**, 163.

Schneir, J., Elings, V. and Hansma, P. K. (1988). *J. Electrochem. Soc.* **135**, 2774.

Schneir, J., Sonnenfeld, R., Hansma, P. K. and Tersoff, J. (1986). *Phys. Rev.* B **34**, 4979.

Schönenberger, C. and Alvarado, S. F. (1989). *Rev. Sci. Instrum.* **60**, 3131.

Schönenberger C. and Alvarado, S. F. (1990a). *Z. Phys.* B **80**, 373.

Schönenberger, C. and Alvarado, S. F. (1990b). *Phys. Rev. Lett.* **65**, 3162.

Schönenberger, C. and Alvarado, S. F., Lambert, S. E. and Sanders, I. L. (1990). *J. Appl. Phys.* **67**, 7278.

Schuster, R., Barth, J.V., Ertl, G. and Behm, R. J. (1991a). *Phys. Rev.* B **44**, 13689.

Schuster, R., Barth, J.V., Ertl, G. and Behm, R. J. (1991b). *Surf. Sci. Lett.* **247**, L229.

Schwarz, U. D., Haefke, H., Jung, Th., Meyer, E., Güntherodt, H.-J., Steiger, R. and Bohonek, J. (1992). *Ultramicroscopy* **41**, 435.

Selloni, A., Carnevali, P., Tosatti, E. and Chen, C. D. (1985). *Phys. Rev.* B **31**, 2602.

Selloni, A., Chen, C. D. and Tosatti, E. (1988). *Phys. Scripta* **38**, 297.

Shechtman, D., Blech, I., Gratias, D. and Cahn, J. W. (1984). *Phys. Rev. Lett.* **53**, 1951.

Shedd, G. M. and Russell, P. E. (1990). *Nanotechnology* **1**, 67.

Shedd, G. M. and Russell, P. E. (1991). *J. Vac. Sci. Technol.* A **9**, 1261.

Shedd, G. M. and Russell, P. E. (1992). *Surf. Sci.* **266**, 259.

Shih, C. K., Feenstra, R. M. and Chandrashekhar, G. V. (1991). *Phys. Rev.* B **43**, 7913.

Shih, C. K., Feenstra, R. M., Kirtley, J. R. and Chandrashekhar, G. V. (1989). *Phys. Rev.* B **40**, 2682.

Shore, J. D., Huang, M., Dorsey, A. T. and Sethna, J. P. (1989). *Phys. Rev. Lett.* **62**, 3089.

Siegenthaler, H. (1992). In: *Scanning Tunneling Microscopy II* (ed. Wiesendanger, R. & Güntherodt, H.-J.), Springer Series in Surface Sciences 28, p. 7 (Berlin, Heidelberg, New York: Springer).

Siegenthaler, H. and Christoph, R. (1990). In: *Scanning Tunneling Microscopy and Related Methods* (ed. Behm, R. J., Garcia, N. & Rohrer, H.), NATO ASI Series E: Appl. Sci. Vol. 184, p. 315 (Dordrecht: Kluwer).

Silver, R. M., Ehrichs, E. E. and de Lozanne, A. L. (1987). *Appl. Phys. Lett.* **51**, 247.

Simmons, J. G. (1963a). *J. Appl. Phys.* **34**, 1793.

Simmons, J. G. (1963b). *J. Appl. Phys.* **34**, 2581.

Slater, C. (1972). *J. Chem. Phys.* **57**, 2389.

Sleator, T. and Tycko, R. (1988). *Phys. Rev. Lett.* **60**, 1418.

Slonczewski, J. C. (1988). *J. Physique* C **8**, 1629.

Slonczewski, J. C. (1989). *Phys. Rev.* B **39**, 6995.

Slough, C. G. and Coleman, R. V. (1989). *Phys. Rev.* B **40**, 8042.

Slough, C. G., Giambattista, B., Johnson, A., McNairy, W. W. and Coleman, R. V. (1989). *Phys. Rev.* B **39**, 5496.

Slough, C. G., Giambattista, B., Johnson, A., McNairy, W. W., Wang, C. and Coleman, R. V. (1988). *Phys. Rev.* B **37**, 6571.

Slough, C. G., Giambattista, B., McNairy, W. W. and Coleman, R. V. (1990). *J. Vac. Sci. Technol.* A **8**, 490.

Slough, C. G., McNairy, W. W., Coleman, R. V., Drake, B. and Hansma, P. K. (1986). *Phys. Rev.* B **34**, 994.

Smith, D. P. E. (1991). *J. Vac. Sci. Technol.* B **9**, 1119.

Smith, D. P. E. and Binnig, G. (1986). *Rev. Sci. Instrum.* **57**, 2630.

Smith, D. P. E., Binnig, G. and Quate, C. F. (1986a). *Appl. Phys. Lett.* **49**, 1641.

Smith, D. P. E., Binnig, G. and Quate, C. F. (1986b). *Appl. Phys. Lett.* **49**, 1166.

Smith, D. P. E., Bryant, A., Quate, C. F., Rabe, J. P., Gerber, Ch. and Swalen, J. D. (1987a). *Proc. Natl Acad. Sci. USA* **84**, 969.

Smith, D. P. E. and Elrod, S. A. (1985). *Rev. Sci. Instrum.* **56**, 1970.

Smith, D. P. E. and Heckl, W. M. (1990). *Nature* **346**, 616.

Smith, D. P. E. and Hörber, J. K. H., Binnig, G. and Nejoh, H. (1990). *Nature* **344**, 641.

Smith, D. P. E., Hörber, H., Gerber, Ch. and Binnig, G. (1989a). *Science* **245**, 43.

Smith, D. P. E., Kirk, M. D. and Quate, C. F. (1987b). *J. Chem. Phys.* **86**, 6034.

Smith, J. R., Bozzolo, G., Banerjea, A. and Ferrante, J. (1989b). *Phys. Rev. Lett.* **63**, 1269.

Smoluchowski, R. (1941). *Phys. Rev.* **60**, 661.

Smolyaninov, I. I., Edelman, V. S. and Zavyalov, V. V. (1991). *Phys. Lett.* A **158**, 337.

Smolyaninov, I. I., Khaikin, M. S. and Edelman, V. S. (1990). *Phys. Lett.* A **149**, 410.

Snyder, E. J., Anderson, M. S., Tong, W. M., Williams, R. S., Anz, S. J., Alvarez, M. M., Rubin, Y., Diederich, F. N. and Whetten, R. L. (1991). *Science* **253**, 171.

Soler, J. M. Baró, A. M., García, N. and Rohrer, H. (1986). *Phys. Rev. Lett.* **57**, 444.

Sollner, T. C. L. G., Goodhue, W. D., Tannenwald, P. E., Parker, C. D. and Peck, D. D. (1983). *Appl. Phys. Lett.* **43**, 588.

Solymar, L. (1972). *Superconductive Tunnelling and Applications* (London: Chapman and Hall).

Sonnenfeld, R. and Hansma, P. K. (1986). *Science* **232**, 211.

Sonnenfeld, R. and Schardt, B. C. (1986). *Appl. Phys. Lett.* **49**, 1172.

Sonnenfeld, R., Schneir, J., Drake, B., Hansma, P. K. and Aspnes, D. E. (1987). *Appl. Phys. Lett.* **50**, 1742.

Sonnenfeld, R., Schneir, J. & Hansma, P. K. (1990). In: *Modern Aspects of Electrochemistry* (ed. White, R. E., Bockris, J. O'M. and Conway, B. E.), Vol. 21, p. 1 (New York: Plenum).

Soto, M. R. (1988). *J. Microsc.* **152**, 779.

Soto, M. R. (1990). *Surf. Sci.* **225**, 190.

Specht, M., Pedarnig, J. D., Heckl, W. M. and Hänsch, T. W. (1992). *Phys. Rev. Lett.* **68**, 476.

Spong, J. K., LaComb, Jr, L. J., Dovek, M. M., Frommer, J. E. and Foster, J. S. (1989a). *J. Physique* **50**, 2139.

Spong, J. K., Mizes, H. A., LaComb, Jr, L. J., Dovek, M. M., Frommer, J. E. and Foster, J. S. (1989b). *Nature* **338**, 137.

Staufer, U. (1990). Ph.D. Thesis, Basel University, Basel, Switzerland.

Staufer, U., Scandella, L., Rudin, H., Güntherodt, H.-J. and García, N. (1991). *J. Vac. Sci. Technol.* B **9**, 1389.

Staufer, U., Scandella, L. and Wiesendanger, R. (1989). *Z. Phys.* B **77**, 281.

Staufer, U., Wiesendanger, R., Eng, L., Rosenthaler, L., Hidber, H. R., Güntherodt, H.-J. and García, N. (1987). *Appl. Phys. Lett.* **51**, 244.

Staufer, U., Wiesendanger, R., Eng, L., Rosenthaler, L., Hidber, H. R., Güntherodt, H.-J. and García, N. (1988). *J. Vac. Sci. Technol.* A **6**, 537.

Stearns, M. B. (1977). *J. Magn. Magn. Mat.* **5**, 167.

Stemmer, A., Engel, A., Häring, R., Reichelt, R. and Aebi, U. (1988). *Ultramicroscopy* **25**, 171.

Stemmer, A., Hefti, A., Aebi, U. and Engel, A. (1989). *Ultramicroscopy* **30**, 263.

Stern, J. E., Terris, B. D., Mamin, H. J. and Rugar, D. (1988). *Appl. Phys. Lett.* **53**, 2717.

Stoll, E. (1984). *Surf. Sci.* **143**, L411.

Stoll, E., Baratoff, A., Selloni, A. and Carnevali, P. (1984). *J. Phys.* C **17**, 3073.

Stoll, E. P. and Gimzewski, J. K. (1991). *J. Vac. Sci. Technol.* B **9**, 643.

Stovneng, J. A. and Lipavsky, P. (1990). *Phys. Rev.* B **42**, 9214.

Stratton, R. (1962). *J. Phys. Chem. Solids* **23**, 1177.

Stroscio, J. A., Feenstra, R. M. and Fein, A. P. (1986). *Phys. Rev. Lett.* **57**, 2579.

Stroscio, J. A., Feenstra, R. M. and Fein, A. P. (1987a). *J. Vac. Sci. Technol.* A **5**, 838.

Stroscio, J. A., Feenstra, R. M. and Fein, A. P. (1987b). *Phys. Rev. Lett.* **58**, 1668.

Stroscio, J. A., Feenstra, R. M. and Fein, A. P. (1987c). *Phys. Rev.* B **36**, 7718.

Stroscio, J. A., Feenstra, R. M., Newns, D. M. and Fein, A. P. (1988). *J. Vac. Sci. Technol.* A **6**, 499.

Stroscio, J. A., First, P. N., Dragoset, R. A., Whitman, L. J., Pierce, D. T. and Celotta, R. J. (1990). *J. Vac. Sci. Technol.* A **8**, 284.

Stupian, G. W. and Leung, M. S. (1987). *Appl. Phys. Lett.* **51**, 1560.

Sugi, M. (1985). *J. Molecul. Electron.* **1**, 3.

Sugihara, K., Sakai, A., Kato, Y., Akama, Y., Shoda, N., Tokumoto, H. and Ono, M. (1991). *J. Vac. Sci. Technol.* B **9**, 707.

Sullivan, T. E. and Cutler, P. H. (1992). In: *Scanned Probe Microscopy* (ed. Wickramasinghe, H. K.), AIP Conf. Proc. 241, p. 111 (New York: AIP).

Sumita, I., Yokotsuka, T., Tanaka, H., Udagawa, M., Watanabe, Y., Takao, M. and Yokoyama, K. (1990). *Appl. Phys. Lett.* **57**, 1313.

Sun, B. N., Boutellier, R. and Schmid, H. (1989). *Physica* C **157**, 189.

Sun, B. N., Taylor, K. N. R., Hunter, B., Matthews, D. N., Ashby, S. and Sealey, K. (1991). *J. Cryst. Growth* **108**, 473.

Suzuki, M. and Fukuda, T. (1991). *Phys. Rev.* B **44**, 3187.

Swartzentruber, B. S., Mo, Y.-W., Kariotis, R., Lagally, M. G. and Webb, M. B. (1990). *Phys. Rev. Lett.* **65**, 1913.

Swartzentruber, B. S., Mo, Y.-W., Webb, M. B. and Lagally, M. G. (1989). *J. Vac. Sci. Technol.* A **7**, 2901.

Synge, E. H. (1928). *Phil. Mag.* **6**, 356.

Tabor, D. and Winterton, R. H. S. (1969). *Proc. Roy. Soc. (London)* A **312**, 435.

Takata, K., Hasegawa, T., Hosaka, S., Hosoki, S. and Komoda, T. (1989a). *Appl. Phys. Lett.* **55**, 1718.

Takata, K., Hosoki, S., Hosaka, S. and Tajima, T. (1989b). *Rev. Sci. Instrum.* **60**, 789.

Takata, K., Kure, T. and Okawa, T. (1992). *Appl. Phys. Lett.* **60**, 515.

Takata, K., Okawa, T. and Horiuchi, M. (1991). *Jpn. J. Appl. Phys.* **30**, L309.

Takata, K., Yugami, J., Hasegawa, T., Hosaka, S., Hosoki, S. and Komoda, T. (1989c). *Jpn. J. Appl. Phys.* **28**, L2279.

Takayanagi, K., Tanishiro, Y., Takahashi, M. and Takahashi, S. (1985a). *J. Vac. Sci. Technol.* A **3**, 1502.

Takayanagi, K., Tanishiro, Y., Takahashi, S. and Takahashi, M. (1985b). *Surf. Sci.* **164**, 367.

Takayanagi, K. and Yagi, K. (1983). *Trans. Jpn. Inst. Met.* **24**, 337.

Tanaka, I., Kato, T., Ohkouchi, S. and Osaka, F. (1990a). *J. Vac. Sci. Technol.* A **8**, 567.

Tanaka, M., Mizutani, W., Nakashizu, T., Morita, N., Yamazaki, S., Bando, H., Ono, M. and Kajimura, K. (1988). *J. Microsc.* **152**, 183.

Tanaka, M., Mizutani, W., Nakashizu, T., Yamazaki, S., Tokumoto, H., Bando, H., Ono, M. and Kajimura, K. (1989a). *Jpn. J. Appl. Phys.* **28**, 473.

Tanaka, M., Takahashi, T., Katayama-Yoshida, H., Yamazaki, S., Fujinami, M., Okabe, Y., Mizutani, W., Ono, M. and Kajimura, K. (1989b). *Nature* **339**, 691.

Tanaka, M., Yamazaki, S., Fujinami, M., Takahashi, T., Katayama-Yoshida, H., Mizutani, W., Kajimura, K. and Ono, M. (1990b). *J. Vac. Sci. Technol.* A **8**, 475.

Tang, S. L., Bokor, J. and Storz, R. H. (1988). *Appl. Phys. Lett.* **52**, 188.

Tang, S. L., Kasowski, R. V. and Parkinson, B. A. (1989). *Phys. Rev.* B **39**, 9987.

Tang, S. L., Kasowski, R. V., Suna, A. and Parkinson, B. A. (1990a). *J. Vac. Sci. Technol.* A **8**, 3484.

Tang, S. L., Kasowski, R. V., Suna, A. and Parkinson, B. A. (1990b). *Surf. Sci.* **238**, 280.

Tanimoto, M. and Nakano, Y. (1990). *J. Vac. Sci. Technol.* A **8**, 553.

Tao, N. J. and Lindsay, S. M. (1991). *J. Appl. Phys.* **70**, 5141.

Taylor, K. N. R., Cook, P. S., Puzzer, T., Matthews, D. N., Russell, G. J. and Goodman, P. (1988). *J. Cryst. Growth* **88**, 541.

Taylor, P. A., Nelson, J. S. and Dodson, B. W. (1991). *Phys. Rev.* B **44**, 5834.

Teague, E. C. (1978). Ph.D. Thesis, North Texas State University, Denton, Texas; reprinted in: *J. Res. NBS* **91**, 171.

Teague, E. C. (1989). *J. Vac. Sci. Technol.* B **7**, 1898.

Teague, E. C. (1992). In: *Scanned Probe Microscopy* (ed. Wickramasinghe, H. K.), AIP Conf. Proc. 241, pp. 371 & 547 (New York: AIP).

Teague, E. C., Scire, F. E., Baker, S. M. and Jensen, S. W. (1982). *Wear* **83**, 1.

Tedrow, P. M. and Meservey, R. (1971). *Phys. Rev. Lett.* **26**, 192.

Tedrow, P. M. and Meservey, R. (1973). *Phys. Rev.* B **7**, 318.

Tekman, E. and Ciraci, S. (1988). *Phys. Scripta* **38**, 486.

Tekman, E. and Ciraci, S. (1989). *Phys. Rev.* B **40**, 10286.

Tekman, E. and Ciraci, S. (1990). *Phys. Rev.* B **42**, 1860.

Terris, B. D., Stern, J. E., Rugar, D. and Mamin, H. J. (1989). *Phys. Rev. Lett.* **63**, 2669.

Terris, B. D., Stern, J. E., Rugar, D. and Mamin, H. J. (1990). *J. Vac. Sci. Technol.* A **8**, 374.

Terris, B. D., Twieg, R. J., Nguyen, C., Sigaud, G. and Nguyen, H. T. (1992). *Europhys. Lett.* **19**, 85.

Tersoff, J. (1986). *Phys. Rev. Lett.* **57**, 440.

Tersoff, J. (1990). In: *Scanning Tunneling Microscopy and Related Methods* (ed. Behm, R. J., Garcia, N. and Rohrer, H.), NATO ASI Series E: Appl. Sci. Vol. 184, p. 77 (Dordrecht: Kluwer).

Tersoff, J. and Hamann, D. R. (1983). *Phys. Rev. Lett.* **50**, 1998.

Tersoff, J. and Hamann, D. R. (1985). *Phys. Rev.* B **31**, 805.

Tersoff, J. and Lang, N. D. (1990). *Phys. Rev. Lett.* **65**, 1132.

Thibaudau, F., Cousty, J., Balanzat, E. and Bouffard, S. (1991). *Phys. Rev. Lett.* **67**, 1582.

Thompson, W. A. (1968). *Phys. Rev. Lett.* **20**, 1085.

Thompson, W. A. and Hanrahan, S. F. (1976). *Rev. Sci. Instrum.* **47**, 1303.

Thompson, W. A. and von Molnar, S. (1970). *J. Appl. Phys.* **41**, 5218.

Thomson, R. E., Walter, U., Ganz, E., Clarke, J., Zettl, A., Rauch, P. and DiSalvo, F. J. (1988a). *Phys. Rev.* B **38**, 10734.

Thomson, R. E., Walter, U., Ganz, E., Rauch, P., Zettl, A. and Clarke, J. (1988b). *J. Microsc.* **152**, 771.

Tokumoto, H., Bando, H., Mizutani, W., Okano, M., Ono, M., Murakami, H., Okayama, S., Ono, Y., Watanabe, K., Wakiyama, S., Sakai, F., Endo, K. and Kajimura, K. (1986). *Jpn. J. Appl. Phys.* **25**, L621.

Tokumoto, H., Miki, K., Murakami, H., Bando, H., Ono, M. and Kajimura, K. (1990a). *J. Vac. Sci. Technol.* A **8**, 255.

Tokumoto, H., Wakiyama, S., Miki, K. and Okayama, S. (1990b). *Appl. Phys. Lett.* **56**, 743.

Toledo-Crow, R., Yang, P. C., Chen, Y. and Vaez-Iravani, M. (1992). *Appl. Phys. Lett.* **60**, 2957.

Tománek, D. and Louie, S. G. (1988). *Phys. Rev.* B **37**, 8327.

Tománek, D., Louie, S. G., Mamin, H. J., Abraham, D. W., Thomson, R. E., Ganz, E. and Clarke, J. (1987). *Phys. Rev.* B **35**, 7790.

Tománek, D., Overney, G., Miyazaki, H., Mahanti, S. D. and Güntherodt, H. J. (1989). *Phys. Rev. Lett.* **63**, 876.

Tománek, D., Zhong, W. and Thomas, H. (1991). *Europhys. Lett.* **15**, 887.

Tonomura, A. (1987). *Rev. Mod. Phys.* **59**, 639.

Tosch, St. and Neddermeyer, H. (1988a). *Phys. Rev. Lett.* **61**, 349.

Tosch, St. and Neddermeyer, H. (1988b). *J. Microsc.* **152**, 415.

Tosch, St. and Neddermeyer, H. (1989). *Surf. Sci.* **211/212**, 133.

Trafas, B. M., Hill, D. M., Benning, P. J., Waddill, G. D., Yang, Y.-N., Siefert, R. L. and Weaver, J. H. (1991a). *Phys. Rev.* B **43**, 7174.

Trafas, B. M., Yang, Y.-N., Siefert, R. L. and Weaver, J. H. (1991b). *Phys. Rev.* B **43**, 14107.

Trautman, J. K., Betzig, E., Weiner, J. S., DiGiovanni, D. J., Harris, T. D., Hellman, F. and Gyorgy, E. M. (1992). *J. Appl. Phys.* **71**, 4659.

Travaglini, G., Rohrer, H., Stoll, E., Amrein, M., Stasiak, A., Sogo, J. and Gross, H. (1988). *Phys. Scripta* **38**, 309.

Trevor, D. J. and Chidsey, C. E. D. (1991). *J. Vac. Sci. Technol.* B **9**, 964.

Trevor, D. J., Chidsey, C. E. D. and Loiacono, D. N. (1989). *Phys. Rev. Lett.* **62**, 929.

Tromp, R. M. (1989). *J. Phys.: Condens. Matter* **1**, 10211.

Tromp, R. M., Hamers, R. J. and Demuth, J. E. (1985). *Phys. Rev. Lett.* **55**, 1303.

Tromp, R. M., Hamers, R. J. and Demuth, J. E. (1986). *Phys. Rev. B* **34**, 1388.

Tromp, R. M., van Loenen, E. J., Demuth, J. E. and Lang, N. D. (1988). *Phys. Rev. B* **37**, 9042.

Tsai, D. P., Jackson, H. E., Reddick, R. C., Sharp, S. H. and Warmack, R. J. (1990). *Appl. Phys. Lett.* **56**, 1515.

Tsao, J. Y., Chason, E., Koehler, U. and Hamers, R. (1989). *Phys. Rev. B* **40**, 11951.

Tsoi, V. S. (1974). *Sov. Phys. – JETP Lett.* **19**, 70.

Tsui, D. C. (1969). *Phys. Rev. Lett.* **22**, 293.

Tsui, D. C., Dietz, R. E. and Walker, L. R. (1971). *Phys. Rev. Lett.* **27**, 1729.

Tsukada, M., Kobayashi, K. and Ohnishi, S. (1990). *J. Vac. Sci. Technol.* A **8**, 160.

Tsukamoto, S., Siu, B. and Nakagiri, N. (1991). *Rev. Sci. Instrum.* **62**, 1767.

Tsuno, K. (1988). *Rev. Solid State Sci.* **2**, 623.

Tsuno, T., Imai, T., Nishibayashi, Y., Hamada, K. and Fujimori, N. (1991). *Jpn. J. Appl. Phys.* **30**, 1063.

Ullah, S., Dorsey, A. T. and Buchholtz, L. J. (1990). *Phys. Rev. B* **42**, 9950.

Uosaki, K. and Kita, H. (1989). *J. Electroanal. Chem.* **259**, 301.

Uosaki, K. and Kita, H. (1990). *J. Vac. Sci. Technol.* A **8**, 520.

Uozumi, K., Nakamoto, K. and Fujioka, K. (1988). *Jpn. J. Appl. Phys.* **27**, L123.

Utsugi, Y. (1990). *Nature* **347**, 748.

van Bentum, P. J. M., Smokers, R. T. M. and van Kempen, H. (1988). *Phys. Rev. Lett.* **60**, 2543.

van Bentum, P. J. M., van de Leemput, L. E. C., Schreurs, L. W. M., Teunissen, P. A. A. and van Kempen, H. (1987). *Phys. Rev. B* **36**, 843.

van Bentum, P. J. M., van de Leemput, L. E. C., Smokers, R. T. M. and van Kempen, H. (1989). *Phys. Scripta* T **25**, 122.

van de Leemput, L. E. C., van Bentum, P. J. M., Driessen, F. A. J. M., Gerritsen, J. W., van Kempen, H., Schreurs, L. W. M. and Bennema, P. (1988a). *J. Microsc.* **152**, 103.

van de Leemput, L. E. C., van Bentum, P. J. M., Schreurs, L. W. M. and van Kempen, H. (1988b). *Physica* C **152**, 99.

van der Walle, G. F. A., Gerritsen, J. W., van Kempen, H. and Wyder, P. (1985). *Rev. Sci. Instrum.* **56**, 1573.

van Hove, M. A., Koestner, R. J., Stair, P. C., Bibérian, J. P., Kesmodel, L. L., Bartos, I. and Somorjai, G. A. (1981). *Surf. Sci.* **103**, 189.

van Hulst, N. F., de Boer, N. P. and Bölger, B. (1991). *J. Microsc.* **163**, 117.

van Kempen, H. (1982). *Can. J. Phys.* **60**, 740.

van Kesteren, H. W., den Boef, A. J., Zeper, W. B., Spruit, J. H. M., Jacobs, B. A. J. and Carcia, P. F. (1991). *J. Appl. Phys.* **70**, 2413.

van Loenen, E. J., Demuth, J. E., Tromp, R. M. and Hamers, R. J. (1987). *Phys. Rev. Lett.* **58**, 373.

van Loenen, E. J., Dijkkamp, D. and Hoeven, A. J. (1988). *J. Microsc.* **152**, 487.

van Loenen, E. J., Dijkkamp, D., Hoeven, A. J., Lenssinck, J. M. and Dieleman, J. (1989). *Appl. Phys. Lett.* **55**, 1312.

van Loenen, E. J., Dijkkamp, D., Hoeven, A. J., Lenssinck, J. M. and Dieleman, J. (1990a). *Appl. Phys. Lett.* **56**, 1755.

van Loenen, E. J., Dijkkamp, D., Hoeven, A. J., Lenssinck, J. M. and Dieleman, J. (1990b). *J. Vac. Sci. Technol.* A **8**, 574.

van Son, P. C., van Kempen, H. and Wyder, P. (1987). *Phys. Rev. Lett.* **58**, 1567.

van Wees, B. J., van Houten, H., Beenakker, C. W. J., Williamson, J. G., Kouwenhoven, L. P., van der Marel, D. and Foxon, C. T. (1988). *Phys. Rev. Lett.* **60**, 848.

Vasile, M. J., Grigg, D. A., Griffith, J. E., Fitzgerald, E. A. and Russell, P. E. (1991). *Rev. Sci. Instrum.* **62**, 2167.

Vázquez, L., Bartolomé, A., Garcia, R., Buendia, A. and Baró, A. M. (1988). *Rev. Sci. Instrum.* **59**, 1286.

Verkin, B. I., Yanson, I. K., Kulik, I. O., Shklyarevski, O. I., Lysykh, A. A. and Naydyuk, Yu. G. (1979). *Solid State Commun.* **30**, 215.

Vieira, S., Ramos, M. A., Buendia, A. and Baro, A. M. (1988a). *Physica* C **153**, 1004.

Vieira, S., Ramos, M. A., Vallet-Regí, M. and Gonzalez-Calbet, J. M. (1988b). *Phys. Rev.* B **38**, 9295.

Villarrubia, J. S. and Boland, J. J. (1989). *Phys. Rev. Lett.* **63**, 306.

Voigtländer, B., Meyer, G. and Amer, N. M. (1991a). *Phys. Rev.* B **44**, 10354.

Voigtländer, B., Meyer, G. and Amer, N. M. (1991b). *Surf. Sci. Lett.* **255**, L529.

Völcker, M., Krieger, W., Suzuki, T. and Walther, H. (1991a). *J. Vac. Sci. Technol.* B **9**, 541.

Völcker, M., Krieger, W. and Walther, H. (1991b). *Phys. Rev. Lett.* **66**, 1717.

Volodin, A. P. and Khaikin, M. S. (1987). *JETP Lett.* **46**, 589.

von Molnar, S., Thompson, W. A. and Edelstein, A. S. (1967). *Appl. Phys. Lett.* **11**, 163.

Wadas, A. (1988). *J. Magn. Magn. Mat.* **71**, 147.

Wadas, A. (1989). *J. Magn. Magn. Mat.* **78**, 263.

Wadas, A. and Grütter, P. (1989). *Phys. Rev.* B **39**, 12013.

Wadas, A., Grütter, P. and Güntherodt, H. J. (1990a). *J. Vac. Sci. Technol.* A **8**, 416.

Wadas, A., Grütter, P. and Güntherodt, H. J. (1990b). *J. Appl. Phys.* **67**, 3462.

Wadas, A. and Güntherodt, H. J. (1990). *Phys. Lett.* A **146**, 277.

Waltman, S. B. and Kaiser, W. J. (1989). *Sensors Actuators* **19**, 201.

Walz, B., Wiesendanger, R., Rosenthaler, L., Güntherodt, H.-J., Düggelin, M. and Guggenheim, R. (1988). *Mat. Sci. Eng.* **99**, 501.

Wan, J.-C., McGreer, K. A., Beauchamp, K. M., Johnson, B. R., Liu, J.-X., Wang, T. and Goldman, A. M. (1990). *Physica* B **165** & **166**, 1505.

Wandass, J. H., Murday, J. S. and Colton, R. J. (1989). *Sensors Actuators* **19**, 211.

Wang, C., Giambattista, B., Slough, C. G., Coleman, R. V. and Subramanian, M. A. (1990a). *Phys. Rev.* B **42**, 8890.

Wang, C., Slough, C. G. and Coleman, R. V. (1991). *J. Vac. Sci. Technol.* B **9**, 1048.

Wang, J., Wu, L.-H. and Li, R. (1989). *J. Electroanal. Chem.* **272**, 285.

Wang, X.-S., Goldberg, J. L., Bartelt, N. C., Einstein, T. L. and Williams, E. D. (1990b). *Phys. Rev. Lett.* **65**, 2430.

Watanabe, M., Dahn, D. C., Blackford, B. and Jericho, M. H. (1988). *J. Microsc.* **152**, 175.

Watanabe, S., Aono, M. and Tsukada, M. (1991). *Phys. Rev.* B **44**, 8330.

Weaver, J. M. R., Walpita, L. M. and Wickramasinghe, H. K. (1989). *Nature* **342**, 783.

Webb, M. B., Men, F. K., Swartzentruber, B. S. and Lagally, M. G. (1990). *J. Vac. Sci. Technol.* A **8**, 2658.

Webb, M. B., Men, F. K., Swartzentruber, B. S., Kariotis, R. and Lagally, M. G. (1991). *Surf. Sci.* **242**, 23.

Weihs, T. P., Nawaz, Z., Jarvis, S. P. and Pethica, J. B. (1991). *Appl. Phys. Lett.* **59**, 3536.

Weimer, M., Kramar, J., Bai, C. and Baldeschwieler, J. D. (1988). *Phys. Rev.* B **37**, 4292.

Weiner, J. S., Hess, H. F., Robinson, R. B., Hayes, T. R., Sivco, D. L., Cho, A. Y. and Ranade, M. (1991). *Appl. Phys. Lett.* **58**, 2402.

Weisenhorn, A. L., Egger, M., Ohnesorge, F., Gould, S. A. C., Heyn, S.-P., Hansma, H. G., Sinsheimer, R. L., Gaub, H. E. and Hansma, P. K. (1991). *Langmuir* **7**, 8.

Weisenhorn, A. L., Hansma, P. K., Albrecht, T. R. and Quate, C. F. (1989). *Appl. Phys. Lett.* **54**, 2651.

Whitman, L. J., Stroscio, J. A., Dragoset, R. A. and Celotta, R. J. (1991a). *Phys. Rev. Lett.* **66**, 1338.

Whitman, L. J., Stroscio, J. A., Dragoset, R. A. and Celotta, R. J. (1991b). *Phys. Rev.* B **44**, 5951.

Whitman, L. J., Stroscio, J. A., Dragoset, R. A. and Celotta, R. J. (1991c). *Science* **251**, 1206.

Wiechers, J., Twomey, T., Kolb, D. M. and Behm, R. J. (1988). *J. Electroanal. Chem.* **248**, 451.

Wierenga, P. E., Kubby, J. A. and Griffith, J. E. (1987). *Phys. Rev. Lett.* **59**, 2169.

Wiesendanger, R. (1987). Ph.D. Thesis, Basel University, Basel, Switzerland.

Wiesendanger, R. (1992). *Appl. Surf. Sci.* **54**, 271.

Wiesendanger, R., Anselmetti, D., Geiser, V., Hidber, H. R. and Güntherodt, H.-J. (1989). *Synth. Met.* **34**, 175.

Wiesendanger, R., Anselmetti, D. and Güntherodt, H.-J. (1990a). *Europhys. News* **21**, 72.

Wiesendanger, R., Bürgler, D., Tarrach, G., Anselmetti, D., Hidber, H. R. and Güntherodt, H.-J. (1990b). *J. Vac. Sci. Technol.* A **8**, 339.

Wiesendanger, R., Bürgler, D., Tarrach, G., Güntherodt, H.-J., Shvets, I. V. and Coey, J. M. D. (1992a). *Surf. Sci.* **274**, 93.

Wiesendanger, R., Bürgler, D., Tarrach, G., Schaub, T., Hartmann, U., Güntherodt, H.-J., Shvets, I. V. and Coey, J. M. D. (1991a). *Appl. Phys.* A **53**, 349.

Wiesendanger, R., Bürgler, D., Tarrach, G., Wadas, A., Brodbeck, D., Güntherodt, H.-J., Güntherodt, G., Gambino, R. J. and Ruf, R. (1991b). *J. Vac. Sci. Technol.* B **9**, 519.

Wiesendanger, R., Eng, L., Hidber, H. R., Oelhafen, P., Rosenthaler, L., Staufer, U. and Güntherodt, H.-J. (1987a). *Surf. Sci.* **189/190**, 24.

Wiesendanger, R., Güntherodt, H.-J., Güntherodt, G., Gambino, R. J. and Ruf, R. (1990c). *Phys. Rev. Lett.* **65**, 247.

Wiesendanger, R., Ringger, M., Rosenthaler, L., Hidber, H. R., Oelhafen, P., Rudin, H. and Güntherodt, H.-J. (1987b). *Surf. Sci.* **181**, 46.

Wiesendanger, R., Rosenthaler, L., Hidber, H. R., Eng, L., Staufer, U., Güntherodt, H.-J., Düggelin, M. and Guggenheim, R. (1988a). *J. Vac. Sci. Technol.* A **6**, 529.

Wiesendanger, R., Rosenthaler, L., Hidber, H. R., Güntherodt, H.-J., McKinnon, A. W. and Spear, W. E. (1988b). *J. Appl. Phys.* **63**, 4515.

Wiesendanger, R., Shvets, I. V., Bürgler, D., Tarrach, G., Güntherodt, H.-J. and Coey, J. M. D. (1992b). *Z. Phys.* B **86**, 1.

Wiesendanger, R., Shvets, I. V., Bürgler, D., Tarrach, G., Güntherodt, H.-J. and Coey, J. M. D. (1992c). *Europhys. Lett.* **19**, 141.

Wiesendanger, R., Shvets, I. V., Bürgler, D., Tarrach, G., Güntherodt, H.-J., Coey, J. M. D. and Gräser, S. (1992d). *Science* **255**, 583.

Wiesendanger, R., Tarrach, G., Bürgler, D. and Güntherodt, H.-J. (1990d). *Europhys. Lett.* **12**, 57.

Wiesendanger, R., Tarrach, G., Bürgler, D., Jung, T., Eng, L. and Güntherodt, H.-J. (1990e). *Vacuum* **41**, 386.

Wiesendanger, R., Tarrach, G., Bürgler, D., Scandella, L. and Güntherodt, H.-J. (1990f). *Mat. Res. Soc. Symp. Proc.* **183**, 237.

Wiesendanger, R., Tarrach, G., Scandella, L. and Güntherodt, H.-J. (1990g) *Ultramicroscopy* **32**, 291.

Wilkins, M. J., Davies, M. C., Jackson, D. E., Roberts, C. J., Tendler, S. J. B. and Williams, P. M. (1992). *Appl. Phys. Lett.* **60**, 1436.

Wilkins, R., Amman, M., Ben-Jacob, E. and Jaklevic, R. C. (1990a). *Phys. Rev.* B **42**, 8698.

Wilkins, R., Amman, M., Ben-Jacob, E. and Jaklevic, R. C. (1991). *J. Vac. Sci. Technol.* B **9**, 996.

Wilkins, R., Amman, M., Soltis, R. E., Ben-Jacob, E. and Jaklevic, R. C. (1990b). *Phys. Rev.* B **41**, 8904.

Wilkins, R., Ben-Jacob, E. and Jaklevic, R. C. (1989). *Phys. Rev. Lett.* **63**, 801.

Williams, C. C., Hough, W. P. and Rishton, S. A. (1989a). *Appl. Phys. Lett.* **55**, 203.

Williams, C. C., Slinkman, J., Hough, W. P. and Wickramasinghe, H. K. (1989b). *Appl. Phys. Lett.* **55**, 1662.

Williams, C. C., Slinkman, J., Hough, W. P. and Wickramasinghe, H. K. (1990). *J. Vac. Sci. Technol.* A **8**, 895.

Williams, C. C. and Wickramasinghe, H. K. (1986a). *Appl. Phys. Lett.* **49**, 1587.

Williams, C. C. and Wickramasinghe, H. K. (1986b). *Microelectr. Eng.* **5**, 509.

Williams, C. C. and Wickramasinghe, H. K. (1990). *Nature* **344**, 317.

Williams, C. C. and Wickramasinghe, H. K. (1991). *J. Vac. Sci. Technol.* B **9**, 537.

Williams, H. J., Foster, F. G. and Wood, E. A. (1951). *Phys. Rev.* **82**, 119.

Williamson, J. B. P. (1967–8). *Proc. Inst. Mech. Eng. London* **182** (3K), 21.

Wilson, I. H., Zheng, N. J., Knipping, U. and Tsong, I. S. T. (1988). *Appl. Phys. Lett.* **53**, 2039.

Wilson, I. H., Zheng, N. J., Knipping, U. and Tsong, I. S. T. (1989). *J. Vac. Sci. Technol.* A **7**, 2840.

Wilson, J. A., DiSalvo, F. J. and Mahajan, S. (1975). *Adv. Phys.* **24**, 117.

Wilson, R. J., Meijer, G., Bethune, D. S., Johnson, R. D., Chambliss, D. D., de Vries, M. S., Hunziker, H. E. and Wendt, H. R. (1990). *Nature* **348**, 621.
Wilson, R. J. and Chiang, S. (1987a). *Phys. Rev. Lett.* **58**, 369.
Wilson, R. J. and Chiang, S. (1987b). *Phys. Rev. Lett.* **59**, 2329.
Wilson, R. J. and Chiang, S. (1988). *J. Vac. Sci. Technol.* A **6**, 398.
Wintterlin, J., Brune, H., Höfer, H. and Behm, R. J. (1988a). *Appl. Phys.* A **47**, 99.
Wintterlin, J., Schuster, R., Coulman, D. J., Ertl, G. and Behm, R. J. (1991). *J. Vac. Sci. Technol.* B **9**, 902.
Wintterlin, J., Wiechers, J., Brune, H., Gritsch, T., Höfer, H. and Behm, R. J. (1989). *Phys. Rev. Lett.* **62**, 59.
Wintterlin, J., Wiechers, J., Gritsch, Th., Höfer, H. and Behm, R. J. (1988b). *J. Microsc.* **152**, 423.
Witek, A. and Onn, D. G. (1991). *J. Vac. Sci. Technol.* B **9**, 639.
Wolf, E. L. (1985). *Principles of Electron Tunnelling Spectroscopy* (New York: Oxford University Press).
Wolf, J. F., Vicenzi, B. and Ibach, H. (1991). *Surf. Sci.* **249**, 233.
Wolfram, T. (ed.), (1978). *Inelastic Electron Tunneling Spectroscopy* (Berlin, Heidelberg, New York: Springer).
Wolkow, R. A. (1992). *Phys. Rev. Lett.* **68**, 2636.
Wolkow, R. and Avouris, Ph. (1988a). *J. Microsc.* **152**, 167.
Wolkow, R. and Avouris, Ph. (1988b). *Phys. Rev. Lett.* **60**, 1049.
Wöll, Ch., Chiang, S., Wilson, R. J. and Lippel, P. H. (1989). *Phys. Rev.* B **39**, 7988.
Wöll, Ch., Wilson, R. J., Chiang, S., Zeng, H. C. and Mitchell, K. A. R. (1990). *Phys. Rev.* B **42**, 11926.
Wolter, O., Bayer, Th. and Greschner, J. (1991). *J. Vac. Sci. Technol.* B **9**, 1353.
Wragg, J. L., Chamberlain, J. E., White, H. W., Krätschmer, W. and Huffman, D. R. (1990). *Nature* **348**, 623.
Wu, X. L. and Lieber, C. M. (1988). *J. Am. Chem. Soc.* **110**, 5200.
Wu, X. L. and Lieber, C. M. (1989a). *J. Am. Chem. Soc.* **111**, 2731.
Wu, X. L. and Lieber, C. M. (1989b). *Science* **243**, 1703.
Wu, X. L. and Lieber, C. M. (1990a). *Phys. Rev.* B **41**, 1239.
Wu, X. L. and Lieber, C. M. (1990b). *Phys. Rev. Lett.* **64**, 1150.
Wu, X. L. and Lieber, C. M. (1991). *J. Vac. Sci. Technol.* B **9**, 1044.
Wu, X. L., Lieber, C. M., Ginley, D. S. and Baughman, R. J. (1989). *Appl. Phys. Lett.* **55**, 2129.
Wu, X. L., Wang, Y. L., Zhang, Z. and Lieber, C. M. (1991). *Phys. Rev.* B **43**, 8729.
Wu, X. L., Zhang, Z., Wang, Y. L. and Lieber, C. M. (1990). *Science* **248**, 1211.
Wu, X. L., Zhou, P. and Lieber, C. M. (1988a). *Phys. Rev. Lett.* **61**, 2604.
Wu, X. L., Zhou, P. and Lieber, C. M. (1988b). *Nature* **335**, 55.
Xhie, J., Sattler, K., Müller, U., Venkateswaran, N. and Raina, G. (1991a). *Phys. Rev.* B **43**, 8917.
Xhie, J., Sattler, K., Müller, U., Venkateswaran, N. and Raina, G. (1991b). *J. Vac. Sci. Technol.* B **9**, 833.
Yamada, H., Fujii, T. and Nakayama, K. (1988). *J. Vac. Sci. Technol.* A **6**, 293.

Yang, Y.-N., Trafas, B. M., Luo, Y.-S., Siefert, R. L. and Weaver, J. H. (1991). *Phys. Rev.* B **44**, 5720.

Yanson, I. K. (1974). *Sov. Phys. – JETP* **39**, 506.

Yanson, I. K. (1983). *Sov. J. Low Temp. Phys.* **9**, 343.

Yata, M., Ozaki, M., Sakata, S., Yamada, T., Kohno, A. and Aono, M. (1989). *Jpn. J. Appl. Phys.* **28**, L885.

Yau, S.-T., Saltz, D. and Nayfeh, M. H. (1990). *Appl. Phys. Lett.* **57**, 2913.

Yeung, K. L. and Wolf, E. E. (1991). *J. Vac. Sci. Technol.* B **9**, 798.

Yoshimura, M., Ara, N., Kageshima, M., Shiota, R., Kawazu, A., Shigekawa, H., Saito, Y., Oshima, M., Mori, H., Yamochi, H. and Saito, G. (1991a). *Surf. Sci.* **242**, 18.

Yoshimura, M., Fujita, K., Ara, N., Kageshima, M., Shioda, R., Kawazu, A., Shigekawa, H. and Hyodo, S. (1990). *J. Vac. Sci. Technol.* A **8**, 488.

Yoshimura, M., Shigekawa, H., Nejoh, H., Saito, G., Saito, Y. and Kawazu, A. (1991b). *Phys. Rev.* B **43**, 13590.

Yoshimura, M., Shigekawa, H., Yamochi, H., Saito, G., Saito, Y. and Kawazu, A. (1991c). *Phys. Rev.* B **44**, 1970.

Young, R. D. (1966). *Rev. Sci. Instrum.* **37**, 275.

Young, R. D. (1971). *Phys. Today*, November, p. 42.

Young, R., Ward, J. and Scire, F. (1971). *Phys. Rev. Lett.* **27**, 922.

Young, R., Ward, J. and Scire, F. (1972). *Rev. Sci. Instrum.* **43**, 999.

Youngquist, M. G., Driscoll, R. J., Coley, T. R., Goddard, W. A. and Baldeschwieler, J. D. (1991). *J. Vac. Sci. Technol.* B **9**, 1304.

Youngquist, M. G., Driscoll, R. J., Coley, T. R., Goddard, W. A. and Baldeschwieler, J. D. (1992). In: *Scanned Probe Microscopy* (ed. Wickramasinghe, H. K.), AIP Conf. Proc. 241, p. 154 (New York: AIP).

Zangwill, A. (1988). *Physics at Surfaces* (Cambridge: Cambridge University Press).

Zeller, H. R. and Giaever, I. (1969). *Phys. Rev.* **181**, 789.

Zener, C. (1934). *Proc. Roy. Soc. London* **145**, 523.

Zettl, A., Bourne, L. C., Clarke, J., Crommie, M. F., Hundley, M. F., Thomson, R. E. and Walter, U. (1989). *Synth. Met.* **29**, F445.

Zhang, H., Hordon, L. S., Kuan, S. W. J., Maccagno, P. and Pease, R. F. W. (1989). *J. Vac. Sci. Technol.* B **7**, 1717.

Zhang, Z., Chen, C.-C., Kelty, S. P., Dai, H. and Lieber, C. M. (1991a). *Nature* **353**, 333.

Zhang, Z., Chen, C.-C., Lieber, C. M., Morosin, B., Ginley, D. S. and Venturini, E. L. (1992). *Phys. Rev.* B **45**, 987.

Zhang, Z., Lieber, C. M., Ginley, D. S., Baughman, R. J. and Morosin, B. (1991b). *J. Vac. Sci. Technol.* B **9**, 1009.

Zhang, Z., Wang, Y. L., Wu, X. L., Huang, J.-L. and Lieber, C. M. (1990). *Phys. Rev.* B **42**, 1082.

Zhong, W., Overney, G. and Tománek, D. (1991). *Europhys. Lett.* **15**, 49.

Zhong, W. and Tománek, D. (1990). *Phys. Rev. Lett.* **64**, 3054.

Züger, O. and Dürig, U. (1992a). *Phys. Rev.* B **46**, 7319.

Züger, O. and Dürig, U. (1992b). *Ultramicroscopy* **42–44**, 520.

Index

625